Advances in Corrosion Control and Materials in Oil and Gas Production

Papers from EUROCORR '97 and EUROCORR '98

EARLIER VOLUMES IN THIS SERIES

Number 1: **Corrosion in the Nuclear Industry**
Prepared by the Working Party on Nuclear Corrosion

Number 2: **Practical Corrosion Principles: a manual of corrosion experiments**
Prepared by the Working Party on Corrosion Education

Number 3: **General Guidelines for Corrosion Testing of Materials for Marine Applications**
Prepared by the Working Party on Marine Corrosion

Number 4: **Guidelines on Electrochemical Corrosion Measurements**
Prepared by the Working Party on Physico-Chemical Methods of Corrosion Testing

Number 5: **Illustrated Case Histories of Marine Corrosion**
Prepared by the Working Party on Marine Corrosion

Number 6: **Corrosion Education Manual**
Prepared by the Working Party on Corrosion Education

Number 7: **Corrosion Problems Related to Nuclear Waste Disposal**
Prepared by the Working Party on Nuclear Corrosion

Number 8: **Microbial Corrosion**
Prepared by the Working Party on Microbial Corrosion

Number 9: **Microbiological Degradation of Materials — and Methods of Protection**
Prepared by the Working Party on Microbial Corrosion

Number 10: **Marine Corrosion of Stainless Steels: Chlorination and Microbial Effects**
Prepared by the Working Party on Marine Corrosion

Number 11: **Corrosion Inhibitors**
Prepared by the Working Party on Inhibitors

Number 12: **Modifications of Passive Films**
Prepared by the Working Party on Surface Science and Mechanisms of Corrosion and Protection

Number 13: **Predicting CO_2 Corrosion in the Oil and Gas Industry**
Prepared by the Working Party on Corrosion in Oil and Gas Production

Number 14: **Guidelines for Methods of Testing and Research in High Temperature Corrosion**
Prepared by the Working Party on Corrosion by Hot Gases and Combustion Products

Number 15: **Microbial Corrosion (Proceedings of the 3rd International EFC Workshop)**
Prepared by the Working Party on Microbial Corrosion

Number 16: **Guidelines on Materials Requirements for Carbon and Low Alloy Steels for H_2S-Containing Environments in Oil and Gas Production**
Prepared by the Working Party on Corrosion in Oil and Gas Production

Number 17: **Corrosion Resistant Alloys for Oil and Gas Production: Guidance on General Requirements and Test Methods for H_2S Service**
Prepared by the Working Party on Corrosion in Oil and Gas Production

Number 18: **Stainless Steel in Concrete: State of the Art Report**
Prepared by the Working Party on Corrosion of Reinforcement in Concrete

Number 19: **Sea Water Corrosion of Stainless Steels — Mechanisms and Experiences**
Prepared by the Working Parties on Marine Corrosion and Microbial Corrosion

Number 20: **Organic and Inorganic Coatings for Corrosion Prevention — Research and Experiences**
Papers from EUROCORR '96

Number 21: **Corrosion-Deformation Interactions**
CDI '96 in conjunction with EUROCORR '96

Number 22: **Aspects of Microbially Induced Corrosion**
Papers from EUROCORR '96 and the EFC Working Party on Microbial Corrosion

Number 23: **CO_2 Corrosion Control in Oil and Gas Production — Design Considerations**
Prepared by the Working Party on Corrosion in Oil and Gas Production

Number 24: **Electrochemical Rehabilitation Methods for Reinforced Concrete Structures – A State of the Art Report**
Prepared by the Working Party on Corrosion of Reinforcement in Concrete

Number 25: **Corrosion of Reinforcement in Concrete — Monitoring, Prevention and Rehabilitation**
Papers from EUROCORR '97

European Federation of Corrosion Publications

NUMBER 26

Advances in Corrosion Control and Materials in Oil and Gas Production

Papers from EUROCORR '97 and EUROCORR '98

Edited by

P. S. Jackman and L. M. Smith

Published for the European Federation of Corrosion by IOM Communications

Book Number 715
Published in 1999 by IOM Communications Ltd
1 Carlton House Terrace, London SW1Y 5DB

IOM Communications Ltd
is a wholly-owned subsidiary of
The Institute of Materials

ISBN 1-86125-092-4

Design and production by
SPIRES Design Partnership

Made and printed in Great Britain

Contents

Part 3 – Martensitic Stainless Steels 217

Part 6 – Corrosion Inhibitors 399

Part 7 – Non-Metallics 439

European Federation of Corrosion Publications

Series Introduction

The EFC, incorporated in Belgium, was founded in 1955 with the purpose of promoting European co-operation in the fields of research into corrosion and corrosion prevention.

Membership is based upon participation by corrosion societies and committees in technical Working Parties. Member societies appoint delegates to Working Parties, whose membership is expanded by personal corresponding membership.

The activities of the Working Parties cover corrosion topics associated with inhibition, education, reinforcement in concrete, microbial effects, hot gases and combustion products, environment sensitive fracture, marine environments, surface science, physico–chemical methods of measurement, the nuclear industry, computer based information systems, the oil and gas industry, the petrochemical industry and coatings. Working Parties on other topics are established as required.

The Working Parties function in various ways, e.g. by preparing reports, organising symposia, conducting intensive courses and producing instructional material, including films. The activities of the Working Parties are co-ordinated, through a Science and Technology Advisory Committee, by the Scientific Secretary.

The administration of the EFC is handled by three Secretariats: DECHEMA e. V. in Germany, the Société de Chimie Industrielle in France, and The Institute of Materials in the United Kingdom. These three Secretariats meet at the Board of Administrators of the EFC. There is an annual General Assembly at which delegates from all member societies meet to determine and approve EFC policy. News of EFC activities, forthcoming conferences, courses etc. is published in a range of accredited corrosion and certain other journals throughout Europe. More detailed descriptions of activities are given in a Newsletter prepared by the Scientific Secretary.

The output of the EFC takes various forms. Papers on particular topics, for example, reviews or results of experimental work, may be published in scientific and technical journals in one or more countries in Europe. Conference proceedings are often published by the organisation responsible for the conference.

In 1987 the, then, Institute of Metals was appointed as the official EFC publisher. Although the arrangement is non-exclusive and other routes for publication are still available, it is expected that the Working Parties of the EFC will use The Institute of Materials for publication of reports, proceedings etc. wherever possible.

The name of The Institute of Metals was changed to The Institute of Materials with effect from 1 January 1992.

The EFC Series is now published by the wholly-owned subsidiary of The Institute of Materials, IOM Communications Ltd.

A. D. Mercer
EFC Series Editor,
The Institute of Materials, London, UK

EFC Secretariats are located at:

Dr B A Rickinson
European Federation of Corrosion, The Institute of Materials, 1 Carlton House Terrace, London, SW1Y 5DB, UK

Mr P Berge
Fédération Europééne de la Corrosion, Société de Chimie Industrielle, 28 rue Saint-Dominique, F-75007 Paris, FRANCE

Professor Dr G Kreysa
Europäische Föderation Korrosion, DECHEMA e. V., Theodor-Heuss-Allee 25, D-60486, Frankfurt, GERMANY

Preface

This EFC publication incorporates papers from the sessions at *EUROCORR '97* and *EUROCORR '98* dealing with Corrosion in Oil and Gas Production. These conference sessions, and attendant workshops, are run by EFC Working Party 13. This Working Party has over 200 members coming from 26 countries throughout the world. Over 80% of members are employed in the oil and gas producing industries, in engineering design houses, in industrial research laboratories or in manufacturing industries. As such the work of EFC Working Party 13 is led by the needs of industry rather than by academia. This is obvious from the content of the 47 papers contained in this publication.

The proceedings are opened by two keynote papers from leading experts in the understanding and control of Corrosion in Oil and Gas Production: Dr S. D. Kapusta, Shell International, The Netherlands and Professor G. Schmitt from Iserlohn University, Germany.

The remaining papers have been separated into six broad topics:

- Carbon and alloy steels
- Martensitic stainless steels
- Corrosion resistant alloys
- Galvanic corrosion
- Corrosion inhibitors, and
- Non-metallic materials.

Within these topics authors have explained new developments in the design of alloys, particularly in martensitic stainless steels, and have improved the understanding of corrosion, and corrosion mitigation, in a wide range of materials from carbon steels to corrosion resistant alloys, reinforced plastics and ceramics. Coming from the need to control corrosion in oil and gas production, the papers deal extensively with corrosion in carbon dioxide, hydrogen sulfide and chloride containing environments.

The contribution made by each of the authors represented here must be gratefully acknowledged. Without their willing help the EUROCORR conferences and this EFC publication would not be possible. The organising committees of *EUROCORR '97* and *EUROCORR '98* are also sincerely thanked.

Finally, thanks are due to Dr L. M. Smith. As previous chairman of Working Party 13 she organised the conference sessions at *EUROCORR '97*. She also reviewed the papers contained in this volume in order to write the Foreword.

P. S. Jackman
Expert Metallurgy Services Limited, United Kingdom
Chairman of EFC Working Party 13 on Corrosion in Oil and Gas Production

Foreword

The role of the material selector can be simply summarised; to choose materials which are safe and cost effective for the application. Both these aims have to be met and many papers in this volume show that these two requirements do not need to be incompatible. Safety need not be bought at a great cost if there is:

- knowledge of the properties of materials
- a rational corrosion model with good assessment of the many controlling parameters
- an assessment of the risks involved in optional material choices
- back up corrosion monitoring in systems where corrosion will arise in service
- a relevant and well managed inspection system.

The first keynote paper expands in detail on these points, illustrating, for operation of equipment, particularly pipelines, in CO_2 and H_2S containing environments, how corrosion rate evaluation, interpretation of monitoring results and establishment of inspection frequencies can all be handled in a probabilistic manner. From such a strategy, confidence can be built in the safe operation of systems, allowing for corrosion in service, based on the combination of a cost effective material choice (carbon steel) with optimised dosing of inhibitors, an effective monitoring system and appropriate inspection.

Such strategies need constant refinement to continuously optimise the design, raise confidence and increase safety. This specifically calls for input into the underlying corrosion model. Paper 2, 3, 5, 8 and 10 refine the understanding of corrosion rates of carbon steels operating in various temperature ranges and gas compositions and when exposed with or without inhibition. For systems operating at higher temperature, where persistency of corrosion product scales significantly influences the corrosion rate, the detailed correlation of fluid flow velocities with cracking of the scale, investigated and reported in Paper 2, is a major advance in mechanistic understanding. Paper 17 describes an experimental set-up to simulate fluid flow in order to investigate its influence on the system corrosivity.

Welds in systems are sites of geometric, fluid flow and metallurgical discontinuity and have been known to result in localised corrosion in both water handling and CO_2-containing production environments. A review of this topic in Paper 6 tries to rationalise the conflicting information established to date, illustrating that microstructure is not the controlling parameter in the propagation of preferential weld attack. Similarly paper 6 suggests that better definition is needed of the role of liquid conductivity and the effects of velocity on localised weld attack and emphasises the complexity of interaction of the many parameters.

One method of reducing the cost of carbon steel systems, that has been considered by a number of operators and steel producers, is by alloying with low levels of chromium (typically around 1%). The influence of chromium is to reduce the overall

CO_2 corrosion rate and thereby reduce the requirement for, or the concentration of, inhibitors. Papers 4, 7, 8 and 9 all investigate this effect. The performance of these chromium-containing steels is shown to be dependent upon the dissolved oxygen in solution (Paper 9) and the presence of hydrogen sulfide and the operating temperature (Paper 8). The key benefit of chromium additions is shown to be in reducing the tendency for severe mesa attack in carbon steels during CO_2 corrosion at worst case temperature and flow conditions (Paper 4), rather than a decrease of the general corrosion rate (Paper 7). With the development of appropriate weld consumables which avoid any localised weld attack it is concluded that these low chromium alloys may now be considered for flowlines and trunk lines with a reduced risk of localised corrosion (Paper 7).

As previously stated, safe use of carbon steel almost inevitably requires the use of corrosion inhibitors to reduce the potential high rates of corrosion. Papers 41 to 45 discuss various aspects of the use of corrosion inhibitors. The formation of corrosion products on the metal surface by pre-corrosion prior to inhibitor injection has a significant influence on the inhibitor performance especially at temperatures above 60°C. The indications are that the inhibitor has a significantly better protection performance on non-corroded surfaces (Papers 41 and 43). Paper 42 investigates the corrosion inhibiting effect of different types of diamine molecules, showing that the carbon chain length influences the reduction in corrosion rate and helping to establish a molecular understanding of the role of inhibitors.

One of the most critical questions regarding inhibition is the ability to control corrosion in the vapour phase of multi-phase pipelines. The rates of corrosion in the vapour space were investigated in Paper 45 considering the influence of oxygen contamination, methanol, and presence of H_2S. Particular blends of inhibitors were found to be more effective than others and their performance in the field was demonstrated.

Performance of inhibitors obviously requires optimum dosing rates which also improves the cost effectiveness of inhibitor application. New electrical resistance probes have been developed (Paper 44) which respond to changes in the corrosion rates in production pipework and allow inhibitor dosing to respond directly to this. The use of these and other corrosion monitoring systems when correctly located (Paper 21) should ideally prevent over-application of chemical products, reducing costs and increasing confidence in corrosion control and, thus, safety.

Resistance of carbon steels to sulfide stress cracking in H_2S-containing environments is critical to safe production operations. Microstructure plays a controlling part in cracking resistance with difference in the cracking resistance being related directly to the type of second phase present (Paper 12) and the grain size (Paper 13). By comparison, corrosion of steels used in very large crude oil carriers which have to withstand exposure to high levels of CO_2, SO_2 and O_2 has been shown to be largely independent of microstructure whilst varying with concentration of acid gases and temperature (Paper 19).

Hydrogen cracking of pipeline steels may also arise from hydrogen present because of low pH environments around buried pipes and/or the influence of cathodic protection. Paper 20 investigates the influence of localised corrosion or other surface defects on the penetration of hydrogen into the steel. Paper 11 establishes a relationship between the critical stress intensity for propagation of hydrogen cracks depending upon the hydrogen concentration generated by cathodic protection, whilst commenting that further work is needed on the possible influence of sulfate reducing bacteria on the cracking risks.

Hydrogen sulfide production by sulfate reducing bacteria was quoted as part of the reason for the failure of a flowline in Paper 18. Other factors included the low flow velocity and lack of pigging which allowed deposits to remain in the line and encourage bacterial activity. This illustrates the need, with carbon steels, to establish appropriate corrosion monitoring, inspection and maintenance systems if the 'cheap option' is to prove cost-effective.

In the context of materials which are safe and cost effective for specific applications a key development of the last 5 years has been increased research activity on the subject of martensitic stainless steels. Correctly selected martensitic stainless steels are corrosion resistant alloys and therefore should give uninterrupted performance for the anticipated service life. The section concerning martensitic stainless steels in this volume effectively presents the current state-of-the-art of this subject.

Improved weldability of martensitic stainless steels has been achieved by modifying the composition, specifically reducing the carbon content to result in reduced hardness in the martensite structure, and balancing the nickel, chromium and molybdenum contents to achieve optimum corrosion resistance and to give fully martensitic structures which can be tempered to give the required strength and toughness. Three papers from Japanese steel makers (Papers 22, 23 and 27) illustrate the development of weldable martensitic stainless steel for flowline applications. The weldable martensitic stainless steels have slightly different chemical compositions, each offering specific properties appropriate for different corrosive applications. A further paper, Paper 24, investigates the application limits of a martensitic stainless steel designed to give improved resistance to corrosion in CO_2- and H_2S- containing environments in (non-welded) downhole tubing.

Selecting the correct grade of martensitic stainless steel does require a detailed study of the properties of the optional alloys available. Papers 28 to 32 illustrate the application of different testing techniques and at the same time use of those techniques to present the evaluation of the properties of these grades of materials particularly in production environments containing hydrogen sulfide gas. Amongst the techniques investigated are studies of the effect of weld restraint (Paper 29) which can have a critical impact upon the performance of welded components in service.

A review of literature on testing various martensitic stainless steels in a number of service environments (Paper 33) combines information from several sources to provide an overview of the impact of environmental parameters on the corrosion performance of martensitic stainless steels falling within different composition ranges. This helps to give a general impression of the safe regimes for service whilst specific comparison of the results given in other papers in this volume indicates that checks on the performance of individual steels may still be needed to safely apply these materials at their limits.

Two papers particularly cover recent uses of extensive quantities of martensitic stainless steels, one in a flowline project (Paper 26) and one in an onshore piping system (Paper 25). The description of the fabrication of the flowline gives some insight into the benefits of selecting the weldable martensitic stainless steel for this particular project and discusses the considerations which led to the selection of welding technique and welding consumable to meet mechanical, corrosion and optimum economic requirements. The use of these materials for the piping system illustrates the technical challenges which had to be met in producing the first longitudinally welded pipe to be applied in service and at the same time procuring the fittings and flanges which were required for the systems. Various problems encountered in making these different products are discussed but it is notable that the martensitic stainless steel was qualified for this demanding application and that it could be successfully re-heat-treated where necessary to recover optimum properties, welded to give required toughness properties and qualified by the certifying authority for safe application for a high pressure system on land.

Giving the wealth of choice open to the material selector in terms of excellent quality steel produced by from several sources, good characterisation of properties made using different types of testing techniques and a growing range of established applications, it may well be anticipated that the use of martensitic stainless steels is one which will be steadily increasing over the coming years.

There is a significant change in performance of many corrosion resistant alloys depending upon the concentration of chloride ions in the solution. Paper 30 investigated the performance of certain martensitic stainless steels and precipitation hardened stainless steels in low chloride environments showing a much increased resistance to corrosion at higher partial pressures of acid gases. At the opposite extreme, the high salinity environments experienced in certain oil fields in Africa present a much higher risk of localised corrosion and in these cases the use of martensitic stainless steels has to be carefully considered as some compositions may pit. In such cases it is more cost effective to select higher alloyed materials like duplex and super duplex stainless steels which have higher corrosion resistance (Paper 34).

As conditions become yet more aggressive particularly with higher levels of hydrogen sulfide it becomes necessary to consider the application of nickel base alloys (Paper 35 to 37). The use of literature reviews can be very helpful in providing a selection of materials which can be considered for different types of applications and Paper 35 provides a useful review of conditions in which cold worked or age hardened nickel based materials have been recommended or validated by corrosion testing for oil country tubular goods application. Paper 36 investigates in more detail the particular performance of Alloy 825 comparing it with the performance of Alloy 28. Paper 37 investigates the influence of grain size on stress corrosion cracking resistance of Alloy G3, an alloy which has been extensively used in some of the most aggressive sour oil and gas production environments.

One cost effective way to make use of corrosion resistant alloys in tubing strings would be to select different tubing materials appropriate for the conditions at each depth in the well (a combination string). Whilst this would be cheaper it is not often carried out because of concerns about galvanic corrosion. Paper 40 investigates the risks of galvanic corrosion between different types of martensitic and duplex stainless steels and nickel alloy materials in production environments, concluding that galvanic corrosion between these alloys would not be serious in oil and gas environment. Nevertheless, galvanic corrosion between a tubing string and the carbon steel casing has resulted in failure in a cold worked duplex stainless steel in the past and this is investigated further in Paper 39. The study emphasises the importance of dynamic plastic straining in test methods for sour service evaluation and states that further tests are necessary in order to define the limits of sour service application.

In the context of materials selection titanium is a material which is regarded as a remedy for corrosion problems in many hostile environments although it may introduce galvanic corrosion problems as it is such a noble metal. Paper 38 discusses practical guidelines for design aimed at avoiding or minimising galvanic corrosion in actual service applications. Further recommendations on the effect of in-service parameters are discussed to optimise the use of titanium in service.

The optimum material for particular applications may obviously require the use of non-metallic materials which are designed to meet particular performance requirements. Paper 47 describes new ceramic-metallic materials giving increased lifetime in component parts of choke valves. Potentially these materials offer improved erosion, erosion-corrosion and toughness properties by optimising the microstructure and the chemical composition of the metallic binder.

Removing the risk of corrosion entirely is a great attraction and glass reinforced epoxy pipelines have proven their successful use for many years as appropriate materials for handling water of various grades. Paper 46 reviews some of this past experience which has led to the establishment of design guidelines and qualification procedures. It also proposes now potential applications which might become more established in the future period.

One approach to improving cost effectiveness of production systems can be through a change in the type of product applied. Coiled tubing was initially used for well servicing and workovers but applications have expanded as tubing diameter, lengths and materials of construction have widened. Materials options now cover a range of non-metallic, metallic and internally and/or externally coated grades of steels so that a cost-effective selection can be made for any given application (Papers 14 and 15).

As stated above, the role of the material selector can be simply summarised, but to successfully select the safe and cost-effective material for the many applications in the oil and gas industry requires detailed knowledge of materials performance under the internal and external corrosion risks and in the stress conditions arising in service. This volume contributes substantially to the knowledge base required.

L. M. Smith
Past Chairman of the EFC Working Party on Corrosion in Oil and Gas Production, 1993–1997

Part 1

Keynote Papers

1
The Materials and Corrosion View of Wet Gas Transportation

S. D. KAPUSTA and B. F. M. POTS
Shell Global Solutions, Shell Research and Technology Centre Amsterdam, Amsterdam, The Netherlands

ABSTRACT
The use of carbon steel for transportation of wet, corrosive gas offers potential savings over the more expensive alternatives, such as use of corrosion resistant alloys or gas drying, but with a higher risk. Shell's approach to assessing the feasibility of using carbon steel for these applications is based on four steps: (1) evaluating the corrosivity of the environment; (2) identifying the best corrosion control option; (3) assessing the risks; and (4) implementing a corrosion management programme to reduce those risks.

This paper describes in more detail the methodology for executing these four steps. The expected corrosion rates are estimated based on predictive models, complemented with laboratory tests and field experience. Corrosion inhibition is one of the most common and versatile methods of corrosion control and it is the focus of this paper. A quantitative approach for assessing corrosion risks is proposed. Some of the data required for this assessment may not be available at the design stage, therefore reasonable 'guesstimates' need to be adopted. The success of carbon steel in corrosive service depends on the implementation of a comprehensive corrosion management programme. The elements of this programme are outlined.

1. Introduction

Transporting wet gas from offshore production facilities for onshore treatment is often an economically attractive alternative to offshore drying, in terms of reduced Capital expenditure (Capex) and sometimes also to allow more flexibility in field development. In some cases (subsea completions, development of marginal fields) offshore drying may not be a technically viable option, and transportation of raw fluids is imperative. The produced fluids may contain significant levels of CO_2, H_2S and acids, which in combination with free water make the pipeline environment potentially very corrosive. Several options for corrosion control of offshore wet gas pipelines are available. The use of corrosion resistant alloys (CRAs) is a technically sound solution, although the costs of materials and the added laying time make this alternative unattractive for long distance, large diameter pipelines.

The use of carbon steel (CS) may offer considerable Capex savings over the more expensive alloys, but it also involves higher operating costs (because of inhibition, inspection, monitoring and staffing), and additional risks. This summarises the main issues to be considered in the design and operation of carbon steel pipelines for corrosive service. The key issues that will be addressed are:

(1) assessing the technical and economic feasibility of carbon steel;

(2) selecting the most cost-effective corrosion control option; and

(3) identifying and managing the corrosion-related risks in operating a wet corrosive gas pipeline.

2. Pipeline Design Considerations

The design of a carbon steel pipeline for corrosive service involves the following steps:

(1) Assessment of the feasibility of using CS;

(2) Determination of the required corrosion allowance;

(3) Assessment of the corrosion risks;

(4) Estimate of the life cycle cost; and

(5) Design of the corrosion control and management programme.

The design of the corrosion control system must be closely integrated with pipeline operations; in particular, the assumptions involved in the design, such as on-line availability of the inhibitor injection system, and the consequences of malfunction of this system, need to be clear to the operators.

3. Carbon Steel or CRA?

Extensive experience with the operation of gas pipelines in corrosive service exists within the Shell Group. Table 1 is a summary of recent pipeline projects involving transportation of corrosive gas. This experience has shown that carbon steel pipelines can be safely operated in very corrosive service if the corrosion control system is properly designed and implemented. Most of this discussion will focus on corrosion inhibition, which is one of the most common and versatile methods of corrosion control. However the same basic ideas also apply to other methods, such as glycol injection, with or without pH control to enhance the formation of a protective scale.

The main technical factors that limit carbon steel use are (1) the corrosivity of the environment, more specifically the temperature of the fluids, and (2) the flow velocity.

Most operators are confident that carbon steel can be inhibited at temperatures below about 120°C. Inhibition is also possible at higher temperatures, but then the inhibitor selection process and the design and operation of the inhibitor injection and the corrosion monitoring systems become extremely critical. The effect of flow on corrosion has been extensively investigated. It is generally agreed that a 'critical velocity' exists that limits the applicability of corrosion inhibitors. There is less

Table 1. *List of recent wet gas/wet oil transportation pipeline projects/prospects*

Case	*D* (in.)	Length (km)	p_{CO_2} (bar)	p_{H_2S} (mbar)	*T* (°C)	Potential corrosion (mm/y)
Troll wet gas	36	2 × 70	0.4	0	50	2
Australia offshore gas	36	140	4	0	70	14
Middle east offshore gas	34	100	3	700	80	10
New Zealand offshore gas	24	36	4	0	60	7
Pacific offshore gas	24	25	5	300	100	17
North Sea offshore gas	12	23	3	8	104	12
Mallard offshore oil	8	14	3.5	5	140	3

agreement on what that velocity is and how it relates to other factors such as inhibitor concentration, temperature, pressure, flow geometry, etc.

A corrosion control system with an on-line availability of close to 100% can be designed and constructed with current technology, for example by full redundancy of all critical components, automatic pipeline shutdown in case of corrosion control system failure, intensive monitoring and maintenance of equipment, etc. The cost of this approach, however, can be high and it needs to be balanced against a reduction of corrosion risks. In most cases, pipeline design is based on less than 100% availability of corrosion control; a value of 95% is often quoted as the maximum achievable following 'normal' equipment and operating procedures.

The economic incentive of a carbon steel pipeline needs to be evaluated for each specific project, as the Capex of CRAs and CS can vary widely depending on external factors and project conditions. The general trend is that life cycle costs of CS will be more attractive for long, large pipelines of relatively short life. This is shown in Fig. 1, where the vertical axis represents the difference in materials costs between carbon steel and clad CRA pipe; the differences become smaller if laying costs are included. On the other hand, the operating costs of CS lines increase almost linearly with lifetime (Fig. 2), but are less sensitive to size. The risk of failure, expressed in terms of cost, needs to be added to any cost comparison between these two alternatives. An example of these calculations is discussed later in this paper.

4. Life Cycle Economics

The main Capex savings can be realised in the use of carbon steel instead of the other, more expensive alternatives. Additional savings are possible by optimising the corrosion allowance, inhibitor injection rate, inspection frequency, staffing of the operation, etc. However, as Fig. 3 shows, these savings also bring about an increase in risks and Operating expenditure (Opex). The exact shape of the cost/risk equation is specific to each project. An optimum balance between these factors results in minimum life cycle costs.

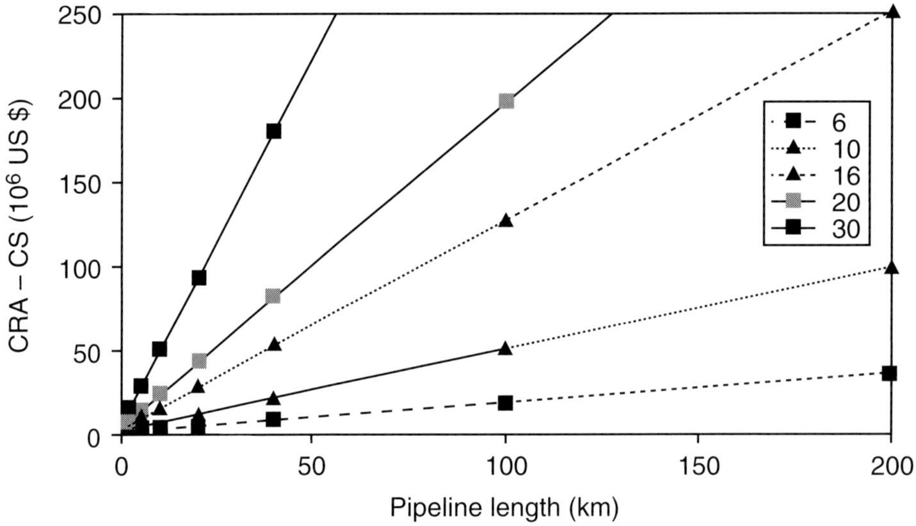

Fig. 1 *Relative cost of clad-CRA vs carbon steel for pipeline diameters of 6–30 in.*

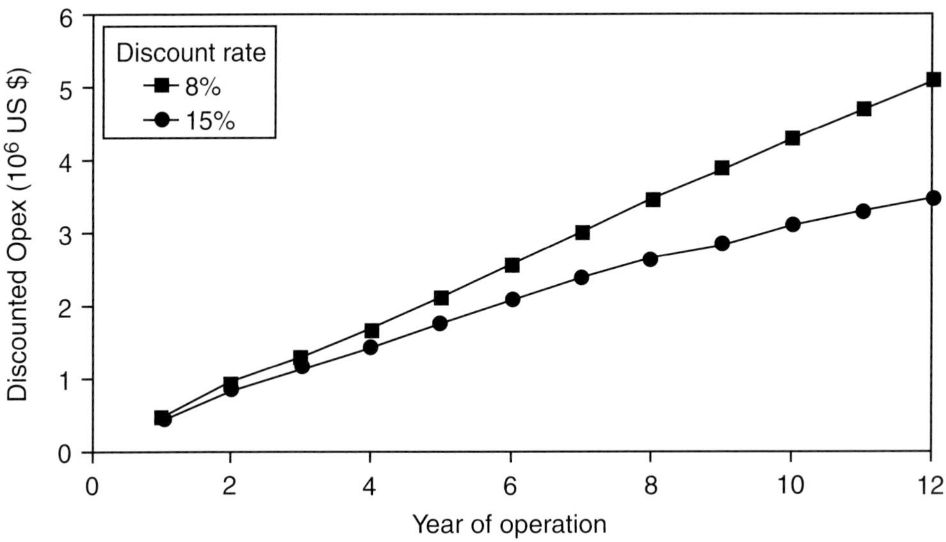

Fig. 2 *Typical operating costs for a 16 in. carbon steel pipeline, including expenses for chemicals, inspection, and personnel.*

5. Risk Analysis

The intention of the following sections is to present the risk assessment methodology. Some of the examples may seem unrealistic, but are given only to illustrate the method. In particular, the distribution of inhibited and uninhibited corrosion rates,

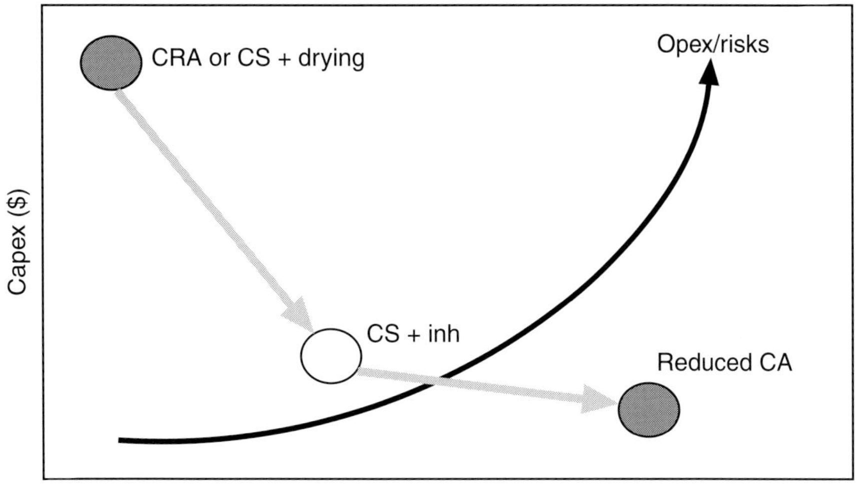

Fig. 3 *Notional Capex, Opex and risks for different corrosion control options. CRA = corrosion resistant alloys; CS = carbon steel; CA = corrosion allowance; inh = corrosion inhibition. The actual cost/risk relationship is project specific.*

and the cost of failure chosen for the risk calculations should be based on a thorough analysis of the system, corrosion mode, consequences of failure, and other data.

Risk is usually defined as the product of the probability and the (economic) consequence of failure. For the purposes of this presentation 'failure' will be defined as a wall loss exceeding the corrosion allowance within the lifetime of the project, without reaching the point of leakage. This 'zero leak' assumption means that an adequate monitoring and inspection programme will be instituted, and corrective actions will be taken to detect excessive wall loss and take appropriate corrective action before leakage.

This paper will focus on issues related to pipeline integrity. However, in addition to the risk of leaks, the use of carbon steel with a corrosion control system introduces another type of risk, related to the operability of the line and the effect of the corrosion control method on downstream processing of hydrocarbons and water. For example, corrosion inhibitors may require periodic pigging of the pipeline to be effective. Pigging will have a direct impact on the size of slugs and of the slug catcher. Corrosion inhibitor may also create environmental problems, for example in case of overboard disposal of produced water. The handling of solids created by corrosion of the pipelines may require special considerations in the design of the slugcatcher and downstream vessels. Field experience has shown that these risks and problems can be managed by proper design and operation of the facilities, if they are identified early in a project.

From a materials and corrosion viewpoint, the transportation of wet, corrosive gas through carbon steel pipelines involves two key processes:

(1) an assessment of the risks; and

(2) design and implementation of a corrosion management programme to reduce those risks to manageable levels.

These two topics will be expanded in the following Sections.

6. Preventing Failures

The three factors that determine the time to failure of a corroding pipeline are:

(1) the corrosion allowance, *CA*;

(2) the corrosion rate, *CR*; and

(3) the time t_r to detect out-of-compliance situations, for example by inspection or monitoring, and take corrective actions, such as repair or replace the line, modify the corrosion control programme, lower the operating pressure, etc.

The corrosion allowance (*CA*) is the difference between the actual pipeline wall thickness (*WT*), and the wall thickness required for pressure containment. The main purpose of the corrosion allowance is to 'buy' sufficient time to detect excessive (= beyond design) wall loss and take the necessary corrective actions to prevent a failure. According to these definitions, the time to failure, t_f, can in principle be calculated as:

$$t_f = CA/CR \tag{1}$$

To prevent a leak,

$$t_r < t_f \tag{2}$$

This equation defines an upper boundary of the initial inspection interval. In reality, the ratio t_r:t_f depends on the risk tolerance of the specific situation, and our confidence in controlling the worst case corrosion rates. The usual range is 2:1 to 1:2. For example, for an uninhibited (worst case) corrosion rate of 10 mm/y and a corrosion allowance of 6 mm, the line could be inspected between 6 months and 1 year into operation, depending on our confidence in the inhibition programme.

These relationships (1) and (2) are at the basis of a corrosion management programme which will include:

(1) a determination of the required corrosion allowance;

(2) design and implement a corrosion control programme;

(3) establish the monitoring procedures to identify out of compliance situations; and

(4) implement a management system to ensure that everything is operated as designed.

7. Corrosion Rate Predictions and Corrosion Allowance Calculations

Traditionally, the corrosion allowance is calculated by multiplying the predicted or worst case corrosion rate by the design life:

$$CA = CR \times N \quad (3)$$

where CA = corrosion allowance in mm, CR = predicted corrosion rate in mm/y, N = design lifetime in years. This 'deterministic' approach is valid when corrosion rates are known or accurately predictable. A more practical approach to defining the required CA is discussed later in this Section.

Several corrosion rate prediction models are available (see Table 2) to determine CR as a function of environmental variables, such as pressure, temperature, flow regime, etc. Some of the models are based exclusively on laboratory tests, while others rely on field experience, or on a combination of field and laboratory data. Until now, a 'head-on' comparison of the relative advantages of the models has not been available; a current joint industry project (JIP) is addressing precisely this gap.

Shell's preferred corrosion prediction tools have evolved from the well known de Waard-Milliams nomogram, to the current program Hydrocor. In addition to the corrosion rate profile, Hydrocor calculates the flow regime, water drop out, effect of cooling, effect of addition of glycol or methanol, scaling tendency, etc. An example of the output screen of the program is shown in Fig. 4. The program can also be used to calculate corrosion rates and the cumulative wall loss over the life of a project, based on the expected production rates and conditions; an example of a multi-year corrosion rate prediction for a 10 km offshore pipeline is shown in Fig. 5.

Hydrocor can also be used to monitor corrosion rates based on pH and iron content of the water phase. An example of its application is the design of the Troll field pipelines (Norway). The pipelines were designed for a 70 year lifetime. Corrosion rates and corrosion allowance were calculated on the basis of Shell's model. Corrosion

Table 2. *Partial list of available CO_2 corrosion prediction model*

Model name	Owner	Comments
Hydrocor	Shell Global Solutions	Integrates corrosion rate prediction with models for multiphase flow, thermodynamics, mass transfer, heat transfer and condensation. Simulates full pipeline. Accommodates more than one corrosion rate model.
SweetCor	Shell Oil Company (USA)	Data base with lab and field data in combination with models of De Waard *et al.* and USL model.
LipuCor	Total	Empirical model based on both lab and field data, including oil pipeline experience.
USL Model	USL	Calculates lifetime of tubing. Includes flash calculations for start of water condensation in tubing and multi-phase flow calculations.
Cormed	Elf	Based on field experience in tubings for (limited) number of countries. Includes effect of organic acids.

CO$_2$ Corrosion in wet gas or oil pipelines — Version 1.05 — ☐ Imperial units ☐ Default values ☐ Elevational profile

Case: M4 test case
Data file: MULTCASE.XLS

OPERATING CONDITIONS

Pressure inlet	bar	110
Temperature inlet	°C	80
Ambient temperature	°C	20
CO$_2$ content gas	mole%	3.5
Pipe inside diameter	m	4.10^{-1}
Gas flow rate	mln Sm^3/d	15
Liquid HC type		NGL ▼
NGL/oil flow rate	m^3/d	800
Water chemistry	No formation water ▼	
Water flow rate	m^3/d	2.00^2
Bicarbonates	ppm	100

CORROSION CONTROL

Alcohol type		MEG ▼
Alcohol concentration	w%	0
Inhibited corrosion rate	mm/y	0.1
Inhibitor availability	%	0

MISCELLANEOUS

Number of pipe nodes		50
Pipeline inclination	°	0
Pipe section length	m	200
Heat transfer coefficient	W/m^{2}K	20
NGL/oil viscosity	Ns/m^2	5.00^{-4}

MAIN OUTPUT

Max BOL corrosion	mm/y	23.78
Max TOL corrosion	mm/y	23.78
Maximum occurs at	m	0
Flow pattern		Annular dispersed
Liquid holdup		1.77%
Mixture velocity	m/s	13.44

Film velocity	m/s	4.95
Relative film length		1.00
Watercut		26.3%
pH		4.96
Fe^{2+}	ppm	0.0
FeCO$_3$ precipitation	mm/y	0.00

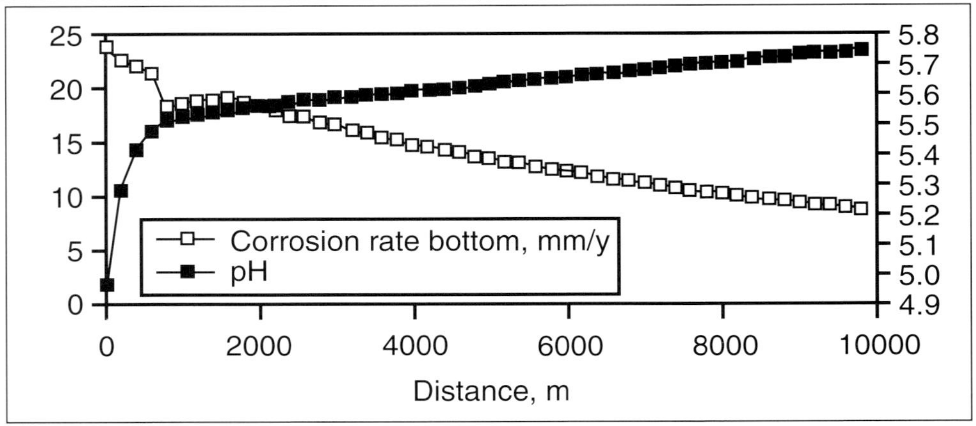

Scale protection factor	1.00
Crude oil protection factor	1.00

Glycol protection factor	1.00
Scale/inhibitor failure	0.00

Notes: Sm^3/d = standard cubic metres per day; MEG = monoethylene glycol; NGL = natural gas liquids; BOL = bottom of line; TOL = top of line.

Fig. 4 *Example of Hydrocor main screen, showing input data and corrosion/pH profile for a 10 km pipeline.*

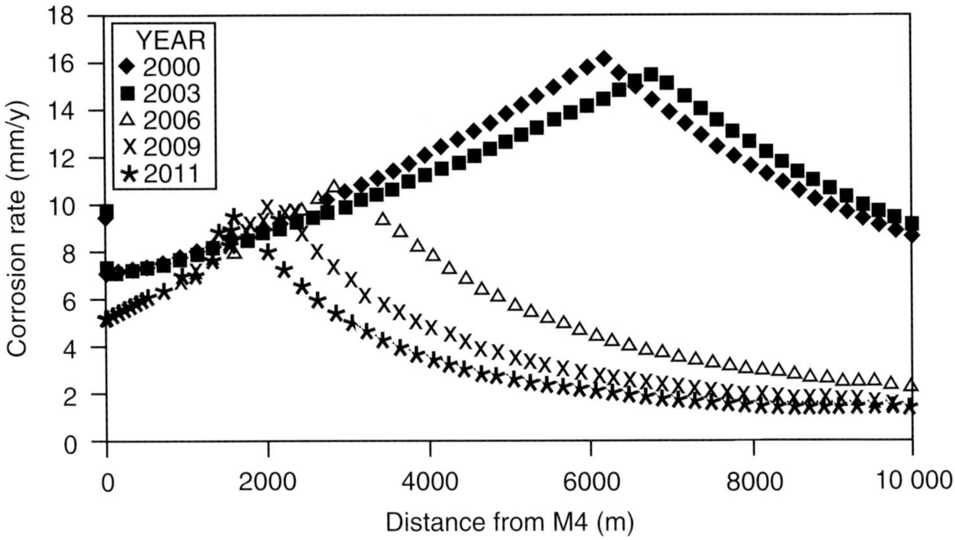

Fig. 5 *Corrosion rate predictions for a 10 km, 16 in. diameter offshore pipeline, for 12 years lifetime. M4 is a new production platform offshore Malaysia connecting to existing central production facilities platform (M3).*

and hydrate control are achieved by injection of glycol. Figure 6 shows an example of the predicted corrosion rates for different glycol injection rates. Recent measurements of dissolved iron and corrosion rates have validated the model predictions.

All models have limitations, either in the range of applicability or in the robustness of the underlying prediction algorithm. The majority of the models are most reliable in low temperature, sweet systems and less reliable for predicting corrosion rates at high temperatures, in the presence of H_2S, in severe flow regimes, or when corrosion inhibitors are used. Figure 7 shows schematically the proven range of application of existing models, on a plot of corrosivity vs temperature. Many of the new and challenging developments fall outside the range of these models. The three major limitations are:

(1) predicting the formation, stability and 'reparability' of protective scales;

(2) predicting the effect of crude oil or condensate in reducing corrosion rates; and

(3) predicting the effectiveness of corrosion inhibitors.

Shell's approach to calculating the required corrosion allowance for an inhibited pipeline is based on the concept of inhibited corrosion rates and inhibitor availability:

$$CA = CR_i \times N \times (1 - D/365) + CR_u \times N \times D/365 \quad (4)$$

where CR_i and CR_u are the inhibited and uninhibited corrosion rates, respectively, and D is the number of days per year that the inhibitor system is NOT available.

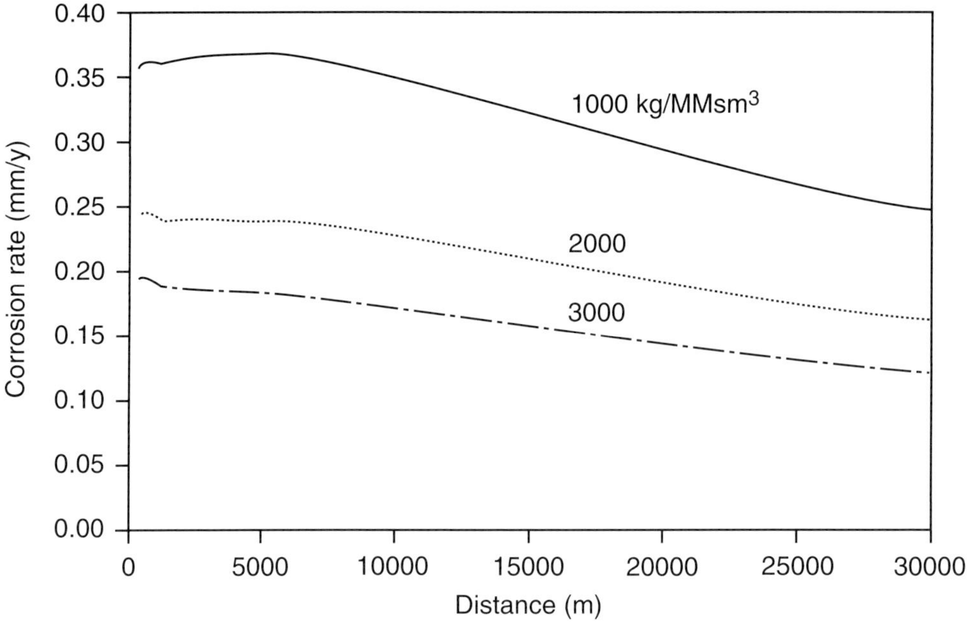

Fig. 6 *Calculated corrosion rates in the Troll pipelines for different glycol injection rates. The target is 0.15-0.20 mm/y for a 70 year lifetime (kg/MMsm^3 = kg per million standard cubic metre).*

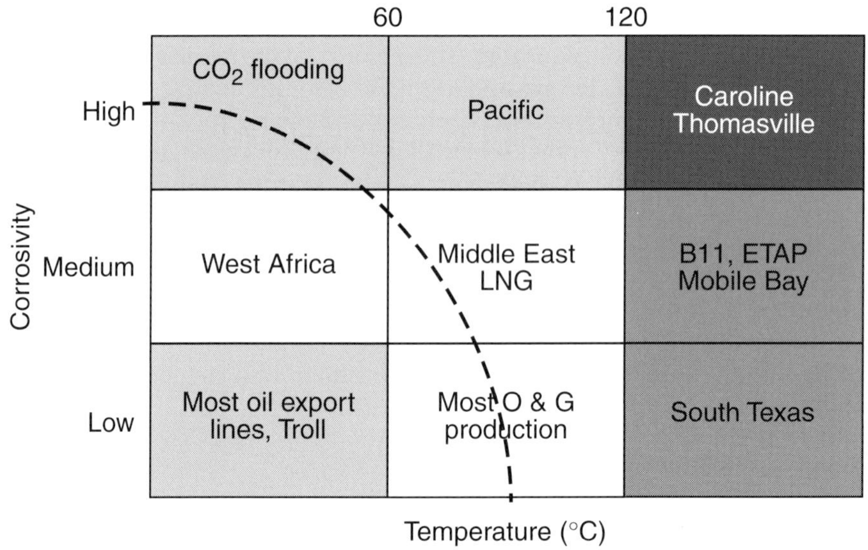

Fig. 7 *Temperature vs corrosivity map showing current and future gas production projects. The dotted line delineates the limit of application of most corrosion prediction models. The shaded areas represent the 'latent corrosivity' of the environment.*

This equation is a simplification of the actual conditions. For example, most inhibitors have a certain persistency, which means that they are not immediately stripped from the surface and continue to protect the pipelines even if not continuously present in the bulk fluids.

Traditionally, CR_i was estimated based on an 'inhibition efficiency' factor determined in laboratory tests. The available information from a number of pipelines, which as been partially summarised in Fig. 8, is that inhibited corrosion rates are fairly independent of the corrosivity of the environment; therefore, the concept of inhibition efficiency is invalid. In laboratory tests, inhibited corrosion rates below 0.1 mm/y are achievable. As will be discussed later, a conservative value of $CR_i = 0.2$ mm/y has been chosen for the corrosion allowance calculations.

When the models are no longer applicable, the Shell approach to estimating corrosion rates is to rely on a combination of selected laboratory testing under the specific conditions of the pipelines, and on field experience in analogous service. Laboratory tests have to be carefully planned and executed in order to represent the field conditions. For example, the water to gas ratio in gas condensate pipeline is low, and may lead to much lower corrosion rates than in typical 'water full' tests due to saturation of the water phase with corrosion products. This and other factors are considered in Shell's test programmes.

In reality, the corrosion allowance is limited by both technical and economic considerations, such as:

(1) maximum pipeline wall thickness that can be welded and laid;

(2) cost of additional steel; and

(3) maximum pit depth that can be inhibited.

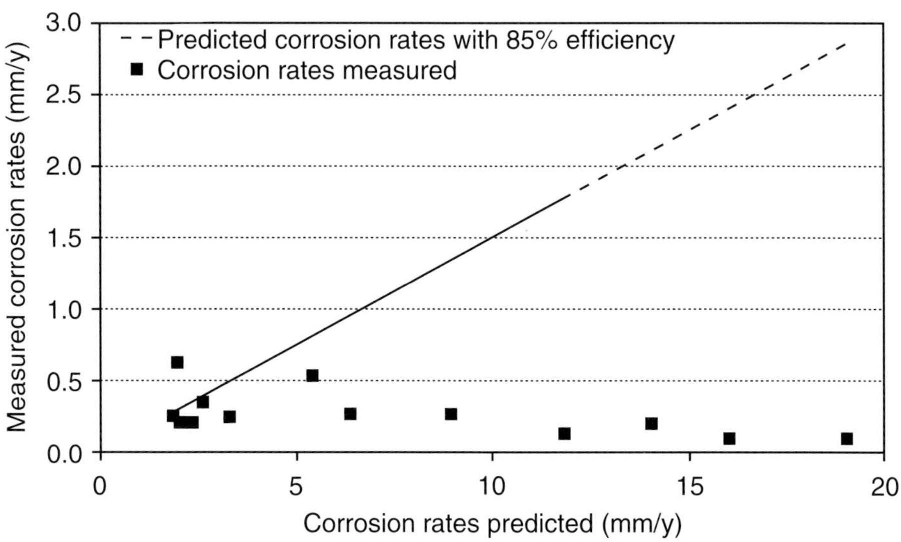

Fig. 8 *Measured vs calculated corrosion rates for existing inhibited wet gas pipelines. The solid line represents an 85% inhibition efficiency, and it clearly overestimates the achievable long term corrosion rates.*

In many cases, the corrosion allowance serves only to provide sufficient reaction time to take corrective actions in case corrosion rates are higher than expected. This means that the corrosion allowance must be allied to the corrosion monitoring and pipeline inspection programme.

Several operators, including Shell, have taken a pragmatic approach to defining the corrosion allowance. A maximum allowance of 6 to 10 mm is often specified, based on corrosivity, inspection plans, pipeline criticality and consequences of failure.

8. Risk Factors

The calculations in the preceding section assume that all parameters in eqn (4) are known, and adopt a single value. In reality, there is considerable uncertainty in the determination of these values, and even in the determination of the environmental conditions that determine corrosivity, such as pressure, temperature, flow and pH. The main uncertainties are:

- the accuracy and confidence level of the corrosion rate, CR_i and CR_u, predictions; that is, what is the confidence that the predicted corrosion rates will not be exceeded, if all systems operate as designed?
- the reliability (= will the inhibitor work as expected?) and downtime of the corrosion control system, which determine D;
- unexpected operation problems, such as higher than expected flow rates, corrosive gas concentrations, etc.

8.1. Accuracy and Confidence Level of Corrosion Rate Predictions

8.1.1. Uninhibited corrosion rates

In addition to temperature and partial pressure of corrosive gases, uninhibited corrosion rates will depend on the formation and stability of a protective scale. In sweet systems, the scale is normally iron carbonate. A lot of effort has been directed to understanding the conditions under which protective scales can form, and more importantly, can self-repair in case of mechanical damage. Shell's model Hydrocor includes an algorithm for predicting scale formation, based on the iron carbonate super-saturation level. The predictions have been supported by experimentation, although less information is available about actual field experience. Laboratory tests have shown, for example, that corrosion rates can fall from 50 to less than 1 mm/y if the dissolved iron concentration in the test solutions increases from 0 to 100 ppm. In a wet gas pipeline, this situation can occur a few metres from the inlet. Corrosion of the production tubing, upstream piping or vessels, if made of carbon steel, may generate a sufficiently high Fe^{2+} concentration to ensure scale protection also at the pipeline inlet. A detailed examination of the production facilities is needed to determine the expected iron concentration at the pipeline inlet.

In practice, we can expect a distribution of corrosion rates, between the 'scaled' and 'non-scaled' values. Intelligent pig inspection results usually show that corrosion rates follow a lognormal distribution, with many low values and a few high values. The underlying cause of this distribution is not obvious, although it probably reflects the complex interaction among all corrosion factors. This type of distribution has been used in the risk assessment calculations, discussed in Section 8.4.

8.1.2. Inhibited corrosion rates

Inhibited corrosion rates lower than 0.1 mm/y can be achieved in laboratory tests. Figure 8 shows that for a large number of inhibited pipelines under a variety of conditions, long term corrosion rates around and below 0.2 mm/y are achievable. For many of these lines, the actual design and operation of the corrosion control system (type of chemical, injection rate, pigging frequency) is unknown. We can take a conservative approach and assume that the system was perfectly designed and operated; this means that an inhibited corrosion rate $CR_i = 0.2$ mm/y is the best that can be achieved in practice. This is the approach taken in applying eqn (4) to most common situations (lower values have been used when sufficient confidence exists about the performance and availability of inhibitors). This again is an approximation. In reality, we expect that corrosion inhibition will have one of two possible outcomes:

(1) the inhibitor will be effective, in which case inhibited corrosion rates will be less than 0.1 mm/y, or

(2) the inhibitor will not be effective, in which case corrosion rates will equal the uninhibited rates.

A lognormal distribution of inhibited corrosion rates has been used in the risk assessment calculations (Section 8.4). An example of this distribution is shown in Fig. 9.

8.2. Reliability and Availability of Corrosion Control System

The parameter used to describe the reliability of the inhibition system is D, the inhibitor downtime. The reliability of the corrosion control system can be calculated from the reliability of its components (pumps, delivery lines, tanks, chemicals, operators) and their interaction. Several programs are available to perform this type of calculation. The system can be designed for close to 100% availability ($D = 0$), for example using full redundancy in all critical components such as pumps, tanks, and valves. The economic consequences of less than full redundancy need to be carefully evaluated.

D depends on:

(i) the time to detect a failure, for example, that inhibitor pumps are not running or the inhibitor tanks are empty; and

(ii) the time to take corrective action, for example, repair the injection pumps.

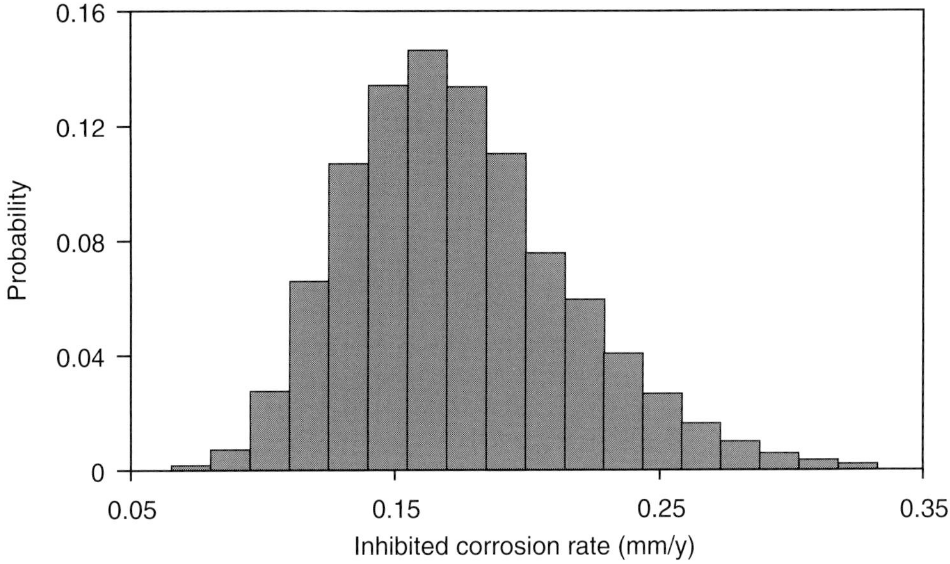

***Fig. 9** Lognormal distribution of inhibited corrosion rates. Mean = 0.2 mm/y, maximum 0.4 mm/y, minimum 0.05 mm/y.*

The corrosion management system should include procedures for minimising these times. For example, to keep *D* below 1 week per year, the operation of the pumps and the level in the corrosion inhibitor tanks needs to be controlled at least twice per week (this approach is similar to that discussed in relation to eqns (1) and (2)). Maintaining this degree of control will normally require some automatic monitoring system, such as 'pump off' and 'level low' alarms. Full stand-by redundancy is recommended for all critical components, for example two pumps interlocked for automatic switch-on.

For the risk assessment calculations in this example, a normal distribution of *D* can be assumed, having a mean of 7 days, a standard deviation of 3 days, a minimum of 1 day, and a maximum of 14 days. This range is fairly typical of several offshore installations. This implies that most shut-downs will be of short duration, between 1 and 7 days. The result of this distribution is graphically shown in Fig. 10.

8.3. Effect of Uncertainty in the Predictions on Corrosion Allowance

The uncertainty in the factors used in eqn (3) lead to a dispersion in the predicted wall loss: the difference between the maximum and minimum predicted wall loss increases, and consequently the required corrosion allowance also increases. As an example, Fig. 11 shows the range of wall loss contained within the 90% confidence level, for increasing uncertainty in the factors in eqn (4), expressed in terms of standard deviation as a percentage of the mean. The average predicted wall loss remains almost constant (= 8 mm), but the upper boundary, which in reality determines the required corrosion allowance, increases with uncertainty to about 12 mm for a 50% standard deviation. If more precise estimates of uninhibited corrosion rates and inhibitor

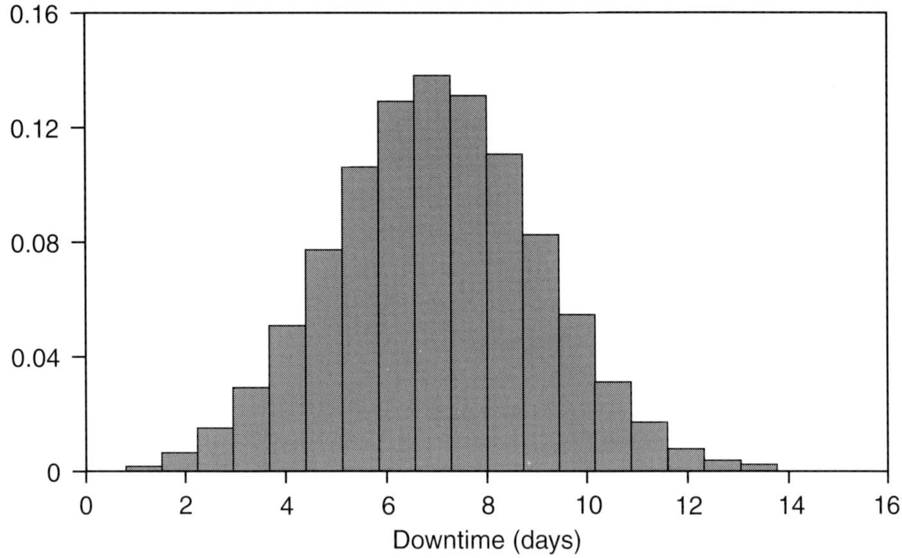

Fig. 10 *Normal distribution of inhibitor downtime. Mean = 7 days.*

system reliability are available, the uncertainty in the calculations and the recommended corrosion allowance could be reduced.

8.4. Probability of Failure

The probability of failure is defined as the probability that the corrosion wall loss will exceed the corrosion allowance during the life of the project. The probabilistic

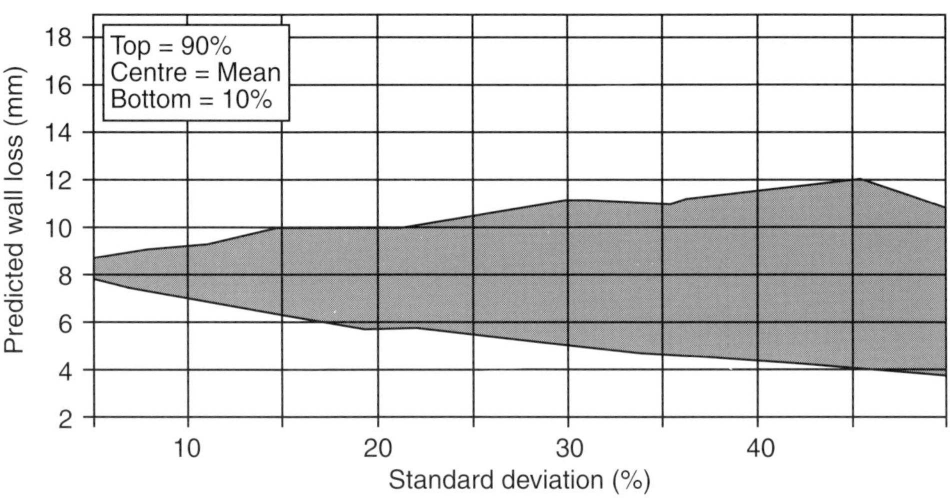

Fig. 11 *Range of predicted wall loss for a 12-year lifetime, calculated using eqn (4), as a function of the standard deviation of the factors used in the equation. The mean values are:* Cr_u *= 12 mm/y,* Cr_i *= 0.2 mm/y,* D *= 7 days.*

wall loss is calculated from eqn (4), using a commercially available risk calculation program. The program takes random samples from the distributions of the parameters in eqn (4), and calculates a randomised wall loss. By repeating (iterating) this procedure a large number of times (100–1000), the distribution of the wall loss is calculated. The probability of exceeding a given corrosion allowance (*CA*) is the integration of the probability distribution between zero and the given value of *CA*. Figures 12 and 13 show examples of these calculations, in the first case as a function of *CA* for several design lifetimes, and in the second case as a function of lifetime for several values of *CA*. For these examples, a minimum *CA* of 8–10 mm is required to reduce the probability of failure to acceptable levels (below 5%).

8.5. Cost of Failure

Cost is one way to represent in monetary terms the consequences of failure. Some of the consequences, for example damage to the environment, loss of human life, and damage to the company's reputation, cannot be fully represented by a cost figure. This means that an actual estimate of the cost of failure is specific to each situation. In the following example, only pipeline replacement costs and loss of production have been considered. This is of course a simplification, but it is consistent with the definition of failure that has been used, which implies that wall loss can be detected and corrected before the pipeline leaks.

Two scenarios have been considered in this example: (i) no failure; and (ii) extensive damage requiring full pipeline replacement. The more common situation, however, is a partial replacement of the most severely corroded sections of the pipeline.

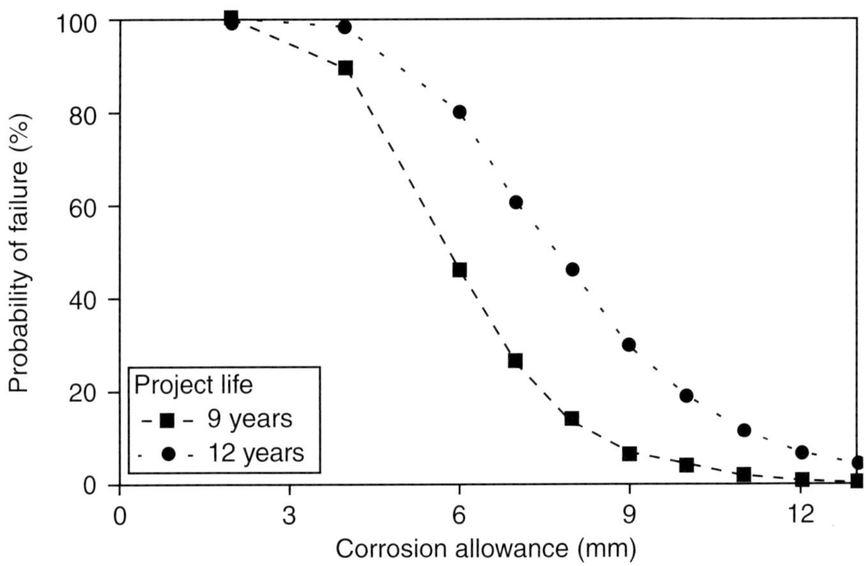

Fig. 12 Dependence of the probability of failure on the corrosion allowance, calculated using eqn (3), a lognormal distribution of corrosion rates and inhibitor downtime, and 2 values of N.

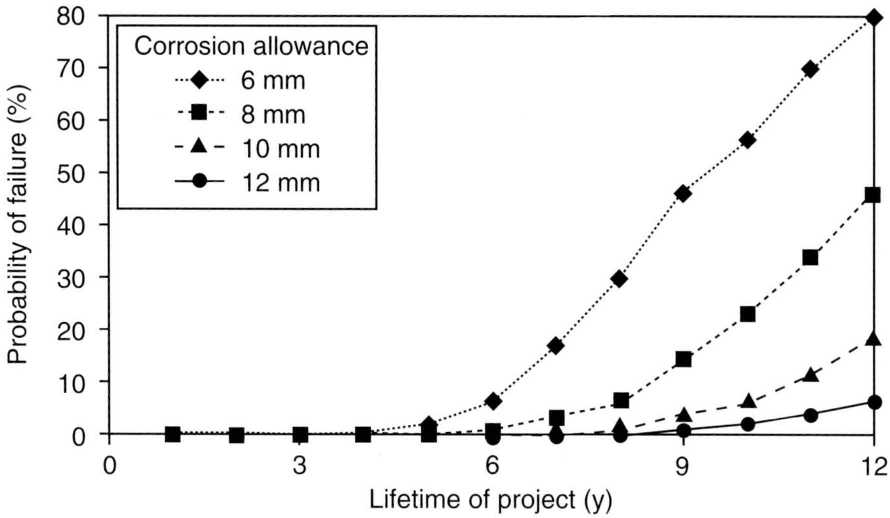

Fig. 13 *Dependence of probability of failure on project lifetime, for different values of corrosion allowance. The assumptions underlying these calculations are the same as for Fig. 11. A minimum corrosion allowance of 8 and 10 mm is required for 7 and 10 years life, respectively, to maintain the probability of failure below 5%.*

The future cost of pipeline replacement was estimated based on the present installation cost, corrected for inflation and discounted to present value. Deferred production is shifted to the last year of operation. In the base calculation, a downtime of 30 days was assumed. This seems reasonable for reconnecting a new pipeline, since the failure scenario assumes replacement in kind before leaking. However, an unexpected failure can result in a much longer downtime. The estimated gas and condensate margin was \$2/MMSCF (millions standard cubic feet) and \$10/bbl (barrel), respectively. In all cases, an inflation rate of 4% and a discount rate of 8% was assumed.

8.6. Risk (= Cost × Probability of Failure) Assessment

We now have all the elements necessary to calculate the risk of failure. The risk is calculated by multiplying the cost of failure (Fig. 14) by the probability of failure (Fig. 13). The results are shown in Fig. 15. The cost of failure is relatively low, maximum around US\$ 700 000. This means that, with these assumptions, the cost of failure makes a small contribution to the NPV (Net Present Value) of the carbon steel plus inhibition option. This option should be preferred for this example. Of course more realistic assumptions should be made on a case by case basis and could result in an opposite conclusion. In particular, the downtime could be considerably longer than 30 days, additional costs could be incurred due to environmental or safety reasons, the uncertainties in the predictions may make the perceived risk unacceptable, etc.

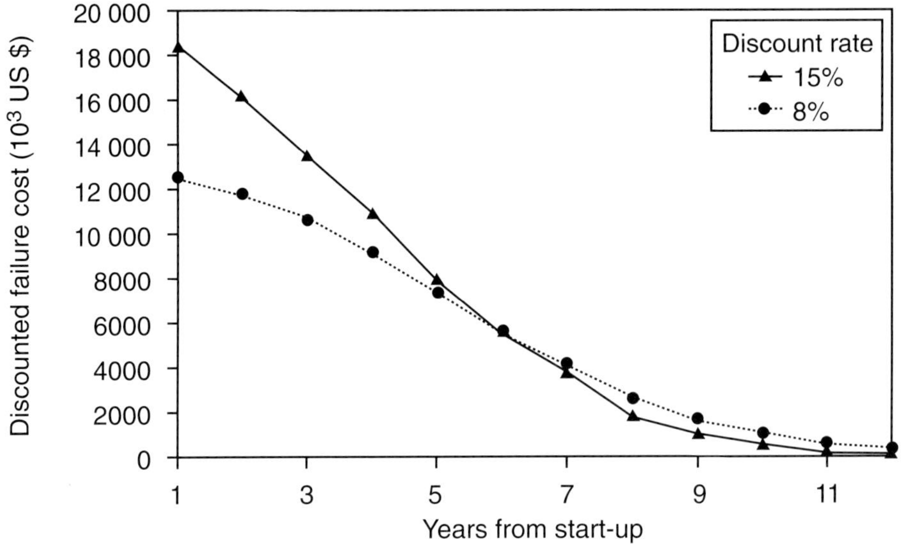

Fig. 14 Net present value of failure, for a full pipeline replacement plus 30 days deferred production.

9. Corrosion Management

In contrast with CRA pipelines, the successful operation of a carbon steel pipeline in wet, corrosive gas service can only be achieved if adequate measures are taken to ensure that the corrosion control programme is operated as designed. This requires

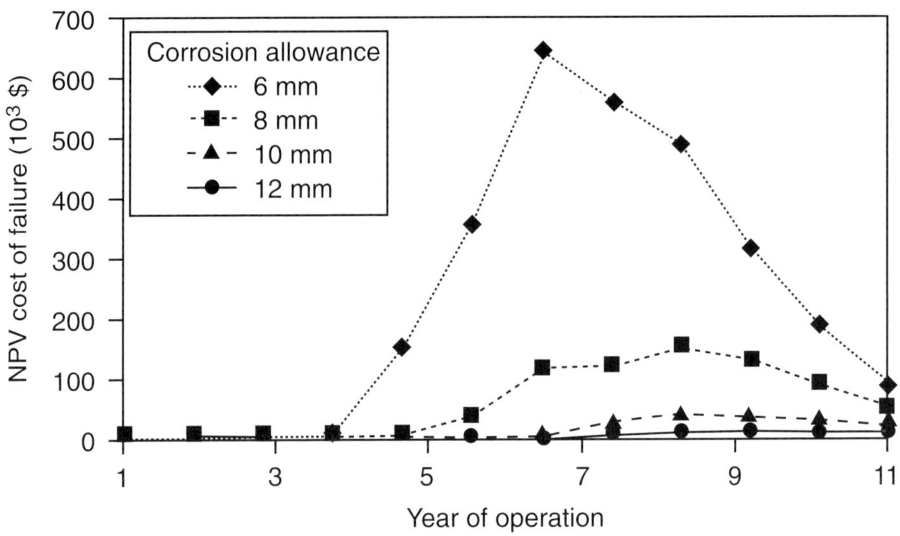

Fig. 15 Risk (= NPV cost of failure) for a 10-km pipeline as a function of life time. A corrosion allowance of 10 mm is the maximum that can practically be recommended. It also minimises the life cycle costs.

a strong commitment on the part of the asset owner, to ensure that the system is operated and maintained as designed. It also requires a clear understanding of the consequences of misoperation of the corrosion control system by all parties involved in the process — this includes operators, maintenance, chemical supplier, corrosion engineer, production chemist, purchasing, etc. The roles and responsibilities of all parties should also be clearly understood, and a 'chain of command' should be established to implement the correct actions in case of problems.

The main elements of corrosion management are:

- proper design of the pipeline and corrosion control systems — this includes the corrosion allowance, selection of corrosion inhibitors, design of the inhibitor injection system;

- implementation of a corrosion monitoring and inspection programme to detect any condition that deviates from the design, for example higher than expected corrosion rates, low inhibitor residuals, failure of the inhibitor injection pumps. This should include the acquisition of relevant production and corrosion data, analysis of the data by trained specialists, quality assurance of all chemicals;

- adequate fall back options to correct problems.

10. Detection of Out-of-Compliance Situations

Early detection of out-of-compliance situations, that is, wall loss or corrosion rates significantly higher than those used in the design of the pipeline, will allow sufficient reaction time to take corrective action before the integrity of the pipeline is affected. In addition to ensuring that pumps, tanks and lines are operational and that chemicals are delivered as specified, corrosion monitoring and pipeline inspection are the two complementary early detection methods.

10.1. Corrosion Monitoring

Corrosion monitoring is an integral part of the corrosion inhibition programme. One of the primary corrosion management measures is sampling and testing of the gas, condensate and water phases. In a highly corrosive environment, on-line sampling and testing should be considered, with feedback via telemetry to the operations control centre.

Traditionally, corrosion monitoring is accomplished using weight loss coupons or electrical resistance probes installed at both ends of the pipeline. Location of the probes is important to obtain significant results. The probes should preferably be flush mounted on the bottom of the line, where they will more likely be water wet. Water 'traps' can also be used to collect water and install corrosion probes.

Coupons and probes provide only single point results, and may be misleading in case of localised (or, as is often the case with CO_2, mesa-like) attack. More accurate measurements can be obtained by installing corrosion spools in the pipeline itself. These spools can be installed, for example, before and after inhibitor injection points to determine the effectiveness of inhibition. Monitoring the spools can be done using permanently installed ultrasonic (US) sensors.

The main disadvantage of these methods is that they provide only spot checks and measure corrosion rates at both ends of the pipeline, but not in the underwater section, where corrosion could be most severe due to higher water ratio, higher velocity, and/or higher condensation rate. A technique for measuring actual corrosion rates and wall loss in a small section of the pipeline itself, which is adaptable for subsea installation, is the FSM (field signature method). This method is similar to the US instrumented spool described above, except that it measures changes in the electrical field when an electrical current is impressed on the pipeline. Location of the spools is very critical for obtaining meaningful information, particularly if the actual operating conditions vary from those assumed during the design. For example, they could be installed in the horizontal run of the pipeline, at least 100 m away from the riser to allow the flow to stabilise.

10.2. Pipeline Inspection

Corrosion monitoring provides only a partial picture of corrosion conditions at the locations where the probes are installed. The pipeline must also be regularly inspected, using intelligent pigs, to determine its overall condition and validate the corrosion rate predictions and monitoring results. Magnetic flux leakage pigs are commonly used. These tools have a sensitivity of about 10% of wall thickness, which for a typical 12.5 mm wall thickness means that detectable pits will be approximately 2 mm deep. More sensitive (around 1 mm detection) ultrasonic pigs are available, but they are also more expensive to run.

The line should be inspected right after construction, to obtain a base line. Assuming that other corrosion monitoring data indicate low corrosion, the first follow-up inspection should be around 1 to 2 years into operation. For Opex estimates, yearly inspections should be assumed.

11. Corrective Actions (Fall-Back Options)

If higher-than-expected corrosion rates are detected through monitoring or inspection, corrective action should be taken to prevent failure. As discussed earlier, the main role of the corrosion allowance is to provide sufficient reaction time to take corrective action before failure. Several fall-back options are available, for example:

(1) increase inhibitor concentration;

(2) change to a more effective inhibitor;

(3) increase pigging frequency;

(4) initiate a batch inhibition programme;

(5) reduce the design life of the pipeline;

(6) downrate the pipeline; or

(7) improve the availability of the corrosion control system.

The choice of corrective action will of course depend on the cause of the problem, and an evaluation of the consequences. It is important that analysis of the monitoring and inspection data, and a 'decision tree' for taking quick corrective action be included in the corrosion management programme. As discussed in Section 9.3, it is important that an experienced and qualified corrosion engineer be assigned the responsibility for evaluating the monitoring and inspection data, and making the necessary recommendations.

12. Conclusions

The use of carbon steel with a corrosion control programme can be a cost effective alternative for wet gas transportation, but it involves additional risks. These risks can be managed, but the success of this approach depends on proper design, a clear understanding of the risks and a strong commitment to the long term operation of the corrosion control system. The Shell Group has successfully operated pipelines in corrosive service for many years with this approach.

2
Modelling the Probability of Flow Induced Localised Corrosion from Critical Hydrodynamic Data and Fracture Mechanics Data of Scales from CO_2 Corrosion of Steel

G. SCHMITT, C. BOSCH and M. MUELLER
Laboratory for Corrosion Protection, Iserlohn University of Applied Science, Frauenstuhlweg 31, D-58644 Iserlohn, Germany

ABSTRACT
Based on the hypothesis that the initiation of flow induced localised corrosion (FILC) is caused by momentum exchange of near-wall turbulence elements with the wall and subsequent fatigue cracking of corrosion product scales, a model has been developed to predict the probability of the onset of FILC by correlating flow intensities in terms of critical wall shear stresses with fracture mechanics data of corrosion product scales. The turbulence elements are quantitatively described by the hydrodynamic properties of the flowing liquid and included into the model with a turbulence coefficient. The effect of corrosion inhibitors on critical wall shear stresses for the initiation of FILC is introduced in the model by a drag reducing coefficient. The validity of the model was proved with experimental data from the system carbon steel/brine/CO_2/inhibitor (critical wall shear stresses without and with inhibitor, fracture stress of iron carbonate corrosion product scales). New experimental results are presented for fracture mechanics properties of iron carbonate scales from CO_2 corrosion of steel (Young's modulus, fracture stress, scale thickness dependent fracture strain, intrinsic growth stress, hardness, scale adhesion). Examples are given on the effect of corrosion inhibitors on critical wall shear stresses for FILC initiation under conditions of sweet and sour gas production.

1. Introduction

The need for the workover of sweet and sour gas production tubings is generally caused by localised corrosion such as pitting and/or flow induced localised corrosion (FILC), often called erosion corrosion. Therefore the understanding of the mechanism of FILC, specifically the initiation of FILC, is essential.

The classic mechanism of FILC (Fig. 1) starts with the scaled metal surface. The roughness of the scale gives rise to microturbulences which cause local thinning of the scale with subsequent formation of pores. Higher corrosion activities at these porous sites result in further local destruction of the scale followed by localised attack at this local anode where scale re-formation is hindered by high local flow intensities, thus stabilising the local anode.

If hydrodynamic forces are responsible for the local destruction of corrosion

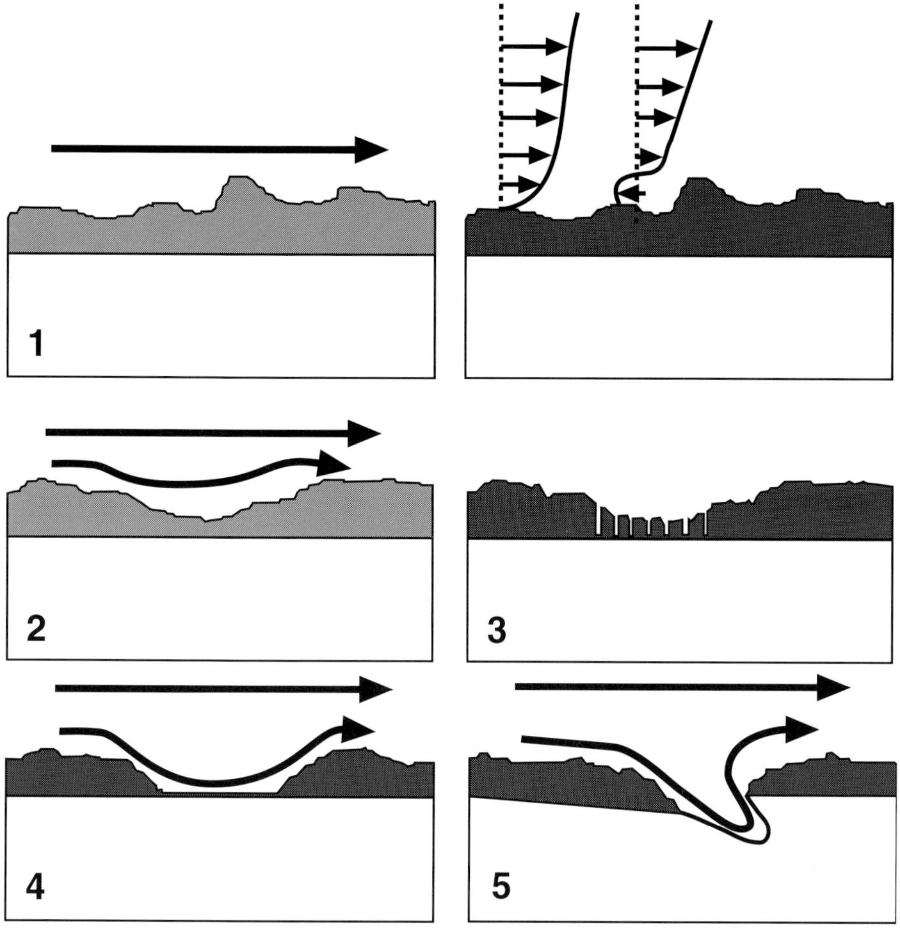

Fig. 1 *Development of flow induced corrosion (FILC); classic mechanism.*

product scales these forces must be high enough to overcome the binding forces in the crystal agglomerates of the scale and the adhesion forces between the metal substrate and the scale. Recent investigations [1,2] have proved, however, that hydrodynamic forces, e.g. expressed in terms of wall shear stresses, are generally several orders of magnitude too small (Table 1) to cause destruction and spalling of protective corrosion product scales and, hence, should not be able to initiate FILC.

On the other hand, there is no doubt on the existence of critical flow intensities above which corrosion product scales are hydrodynamically destroyed [3,4]. Our proposed hypothesis is that this is due to microturbulences in the near-wall region which exchange momentum with the wall and cause fatigue cracking of the corrosion product layer, which is considered as a porous crystal agglomerate. It is assumed that critical wall shear stresses result in the production of near-wall turbulence elements (vortices) which are critical in size, frequency, and energy to cause fatigue cracking of the scale.

Table 1. *Hydromechanics vs fracture mechanics*

Forces	Magnitude (Pa)
Wall shear stress • in typical technical systems • in regions of high flow intensities, e.g. at flow disturbances	 10^0–10^2 10^4–10^5
Adherence of corrosion product scales to the metal substrate	10^6 to 3×10^7
Fracture stresses of corrosion product scales • high porosity scales • low porosity scales	 10^7–10^8 10^8–10^9

This prompted us to develop a consistent concept to understand and quantify critical hydrodynamic and critical scale parameters, and to produce a probabilistic model that allows prediction of critical environment and materials related conditions for the onset of FILC under service conditions.

It should be emphasised that the model described below relates at this stage of the development only to the initiation of FILC by hydrodynamic destruction of 'smooth' surfaces with 'uniform' scales (initiation step type 3 in Fig. 2). Not considered here are defects in the scale due to flow independent pitting (e.g. due to selective adsorption [5,6], insufficient inhibition [5,6] and/or parent material inhomogeneities (initiation step 1 in Fig. 2) or flow independent cracking and spalling of scales (e.g. due to intrinsic compressive growth stresses [2,7,8] (initiation step type 2 in Fig. 2), although initiation steps 1 and 2 can also be influenced by flow. All three steps have in common that they produce unprotected surface sites where deep or shallow pitting will occur. At these pits the flow is disturbed. These surface imperfections create local turbulences which prevent scale re-formation above critical flow intensities. Under these conditions the erosion attack will propagate (Fig. 2).

The following theoretical considerations are general. However, to date, experimental verification has been demonstrated only for scales from CO_2 corrosion of steel.

2. Boundary Conditions for Flow Induced Scale Destruction

In order to understand the effect of flow on the initiation step of flow induced scale destruction one has to describe the pattern of turbulent flow, specifically in the near-wall region.

The current theories of wall turbulences consider the existence of turbulent bursts, ejections, sweeps and wall streaks [9–13]. These turbulence elements in the near-wall region of the turbulent boundary layer could cause fatigue cracking of scales due to local pressure changes.

If the scale (corrosion product layer) is considered as a crystal agglomerate with pores and voids, it makes sense to assume that a frequently repeated impact of turbulence elements onto this crystal agglomerate can cause a loosening of the crystals in the scale agglomerate and initiate destruction of the scale. This idea is sketched in Fig. 3 in a very simple two-dimensional model. The crystal agglomerate is symbolised

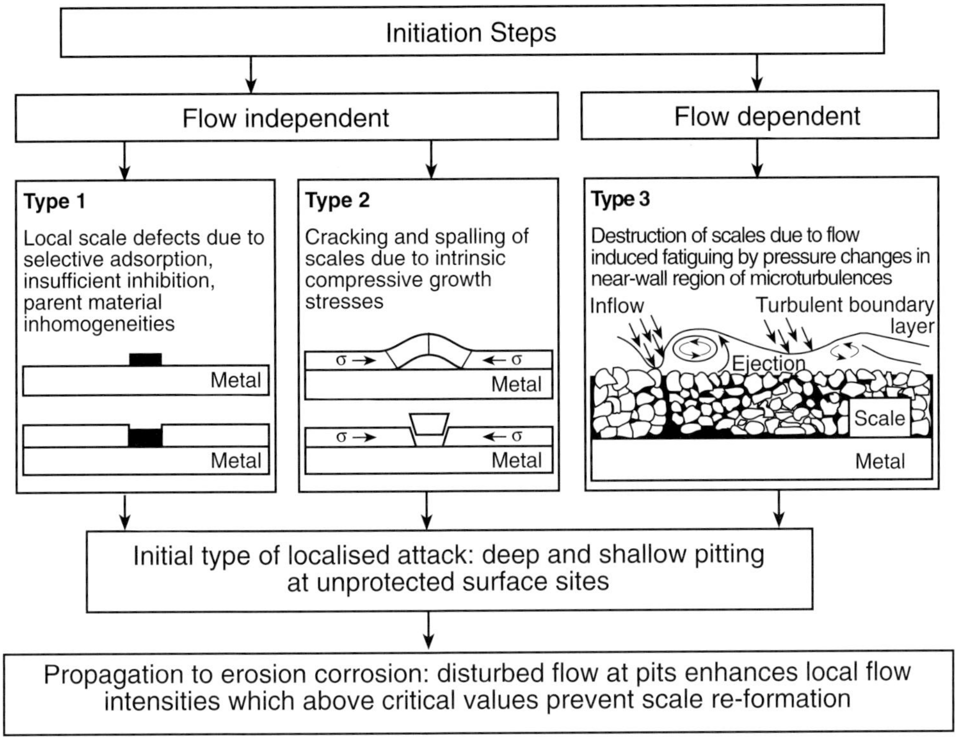

Fig. 2 *New concepts of FILC initiation.*

as an arrangement of small spheres sitting in a box with strong walls. A turbulence element is symbolised by a small piston that repeatedly impacts onto the surface of this arrangement of spheres. In order to loosen the spheres the impact of one hit must exceed a critical energy value, the magnitude of which depends on the adhesion and friction forces between the spheres and also on the impact surface area. The smaller the impact surface area the easier it is to loosen the arrangement of spheres. At a given impact intensity of the piston and given mechanical properties of the sphere arrangement there will exist a critical impact surface area above which no loosening of the sphere arrangement is possible (Fig. 4).

A predictive model must therefore correlate the mechanical properties of the scale as a crystal agglomerate (Figs 3, 4) with the hydrodynamic impact exerted onto the scale by turbulence elements (Figs 5, 6). For that purpose the turbulence intensity I_{turb} is introduced.

The turbulence intensity I_{turb} can be defined as the intensity of momentum transfer by near-wall turbulence elements (vortices) to the wall according to eqn (1):

$$I_{turb} = \frac{P_V}{T_c \cdot A_V} \tag{1}$$

with I_{turb} = turbulence intensity (Pa), P_V = momentum transferred by the turbulence elements to the wall (single burst) (kg ms^{-1}) and T_c = time between single bursts (s).

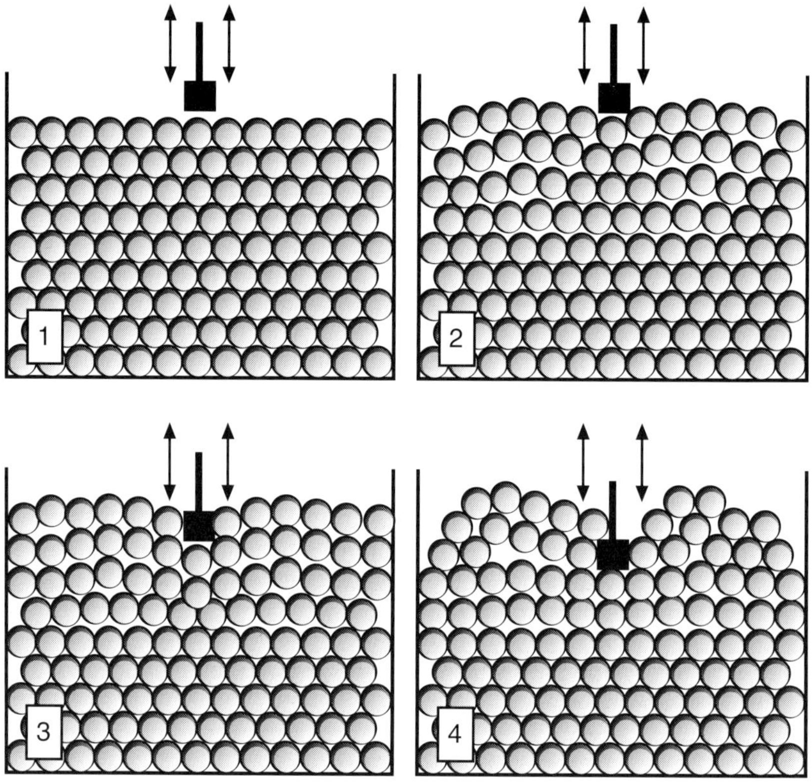

Fig. 3 *Loosening of scale crystal agglomerates by local impact (two-dimensional model).*

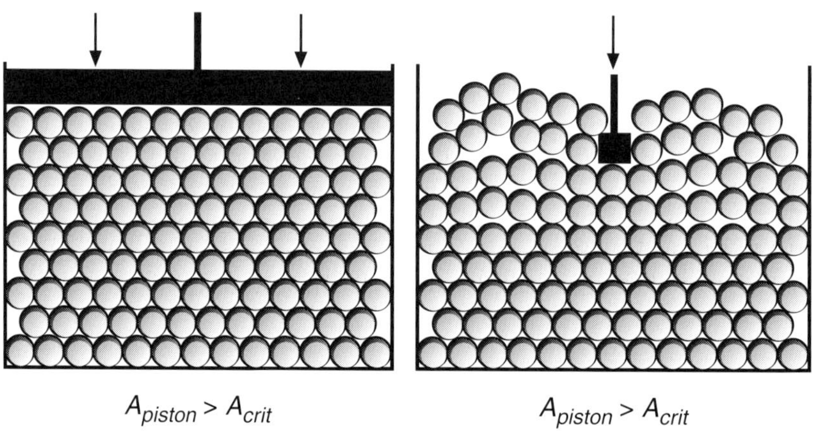

Fig. 4 *Critical surface area for destructive penetration into scales (two-dimensional model).*

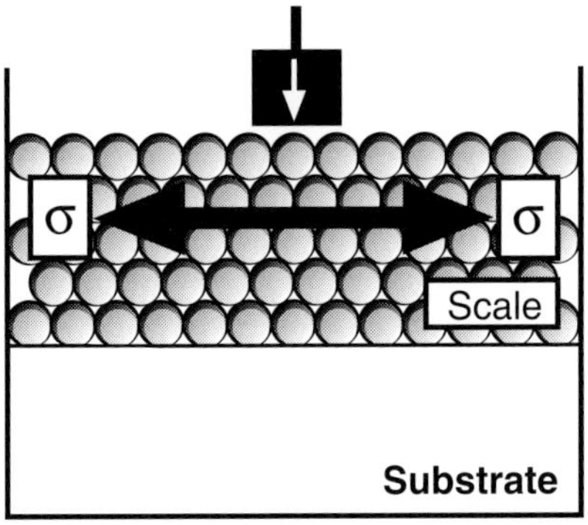

Fig. 5 *Tensile stress in corrosion product scale exerted by impact of turbulence elements.*

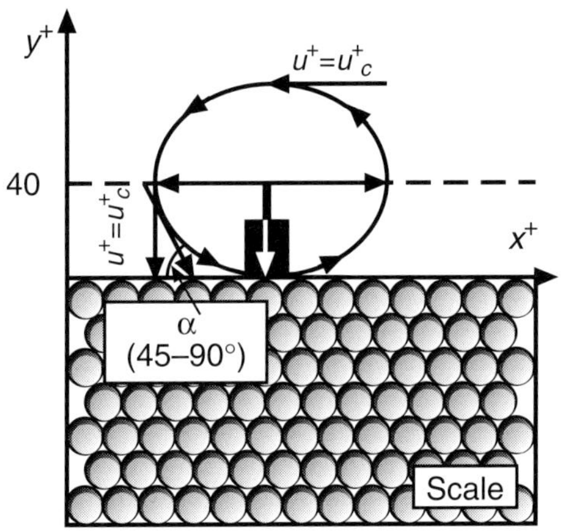

Fig. 6 *Model of a near-wall turbulence element (vortex).*

$$T_c = T_c^+ \frac{\nu \cdot \rho_V}{\tau_w} \tag{2}$$

with T_c^+ = dimensionless time between single bursts, ρ_V = density of the vortex (kgm^{-3}), ν = kinematic viscosity of the vortex (m^2 s^{-1}), τ_w = wall shear stress (Pa) and A_V = surface area impacted by a single burst (m^2).

In the porous corrosion product scale an extrinsic stress is built up as a consequence of the momentum transfer by the turbulence elements (Fig. 5). Scale fracture will occur if the turbulence intensity exceeds the fracture stress σ_y of the corrosion product scale (eqns (3),(4)):

$$I_{turb} \geq \sigma_y = E_{SC} \cdot \varepsilon_{SC} = E_0 (1-p)^3 \cdot \varepsilon_{SC} \tag{3}$$

with E_{SC} = Young's modulus of the scale (Pa), ε_{SC} = critical fracture strain, E_0 = Young's modulus (single crystal) (Pa) and p = porosity of the scale.

$$\frac{P_V}{T_c \cdot A_V} \geq \sigma_y \tag{4}$$

An additional condition for scale fracture is that the impact area of a single burst is small enough, i.e. (sub)critical to yield fatigue destruction of the scale. This (sub)critical impact area can be estimated from eqn (5):

$$A_{V(crit)} \leq \frac{P_V}{T_c \cdot \sigma_y} \tag{5}$$

The momentum transferred by a single burst to the wall is given by:

$$P_V = m_V \cdot u_c = V_V \cdot \rho_V \cdot u_c \tag{6}$$

with m_V = mass of the near-wall turbulence element (kg) and u_c = characteristic streamwise vortex velocity (ms^{-1}).

$$u_c = u_c^+ \sqrt{\tau_w / \rho_V} \tag{7}$$

with u_c^+ = dimensionless characteristic vortex velocity and V_V = volume of the vortex (m^3).
In the literature [10,12–17] direct information is given concerning the wall distance of formation, size and streamwise velocity of a near-wall turbulence element:

- At $y^+ = 40$ the most energetic streamwise fluctuations are characterised by streamwise dimensions of $\lambda^+ \approx 100$ [14–17] (λ^+ = dimensionless streamwise eddy diameter).

- These turbulence elements exchange momentum with the wall. The dimensionless diameter of the momentum transfer area is about $x^+ \approx 100$ [10,13,17].

- The impact angle with the wall is about 45° to 90° [10].

- The large eddies travel downstream with a speed characteristic of the boundary layer [15].

With this information a near-wall turbulence element can be modelled as an ellipsoidal vortex (Fig. 6). The volume of the vortex is therefore given by:

$$V_V = \tfrac{4}{3}\pi(x_c^+)^2 \frac{\nu^2 \rho_V}{\tau_w} y_c^+ \frac{\nu}{\sqrt{\tau_w / \rho_V}} \tag{8}$$

with x_c^+, y_c^+ = dimensionless characteristic radii of the ellipsoidal vortex.

Including the momentum of a single burst (eqns 6–8) and the time between single bursts (eqn (2)) into eqn (5), the critical impact area becomes:

$$A_{V(crit)} \leq \frac{P_v}{T_c \cdot \sigma_y} = \frac{4\pi(x_c^+)^2 y_c^+ u_c^+ \nu^2 \rho_V}{3T_c^+ \cdot \sigma_y} \tag{9}$$

On the other hand the critical impact area $A_{V(crit)}$ is given by the circular projection area of the ellipsoidal vortex on the smooth scale surface:

$$A_{V(crit)} = \pi \cdot a^2 = \frac{\pi(x_c^+)^2 \nu^2 \rho_V}{\tau_{w(crit)}} \tag{10}$$

with a = radius of the projection area (m).

By equating eqn (9) with eqn (10) an expression is obtained which correlates the critical wall shear stress with the fracture stress of the corrosion product layer:

$$\tau_{w(crit)} = \frac{3T_c^+}{4y_c^+ u_c^+} \cdot \sigma_y \tag{11}$$

The parameters T_c^+, y_c^+ and u_c^+ represent the turbulence elements of the flow and can be combined to a turbulence coefficient K_V:

$$K_V = \frac{3T_c^+}{4y_c^+ u_c^+} \tag{12}$$

Introducing data from the literature, T_c^+ e.g. =100 [17], $y_c^+ = 40$ [14–17] and using the universal velocity profile [18]:

$$u_c^+ = 2.5 \ln y_c^+ + 5.5; \quad 40 \leq y_c^+ \leq 100 \tag{13}$$

a value of $K_V = 0.127$ is obtained.

3. Drag Reduction by Surface Active Compounds (SAC)

It has been known for about 50 years that SACs can reduce the wall shear stress of flowing liquids even if added only in minute, ppm, concentrations [19]. This 'drag reducing' effect was reported primarily for linear, randomly-coiling macromolecules (polyethoxylates, polyacrylamides) [20]. However, also detergents like cetyltrimethylammonium bromide [21] and other quaternary ammonium compounds [22] were found to exert drag reducing effects which in contrast to the macromolecular additives are significantly less susceptible to mechanical degradation processes. In our own investigations 1 mmolL^{-1} of corrosion inhibitors like dodecylamine, octadecylamine, also in combination with dodecylalcohol tetraethoxylate (C_{12}-4EO), proved to significantly reduce the wall shear stress of a brine/hydrocarbon (90/10 vol/vol) film driven by an 18.4 ms^{-1} CO_2 gas flow under annular mist flow conditions [23,24]. Therefore, part of the inhibitive action of certain inhibitors must be attributed to its drag reducing effect. The C_{12}-ethoxylate with much less drag reducing power exhibited also a much lower inhibitive efficiency. The high protectivity of an equimolar mixture of dodecylamine with the ethoxylate is believed to be due to a combination of cooperative adsorption and drag reducing effect. Evidence for the drag reducing action of some other corrosion inhibitors was also observed in CO_2 corrosion of steel under full bore flow conditions [3,25–27].

Although the drag reducing effect has been explained mostly by coil-stretch transition of macromolecules in velocity gradients [28,29], there is now clear experimental evidence that adsorbed species and SAC aggregates in the near-wall boundary layer of the flowing liquid play a significant role in decreasing wall shear stresses [23,24,30,31] and can also increase critical flow intensities of FILC [4]. The combination of drag reducing with inhibiting effects can open a new door in inhibitor development for highly turbulent flow systems.

3.1. Aggregation of SAC

Surface active compounds usually aggregate as lamellar or spheroidal micelles. Spheroidal micelles (Fig. 7) are usually formed by aggregation of 10 to 100 monomers and reach diameters of 3–6 nm [32]. Above critical monomer concentrations and shear rates rod-like micelles are formed by aggregation of spheroidal micelles; e.g. tetradecyltrimethylammonium bromide forms rod-like micelles of about 250 nm in length [33].

The formation of rod-like micelles above critical SAC concentrations and shear rates is a precondition for the drag reducing effect [34,35]. The flow itself orients the

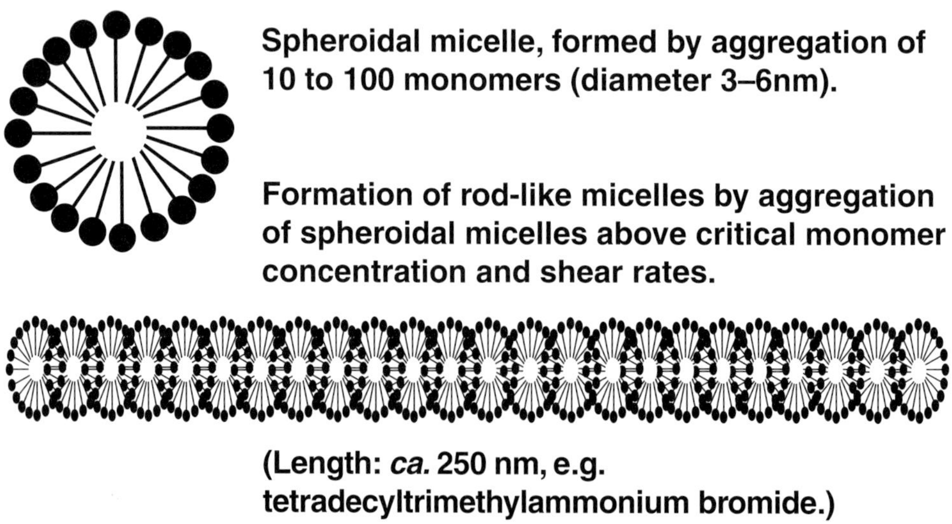

Fig. 7 Aggregation of surface active compounds (SAC) and formation of micelles.

rod-like micelles in the mean flow direction and forms shear induced structures (SIS). A model proposed by Hoffmann [36] is shown in Fig. 8. The macroscopic consequence for pipe flow is a damping effect on the radial turbulent fluctuations. The characteristic length of the near-wall turbulence elements is increased while the burst frequency decreases [37].

3.2. Modelling SAC Effects

In the presence of surface active compounds (SAC) the momentum P_V transferred by a single burst (eqn (6)) is strongly reduced. This can be modelled by introduction of a drag reducing coefficient K'_{SAC}.

$$P_V = K'_{SAC} \cdot V_V \cdot \rho_V \cdot u_c \tag{14}$$

with K'_{SAC} = drag reducing coefficient related to SAC effects on the momentum transfer of a single burst ($0 < K'_{SAC} < 1$), V_V = volume of the vortex (m^3), ρ_V = vortex density (kgm^{-3}) and u_c = characteristic vortex velocity (ms^{-1}).

As stated above, the streamwise characteristic length of the near-wall turbulence elements is increased by SAC. This affects the critical impact area (eqn (10)). The increase of $A_{V(crit)}$ can be described by use of an interaction coefficient K''_{SAC} which takes into account the effects of SAC on vortex geometry:

$$A_{V(crit)} = \pi \cdot (aK''_{SAC})^2 \tag{15}$$

with a = radius of the projection area (m).

$$a = \frac{\nu \cdot x_c^+}{\sqrt{\tau_w / \rho_V}} \tag{16}$$

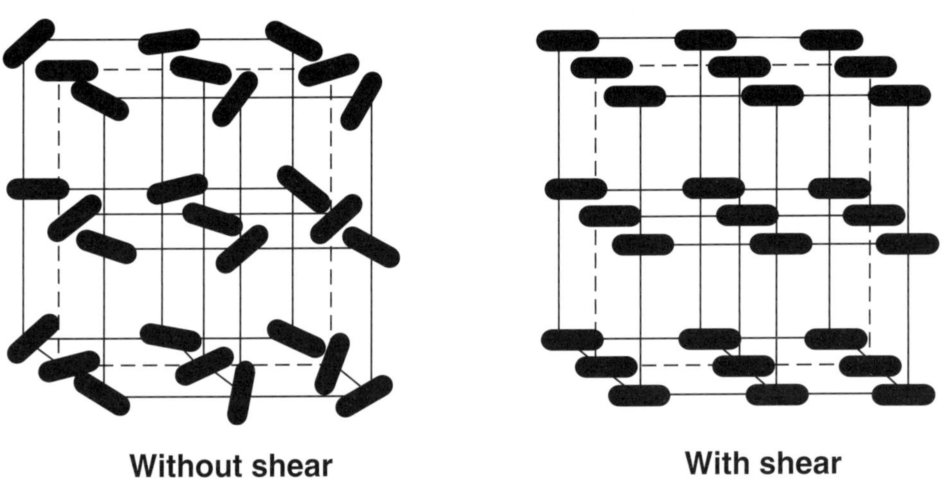

Fig. 8 Effect of flow on SAC aggregates: orientation of rod-like micelles under influence of shear [36].

The effect of SAC on the impact area is demonstrated in Fig. 9. The oriented rod-like micelles cause an increase in vortex geometry, i.e. an increase of the impact area of single bursts in combination with the reduction of the rotation velocity $|u|$ of near-wall vortices. Therefore the momentum exchange of the turbulence elements with the wall perpendicular to the mean flow direction is reduced.

By use of eqn (9) and eqn (14) the critical impact area $A_{V(crit)}$ can be modelled as follows:

$$A_{V(crit)} \leq \frac{P_V}{T_c \cdot \sigma_y} = \frac{K'_{SAC} \cdot V_V \cdot \rho_V \cdot u_c}{T_c \cdot \sigma_y} = K'_{SAC} \frac{4y_c^+ u_c^+}{3T_c^+} \cdot \pi (x_c^+)^2 \nu^2 \rho_V \cdot \frac{1}{\sigma_y}$$

$$= K'_{SAC} \cdot \frac{1}{K_V} \cdot \pi (x_c^+)^2 \nu^2 \rho_V \cdot \frac{1}{\sigma_y} \tag{17}$$

On the other hand $A_{V(crit)}$ is given by eqn (15):

$$A_{V(crit)} = \pi (a \cdot K''_{SAC})^2 = (K''_{SAC})^2 \cdot \pi (x_c^+)^2 \nu^2 \rho_V \cdot \frac{1}{\tau_{w(crit)}} \tag{18}$$

From eqn (17) and (18) a relation can be obtained between the critical wall shear stress and the fracture stress of the scale with regard to SAC effects:

$$\frac{K'_{SAC}}{K_V \cdot \sigma_y} = \frac{(K''_{SAC})^2}{\tau_{w(crit)}} \tag{19}$$

$$\tau_{w(crit)} = K_V \cdot \frac{(K''_{SAC})^2}{K'_{SAC}} \cdot \sigma_y = \frac{K_V}{K_{SAC}} \cdot \sigma_y \tag{20}$$

with K_{SAC} = drag reducing coefficient and K_V = turbulence coefficient.

The drag reducing coefficient K_{SAC} can be evaluated by introducing experimental data for critical wall shear stresses and fracture mechanics data of the corrosion product scales formed under these conditions.

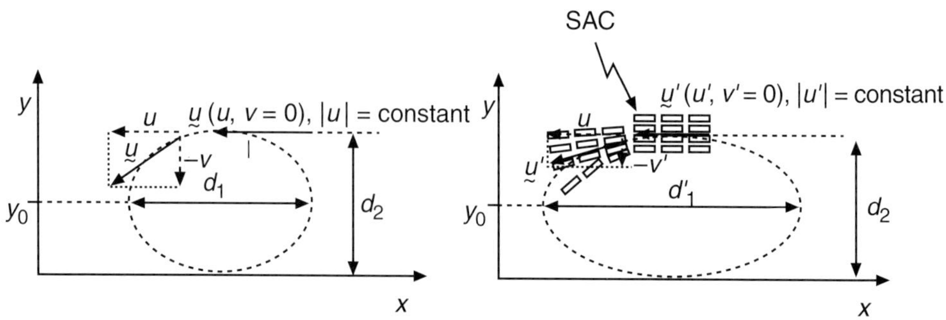

***Fig. 9** Effect of SAC on vortex geometry.*

In the absence of SAC the critical wall shear stress is given by eqn (11a):

$$\tau_{w(crit)} = K_V \cdot \sigma_y \tag{11a}$$

$$K_V = \frac{\tau_{w(crit)}}{\sigma_y} \tag{21}$$

In the presence of SAC the critical wall shear stress can be estimated using eqn (22):

$$\tau_{w(crit,SAC)} = \frac{K_V}{K_{SAC}} \cdot \sigma_{y(SAC)} \tag{22}$$

from which

$$K_V = \frac{\tau_{w(crit,SAC)} \cdot K_{SAC}}{\sigma_{y(SAC)}} \tag{23}$$

Therfore K_{SAC} can be evaluated from experimental data in the following way:

$$K_{SAC} = \frac{\tau_{w(crit)}}{\tau_{w(crit,SAC)}} \cdot \frac{\sigma_{y(SAC)}}{\sigma_y} \tag{24}$$

Equation (20) can be applied to correlate turbulence effects in terms of critical wall shear stresses with the fracture stress of corrosion product scales for the case in which the energy of a single burst is high enough to cause a microcrack in the scale. Consequently the boundary condition for the crack growth is given by:

$$\frac{E_{V,kin}}{A_V} > \frac{1}{2}\frac{\rho_V \cdot V_V \cdot u_c^2}{A_V} > 2\gamma_0(1-p)^3 \tag{25}$$

with $E_{V,kin}$ = kinetic energy of a single turbulence element (J), A_V = critical impact area (m), γ_0 = surface fracture energy (single crystal) (Jm^{-2}) and p = porosity.

In this case a repeated impact of single bursts would result in growth of the crack length and finally cause a through-scale crack.

Dependent on the burst frequency n_c the time for the development of a through-scale crack can be estimated. With the burst frequency

$$n_c = \frac{1}{T_c} = \frac{\tau_w}{T_c^+ \cdot \nu \cdot \rho_V} \tag{26}$$

and the number of bursts n needed for through-scale cracking

$$n \cdot \sigma_{single\ burst} = \sigma_y \tag{27}$$

$$n = \frac{\sigma_y}{\sigma_{single\ burst}} \tag{28}$$

with

$$\sigma_{single\ burst} = \tau_w \cdot \frac{K_{SAC}}{K_V} \tag{29}$$

where τ_w = supercritical wall shear stress (in Pa).

The time T_{crack} for through-scale cracking is given by

$$T_{crack} = \frac{n}{n_c} \tag{30}$$

For a given set of experimental data (Table 2) obtained from CO_2 corrosion experiments with low alloy steel a through-scale cracking time of 21 min was calculated according to eqn (30). It is interesting to note that in the relevant jet cell experiment [38] which was performed under video control, the time between the start of the experiment and the beginning of scale spalling was about 20 min. This gives some confidence that the approach outlined above yields relevant results and can be used in further modelling.

4. Generation of Data of the Fracture Mechanics Properties of Corrosion Product Scales, Demonstrated for Iron Carbonate Scales from CO_2 Corrosion of Steel

To assess the fracture mechanical properties of corrosion product scales information is needed on the Young's modulus E, the yield stress σ_y, the critical fracture strain, ε_{crit}, the intrinsic growth stress, σ_{intr}, the limiting scale thickness, h_y, the surface fracture energy, γ, and the adhesion force of the scale to the metal substrate.

In previous work [2,7,8] some of these data were estimated for iron carbonate scales formed by CO_2 corrosion of steel from theoretical considerations and literature data. The present work generated new experimental data which will be outlined below.

Table 2. *Estimation of time for through-scale cracking (data from CO_2 corrosion of steel)**

Turbulence coefficient K_V	0.127
Drag reducing coefficient K_{SAC}	0.115†
Fracture stress σ_y (GPa)	1.08
Single burst stress $\sigma_{single\ burst}$ (Pa)	172
Burst frequency n_c (s^{-1})	5022
Number of bursts n needed through-scale cracking	6.3×10^6
Time for through-scale cracking (min)	21

* 1M NaCl, 1 mmol L^{-1} MNQB(see section 4.1), 5 bar CO_2, 80°C, τ_{crit} = 190 Pa, 96 h.
† without inhibitor: τ_{crit} = 1 Pa, σ_y = 0.049 GPa (from Fig. 18, p.44).

4.1. Origin of Iron Carbonate Scales

The iron carbonate scales investigated in this study were formed by corrosion of low alloy C–Mn steels (chemical composition in mass%: C 0.36–0.37; Si 0.19–0.25; Mn 1.26–1.32; Cr 0.03–0.11; Ni 0.01–0.11; Mo 0.01–0.07; Cu 0.03–0.09; P 0.012–0.017; S 0.005–0.024; Al 0.005–0.074) with different microstructure (as-rolled ferritic–pearlitic (API 5CT J55) and quenched-and-tempered martensitic–bainitic (API 5CT C75 and C90) microstructure) in 1M NaCl solution under 1 and 5 bar CO_2 at 80°C. The effect of inhibitors was tested using 10^{-3}M N-(2-methylenenaphthyl)quinolinium bromide (MNQB).

4.2. Experimental Methods

4.2.1. Fracture stress and fracture strain

The fracture stress and fracture strain of iron carbonate scales were directly measured by 4-point loading tests (4-PLT) (Fig. 10) with a constant bending rate of 10 $\mu m s^{-1}$. The test coupons (60 × 18 × 2 mm) were covered with the carbonate scale only on the outer side and only on an area of 40 × 18 mm. The rest of the surface had been protected during the scale-forming corrosion experiment by an easily removable organic coating. Acoustic emission (8900 Locan, Fischer-Pierce & Waldburg GmbH, Germany) was used to identify the onset of scale cracking. A typical acoustic emission plot of a 4-PLT is given in Fig. 11. The scale-covered coupon surface was investigated by scanning electron microscope (SEM) before and after the 4-PLT to prove that the scale cracked only during the test and to identify the cracking mode.

The fracture stress and fracture strain were calculated according to eqn (31) and (32) [39].

$$\sigma = \frac{12E \cdot t \cdot y}{3H^2 - 4A^2} \text{ (MPa)} \tag{31}$$

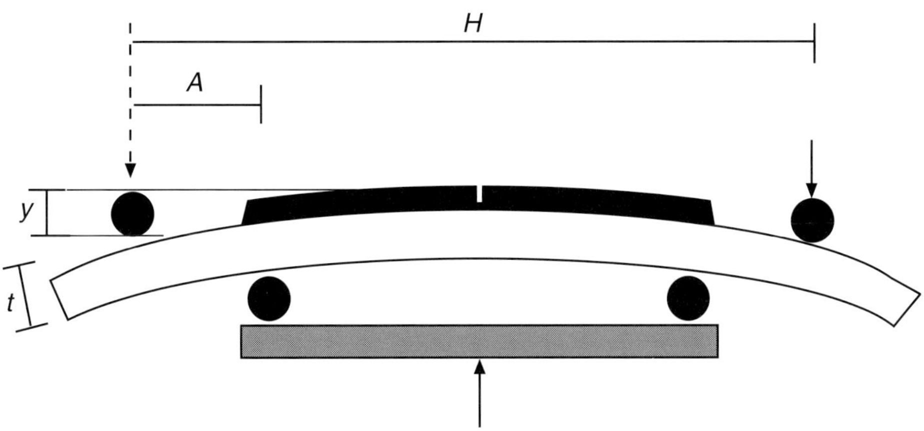

Fig. 10 *4-point loading test.*

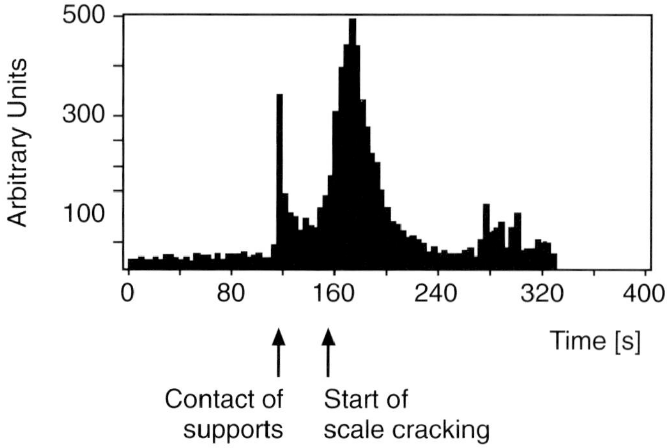

Fig. 11 *Acoustic emission plot.*

$$\varepsilon = \frac{12 \cdot t \cdot y}{3H^2 - 4A^2} \tag{32}$$

In these equations σ is the maximum tensile stress (Pa), E the Young's modulus (GPa), t the thickness of the specimen (including the scale) (m), y the maximum deflection between the outer supports (m), H the distance between the outer supports, A the distance between the inner and outer supports, and ε the fracture strain.

Fracture stress and fracture strain testing was performed on iron carbonate scales with thicknesses ranging from 2 to 100 μm. These scales were produced on J55 steel samples by CO_2 corrosion under standard conditions at 5 bar CO_2 in the absence and presence of the inhibitor MNQB and with exposure times between 3 and 120 h.

4.2.2. Intrinsic growth stress

Intrinsic growth stresses of iron carbonate scales result from the fact that carbonate crystals growing on the metal lattice have a higher space demand than the metal lattice. They were measured by two methods:

(i) by direct dilatometric measurement, and

(ii) by X-ray diffraction.

Dilatometric method. Steel strips (200 × 10 × 0.5 mm) were fixed in an internal stress meter originally developed to measure *in situ* the intrinsic stress of electroplated metal coatings [40] (Fig. 12). The intrinsic stress produced by the growing coating or scale results in a lengthening (compessive stress) or a shortening (tensile stress) of the steel strip which was measured directly with an accuracy of 1 μm. Equipment was developed for direct dilatometric measurements at ambient pressure and elevated pressures and temperatures.

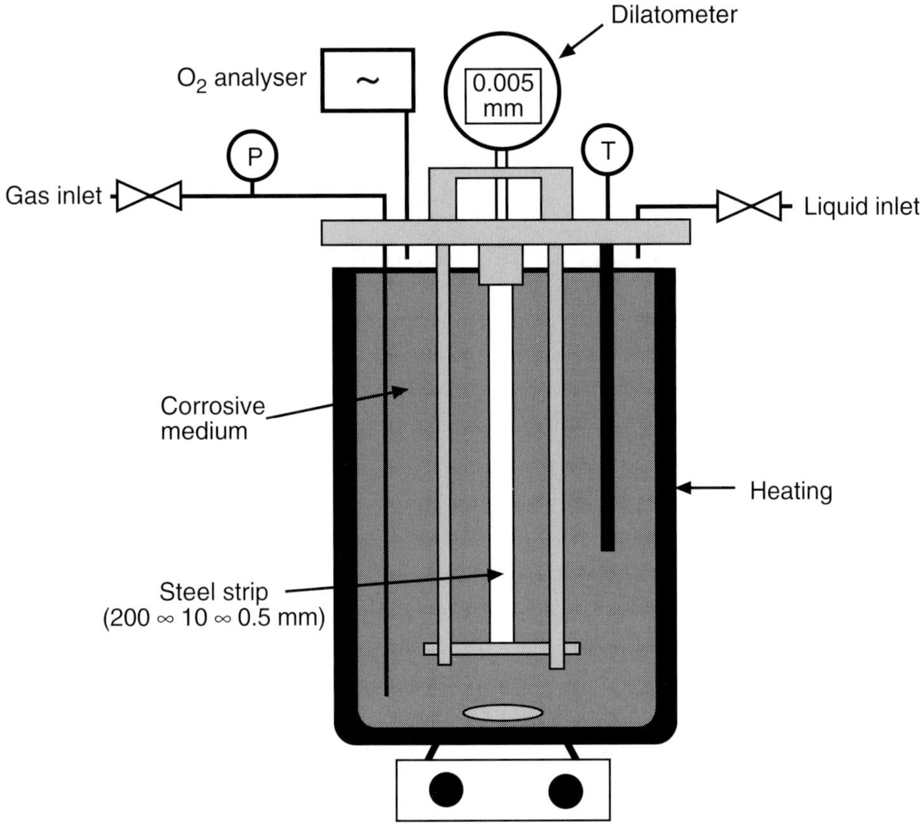

Fig. 12 *Experimental setup for dilatometric measurements.*

The intrinsic growth stress of the scale was calculated according to eqn (33) [40].

$$\sigma_{intr} = \frac{E \cdot d \cdot \nu}{2l \cdot t} \text{ (MPa)} \tag{33}$$

where σ_{intr} is the intrinsic growth stress in the scale, E the Young's modulus of the steel strip (210 GPa), d the thickness of the steel strip (0.5 mm), ν the shortening or lengthening of the steel strip, l the length of the steel strip (200 mm) and t the thickness of the scale. In the presence of 1 mmol L^{-1} of the inhibitor MNQB even after 96 h exposure no indication of a dilatation of the steel strip could be observed within the accuracy limit of 1 μm.

This is obviously due to the fact that the scale thickness (max. 12 μm after 96 h) is too low to produce a measurable dilatation. This made it impossible to decide with this method whether the thickness related intrinsic stress in scales formed in the presence of an inhibitor is the same or lower than in scales formed in uninhibited corrosion systems.

X-ray diffraction method. The intrinsic stress in iron carbonate scales on steel coupons (40 × 20 × 3 mm) was also measured with a Ψ-goniometer by X-ray diffraction [41,42]. The intrinsic stress can be calculated from eqn (34):

$$\sigma_{intr} = m \cdot \frac{E}{1+\nu} \text{ (MPa)} \tag{34}$$

where E is the Young's modulus, ν the Poisson number and m the slope in the $D_{\{100\}}/\sin^2\psi$ plot.

It was proved by X-ray diffraction (Fig. 13) that the scales formed on the C75 grade carbon–manganese steel at 80°C in 1 M NaCl under 5 bar CO_2 consisted of siderite ($FeCO_3$). The $D_{\{100\}}/\sin^2\psi$ plot of the line at 2 Θ = 137° (marked in Fig. 13) is shown in Fig. 14. The scatter of the data points is caused by the different sizes of the iron carbonate crystals. With E_{scale} = 130 GPa and ν = 0.3 an intrinsic compressive stress of –90±40 MPa was obtained for a scale grown during an exposure time of 96 h under corrosion conditions given above.

4.2.3. Young's modulus, Vickers hardness

These fracture mechanics data can be obtained from the micro-indentation method [43–48]. Force/penetration and force/relief curves were measured on polished iron carbonate scales with a Vickers-indenter at a maximum force of 150 mN. In brittle materials the indentation results in cracks starting from the corners of the Vickers indentation (Fig. 15). The length of the cracks indicates the degree of intrinsic stress. The crack length was measured on SEM photomicrographs of the indentation.

The evaluation of the characteristic data of micro-indentations yields the Young's modulus according to eqn (35):

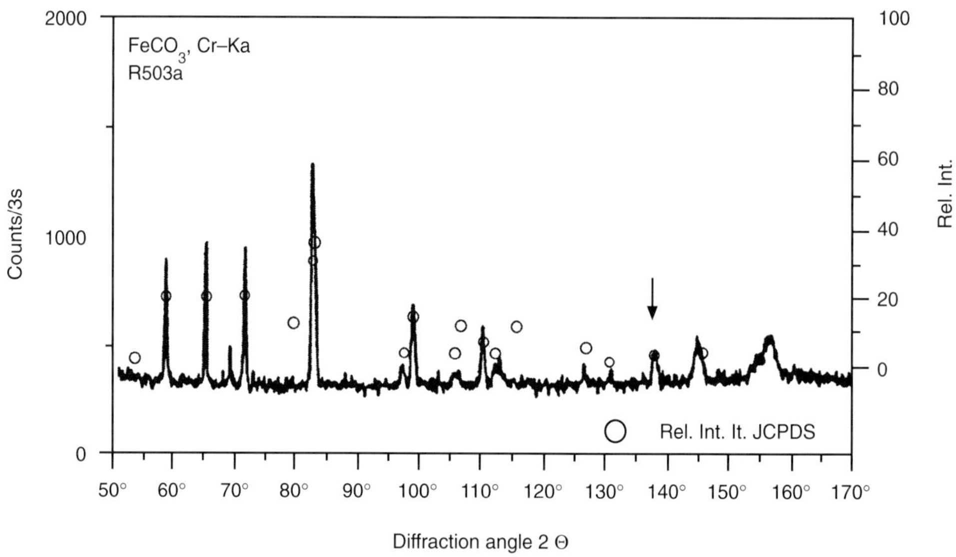

Fig. 13 *X-ray diffraction plot of iron carbonate scale.*

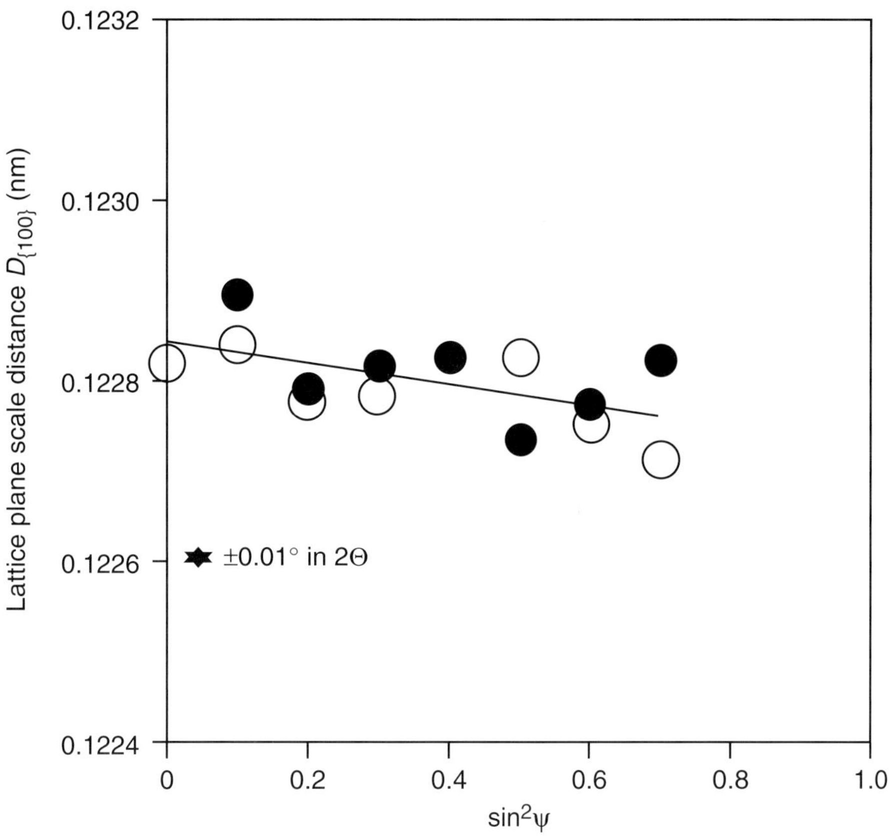

***Fig. 14** Lattice plane scale distance vs sin^2 ψ at 2Θ =137°.*

$$\frac{E_{SC}}{(1-\nu^2)} = \frac{F \cdot \sqrt{\pi}}{4 \cdot \tan(68°) \cdot h_p \cdot \frac{dh}{dF} \cdot F_{\max}} \text{ (GPa)} \tag{35}$$

where ν is the Poisson number, F the indentation force, $h_p = h_{ep} - h_e$ the penetration depth in the plastic zone and dh/dF the penetration intensity.

The Vickers hardness H_v is calculated according to eqn (36):

$$H_v = 0.1891 \cdot F \cdot d^{-2} \text{ (MPa)} \tag{36}$$

where d is the diagonal of the indentation.

The average value for the Young's modulus was 125 GPa, i.e. close to the value estimated from theoretical considerations (150 GPa) [7]. The Vickers hardness ranged between 220 and 650 MPa.

4.2.4. Adhesion

Adhesion measurements were performed with round C75 coupons (dia. 6 × 2 mm)

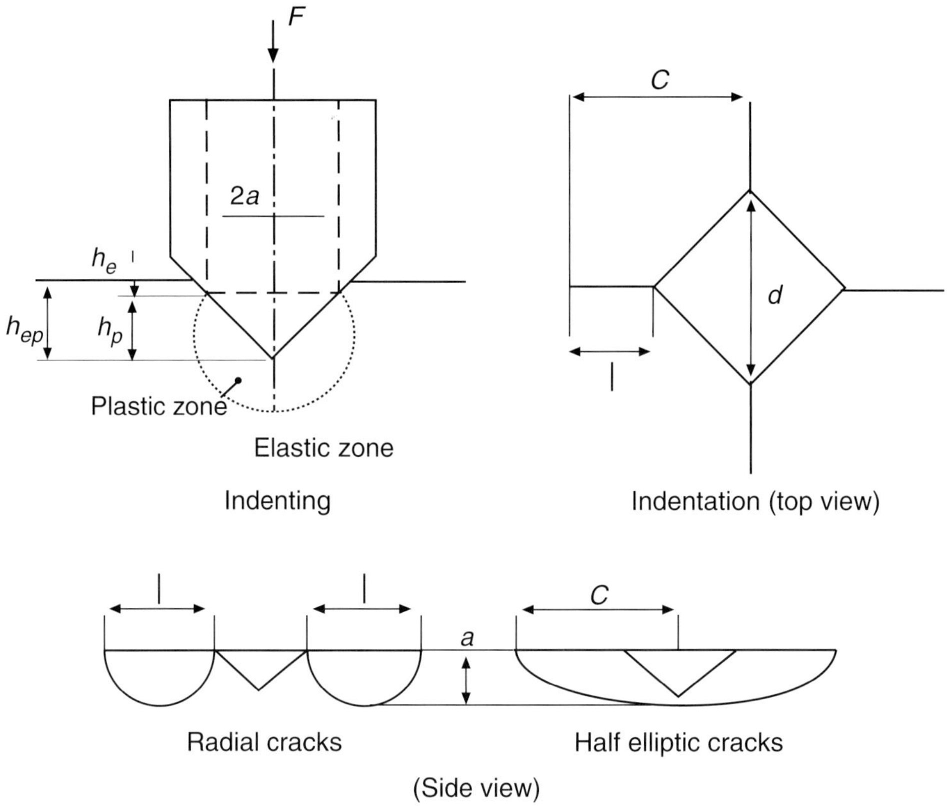

***Fig. 15** Initiation of cracks in brittle materials by microindentation.*

corroded 96 h at 80°C in 1M NaCl solution under 5 bar CO_2. A piston was glued onto the scale (Fig. 16) and the force needed to tear the piston from the coupon (Fig. 17) was measured.

It was ascertained by microscopic and SEM investigations that the materials had separated between the scale and the base material. The surface related adhesion of iron carbonate scales produced under the above described conditions ranged between 8.4 and 18.3 MPa.

4.3. Fracture Mechanics Data

A comparison of estimated and measured fracture mechanics data is shown in Fig. 18. All fracture strain data generated from 4-PLT experiments were plotted against the scale thickness in a failure mode diagram. The three different boundary lines represent three different defect intensities in the scales reducing the effective fracture surface energy from 100% (3 Nm^{-1}) to 50 and 20%. The Young's modulus E in these calculated diagrams was assumed to be 130 GPa. According to data measured in this work a Young's modulus of 125 GPa was used to calculate the fracture strains from 4-PLT data.

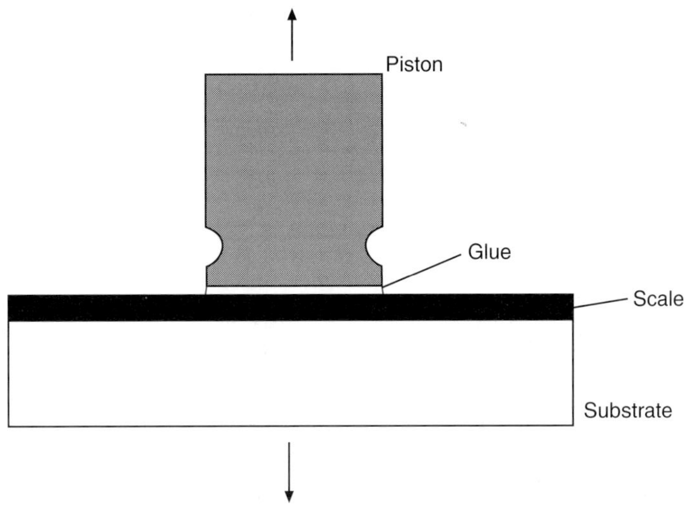

Fig. 16 *Arrangement for adhesion measurements.*

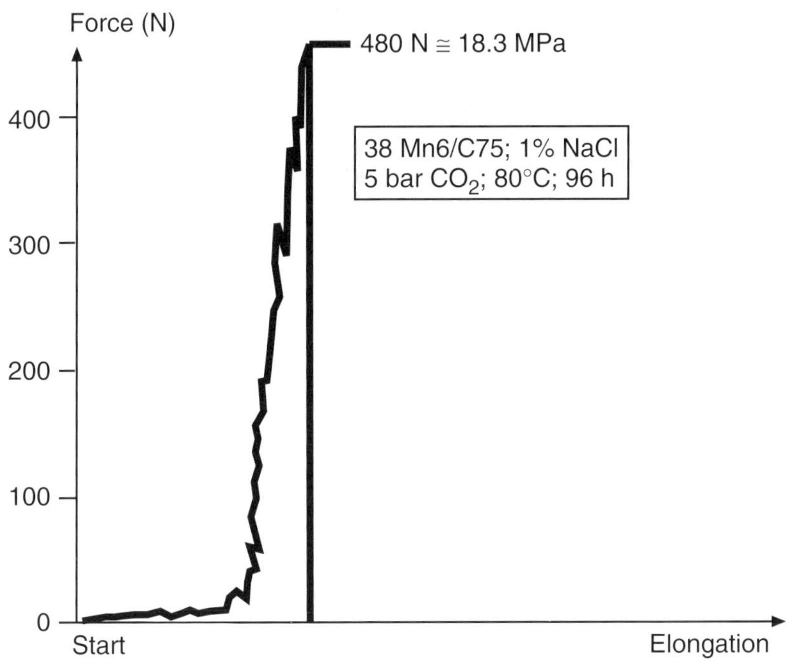

Fig. 17 *Force vs elongation plot from adhesion measurements.*

The experimental data in Fig. 18 match closely with the calculated range. Only at higher thicknesses (>30 μm) is some deviation observed. The maximum scale thickness obtained in the presence of the inhibitor MNQB was approximately one order of magnitude lower than the maximum scale thickness produced in the absence of an inhibitor.

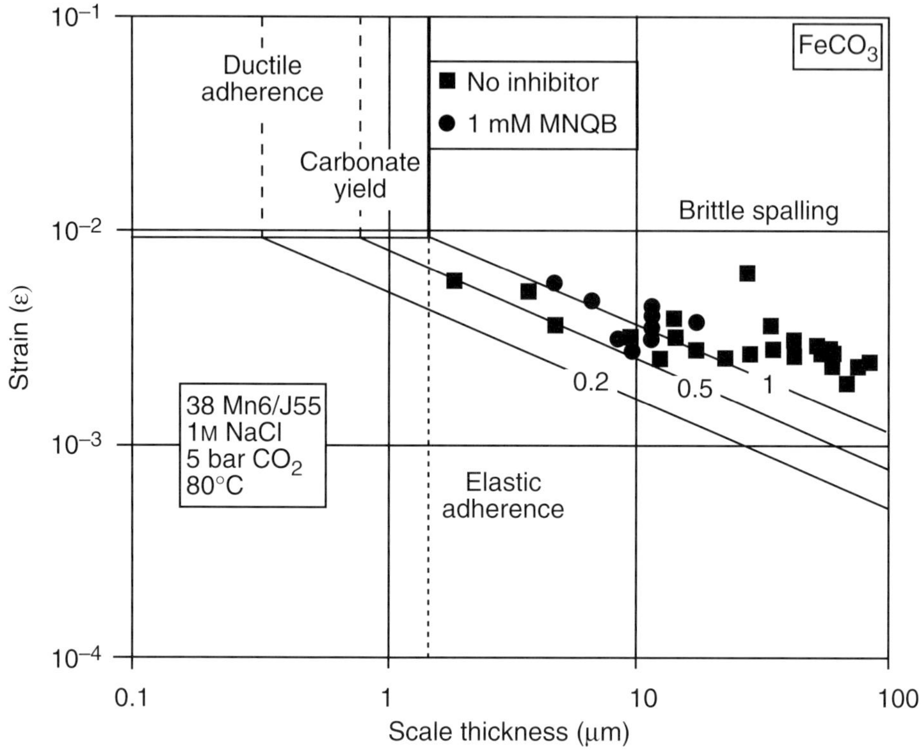

Fig. 18 Failure mode diagram for iron carbonate scales.

The results on the thickness dependent fracture strain of iron carbonate scales obtained with the 4-PLT method match with the data obtained from scale defect structure calculations published earlier [2]. This gives confidence to use these data in future model calculations.

The fracture mechanical data now available for iron carbonate scales on low alloy steel are listed in Table 3. They match closely with values from theoretical calculations.

5. Critical Wall Shear Stresses for FILC Initiation

Comparison of the data for the fracture stress or adhesion of iron carbonate scales with wall shear stresses produced in flowing technical media (10^0–10^5 Pa) (Table 1) [1] clearly indicates that hydrodynamic forces expressed in terms of wall shear stresses are orders of magnitude too small to cause destruction and spalling of scales from CO_2 and H_2S corrosion of carbon steels and to initiate FILC. However, critical wall shear stresses for the initiation of FILC exist [3,25–27] and have been evaluated under sweet (CO_2) and sour gas (H_2S) corrosion conditions using the high pressure submerged jet technique [4,8,38].

It is of high technical interest to evaluate the effect of inhibitors and their concentration on critical wall shear stresses, because this would permit the inhibitor concentration to be minimised for a given gas production rate or to optimise the gas

Table 3. *Fracture mechanical data of iron carbonate scales from CO_2 corrosion of steel*

Parameter	Estimated	Measured
Surface fracture energy γ (Nm^{-1})	3.0	No
Young's modulus *E* (GPa)	150	125[a]
Fracture stress σ_y (MPa)	1400 (single crystal)	75–100[b,c] 230–720[d]
Fracture strain ε_y ($\times 10^{-3}$)	9.33	1.8–5.6[d]
Intrinsic stress σ_{sc} (MPa)	– (200–220)	– (20–50)[e] –90 ± 40[f]
Limiting scale thickness *h* (m)	1.46×10^{-6}	No
Critical strain ε_{crit}	2.5×10^{-4}	5.3×10^{-4}[b]
Adhesion σ_{Ad} (MPa)	No	8.4–18.3[g]
Hardness (Vickers) H_v (MPa)	No	220–650[a]

[a] Micro-indentation method. [b] Micro-3-point-bend measurement. [c] Scale thickness 80 μm; 0.2 γ. [d] 4-point loading test. [e] Dilatometric measurement. [f] X-ray diffraction. [g] Tensile test.

production rate with an appropriate inhibitor concentration. This situation is demonstrated in Fig. 19 [4] which summarises jet impingement results from a corrosion system that exists in many north German sour gas fields where monoethylamine (MEA) injection is used to keep elemental sulfur, from sulfur-containing formations, in solution. The corrosion system investigated consisted of carbon steel in 5M MEA-solution with 60 gL^{-1} polysulfide under 22 to 42 bar H_2S. The results in Fig. 19 were obtained at 110°C after 93–101 h exposure.

It can be seen that a concentration of 300 ppm commercial inhibitor (based on carbonamides from polyamines) was not sufficient to increase the critical wall shear stress from values below 10 Pa to values in the range of 20 Pa. This could be achieved only with increasing the inhibitor concentration to at least 900 ppm. Further increase of the inhibitor concentration to 2700 ppm yielded no significant increase of the critical wall shear stress at this temperature.

Attempts to measure critical wall shear stresses in inhibitor-free CO_2 corrosion systems of carbon steel were unsuccessful because flow independent pitting occurred under these conditions. Thus, scale destruction was observed already under stagnant conditions. However, if pitting was prevented by addition of corrosion inhibitors, critical wall shear stresses could be measured.

Figure 20 gives an example for experiments with the inhibitor NMQB applied in concentrations of 0.1 and 1.0 mmol L^{-1}. In the presence of 0.1 mmol L^{-1} NMQB the critical wall shear stress ranges in the order of 25 Pa and higher, indicated by a steep increase of the mass loss rate. Increasing the inhibitor concentration to 1 mmol L^{-1}. increases the critical wall shear stress to values above 250 Pa.

The iron carbonate scales obtained in the experiments summarised in Fig. 20 were investigated in the SEM. It was observed that the surface-related number of cracks and their width increased with increasing wall shear stress, i.e. with increasing flow intensity. For a rough estimation of the effect of the flow intensity on the cracking

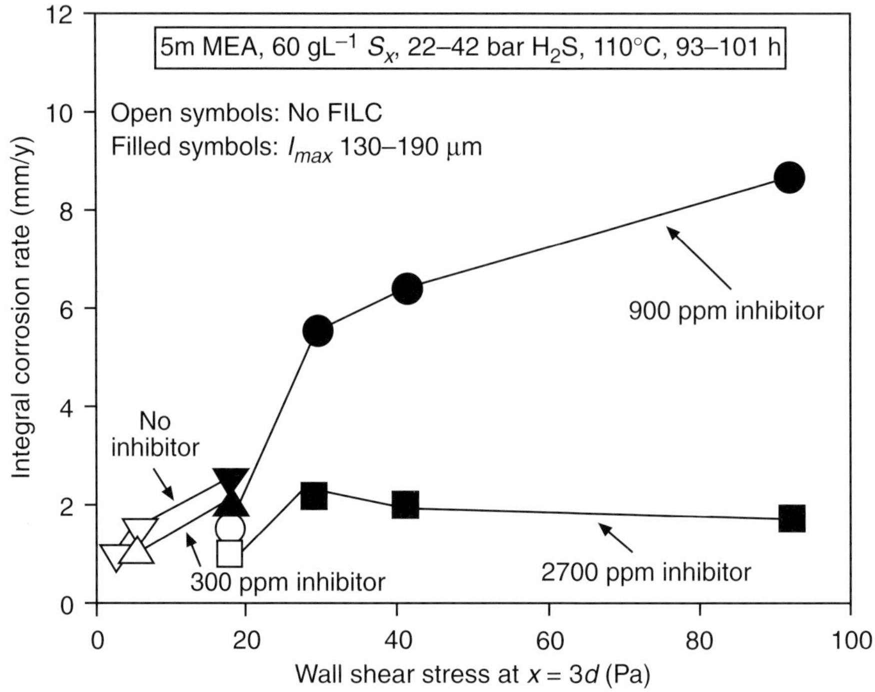

Fig. 19 Effect of flow and inhibitor concentration on the H_2S corrosion of low alloy steel (I_{max} = maximum penetration depth).

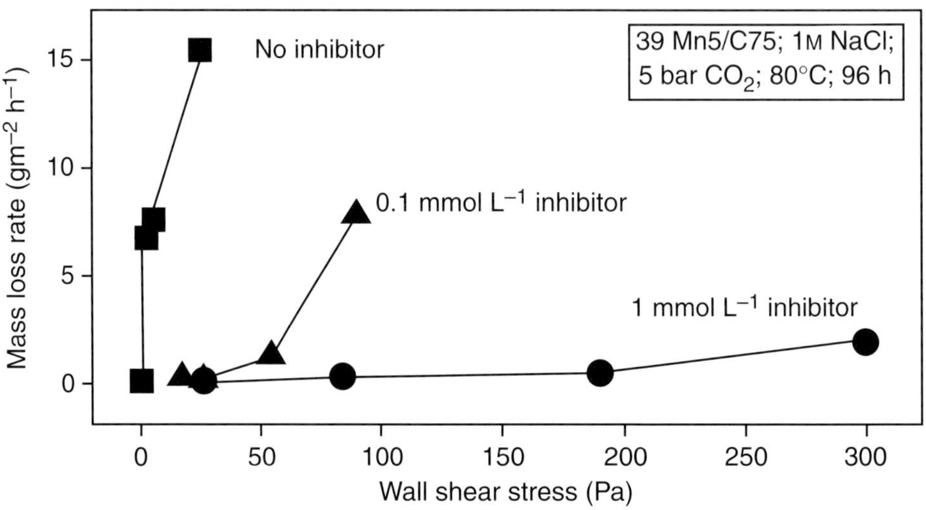

Fig. 20 Effect of flow and inhibitor (MNQB) concentration on the CO_2 corrosion of low alloy steel.

tendency of the carbonate scales the surface areas of all crack lines on a unit surface area were related to the total unit surface area for a given corrosion experiment. Such crack area ratios were plotted in Fig. 21 for uninhibited and inhibited corrosion conditions relative to the flow intensity expressed in terms of wall shear stresses. The crack area ratios in uninhibited or poorly inhibited systems steeply increase with increasing wall shear stress. With increasing inhibitor concentration the slope decreases.

The increase of the cracking tendency of the carbonate scales with increasing flow intensity supports the hypothesis of fatigue cracking of scales by momentum transfer of turbulence elements in the near-wall region having critical sizes, frequencies and energy contents. Thus, wall shear stresses become critical when the magnitude of single impacts (bursts) and the range of burst frequencies match with the fracture mechanics properties of the scale, which in turn depend on the environmental conditions of the corrosion reaction.

6. Summary

According to present day knowledge it must be assumed that the initiation of flow induced localised corrosion (FILC) is caused by hydrodynamic forces which mechanically destruct protective corrosion product scales. However, there are two experimental findings that seem to contradict each other:

(i) Wall shear stresses, as a measure of hydrodynamic forces, were found to be

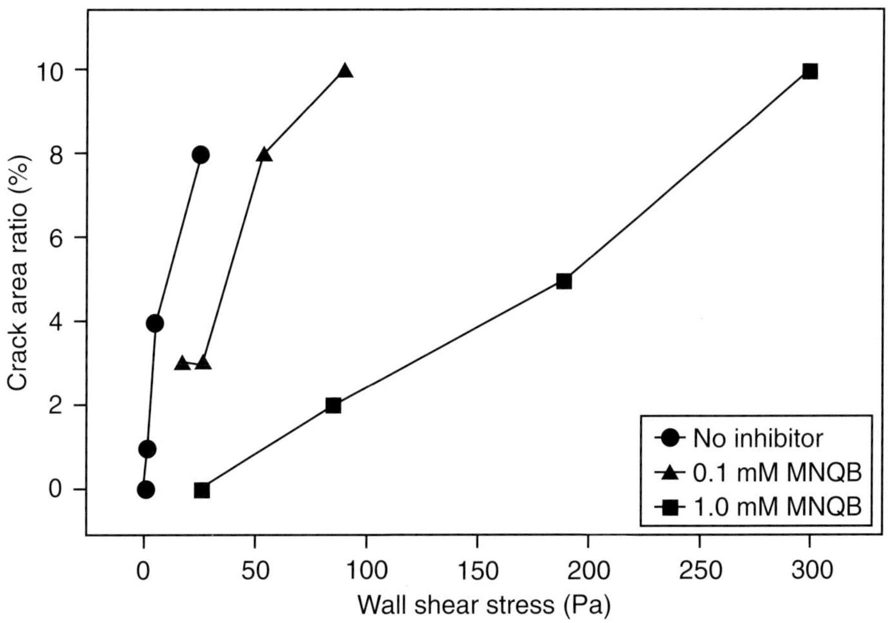

***Fig. 21** Effect of flow intensity on crack area ratio of iron carbonate scales.*

orders of magnitudes smaller than fracture stresses of corrosion product scales or adhesion forces between scale and the base material.

(ii) It is possible to measure critical wall shear stresses above which FILC is initiated.

The following hypothesis would solve this problem:

Turbulence elements, created in the near-wall region of the turbulent boundary layer, exchange momentum with the wall and reach critical sizes, frequencies and energy values at critical wall shear stresses to cause fatigue cracking of scales.

Based on this hypothesis a model has been developed that allows the prediction of the probability of the onset of FILC by correlating flow intensities in terms of critical wall shear stresses with fracture mechanics data, specifically with the fracture stress of corrosion product scales.

The turbulence elements are quantitatively described by the hydrodynamic properties of the flowing liquid which are included in the model with a turbulence coefficient.

The effect of corrosion inhibitors on critical wall shear stresses for the initiation of FILC was introduced in the model by a drag reducing coefficient. This coefficient is a new criterion to characterise the efficiency of a corrosion inhibitor.

The validity of the model was proven with experimental data from the system carbon steel/brine/CO_2/inhibitor including critical wall shear stresses without and with inhibitor, measured fracture stresses of iron carbonate scales. Using the new model it was possible to calculate for a given set of corrosion conditions the time to FILC initiation which matched closely with the experimentally measured time to start of scale spalling.

New experimental results were obtained for fracture mechanics properties of iron carbonate scales from CO_2 corrosion of carbon steel (Young's modulus, fracture stress, scale thickness dependent fracture strain, intrinsic growth stress, hardness, scale adhesion).

Examples are given on the effect of corrosion inhibitors on critical wall shear stresses for FILC initiation under conditions of sweet and sour gas production.

The increase of the cracking intensity of iron carbonate scales in inhibited CO_2 corrosion systems with increasing wall shear stress supports the hypothesis of fatigue cracking of scales by momentum exchange of turbulence elements in the near-wall region having critical sizes, frequencies and energy contents.

The magnitude of critical wall shear stresses depends on the inhibitor concentration. Thus, the inhibitor concentration determines the magnitude of critical wall shear stresses and, hence, the critical flow rate below which corrosion damages due to FILC can be prevented in a given flow system. This is important to determine, for example, the critical production rate in inhibited carbon steel tubulars for oil and gas production.

The probabilistic model for FILC initiation will be improved in future work to widen its range of application.

References

1. G. Schmitt and T. Gudde, Local mass transport coefficients and local wall shear stresses at flow disturbances, *Corrosion '95*, Paper 102, NACE International, Houston,Tx, 1995.
2. G. Schmitt, T. Gudde and E. Strobel-Effertz, Fracture mechanical properties of CO_2 corrosion product scales and their relation to localized corrosion, *Corrosion '96*, Paper 9. NACE International, Houston, Tx, 1996.
3. G. Schmitt, T. Simon and R.H. Hausler, CO_2 erosion corrosion and its inhibition under extreme shear stresses II. Performance of inhibitors, *Corrosion '93*, Paper 86, NACE International, Houston,Tx,
4. G. Schmitt, C. Bosch, U. Pankoke, W. Bruckhoff and G. Siegmund, Evaluation of critical flow intensities for FILC in sour gas production, *Corrosion '98*, Paper 46, NACE International, Houston, Tx, 1998.
5. S. Feinen, PhD Thesis, Aachen University of Technology, 1998.
6. G. Schmitt and S. Feinen, Media effects in pitting of iron and steel during CO_2 corrosion, *Proc. Int. Conf. on Problems of Corrosion and Corrosion Protection of Materials*, 9–11 June, 1998, Lviv, Ukraine, p.13–16. Publ. Karpenko Physico–Mechanical Institute of the National Academy of Ukraine, Lviv, Ukraine.
7. G. Schmitt, Hydrodynamic limitations of corrosion inhibitor performance, *Proc. 8th Europ. Symp. on Corrosion Inhibitors (8 SEIC)*; Ann. Univ. Ferrara, N.S, Sez. V, Suppl. N.10, 1995, p.1075.
8. G. Schmitt *et al.*, Initiation of flow induced localized corrosion in oil and gas production — Hydrodynamic forces vs mechanical properties of corrosion product scales, *Proc. 13th Int. Corrosion Congr.*, Melbourne, Australia, 25–29 Nov., 1996, Vol. II, Paper 150. Publ. Australasian Corrosion Association Inc., Clayton, Australia.
9. M. T. Landahl and E. Mollo-Christensen, *Turbulence and Random Processes in Fluid Mechanics*, Cambridge University Press, 1987, pp.111–120.
10. A. K. Praturi and R. S. Brodkey, A stereoscopic visual study of coherent structures in turbulent shear flow, *J. Fluid. Mech.*, 1987, **89** (2), 251–272.
11. G. R. Offen and S. J. Kline, A proposed model of the bursting process in turbulent boundary layers, *J. Fluid. Mech.*, 1975, **70**, 209–228.
12. S. K. Robinson, Coherent motions in the turbulent boundary layer, *Ann. Rev. Fluid Mech.*, 1991, **23**, 601–639.
13. A. S. Thomas and M. K. Bull, On the role of wall pressure fluctuations in deterministic motions in the turbulent boundary layer, *J. Fluid. Mech.*, 1983, **128**, 283–322.
14. H. L. Grant, The large eddies of turbulent motion, *J. Fluid. Mech.*, 1958, **4**, 149.
15. D. J. Tritton, Some new correlation measurements in a turbulent boundary layer, *J. Fluid. Mech.*, 1967, **28**, 439.
16. W. R. B. Morrison and R. E. Kronauer, Structural similarity for fully developed turbulence in smooth tubes, *J. Fluid. Mech.*, 1969, **39**, 117.

17. S. L. Lyons, S. Nikolaides and T. J. Hanratty, The size of turbulent eddies close to a wall, *AIChE J.*, 1988, **34**, 938.
18. T. von Karman, Analogy between fluid friction and heat transfer, *Trans. ASME*, 1939, **61**, 705.
19. B. A. Toms, Some observations on the flow of linear polymer solutions through straight tubes at large Reynolds numbers, *Proc. 1st Int. Congr. on Rheology*, Vol.II, pp.135–141, North Holland, Amsterdam, 1948.
20. P. S. Virk, Drag reduction fundamentals, *AIChE J.*, 1975, **21**, 625–656.
21. A. White, Flow characteristics of complex soap systems, *Nature*, 1967, **214** (5).
22. J. Weber and K. Knopf, *Werkst. und Korros.*, 1991, **42**, 348.
23. G. Schmitt and W. Buecken, Effect of corrosion inhibitors and surfactants on the fluid dynamic properties of multiphase annular mist flow in carbon dioxide corrosion systems, *Corrosion '92*, Paper No. 6, NACE International, Houston, T, 1992.
24. G. Schmitt and W. Buecken, Performance of inhibitors under high turbulent flow, *Proc. 12th Scand. Corrosion Congr. & EUROCORR '92*, 31 May–4 June 1992, Dipoli, Espoo, Finland.
25. T. Simon, Inhibition of flow induced localized corrosion under extreme wall shear stresses in CO_2 containing media (in German), PhD Thesis, Aachen University of Technology, Germany, 1993.
26. G. Schmitt, T. Simon and R. H. Hausler, CO_2 erosion corrosion and its inhibition under extreme shear stress — I. Development of methodology, *Corrosion '90*, Paper No. 22, NACE International, Houston, T, 1990.
27. G. Schmitt, Flow induced localized Corrosion — A challenge for corrosion inhibition in oil and gas production, *Proc. 15th Ann. Conf. On Corrosion Problems in Industry*, 9–10 Oct., 1996, Alexandria, Egypt, p.1–15, The Egyptian Corrosion Society, Cairo, Egypt.
28. P. G. de Gennes, Coil-stretch transition of diluted flexible polymers under ultrahigh velocity gradients, *J. Chem. Phys.*, 1974, **60**, 5030.
29. A. Ambari, C. Deslouis and B. Tribollet, Coil-stretch transition of macromolecules in laminar flow around a small cylinder, *Chem. Eng. Commun.*, 1984, **29**, 63–78.
30. P. Peyser, R. C. Little and C. R. Singleterry, Study of drag reduction and absorption of polymer solutions, Naval Research Laboratory, Washington, D.C., NRL-Rep. 7222, Jan. 1971.
31.J. Wu, Surface containing of polymer solutions and pulsative ejection, *Nature, Phys. Sci.*, 1971, **231** (24), June 14.
32. H. G. Elias, *Makromolekuele*, Hüthig and Wepf, 5th edn, Basel, 1990.
33. P. Lindner *et al.*, Drag reducing surfactant solutions in laminar and turbulent flow investigated by small-angle neutron scattering and light scattering, *Prog. Colloid Polymer Sci.*, 1990, **81**, 107.
34. R. Heen, Untersuchung der Orientierung und Assoziatbildung von Tensiden und Polymerlösungen, PhD Thesis, University of Dortmund, 1993.
35. W. Pfeiffer, F. Henkel, E. Sackmann, W. Knoll and D. Richter, Local dynamics of lipid bilayers studied by incoherent quasi-elastic neutron scattering, *Europhys. Lett.*, 1989, **8**, 201.
36. H. Hoffmann, G. Platz, H. Rehage, W. Schorr and W. Ulbricht, Viskoelastische Tensidlösungen, *Ber. Bunsenges. Phys. Chem.*, 1981, **85**, 255.
37. H. Giesekus *et al.*, "Rheology", *Fortschr. Verfahrenstechnik*, 1985, **23**, 3.
38. M. Mueller, PhD Thesis, Aachen University of Technology, 1999.
39. EN ISO 7539-2: 1995, Corrosion of Metals and Alloys – Stress Corrosion Testing — Part 2: Preparation and Use of Bent-Beam Specimens, Beuth-Verlag, Berlin, 1995.
40. A. Dvorak and L. Vrobel, A new method for the measurement of internal stress in electrodeposits, *Trans. Inst. Met. Finish.*, London, 1971, **49**, 153–155.
41. G. Faninger, V. Hauk, E. Macherauch and U. Wolfstieg, Recommodations for the practical use of X-ray stress determination methods (in iron materials), *Haerterei-Tech. Mitt.*, 1976, **31**, p. 1–124.

42. V. Hauk, *Structural and Residual Stress Analysis by Nondestructive Methods*, Elsevier Science, New York, 1997.
43. H.-D. Tietz, *Technische Keramik*, VDI Verlag, 1994.
44. D. B. Marshall and B. R. Lawn, *Indentation of Brittle Materials, in Microindentation Techniques in Materials Science and Engineering*, P. J. Blau and B. R. Lawn eds, ASTM STP 889, Am. Soc. for Testing Materials, West Conshohocken, Pa, 1986, p.26–45.
45. P. Chantikul, G. R. Anstis, B. R. Lawn and D. B. Marshall, A critical evaluation of indentation techniques for measuring fracture toughness: 1. Direct crack measurements, *J. Am. Ceram. Soc.*, 1981, **64**, 533–538.
46. G. R. Anstis, P. Chantikul, B. R. Lawn and D. B. Marcal, A critical evaluation of indentation technique for measuring fracture toughness: 2. Strength Method, *J. Am. Ceram. Soc.*, 1981, **64**, 539–543.
47. R. W. Steinbrech *et al.*, Indentation studies on Y_2O_3-stabilized ZrO_2: 1. Development of indentation-induced cracks, *J. Am. Ceram. Soc.*, 1994, **77**, 1185–1193.
48. R. W. Steinbrech *et al.*, Indentation studies on Y_2O_3-stabilized ZrO_2: 2. Toughness determination from stable growth of indentation-induced cracks, *J. Am. Ceram. Soc.*, 1994, **77**, 1194–1201.

Part 2

Carbon and Low Alloy Steels

3
Controlling Risk Through Prediction of Degradation Mechanism and Failure Modes in Pipelines

J. D. A. EDWARDS and T. SYDBERGER
Det Norske Veritas, Høvik, Norway

ABSTRACT
Materials selection for transport of CO_2-containing oil and gas is usually based on either 'expensive' corrosion resistant alloys or 'cheaper' carbon steel. Use of the latter material generally requires prediction of the corrosion rate inclusive of the effect of corrosion control treatments if used. Probabilistic methods can be used to evaluate the confidence in these rates. Corrosion damage will interact with service stresses and hence affect the failure mode of the pipe and therewith the consequences of failure. Once inspection data become available the rates and uncertainties can be revised. When all such uncertainties are accounted for, the materials selection can be based on the lowest lifetime cost option including the Risk cost as well as capital expenditure and operating expenditure.

1. Introduction

Carbon dioxide corrosion remains a subject of high concern in securing future extraction and supply of oil and gas. When new equipment is to be built, the possibility of CO_2-corrosion is one of the primary factors that affects materials selection. Generally, stainless steels or CRAs (Corrosion Resistant Alloys) offer a more-or-less guaranteed resistance to corrosion. However, the high cost of such materials means that the cheaper alternative of carbon or low-alloy steels (C-steel) must be shown to be unsuitable before CRAs are selected. The actual cost of the C-steel is strongly affected by the operational phase, and is not always considered in detail. This paper addresses some of the aspects that have not usually been considered in the selection philosophy of such materials viz. the difference in 'Risk' that is inherent in the selection of C-steel as opposed to CRAs.

It should be appreciated that 'Risk' is a function of both the probability of an event occurring and the consequences of the event. The term 'Probability' infers that a quantitative value is used. However, current state-of-the-art analyses are rather more semi-quantitative, so the term 'likelihood' is preferred to avoid misconception of accuracy. Thus, in the present context, Risk is expressed as *the product of the likelihood of corrosion failure and the cost consequences of the failure*. In the case of CO_2-corrosion, 'failure' may take several forms, including a leakage, or 'loss of compliance with mandatory codes'. Risk control involves controlling either or both of the consequences and likelihood. In the case of CRA vs C-steel, a CRA option would have a low Risk, whilst the C-steel has a Risk that is governed by the measured or perceived corrosivity of the prevalent conditions including any corrosion mitigation activities.

A quantitative Risk analysis requires that the failure likelihood is defined as a true probability value, (such as there is a 10^{-4} probability per annum that failure will occur), whilst the consequences should be expressed in tangible terms such as money. Once these parameters are identified the operator would be in a position to make an optimum' materials selection based on the cheapest lifetime solution accounting for the Risk cost as well as investment (Capex) and operational (Opex) costs. This concept is not always readily accepted as there is a strong tendency to relate 'Risk' to 'Safety' (i.e. death and injury to personnel). Any suggestion that a higher Risk solution has been selected merely to save money may not be 'morally' acceptable in some quarters. In practice the Risk cost comprises both the 'safety' aspect and the 'business interruption/reflection' cost though this is not always stated explicitly. In a number of cases analysed by Det Norsk Veritas (DNV) for offshore installations and pipelines, the potential business interruption costs have outweighed the safety costs by several orders of magnitude. This observation leads to the conclusion that if the business interruption costs are under control, then there should be no concern for safety. The incentive for basing corrosion control on Risk is straightforward business acumen; there is no financial advantage in using CRAs if C-steels will suffice and are cheaper. Admittedly, the economic calculations are not straightforward, alternatives must address long term accounting, and include changes in monetary values, and variation in field operation. Generally, these are affected by project specific requirements and such financial advice has to be included in the analysis. A successful evaluation should result in a Risk based corrosion-control that minimises the total lifetime costs.

2. Consequences

The consequences of failure may be considered to comprise three factors: safety, business losses and environmental damage. Ultimately all three can be expressed in monetary terms. For pipelines, the safety aspect is most pertinent in the vicinity of installations and at landfalls. This is reflected in contemporary rules [1] where analyses of pipelines allow required safety levels to be stipulated for differing applications. Similarly, significant environmental impact is to be expected from a leak occurring on a landfall in sensitive ecosystems, rather than in a remote subsea location. Not withstanding the safety and environmental issues, the enormous value of the transported commodity, particularly in export or major trunk lines, implies significant financial losses should a failure occur. It is important to be able to differentiate between the expected failure modes so that the consequences can be addressed accurately. A small 'pin-hole' leak may be patched temporarily, or be by-passed, whereas a full breach would require immediate and long term shutdown. 'Failure' is also used when a maximum allowable wall thickness has been consumed. Even though no loss of containment occurs with this event, the line may have to be written-off, and a new line built if warranted. If inspection and/or monitoring is able to detect the reductions in wall thickness in due time, then business losses can be minimised as a new line can be designed and built before the original line has to be closed. It should be apparent from the above comments that the consequences of a failure are a feature of the failure mode, which in turn can be derived from knowledge of the degradation mechanisms and operational loads expected to be active in the line.

3. Likelihood

There are many parameters that must be evaluated as part of the decision for what is the 'likely' mode(s) and associated corrosion rate. Typically this will be based on one or more of the several models that have been developed, mostly from modifications of the 'De Waard and Milliams' equation published in 1975 [2–5]. Despite the widespread use of these models, there are no recognised guidelines to describe how their results should be applied to practical situations. Commendably though, the EFC Oil and Gas Working group has developed a document on this subject [7], whilst the NORSOK Standard [6] states that "evaluation of CO_2 corrosion should be based on . . .a corrosion prediction model". The reliability of the results, and relation to individual and significant variations in materials type [8], corrosion morphology and any chemical treatment [9], etc. are not discussed, yet these all affect how well the model will match up to real-life. Irrespective of whether the corrosion engineer is conscious of this or not, the uncertainty of these parameters, and any engineering judgement that may be exercised when determining a corrosion rate, heavily affect Risk associated with operation of the equipment. This applies during both phases of design and service.

3.1. Design

Materials selection at the design stage demands greatest faith in the predictive model of CO_2-corrosion rates and the efficiency of any corrosion control (e.g. inhibitor). Input parameters are often very 'preliminary'. Predicted operational data is usually sparse and based on thermodynamic modelling where essential data such as CO_2 content, pressure and temperature may be given as ranges, or as 'maximum' values without further qualification of the certainty involved. The traditional 'safe' approach is to select the worst case combination. The resultant corrosion rate is used to evaluate alternative materials selection options, which can be generalised as follows:

- Very low corrosion rate: use C-steel, possibly with a corrosion allowance
- Moderate corrosion rate: use C-steel with a corrosion allowance, and chemical treatment (inhibitor),
- High corrosion rate: use CRAs.

The difficulty of course, is to decide which category a case belongs to. To some extent the NORSOK Standard [6] attempts this for pipelines when it states that CRAs should be used when the corrosion allowance exceeds 10 mm. (This includes an option to use a 'conservative' design value of '85% efficient' for an inhibitor treatment). The challenge is to quantify the uncertainties involved in the prediction. Although much work has been expended in developing CO_2 corrosion models, based primarily on laboratory data, any corrosion rate value returned is heavily affected by the reliability of the inhibitor treatment when used. One approach has been to correlate the model to real-life using engineering judgement and field data [10]. Here the inhibitor performance was defined for a few relevant cases, and sensitivity analyses performed to assess the impact of the uncertainty in judgement on the overall result. Some more work along these lines would be beneficial, but field data are not readily available.

If principles of Risk are applied to the materials selection, it is first necessary to consider what may happen if corrosion in excess of that accommodated by the corrosion allowance occurs. Potential failure modes result from interaction of the corrosion damage and the loads experienced during service (both operational and accidental). These have been discussed in detail elsewhere [11], and for tubular components the tensile hoop stresses are often the dominating load. In the absence of stratified flow, 'pitting' and circumferential grooving at welds are the most predominant damage modes for CO_2 corrosion. Such localised attack does not interact significantly with the hoop stresses, and with wall thicknesses typical of tubular components, small pits will not normally affect structural integrity even when their depth approaches the wall thickness. These types of defects can usually be detected by inspection in due time. Further, if this damage continues unabated the result will be a 'small' leak rather than a larger 'burst' that would occur with larger defects, i.e. corrosion that exceeds the corrosion allowance can be acceptable for such localised damage modes.

Traditional design with C-steel usually strives to avoid any corrosion, uniform or localised, that exceeds the thickness of the corrosion allowance. In contrast the Risk based approach accepts that some corrosion may exceed the corrosion allowance, and indeed may allow a leak to occur. The condition is that the probability of leakage and the consequences of the leak are estimated, and that the resultant Risk is accepted. Typically the Risk can be expressed in monetary terms and weighed against 'risk reducing' costs, such as using a thicker corrosion allowance, chemical treatment, more inspection, or a switch to CRAs.

3.2. Service

When equipment has been designed, a service limit state can be defined as the 'Maximum Accepted Risk'. The CO_2 corrosion model can be applied to the prediction of the lifetime of a given component in terms of the time to reach the Maximum Accepted Risk. This is illustrated schematically in Fig. 1, where it can be seen how the uncertainty surrounding the predictive model, and hence the uncertainty in Risk, increases as the model is used for the longer forecasts. However, once in service, many uncertainties are removed: the material has been fixed, and actual flow data are available, and in most cases it is possible to monitor corrosion rates and inspect for damage. Preferably, the corrosion predictions should be reviewed and revised after commissioning using these field data.

Figure 2 is a modification of Fig. 1 and shows how the Risk is reduced when more accurate data are used. Any further modification to this Figure over time is strongly dependent on the inspection and confidence in inspection findings.

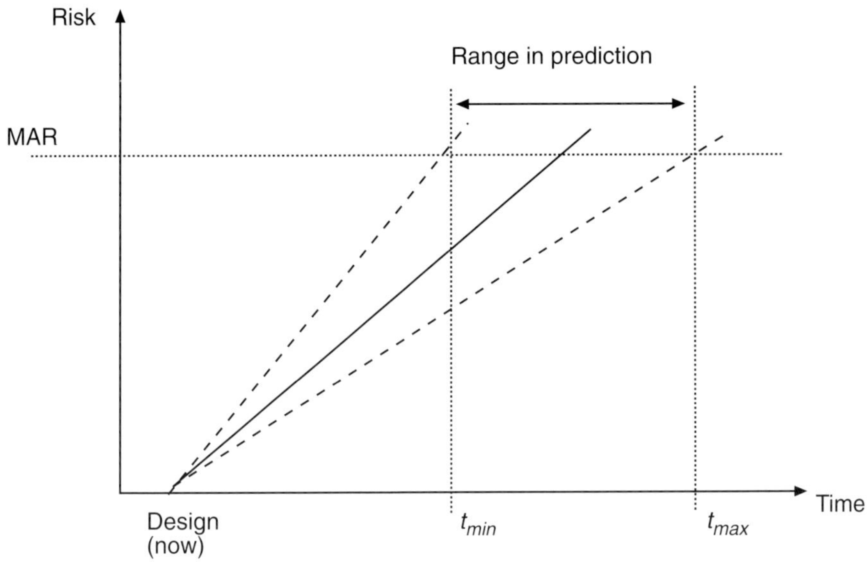

Fig. 1 *Development of Risk over time due to a corrosion process based on data available at design. The range in time to attain Maximum Acceptable Risk (MAR) results from the uncertainty in data and the model.*

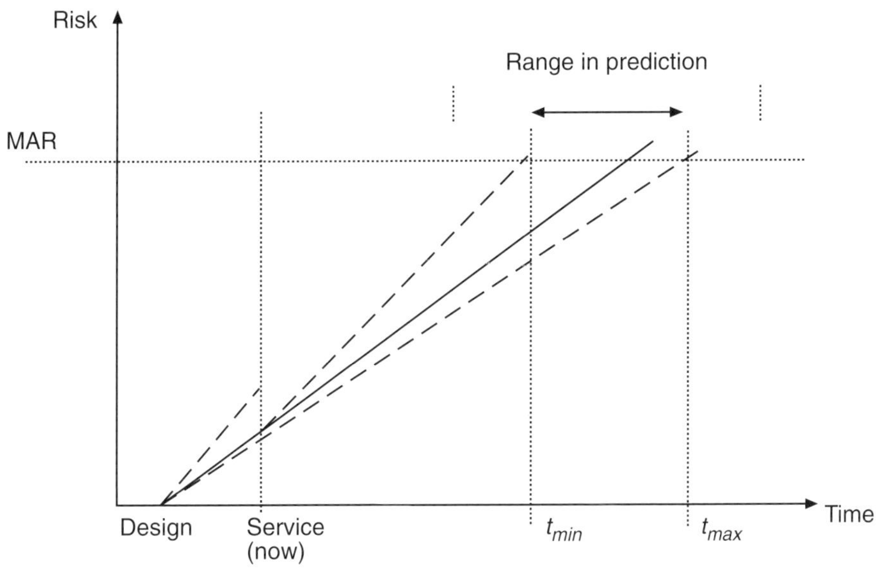

Fig. 2 *Modification of Fig. 1 illustrating how the use of service data shortly after commissioning, provides more certain prediction, and reduces the range in time to reach MAR.*

3.3. Inspection

It should be apparent from the above that inspection represents a most important part of controlling the Risk associated with CO_2 corrosion. In general, the Risk is lower the more often inspection is carried out, assuming that inspection tools have adequate detection capabilities. In cases where inspection reveals significantly more corrosion damage than anticipated — for example in a 'non-corrosive' dry-gas system – the Risk is reduced because the condition is known and remedial action can be planned and scheduled , i.e. the Risk is controlled.

Preferably, inspection does not reveal such surprises, and the measured findings can be used to adjust the predictions and hence the Risk development over time. This is illustrated in Fig. 3, where the measured value is used to derive a time-averaged corrosion rate that is used instead of the modelled version. It will be noted that the measured point in Fig. 3 does not give an absolutely correct value. This is because there is always some uncertainty associated with the inspection method due to its sizing capability, whilst the overall efficacy is affected by the amount of area actually covered by the inspection (expressed as percentage coverage). Measurement of wall thicknesses at pre-determined locations can be considered to provide accurate corrosion rates and damage sizing, but only at the point where they are deployed. If a mechanism is very well understood, then it is possible that measured indications of damage in one point can be correlated to corrosion rates elsewhere in the system. Although such correlations can be identified in CO_2 systems together with local 'hot-spots', it is difficult to rely on this approach in general. Inspection pigging of pipelines represents an alternative method; here a large surface area is inspected, but with a sizing accuracy and detection limit that is less than that from local monitoring.

For most inspection methods, the ability to discover damage of a given size is expressed by the 'Probability of Detection' (POD). If there is 100% coverage then the

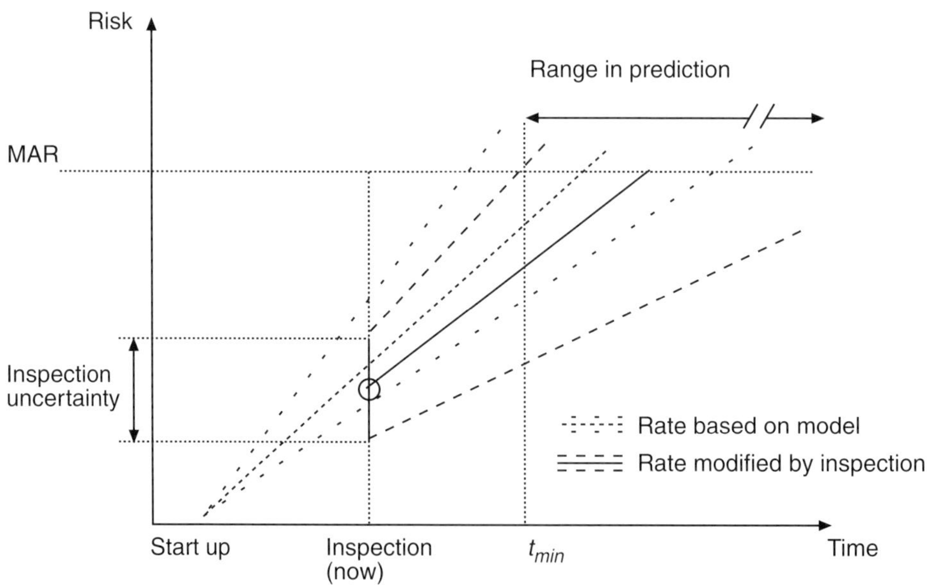

***Fig. 3** An inspection point is shown with a range representing the uncertainty due to probability of detection. Forecast from the predictive model is modified to account for the inspection findings. Here the revised range in prediction of time to attain MAR is increased, but the minimum 'lifetime' is longer.*

POD is limited by the physical detection limits, whereas a 'spot check' will be limited by both the physical detection limits of the method and the probability that the method has actually 'covered' an area with a defect. With the exception of 100% coverage, the actual reliability of the inspection procedure is difficult to quantify and there can be considerable uncertainty accompanying any results so derived. It is then possible to review the inspection results together with the predictive model, and make a new prediction based on the evaluation of these two factors. For example, if there is mutual agreement between the likelihood of corrosion damage predicted by the 'model', and the findings of the inspection, then the confidence in the rate will be increased and Risk reduced. This is illustrated in Fig. 3 where the improved result is used to extend the inspection period. In general, this routine can be repeated following each inspection. It would be hoped that each inspection will increase the confidence in the prior inspections, and the inspection interval can be increased as predictions become more accurate.

4. Cost Savings

It was mentioned earlier that there is a business incentive for using Risk to control corrosion. It is difficult to quantify potential gains in general terms, but worked examples in literature for pipelines [9,11] suggest that Risk based materials selection (i.e. using CRAs vs C–Mn steel) may reduce lifetime costs by about 30–50% depending on conditions. Even in cases where C-steel is selected there can be significant costs associated with the 'corrosion allowance'. For example, a 3 mm corrosion allowance on a 500 km 40 in. long export line of C-steel may incur an additional 70M US dollars in materials and fabrication: As a 'dry' gas line the likelihood of corrosion is low and with the anticipated damage morphology as localised pitting, the Risk is also expected to be low with the 3 mm allowance offering minimal benefit. A Risk analysis should show whether or not the extra allowance is a actually a good investment.

5. Conclusions

The concept of Risk is expressed as a function of Consequences of a Failure and the Likelihood of the Failure. Failure must address the corrosion damage morphology and its possible interaction with loads.

Risk control starts at the design stage by being aware of the uncertainties involved in predictive models and in the input parameters used in these models. Today, the traditional 'safe' approach is to select the worst case combination of expected conditions in combination with a 'worse case' or 'conservative' predictive model.

Risk based analyses involve a move away from deterministic corrosion rates that are frequently expressed as 'worst case' or 'conservative' values, towards distributions where terms such as 'most likely', or '0.9 probability that corrosion will not exceed given value', are preferred. This is not a straightforward change. Even with CO_2-corrosion, which is probably one of the most extensively 'modelled' corrosion phenomena, there is significant uncertainty involved in transposing the laboratory data to real-life. A modification to these models to render corrosion rates as

'Probability Distributions' would be useful in describing the uncertainties, and would hence be a valuable contribution to Risk based calculations.This implies correlating laboratory models to real-life operational data.

With Risk expressed in monetary terms, alternative materials selection options can be compared for their intended lifetime. The Risk cost should be considered together with Capex and Opex, including Risk control measures such as, any chemical treatments, increased wall thickness, and inspection costs.

References

1. DNV Rules for Submarine Pipeline Systems (1996).
2. C. de Waard and D. E. Milliams, "Carbonic acid corrosion of steel", *Corrosion*, 1975, **31**, 950–69.
3. C. de Waard and D. E. Milliams, "Predictive model for CO_2 corrosion engineering in wet natural gas pipelines", *Corrosion*, 1991, **47**, (12) 976.
4. C. De Waard and D. E. Milliams, "Prediction of CO_2 corrosion of carbon steel", *Corrosion '93*, Paper No. 69. NACE International, Houston, Tx, 1993.
5. C. de Waard, U. Lotz and A Dugstad, "Influence on liquid flow velocity on CO_2 corrosion. A Semi-empirical Model", *Corrosion '95*, Paper No. 128. NACE International, Houston, Tx, 1995.
6. NORSOK Offshore Standard M-DP-001. "Design Principles Materials Selection", Oleindustriens Landsforening, Stavanger, Rev 1. 1995.
7. *CO_2 Corrosion Control in Oil and Gas Production* (M. B. Kermani and L. M. Smith, Eds). Publication No. 23 in European Federation of Corrosion Series. Published by The Institute of Materials, London, 1997, 56pp.
8. J. Crolet and W. Wilhelmsen, " Influence of a layer of undissolved cementite on the rate of CO_2 corrosion of carbon steel". *Corrosion '94*, Paper No. 4. NACE International, Houston, Tx, 1994.
9. S. Olsen, T. Blygstad and T. Wenn, "Flow effects on corrosion inhibitors in topside flowlines". *Corrosion '95*, Paper No. 129. NACE International, Houston, Tx, 1995.
10. T. Sydberger, J. D. Edwards and K. J. Mork, "A probalistic approach to the prediction of CO_2 corrosion and its application to life cycle cost analyses of oil and gas equipment." *Corrosion '95*, Paper No. 65. NACE International, Houston, Tx, 1995.
11. T. Sydberger, J. D. Edwards and O. Bjornoy, "Modes of CO_2 Corrosion damage: consequences for corrosion control strategies", *Corrosion '96*, Paper No. 28. NACE International, Houston, Tx, 1996.
12. J. D. Edwards, T. Sydberger and K. J. Mork, "Reliability based design of CO_2 corrosion control". *Corrosion '96*, Paper No. 29, NACE International, Houston, Tx, 1996.

4
Effect of Chromium on Mesa Corrosion of Carbon Steel

R. NYBORG, A. DUGSTAD and P.-E. DRØNEN
Institute for Energy Technology, (IFE), Norway

ABSTRACT
The factors affecting local breakdown of protective corrosion films in CO_2 corrosion of carbon steel have been studied in flow loop experiments at 40–80°C and pH 5.8. Above 60°C corrosion films with good protective properties can form under these conditions, but mesa attack with very high local corrosion rates can occur especially at high flow rates. The presence of 0.5% chromium in the steel was found to reduce significantly the tendency for severe mesa attack in carbon steels during CO_2 corrosion.

1. Introduction

The application of carbon steel in oil and gas pipelines and production tubulars depends to a large degree on either the formation of protective corrosion product films or the use of corrosion inhibitors. Above 60°C and at pH values higher than 5.5 iron carbonate corrosion films with good protective properties can form in wet CO_2 systems, but local breakdown of the corrosion films may result in rapid local attack or mesa corrosion attack, especially at high flow rates. In these mesa corrosion areas the local corrosion rates can become very high.

Carbon dioxide corrosion of carbon steel is very dependent on water chemistry, and it is important to differentiate between conditions leading to protective film formation and the so-called 'worst case corrosion' where no protective films are formed. The experiments presented here were performed at The Institute for Energy Technology during the two multi-client projects Kjeller Sweet Corrosion III and IV, which concentrated on conditions where protective corrosion films can form. Flow loop experiments were performed at 40–80°C and pH 5.8. In this paper the focus is on the factors affecting initiation of mesa corrosion and particularly the effect of chromium content of the steel. In the earlier projects Kjeller Sweet Corrosion I and II experiments under conditions giving no or little corrosion films were performed [1–3].

2. Experimental

Flow loop experiments were carried out in an 80 mm i.d. high pressure one-phase water flow loop made of duplex stainless steel. The steel specimens were tested as flat 50 × 60 × 2.5 mm coupons mounted along the diameter of the flow channel and separated by PTFE spacers. Three to four specimen racks with different flow rates were used, and up to twelve specimens could be mounted in each specimen rack. This enabled testing of a large number of different steels at different flow rates from 0.1 to 7 ms^{-1} in one loop experiment. The specimens were electrically insulated from each other and from the loop, and some of the specimens had electrical connections enabling electrochemical measurement to be conducted. Corrosion rates were measured for all specimens by weight loss and by measurement of the depth of mesa attack or pits. The corrosion rates of some of the specimens were followed during the experiment by linear polarisation resistance measurements (LPR) and by a radioactive technique where some of the steel specimens were neutron activated and the decrease in activity resulting from metal loss was measured with a scintillation counter [4].

Most experiments were conducted at 80°C, with some at 40°C and a few at 100–150°C. All experiments were with 1.8 bar CO_2 partial pressure and pH values between 5.5 and 5.8. The desired pH value was obtained by addition of sodium bicarbonate. Distilled water was used in all the experiments. Some experiments were done with 3% NaCl in the water and some without salt. The loop pressure was kept constant at 8–12 bar by using a pressuriser half filled with water and kept at a higher temperature than the loop. Corrosion products were precipitated out in the pressuriser due to the lower iron carbonate solubility at higher temperature. The level of ferrous ions (Fe^{2+}) was kept at the desired level by adjusting the flow of loop water through the pressuriser. The content of ferrous ions was kept above the iron carbonate solubility, thus enabling precipitation of iron carbonate films on the specimen. The Fe^{2+} content was usually 5–40 ppm in 80°C experiments and 100–300 ppm at 40°C. The duration of the experiments was two to three weeks. Seven different steels were tested, three carbon steels without addition of chromium or nickel, two steels with 0.5–1% Cr and two steels with 0.5–1% Ni. The chemical composition of the steels is shown in Table 1.

3. Results

Several aspects of CO_2 corrosion of carbon steels were studied, including the formation and breakdown of protective corrosion films, effect of supersaturation of iron carbonate, effect of small amounts of Cr and Ni in the steel, the influence of

Table 1 Chemical composition and microstructure of the tested steels

Material	C	Si	Mn	S	P	Cr	Ni	Cu	Microstructure
St-52 batch 1	0.18	0.34	1.50	0.017	0.023	0.08	0.03	0.010	Ferritic–pearlitic
St-62 batch 2	0.15	0.18	1.57	0.011	0.014	0.03	0.04	0.015	Ferritic–pearlitic
Low C steel	0.04	0.34	1.34	0.001	0.008	0.03	0.06	0.11	Ferritic
16Mn Cr5	0.18	0.31	1.08	0.029	0.010	0.87	0.09	0.12	Ferritic–pearlitic
API 5LX-X60	0.06	0.24	1.12	0.003	0.008	0.53	0.02	<0.01	Quenched and tempered
API 5LX-X65	0.20	0.32	1.6	0.009	0.003	0.02	0.75		Ferritic–pearlitic
API 5LX-X60	0.05	0.25	1.32	0.003	0.013	0.02	1.00	0.1	Quenched and tempered

prehistory, different steel surfaces and precorrosion, and the effect of Ca^{2+} or acetic acid in the water. Only some of the results are presented here, with emphasis on the effect of Cr on the occurrence of mesa attack.

Loop experiments at 80°C, 1.8 bar CO_2 and pH 5.8 resulted in severe mesa corrosion attack at a high flow rate of 7 ms^{-1}. The mesa attacks were deep and flat-bottomed and usually without any corrosion film in the bottom, as seen in Fig. 1. The local corrosion rate in the mesa attacked areas was very high, typically 20–40 mm/y. The areas which were not mesa attacked usually had a 50–100 μm thick corrosion film consisting mostly of iron carbonate. In these areas the corrosion rate was in the range 0.2–2 mm/y. Mesa attack occurred in the carbon steels without addition of chromium or nickel and in the steels with 0.5–1% Ni. Deep, film-free mesa attack did not occur in the two steels containing 0.5–1% chromium. In some of the experiments a shallow, flat-bottomed mesa-like attack occurred in some of the chromium containing steel specimens. This attack was covered with corrosion film in the bottom, as seen in Fig. 1.

In situ corrosion rate measurements with LPR measurements showed that the specimens without mesa attack usually had high corrosion rates at the start of the exposure, while the corrosion rate decreased rapidly during the first day or two due to precipitation of protective corrosion films. The corrosion rates remained very high throughout the experiment for specimens with deep, film-free mesa attack. The LPR measurements for the 0.5% Cr steel specimens with a shallow, film-covered local attack as seen in Fig. 1 showed that the corrosion rate in these specimens decreased gradually during a period of two weeks towards the same level as specimens without any mesa attack, as a result of reformation of protective films in the bottom of the local attack.

Measured corrosion rates in loop experiments at 80°C, 1.8 bar CO_2 and pH 5.8 are shown as function of liquid flow rate in Fig. 2. The group of very high corrosion rates above 10 mm/y represent specimens with deep, film-free mesa attack. The corrosion rates for these specimens are calculated from the depth of the mesa attack. For specimens without local attack weight loss data are used. The corrosion rate in these specimens decreased during the experiment due to protective film formation, and final corrosion rates are lower than the weight loss data shown. The purpose of this figure is primarily to show when mesa attack occurs.

Corrosion rates in experiments at 40°C, 1.8 bar CO_2 and pH 5.8 are shown in Fig. 3. At this lower temperature the specimens suffered either uniform attack with little

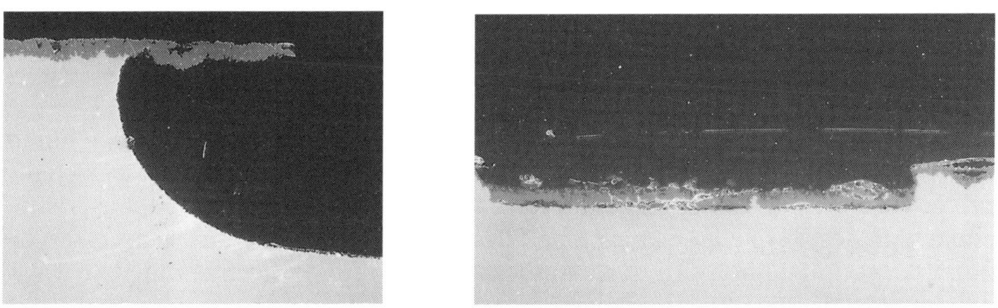

Fig. 1 *Mesa attack in carbon steel without chromium* (left) *and with 0.5% Cr* (right) *×20 magnification.*

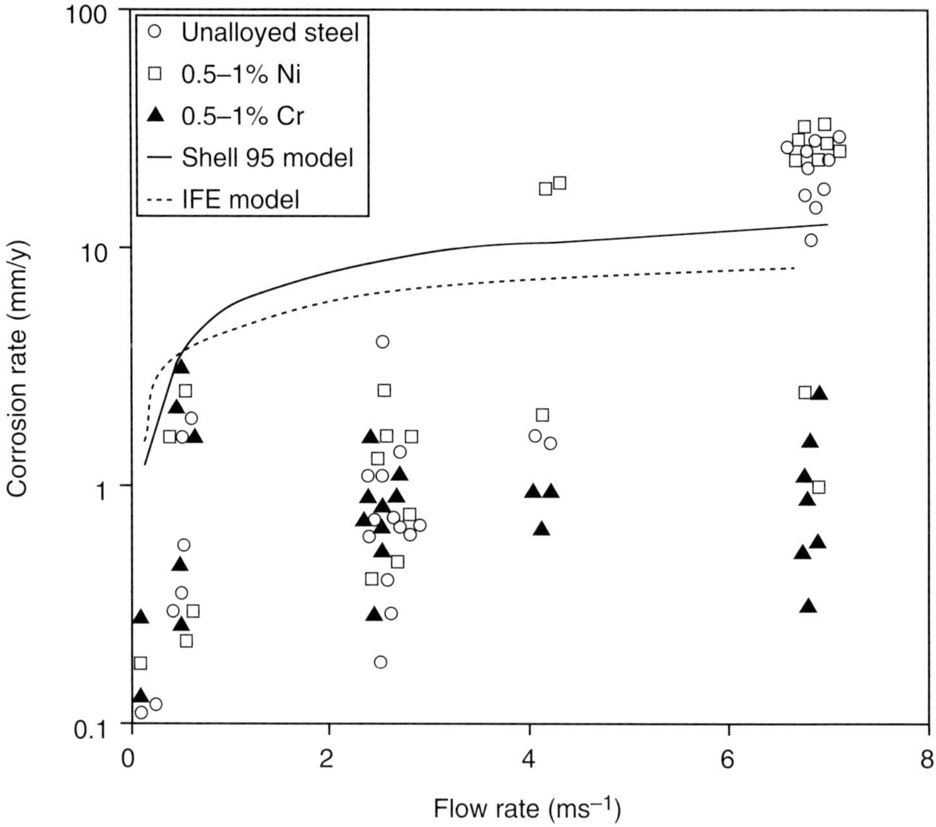

Fig. 2 Corrosion rates in loop experiments at 80°C, 1.8 bar CO_2 and pH 5.8.

film formation or mesa attack without corrosion film in the bottom. This was the case also for the chromium containing steels. The mesa attack was not so deep as in the experiments at 80°C, and the corrosion rates in areas without mesa attack were higher than at 80°C. The corrosion films formed at 40°C were much more porous and less protective than the films formed at 80°C because the iron carbonate precipitation rate is much lower at 40°C than at 80°C.

4. Discussion

Mesa attack is a localised attack which can occur when a protective corrosion film is broken down locally. A prerequisite for mesa attack is therefore that a partially protective corrosion film is formed. The precipitation rate of iron carbonate is slow, and a high degree of supersaturation of iron carbonate in the water is necessary in order to precipitate iron carbonate in the corrosion film. Protective films are easily formed at 80°C when the pH is above 5.

The experiments reported here have shown that carbon steels without chromium addition suffer mesa corrosion easily at 80°C and pH 5.8 when the flow rate is above

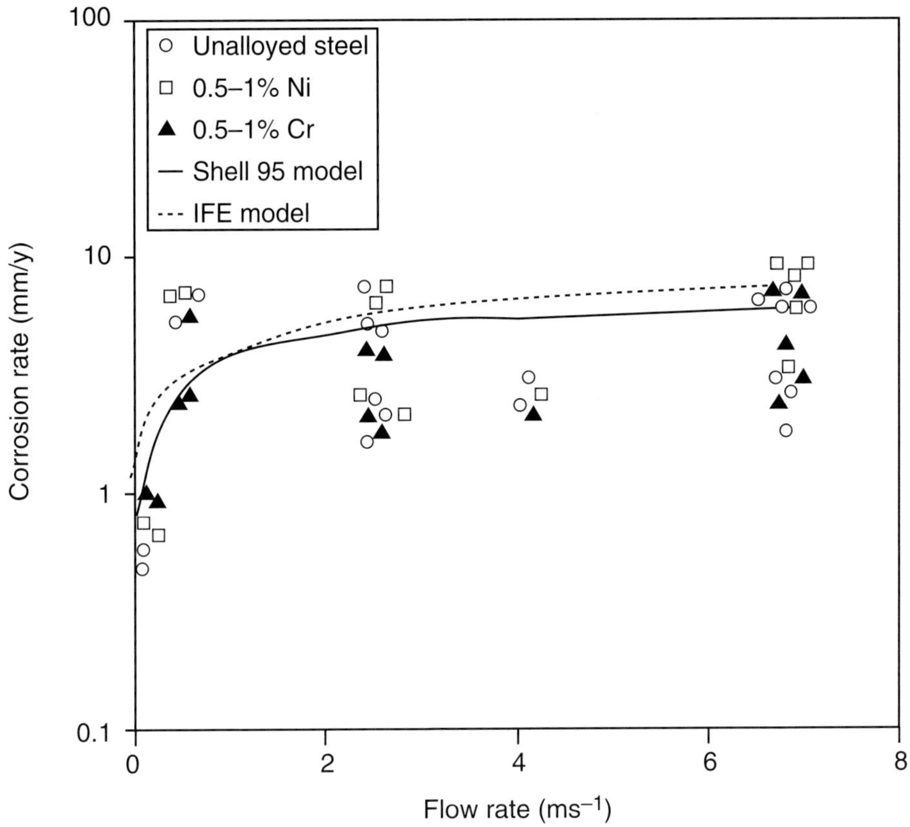

***Fig. 3** Corrosion rates in loop experiments at 40°C, 1.8 bar CO_2 and pH 5.8.*

2.5 ms^{-1}. At lower flow rates localised attack can occur but protective corrosion films are easily reformed, thereby lowering the corrosion rate in the locally corroded areas. At 4–7 ms^{-1} flow rate the experiments have shown that protective corrosion films are not reformed in the areas with mesa attack as long as the flow is maintained, resulting in deep mesa attack with local corrosion rates in the order of 20 mm/y. The corrosion rate in the mesa attacked area can even be higher than in the 'worst-case' situation without any corrosion films, possibly due to a galvanic cell between the film-free metal in the bottom of the mesa attack and the film-covered metal outside the mesa attack. This is illustrated by the lines in Fig. 2, representing the Shell model [5] and the IFE model [1] for CO_2 corrosion without protective corrosion films. Both these models have been fitted to the large number of loop experiments in the Kjeller Sweet Corrosion II project at IFE [1].

Deep, film-free mesa attack was not found in the steels with 0.5–1% chromium. While localised attack can initiate also in these steels, protective corrosion films seem to reform more easily. This ability to reform protective films makes localised attack less dangerous in chromium-containing steels than in plain carbon steels. Addition of 0.5–1% nickel did not seem to have any pronounced effect on the occurrence of mesa attack in these experiments.

The corrosion rates in specimens without mesa attack at 80°C are well below the predicted worst case corrosion rates given by the lines in Fig. 2, showing that the corrosion films have a protective effect. The corrosion films formed at 40°C offer little protection, as seen in Fig. 3.

Chromium is commonly used to improve the corrosion resistance in wet CO_2 environments, and several oil companies have started to use 0.5% Cr steel in pipelines. Work at Sumitomo [6,7], Kawasaki [8], and IFE [2] shows a good effect of small amounts of chromium in CO_2 saturated water at temperatures below 90°C. It is a general view that chromium does not reduce the corrosion rate significantly on a film-free steel surface and that the effect is attributed to better protective properties of the corrosion film. It is suggested that Cr is enriched in the iron carbonate film and makes it more dense and stable. Kimura has reported [8] that mesa attack did not occur in carbon steels containing more than 0.25% Cr in rotary immersion tests at 80°C and pH 6.0. This is in accordance with the experiments reported here. At higher temperatures the effect of Cr seems to be more unclear. Current experiments at IFE in the Kjeller Sweet Corrosion V project includes similar loop experiments at 100–150°C.

5. Conclusions

Carbon steel without chromium addition can suffer severe mesa corrosion attack at 80°C and pH around 5.8 when the flow rate is higher than 2.5 ms^{-1}. At lower flow rates protective corrosion films can reform in areas with localised attack. The corrosion rate in the mesa attacked area can be higher than in a situation without any corrosion films, possibly due to a galvanic cell between the film-free bottom of the mesa attack and the film-covered metal on the outside.

Addition of 0.5–1% chromium prevents the occurrence of deep mesa attack at 80°C and pH 5.8. While localised attack can initiate also in these steels, protective corrosion films seem to reform more easily in the chromium-containing steels. This ability to reform protective films makes localised attack less dangerous in chromium-containing steels than in plain carbon steels.

6. Acknowledgements

The experiments described in this paper were carried out in the two multi-client research projects Kjeller Sweet Corrosion III and IV financed by Elf, BP, Shell, Total, Mobil, Phillips Petroleum, Statoil, Norsk Hydro, Saga Petroleum and Kawasaki Steel. The authors wish to thank these companies for their technical and financial support.

References

1. A. Dugstad and L. Lunde, "Parametric study of CO_2 corrosion of carbon steel", *Corrosion '94*, Paper No. 14, NACE International, Houston, Tx,1994.

2. A. Dugstad, L. Lunde and K. Videm, "Influence of alloying elements upon the CO_2 corrosion rate of low alloyed carbon steels", *Corrosion '91*, Paper No. 473, NACE, Houston, Tx, 1991.
3. K. Videm and A. Dugstad, "Effect of flow rate, pH, Fe^{2+} concentration and steel quality on the CO_2 corrosion of carbon steels", *Corrosion '87*, Paper No. 42, NACE, Houston, Tx,1987.
4. A. Dugstad and K. Videm, "Radioactive techniques for corrosion monitoring", *Corrosion '89*, Paper No. 159, NACE, Houston, Tx, 1989.
5. C. de Waard, U. Lotz and A. Dugstad, "Influence of liquid flow velocity on CO_2 corrosion: a semi-empirical model", *Corrosion '95*, Paper No. 128, NACE International, Houston, Tx, 1995.
6. A. Ikeda, M. Ueda and S. Mukai, "CO_2 corrosion behaviour and mechanism of carbon steel and alloy steel", *Corrosion '83*, Paper No. 45, NACE, Houston, Tx, 1983.
7. M. Ueda and A. Ikeda, "Effect of microstructure and Cr content in steel on CO_2 corrosion", *Corrosion '96*, Paper No. 13, NACE International, Houston, Tx, 1996.
8. M. Kimura, Y. Saito and Y. Nakano, "Effects of alloying elements on corrosion resistance of high strength linepipe steel in wet CO_2 environment", *Corrosion '94*, Paper No. 18, NACE International, Houston, Tx, 1994.

5
Formation of Protective Corrosion Films During CO_2 Corrosion of Carbon Steel

A. DUGSTAD
Institute for Energy Technology, (IFE) Norway

ABSTRACT
For many oil and gas pipelines the use of carbon steel is limited by CO_2 corrosion. The parameters affecting formation of protective corrosion films have been studied in a large number of flow loop experiments performed with 2 bar CO_2 partial pressure and pH 5.8 at 40–120°C and flow rates 0.1–7 ms^{-1}. The importance of dissolved corrosion products, temperature, pH and flow rate has been studied in detail. The risk for local attack during operation is also dependent upon the previous history of the steel surface. Periods of shutdown or oil wetting may facilitate formation of protective films because corrosion products accumulate at the steel surface. The presence of mill scale or rust from tube manufacturing and storage may also enhance the protectiveness of the corrosion film and reduce the risk for local attack. Periods with stagnant conditions during the loop experiments were shown to increase the protectiveness. The influence of the history of the steel surface can be one explanation for the tendency to lower corrosion in the field compared to laboratory experiments.

1. Introduction

The successful application of carbon steel in oil and gas pipelines and production tubulars depends to a large degree on either the formation of protective corrosion product films or the use of corrosion inhibitors. Protective iron carbonate films can form in wet CO_2 systems at all temperatures provided that the pH and the dissolved iron carbonate concentration in the bulk is sufficiently high. The growth of iron carbonate is a slow temperature-dependent process. A high supersaturation of Fe^{2+} and CO_3^{2-} is therefore necessary in order to form a protective film. Once the film is formed, however, it will remain protective at a much lower supersaturation.

Protective film formation is accelerated by all measures that restrict the transport of reaction products from the surface and which can anchor the corrosion product. Freshly ground and continuously exposed specimens are therefore much more susceptible to high corrosion rates than real pipelines, where the surface is covered by oxide layers grown during rolling and storage and carbonate layers formed during stagnant periods and periods with low water wetting when the surface is water wet but the water is replenished at a low rate.

The initiation, growth and stability of corrosion product films has been studied extensively at The Institute for Energy Technology (IFE) during the three multi-client projects Kjeller Sweet Corrosion III, IV and V [1,2] and the ESI project [3]. Some of the results obtained in the KSC-III project carried out in the period 1990 to 1993 are presented and discussed in this paper. The project is particularly focused on how operational parameters like stagnant periods and periods with low water-wetting affect the protective properties of the corrosion product film.

2. Experimental

The flow loop experiments were carried out in an 80 mm i.d. high pressure one-phase water flow loop made of duplex stainless steel (IR10). The steel specimens were tested as flat $50 \times 60 \times 2.5$ mm coupons mounted along the diameter of the flow channel and separated by PTFE spacers. Three to four specimen racks with different flow rates were used, and up to twelve specimens could be mounted in each specimen rack. This enabled testing of a large number of different steels at different flow rates from 0.1 to 7 ms^{-1} in one loop experiment. The specimens were electrically insulated from each other and from the loop, and some of the specimens had electrical connections enabling electrochemical measurements to be made. Corrosion rates were measured for all specimens by weight loss and by measurement of the depth of mesa attack or pits. The corrosion rates of some of the specimens were followed during the experiment by linear polarisation resistance measurements (LPR) and by a radioactive technique where some of the steel specimens were neutron activated and the decrease in activity resulting from metal loss was measured with a scintillation counter [4].

Experiments were performed at 40, 60, 80 and 120°C. The CO_2 partial pressure varied from 1.8 to 2.6 bar and pH values between 5.5 and 6.0. The desired pH value was obtained by addition of sodium bicarbonate. Distilled water was used in all the experiments. Some experiments were done with 0.1% NaCl in the water and some without salt. Only the radioactive corrosion monitoring technique was used in the experiments without salt. The loop pressure was kept constant at 8–12 bar by using a pressuriser half filled with water and kept at a higher temperature than the loop. Corrosion products were precipitated out in the pressuriser due to the lower iron carbonate solubility at higher temperature. The level of ferrous ions was kept at the desired level by adjusting the flow of loop water through the pressuriser. The Fe^{2+} content was usually 5–40 ppm in experiments above 80°C and 50–300 ppm at lower temperatures. The duration of the experiments was two to three weeks. A large number of carbon steels have been studied and the chemical composition of those discussed in this paper are shown in Table 1.

Under field conditions the pipe wall can be exposed periodically to the water and

Table 1. *Chemical composition and microstructure of the tested steels. Descriptive names of the steels have been given in the first column for easy identification in the text.*

Material		C	Si	Mn	S	P	Cr	Ni	Cu	Microstructure
St52	St-52	0.15	0.18	1.57	0.011	0.014	0.03	0.04	0.015	Ferritic–pearlitic
Cr0.87	16Mn Cr5	0.18	0.31	1.08	0.029	0.010	0.87	0.09	0.12	Ferritic–pearlitic
Cr0.53	API 5LX-X60	0.06	0.24	1.12	0.003	0.008	0.53	0.02	<0.01	Quenched and tempered
Ni 1.0	API 5LX-X60	0.05	0.25	1.32	0.003	0.013	0.02	1.00	0.01	Quenched and tempered

the oil (condensate) phase. When the oil phase replaces the water phase, the steel surface becomes either oil wet or it continues to be wetted by a thin water film embedded in the corrosion product layer. A similar water film will be present if the pipe is drained. As long as the thin water films are present the corrosion attack will continue, but at a much lower rate since the water is replenished at a very low rate or not at all. Under these conditions, corrosion products are forced to precipitate out on the surface, and this facilitates protective film formation significantly. The described scenario was simulated with so-called 'dry periods' during the exposure. The flow loop was drained in these periods and the steel surface was kept wet by exposing the specimens in a CO_2 atmosphere with high humidity.

3. Results

A survey of the corrosion rates obtained in the various loop experiments is given in Table 2. Four types of results can be distinguished. The corrosion rates given in the first column of results are obtained on fresh ground steel specimens which were continuously exposed at low iron carbonate supersaturation. Protective films did

Table 2. *Survey of the results obtained from various experiments carried out at 1.8–2 bar CO_2 partial pressure and pH 5.7–5.9. Corrosion rates are measured by weight loss (W), LPR and radioactive techniques (Ra). The supersaturation S is indicated in the table*

Temp.	Flow	Steel	Corrosion rate mm/year			
			Fresh surfaces S = 1–3	Fresh surfaces S = 10–100	Dry period 2–5 days	Pre-corroded surfaces
	2.5	St52B	2.5 W			
120	6.8	St52B	2.1 Ra			0.05 Ra
		Cr0.87		0.4 LPR		
80		Ni1.0		<<1 LPR		
	2.5	St52B	8** depth	<0.3 Ra		
		Cr0.53		0.2 LPR		
		Cr0.87		1.8 LPR	0.11 LPR	
60		Ni1.0		1.0 LPR	0.1* LPR	
	2.5	St52B	7.5 Ra	4.3 LPR	0.3** depth	
		Cr0.53		1.1 LPR	<0.1 LPR, Ra	
		Cr0.87		3.2 W	<0.1* LPR	0.06 LPR
		Ni1.0		1.7 W	<0.01 LPR	<0.01* LPR
	2.5	St52B	4.4 Ra	3.0 Ra	0.4 LPR	<0.01 LPR
40		Cr0.53		2.0 Ra	<0.1* LPR	<0.1 LPR
		Cr0.87		3.8 W	0.3* LPR	<0.1 LPR
		Ni1.0		2.0 W	0.8 LPR	<0.1* LPR
	6.8	St52B	6.3 Ra	5.8 Ra	0.5 LPR	<0.01 LPR
		Cr0.53		2.1 Ra	0.4 LPR	<0.1 LPR

* Small pits. ** mesa atacked areas, corrosion rate calculated from depth measurement.

not form easily at 40 and 60°C under these conditions and the corrosion rates obtained were therefore still high after a few weeks exposure.

When the temperature was increased to 80°C, however, protective films started to form as the corrosion precipitation process accelerated. Only parts of the surface obtained protection in the reported experiment and a mesa attack did therefore develop. Mesa corrosion is characterised by a deep, more or less film free, attack surrounded by a protected surface. Mesa attacked specimens are indicated with two asterisks in the table. The film formed at 120°C gave protection, but less so than at 80°C, no localised attack was seen.

When supersaturation is increased iron carbonate films form more easily. Better protection was therefore observed at all temperatures, as shown by the results in column 2 in Table 2, where the supersaturation was higher. No mesa attack developed at 80°C and 2.5 ms^{-1} flow velocity. At higher flow velocities (4.1 and 6.8 ms^{-1}), however, mesa attack was also observed in these experiments (not included in the table). At lower temperatures, more film accumulated on the surface, but the film was very porous and gave low protection. These films can be very thick as shown in Fig 1(a). The development of thick, non-protective porous films is not only a laboratory effect, as similar films can be seen on steel exposed in the field (see Fig. 1(b)). The films can grow for months without giving protection unless the steel is exposed to stagnant or 'wet' conditions. After a stagnant or wet period of a few days, corrosion rates are usually reduced 5–100 times, as shown by the results in column 3 in Table 1. This is interesting as it can explain the large gap often seen between laboratory results and corrosion rates measured in the field. Laboratory experiments have traditionally been conducted with continuously exposed fresh ground steel samples and very conservative worst case results have therefore been obtained. A corrosion survey carried out by Elf showed that high corrosion rates were seldom experienced at pH values higher than 5.6 [5]. This is also the case in the laboratory experiments when simulation of operational parameters like shut down and pre-formed oxide and carbonate films are included in the test procedure.

When protective films are once formed they usually remain protective if the supersaturation in the bulk is reasonable high. This is illustrated in column 4 in Table 1, where pre-corroded steels have been exposed. Corrosion rates are generally very low. Some small pits were seen now and then and it can be speculated whether they were still active when the experiments were stopped. Scanning electron microscopy investigation showed that many of them were filled with corrosion products. It is therefore assumed that they would obtain protection with time.

4. Discussion

All the experiments discussed above were conducted under supersaturated conditions. That means that iron carbonate can precipitate and deposit on the surface if it is not removed progressively by the flow. Whether it deposits or not depends amongst other factors on how it is anchored to the surface. If a massive precipitation takes place a dense iron carbonate film can form on most steel surfaces. At lower precipitation rates, however, the formation of an iron carbonate film depends on the presence of a skeletal substrate which can strengthen the film during the initial growth process. This substrate can be iron carbide (steel St52) as shown in Fig. 1, chromium oxides or carbides (steel Cr 0.87 and Cr 0.53) or other constituents accumulated on the steel. Iron carbonate precipitation can then fill up the porous skeletal substrate. Since small differences in the steel composition and the microstructure can affect the anchoring properties, it is not surprising that large differences in the performance of similar carbon steels has been seen in both field applications and laboratory work.

When iron carbonate precipitates on the steel surface it is the morphology and the density of the corrosion film which controls the transport of different species through the film. Both laboratory and field experiments have shown that thick, non-protective

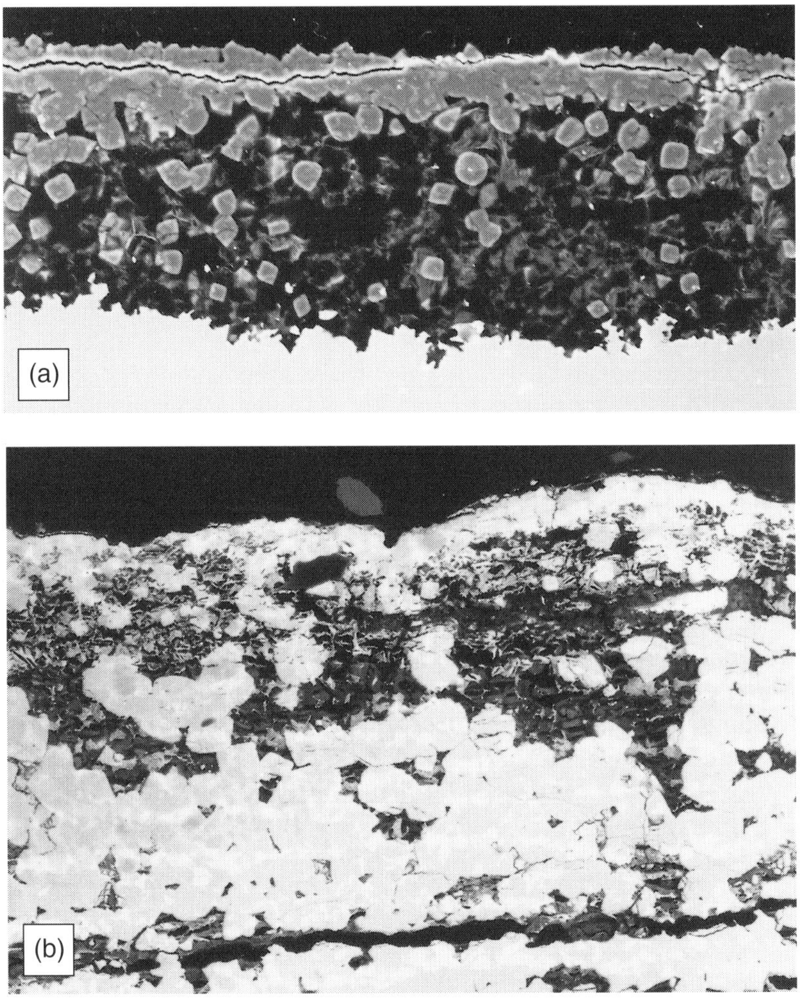

Fig. 1 *(a) Cross-section of a corrosion product film formed in a laboratory experiment carried out at 60 °C, pH 5.8 and 5–10 times supersaturation. A carbide film partly filled with iron carbonate can be seen (×450 magnification). (b) Cross-section of a corrosion product film formed on a weight loss coupon exposed 3 months in a flow line in the North Sea. Temperature 60 °C and 1 bar CO_2 partial pressure. A carbide film partly filled with iron carbonate can be seen. The film was more than 1 mm thick (×120 magnification).*

films containing large amounts of iron carbonate can form. This is particularly the case at low temperatures (20–60°C). The loop experiments have shown that such films can be sealed by denser carbonate precipitates and the corrosion rate reduced substantially by draining the system or by exposing the steel to stagnant conditions.

Once a reasonably dense corrosion product film has formed it can be speculated whether it is the reduced transport through the iron carbonate film or the formation of a passive film which reduces the corrosion rate [6]. Experiments carried out at IFE in collaboration with Elf have shown that the steel shows passive-like behaviour at high pH. High pH values can be obtained close to the steel when the transport of reactants and corrosion products is restricted by the deposited corrosion products.

5. Conclusions

Protective iron carbonate films can form in wet CO_2 systems at all temperatures provided the pH and the dissolved iron carbonate concentration in the bulk is sufficiently high. The growth of iron carbonate is a slow temperature-dependent process. A high supersaturation of Fe^{2+} and CO_3^{2-} is therefore necessary to form a protective film, particularly at low temperatures. Once the film is formed, however, it will remain protective at a much lower supersaturation.

Protective film formation is accelerated by all measures which restrict the transport of reaction products from the surface. Freshly ground and continuously exposed specimens are therefore much more susceptible to high corrosion rates than real pipelines, where the surface is covered by oxide layers grown during rolling and storage and carbonate layers formed during stagnant periods and periods with low water wetting. Corrosion films can grow for months without giving protection unless the steel is exposed at stagnant or 'wet' conditions. After a stagnant or wet period of a few days, corrosion rates are usually reduced 5–100 times because corrosion products are accumulated at the steel surface and protective corrosion films more easily formed.

The introduction of pre-formed film and stagnant and dry periods in laboratory experiments gives a better simulation of the conditions in the field. Laboratory experiments have traditionally been conducted with continuously exposed fresh ground steel samples and very conservative worst case result have therefore been obtained. The influence of the history of the steel surface can be one explanation for the tendency to lower corrosion rates in the field compared to laboratory experiments.

6. Acknowledgements

The experiments described in this paper were carried out in the multi-client research project Kjeller Sweet Corrosion III financed by Elf, BP, Shell, Total, Mobil, Phillips Petroleum, Statoil, Norsk Hydro, Saga Petroleum and Kawasaki Steel. The author wish to thank these companies for their technical and financial support.

References

1. A. Dugstad. L. Lunde and K. Videm, "Influence of alloying elements upon the CO_2 corrosion rate of low alloyed carbon steels", *Corrosion '91*, Paper No.473, NACE, Houston, Tx, 1991.
2. A. Dugstad and L. Lunde, "Parametric study of CO_2 corrosion of carbon steel", *Corrosion '94*, Paper No. 14, NACE International, Houston, Tx, 1994.
3. J. L. Crolet, S. Olsen and W. Wilhelmsem, "Observation of multiple steady states in the CO_2 corrosion of carbon steel", *Corrosion '95*, Paper No. 127, NACE International, Houston, Tx, 1995.
4. A. Dugstad and K. Videm, "Radioactive techniques for corrosion monitoring", *Corrosion '89*, Paper No. 159, NACE, Houston, Tx, 1989.
5. J. L. Crolet and M. Bonis, "Prediction of the risk of CO_2 corrosion in oil and gas wells", *SPE Production Engineering*, 1991, **6**, (4), 449-453.
6. J. L. Crolet, N. Thevenot and S. Nesic, "Role of conductive corrosion products on the protectiveness of corrosion layers", *Corrosion '96*, Paper No. 4, NACE International, Houston, Tx, 1996.

6
On the Effect of Microstructure in CO_2 Corrosion of Carbon Steel Welds

R. ANDREASSEN and J. ENERHAUG*
Hogskolen i Narvik, Norway
*Statoil Research Centre, Norway

ABSTRACT

Oil companies have experienced considerable corrosion in weld zones of hydrocarbon piping systems made of carbon steels. In CO_2-containing environments corrosion failures are often explained by the varying tenacity of protective corrosion scales on different carbon steel microstructures. Preferential corrosion of weld metal and heat affected zone (HAZ) has mostly been explained by galvanic effects caused by differences in alloy content and microstructure. This paper reviews opinions on how the microstructure affects corrosion on carbon steel materials for top side production piping, but because only a few sources deal specifically with these materials carbon steel in general and pipe line materials have also been reviewed. Possible hypotheses for the effects of microstructure are discussed.

In the experimental part metallographic studies were carried out by nital etching and colour etching on corroded welds from top side production piping. Coupons of uncorroded welds were used as references. Certain theories on the tenacious nature of the corrosion scale are challenged and discussed against the background of the metallographic studies. Initiation seems to be important for weld corrosion. Propagation seems to have the nature of localised corrosion that is not significantly affected by the microstructure. The result of the metallographic study lends support to the hypothesis that CO_2 corrosion may propagate beneath a corrosive scale.

1. Introduction

Carbon dioxide corrosion of carbon steels is a major challenge in the oil and gas industry as carbon steels are widely used because of their strength and cheapness. Extensive research over many years has been conducted to understand this form of corrosion and to estimate corrosion rates [1–5] that may vary by more than an order of magnitude from one situation to another. The use of carbon steels in this environment relies upon a protective scale of mainly iron carbonate, $FeCO_3$, although the protectiveness varies with conditions in the fluids such as temperature, saturation [5–7], H_2S content [8,9] and corrosion history [4,5,10]. Corrosion resulting from CO_2 has been videotaped *in situ* [11]. A physically based model proposes that corrosive conditions may arise beneath a precipitated corrosion scale [12,13].

This paper targets the corrosion found in welds on topside low carbon steel piping. Because only few references treat this specifically, it has been necessary to study the literature on CO_2 corrosion in general, including pipeline steels. The metallographic studies were carried out on low carbon welds from flowlines.

Characterising the structures in carbon steel welds is seldom an easy task. The degree of precision and minuteness must be related to the field of investigation. In the classical description of the austenite to ferrite/carbide transformation it is assumed that the material is pure carbon steel with known thermal history. The applied material standards will usually specify performance and certain chemical requirements only. The structure in CO_2 exposed base metals in the oil industry varies with their production routes, e.g. normalised to ferritic/pearlitic steels, QT (quenched-and-tempered) or TMCP (Thermo-mechanically Controlled Process). Forged flange material studied in this paper contains Nb (0.02%) that enhances the formation of bainite even after normalising heat treatment. In corrosion tests with X65-steels the carbon content has been seen to vary from 0.31% [3] (QT-type) to 0.06% [14] (TMCP-type).

In weld zones the thermal history varies with the distance from the weld metal and the structure becomes very complex. For weld qualification purposes the optical microscope is most frequently used for structural description. Various structure classification systems that differ from the classical have been suggested, and disagreements as well as confusion on the designations of ferrite–carbide aggregates have been noticed [15]. In this work a SINTEF modification is used [16]. The structures are divided into (i) martensite and lower bainite, (ii) sideplate–ferrite, including Widmanstätten ferrite and upper bainite, (iii) acicular ferrite, (iv) grain boundary ferrite and polygonal ferrite and (v) pearlite. In multi-run welds, the heat treatment from the subsequent run affects the previous run and breaks up the coarse grained structures in the HAZ (heat affected zone) and weld metal. Parts of the metal are polygonised (normalised) and the columnar texture in the weld metal is obscured or becomes less visible.

2. Extracts from Literature Survey of CO_2 Weld Corrosion

Alloying elements are known to affect the nobleness of the steel depending on the environment. Nickel is assumed to be favourable by some workers [17,18] while others find no beneficial effects [4]. Also copper and molybdenum are found to have positive effects on both weld metal and HAZ [17]. Chromium (about 0.5%) is found to strengthen the protectiveness of iron carbonate [3,4,19].

It is generally agreed that cementite (Fe_3C) is cathodic in CO_2 environments and enhances the corrosion galvanically as it (the cementite) is exposed by the corrosion of ferrite [7]. In CO_2 corrosion tests of welds, galvanic effects are demonstrated [17] but it has not been possible to explain the observed corrosion rates by measuring galvanic currents between parts of split weldments. Some workers generally suppose that preferential CO_2 corrosion at welds is caused by minor compositional and/or microstructural differences across the weldment. There are indications that inhibitors may exacerbate this kind of corrosion [20]. Initiation of pitting is believed to arise where inclusions and particles form microgalvanic cells, sulfides being among the more important. High density of inclusions is in itself considered to make a metal more anodic. There is no general agreement on how the different cementite forms affect CO_2 corrosion [23].

The metallurgical condition of the steel is often believed to have an effect on how firmly the corrosion scale sticks to the surface [21] and it is a common opinion that lamellar pearlite provides a better grip than particle pearlite [4,7,22,23] because of its protruding cementite lamellae. Remnants of the pearlite bands have been found in the corrosion scale [11,22]. It is claimed that iron carbonate is less tenacious and less crystalline on spheroidised and QT-steel, and that the corrosion scale forms more dense crystals on normalised steel [22]. Some workers have found a X52-type steel more susceptible to corrosion in QT condition than in normalised condition [23]. Others have found contradictory results with QT-treatment of two different steels to the same specification (St52) [4]. Bainitic structures are found to become more anodic than ferritic/pearlitic steel [17]. In one case it is found that fine grained ferrite corrodes slower thall coarse grained [23]. In another investigation one low carbon X65-steel (TMCP) appears more anodic than a 'St52' steel (ferritic/pearlitic) [14].

Taking all theories and opinions in the surveyed literature into account, they lead to the following working hypotheses:

1. Due to differences in the tenacity of iron carbonate, corrosion should be more severe in low temperature HAZs and other areas with spheroidised or particulate cementite. Corrosion should be less severe in areas with lamellar cementite, such as pearlite.

2. Different ferrite structures may provide different tenacity for the corrosion scale so giving variations in the corrosion.

3. Due to the banded texture of hot rolled carbon steels galvanic effects should cause differences in the corrosion in areas where the bands break the surface. Ferrite should be more severely attacked.

4. Various fine grained structures of intragranular ferrite may render the weld metal and high temperature HAZ more anodic.

3. Results and Discussion of the Metallographic Investigations

Metallographic investigations were carried out on corroded welds. Uncorroded welds were included as references. The welds are listed in Table 1. The corroded weldments come from flowlines transporting unstabilised water and CO_2-containing oil production fluids. During replacement and field cut out the specimens were not given any special treatment to ensure minimum damage to the corrosion scale. The base metals seem similar to the ASTM A106 GrB and ASTM A105 type materials used in other cases where specifications were known.

The corroded surfaces can broadly be divided into three different types.

Smooth: no corrosion scale is visible.

Jagged: the surface is jagged or heavily curved; corrosion scale is often seen.

Rugged: the surface is rugged or fringed, often with remnants of grains as if

Table 1. *Chemical composition and microstructure of the tested steels. Descriptive names of the steels have been given in the first column for easy identification in the text*

Specimen	Base metal 1	Base metal 2	Weld metal alloying	Mose severe corrosion
W1	BM1-F-A	BM2-HR-A	Unalloyed	WM + BM
W2	BM1-F-A	BM2-HR-A	1 Ni	HT HAZ, both sides
W3	BM1-F-A	BM2-HR-A	1 Ni	HT HAZ, BM2-side
W4	BM1-F-A	BM2-HR-A	1 Ni	HT HAZ, BM2-side
W5	BM1-F-A	BM2-HR-B	Unalloyed	LT/IC HAZ BM2-side
W6	BM2-HR-C	BM2-HR-C		Not exposed

Key WM: Weld metal; BM: Base metal; F: Forged; HR: Hot Rolled; HT: High Temperature; LT: Low Temperature; IC: Inter Critical.

Alloys Forged materials are similar to known ASTM A105 type.
Typical analysis: 0.2 C, 1.3 Mn, 0.25 Si, 0.1 Cr, 0.02 Nb.
Hot rolled materials are similar to known ASTM A106 GrB type.
Typical analysis: 0.17 C, 1 Mn, 0.25 Si, 0.2 Cu.

Weld Metal *Typical analysis:* 0.1 C, 0.25 Si, 1.3 Mn in addition to elements specified in the table.

Alloy contents are given in wt%.

the grain boundaries are preferentially attacked; corrosion scale is present and fills the space between the grain remnants.

The adhesion of corrosion scale is often explained as an anchoring effect of lamellar pearlite. However, microexamination cannot prove any differences in the degree of corrosion on surfaces where the pearlite bands reach the metal surface. On the specimen W5 it is found that the deepest attack is in the low temperature HAZ with spheroidised pearlite, while in another clock (angular) position on the same weld, the deepest attack is in the normalised zone containing fine lamellar pearlite. In the forged base metals, where lamellar pearlite is not present, the deepest attack is, in one case, in the high temperature HAZ, while in the other case it is in the subcritical–intercritical HAZ.

On specimen W1 the pearlite bands/carbon rich bands can be followed from the steel into the corrosion scale along with traces of slender MnS phase (see microphotograph, Fig. 1). Assuming that the stratified cracked layers of corrosion scale in W1 express the propagation of corrosion, it can be seen that the corrosion rate appears to be quite even. There are no observations that indicate the establishment of steady galvanic effects close to bands of pearlite or other ferrite carbide aggregates. These effects must at least have been insignificant and overridden by other local cell reactions. Other workers have also noticed the lack of correlation between corrosion attack and pearlite bands [7].

Deoxidation particles are present in weld metal. In the unalloyed weld of W1, the corrosion has affected weld metal as well as base metal. It seems that the corrosion obscures any differences in susceptibility to initiation of corrosion between base metal

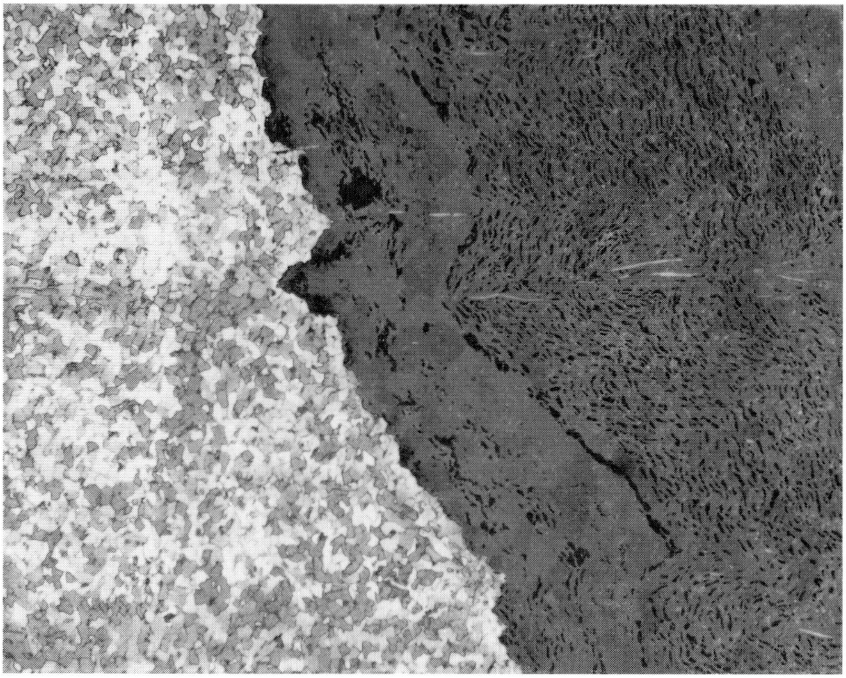

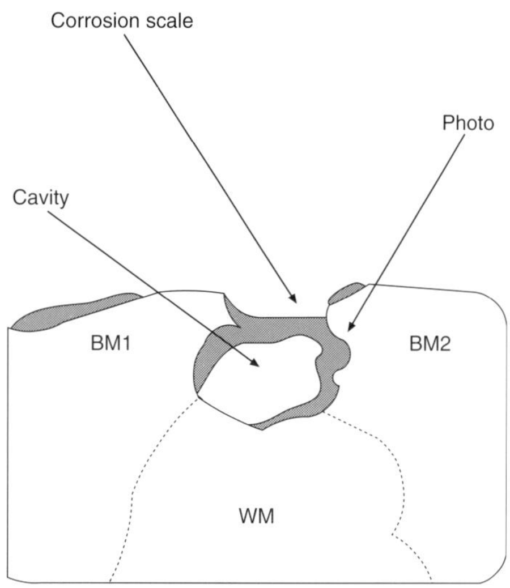

***Fig. 1** Microphotograph of W1,200× magnification, colour etched. Stratified corrosion scale. Carbon rich band from former pearlite band follows MnS phases horizontally in the middle of the photo. The corrosion has reached polygonised ferrite in the HAZ.*

and weld metal. On the specimens the ferrite grain sizes vary from 3–5 μm in polygonal ferrite in HAZ and weld metal up to 15–30 μm in normalised hot rolled base metal. The various exposed ferrite shapes include polygonal, fine grained sideplate ferrite and acicular ferrite. The microstructural constituents studied near corrosion sites on the specimens include both high angle grain boundaries in polygonal and acicular ferrite and low angle grain boundaries combined with high aspect ratios in side plate ferrite. On the investigated weldments unambiguous indications that the ferrite shape or grain size affect the corrosion propagation are not found. This conclusion is even more obvious when it is considered that deep corrosion in HAZ or weld metal must penetrate different microstructures.

Frequently the most severe corrosion sites on the specimens are covered with thick crystalline looking corrosion scale. It is assumed that the cracks in the thick scale appeared when the scale dried out. Crolet *et al.* [11,13] have hypothesised that a corrosion layer may establish a corrosive environment with its own electrolyte, and that both protective and corrosive scales may be found within the same chemical system. Observations on the investigated welds seem to lend support to these hypotheses. Also Nesic and Lunde [7] observed cavities and pits filled with corrosion scale. They presumed it was protective or, tentatively, that there was some kind of crevice corrosion mechanism that established a local galvanic effect. Joosten *et al.* [11] noticed from their video recording of progressing CO_2 corrosion that significant interactions must take place in the interface between metal and corrosion scale.

4. Conclusions

1. Working hypotheses based on the literature survey have been discussed against metallographic observations on corroded weldments.

2. The metallographic studies cannot support the theories stating that pearlite, spheroidised pearlite or other carbide aggregates affect the propagation of CO_2 corrosion under field conditions.

3. The metallographic studies do not indicate that the propagation of CO_2 corrosion in weld zones is associated with any specific microstructural constituent alone.

4. It is suggested as a hypothesis that weld corrosion in CO_2 corrosion under field conditions is nucleated at various metallurgical discontinuities found in HAZ or weld metal. The individual microstructural constituents do not affect differently the propagation of corrosion.

5. During the metallographic studies there were observations that lend support to the hypothesis stating that corrosive conditions may occur under corrosion scale from CO_2 corrosion.

References

1. "Advances in CO_2 Corrosion", Vol. I, 1984 , P. A. Godard; P. A. Burke, B. S. Asphahani and B. S. Wright (Eds) "Advances in CO_2 Corrosion", Vol. II, 1986, Selected Papers from *Corrosion '84* and *'85*, NACE International, Houston, Tx, 1984/85.
2. C. de Waard and U. Lotz, "Prediction of CO_2-corrosion of carbon steel", *Corrosion '93*, Paper No. 69, NACE International, Houston, Tx, 1993.
3. A. Dugstad and K. Videm, "KSC II, Kjeller Sweet Corrosion II, Final Report", Institutt for Energiteknikk, Kjeller, Norway, 1990.

4. A. Dugstad, H. Hemmer, H. Kooyman and W. Wilhelmsen, "KSC III, Kjeller Sweet Corrosion III, Final Report", Institutt for Energiteknikk, Kjeller, Norway,1994.
5. J. L. Crolet,"Which CO_2 corrosion, hence which prediction", *Progess in the Understanding and Prevention of Corrosion*, J. Costa and A. D. Mercer (Eds) (Proc. 10th European Corrosion Congress, 1993) Vol 1, 473–497. The Institute of Materials, London, 1993.
6. A. Dugstad, "The importance of $FeCO_2$ supersaturation on the CO_2 corrosion of carbon steels", *Corrosion '92*, Paper No. 14, NACE International, Houston, Tx, 1992.
7. S. Nesic and L. Lunde, "Carbon dioxide corrosion of carbon steel in two-phase flow", *Corrosion*, 1994, **50** (9), 717–727.
8. R. G. Asperger, "Iron sulfide pit initiation, propagation and passivation", *Corrosion '93*, Paper No. 223, NACE International, Houston, Tx, 1993.
9. K. Videm and J. Kvarekval, "Corrosion of carbon steel in CO_2 saturated aqueous solutions containing small amounts of H_2S", *Corrosion '94*, Paper No. 12, NACE International, Houston, Tx, 1994.
10. D. Harrop, S. Webster, A. J. McMahon and G. J. Partridge, "CO_2 Corrosion: A testing problem with a predictive solution", Progress in the Understanding and Prevention of Corrosion, J. Costa and A. D. Mercer, Eds. (Proc. 10th European Corrosion Congress, 1993), Vol. 1, 473–497. The Institute of Materials, London, 1993.
11. M. W. Joosten, T. Johnsen, A. Dugstad, T. Walmann, T. Jossang, P. Meakin and J. Feder, "*In Situ* observations of localized CO_2 corrosion", *Corrosion '94*, Paper No. 3, NACE International, Houston, Tx, 1994.
12. J. L. Crolet, S. Olsen and W. Wilhelmsen, "Influence of a layer of undissolved cementite on the rate of the CO_2 corrosion of carbon steel", *Corrosion '94*, Paper No. 4, NACE International, Houston, Tx, 1994.
13. J. L. Crolet, S. Olsen and W. Wilhelmsen, "Observations of multiple steady states in the CO_2-corrosion of carbon steel", *Corrosion '95*, Paper No. 127, NACE International, Houston, Tx, 1995.
14. S. Nesic, J. Postlethwaite and S. Olsen, "An electrochemical model for prediction of CO_2 corrosion", *Corrosion '95*, Paper No. 131, NACE International, Houston, Tx, 1995.
15. Ø. Grong, *Metallurgical Modelling of Welding*. The Institute of Materials, London, 1994.
16. O. M. Akselsen and Ø.Grong, Offprint in Norwegian, SINTEF/NTH, Trondheim, Norway.
17. S. Endo, M. Nagae, S. Fujita and O. Hirano, "Preferential corrosion of weldment of carbon steel in CO_2 containing environment", 1992 OMAE Volume III-A, Mater. Eng. ASME, 1992, pp. 95–102.
18. C. A. Wright and K. Hladky, "Preferential Corrosion of Nickel Containing Weldments in Carbon Dioxide Containing Hydrocarbon Systems", Capcis Ltd. Report, March, 1994.
19. M. Murakami, S.I. Deshimaru, M. Kimura, T. Hatomura and T. Yamaura, "Development of high strength UOE pipe with superior CO_2 corrosion resistance", 1992 OMAE Volume V-B, Pipeline Technology, ASME, 1992.
20. J. W. Palmer, J. L. Dawson, T. Ulrich and A. N. Rothwell, "Inhibition of weld corrosion under flowing conditions — The development of test procedures", *Corrosion '93*, Paper No. 110, NACE International, Houston, Tx, 1993.
21. J. L. Dawson, C. C. Shih and P. K. N. Bartlett, "Models and predictions of CO_2 corrosion and erosion — corrosion under flowing conditions", *Progress in the Understanding and Prevention of Corrosion*, J. Costa and A. D. Mercer, Eds (Proc. 10th European Corrosion Congress, 1993), Vol. 1, 473–197. The Institute of Materials, London, 1993.
22. C. A. Palacios and J. R. Shadley, "Characteristics of corrosion scales on steels in a CO_2-saturated NaCl brine", *Corrosion*, 1991, **47**, (2), 122–127.
23. B. Mishra, S. Al-Hassan, D. L. Olson and M. M. Salama, "Prediction of microstructural effects on corrosion of linepipe steels in CO_2-brine solution", *Corrosion '93*, Paper No. 90, NACE International, Houston, Tx, 1993.

7
Influence of Chromium Addition up to 1% on Weight Loss Corrosion of Line Pipe Steels in Wet CO_2 Environments

R. POPPERLING*, Y. M. GUNALTUN, C. LINNE† and J. M. RIVREAU
Total, Paris, France
*Mannesmann, Duisburg, Germany
†Vallourec, Aulnoye, France

ABSTRACT

Corrosion tests on welded and non-welded chromium-containing steels (Cr < 1%) were conducted in a 0.1% NaCl solution saturated with 1 bar CO_2 at temperatures between 20 and 90°C under dynamic conditions. The flow rate of the test environment was 1 ms^{-1}. The data measured each day included weight loss of the specimens and the iron content and the pH of the test solution. The corrosion rates increased with increasing temperature. The results at higher temperatures indicate that the corrosion rate is determined by the limited solubility of iron in the test solution. The maximum possible corrosion rate at these temperatures is probably higher than that measured. No preferential corrosion was observed on the weldments.

Despite the effect of the limited iron solubility on the measured corrosion rate, it can be concluded from the results that the general corrosion rate is independent of the chromium content of the steel. In other words, contrary to previous published findings, chromium additions in the range studied do not improve the general corrosion resistance of line pipe steels in CO_2 environments.

1. Introduction

According to several published data the addition of about 0.5% chromium to standard carbon steel reduces significantly the corrosion rates in wet CO_2 environment. As early as 1983, a study published by Ikeda *et al.* [1] showed that corrosion rates could be reduced by up to 50% if the chromium content is about 0.5% and up to about 75% if the chromium content is around 1%. In 1991, reports issued by Kawasaki [2,3] and NKK (Nippon Ko Kan) [4] confirmed the results of Sumitomo [1].

According to a Vallourec report [5] issued in 1993, the influence of chromium additions is noticeable if the corrosion rate is above 1 mm/y. The tendency to form localised corrosion is low with chromium additions but the iron content of the test solution is important. Some studies carried out by IFE (Institute for Energy Technology Norway) [6] confirmed Vallourec findings: in some conditions where steels without chromium suffered severe mesa attack (corrosion rates between 12 and 15 mm/y), the corrosion rate of 0.5% chromium steel remained below 1.5 mm/y and no mesa attack was observed. The uniform corrosion rate of the 0.5% chromium steel was similar to that measured on normal carbon steel if no mesa attack takes place. It was established that the main advantage of 0.5% chromium steels is not the reduction of general corrosion rate but the resistance to mesa attack in severe flow and corrosive conditions.

Finally, the Kawasaki [4], Vallourec and IFE findings are in good agreement: i.e. the steels with no chromium are subject to mesa attack but chromium additions between 0.25% and 1% prevent mesa attack.

Some other tests carried out by IFE in solutions with very low iron contents showed that flat samples of 0.5% chromium steels are not subject to mesa corrosion although mesa attack can be observed at obstacles (weldments, upwards steps etc.). However, even with mesa attack, the penetration rates of 0.5% chromium steels are significantly lower than those of steels with no chromium (4.2–5.8 compared to 11.6–18.7 mm/y).

The main problem with CO_2 corrosion is its localised aspect, especially mesa corrosion (the corrosion allowances for the flowlines of some new fields are as high as 10 mm due to the high corrosivity). Based on the above results, it can be concluded that with the addition of 0.5% chromium the penetration rate and the risk of localised corrosion can be significantly reduced compared to steels with no chromium, particularly in cases of high corrosivity with risk of mesa attack.

Even though good corrosion behaviour of 0.5% chromium steels was established, the behaviour of welded 0.5% chromium pipes was not known. Before using these steels for a field application it was necessary to study the mechanical and corrosion behaviour (preferential corrosion) of weldments to guarantee the continuity of the mechanical properties and corrosion resistance in these areas. In fact, some of the recently developed welding consumables necessary for manual welding and repairs were not available on the market when this study started, and so their behaviour was not known.

Subsequently, a collaborative research programme was established by TOTAL to verify the effect of low chromium additions on the weld behaviour and corrosion control by corrosion inhibitor injection. The welding of pipe and plate samples was carried out by ETPM/SERIMER. The experimental work described below was carried out at CEV (centre d'études Vallourec) and MFI (Mannesmann Forschungs-Institut) on seamless line pipes and welded plates respectively. The inhibition tests carried out in the TOTAL laboratory will be published separately.

2. Materials and Mechanical Test Results

2.1. Seamless Line Pipe

A quenched and tempered API 5L X65 commercial pipe 168.3 mm dia. × 12.5 mm w.t. was sampled to produce simulated girth welds with modified consumables since the, usual unalloyed cellulosic electrodes lead to poor impact properties and generate galvanic effects that amplify the corrosion phenomena in the weldment. Matching the parent pipe and consumable appeared to be a solution and so modified electrodes were necessary to carry out manual welding successfully (cellulosic and basic coated electrodes, SMAW (shielded metal arc welding) technique) and wires for the automatic welding (GMAW (gas metal arc welding) technique).

Chemical analysis was conducted on the parent metal, the consumable, the weld bead and the heat affected zone (HAZ) using the glow discharge spectrometer technique. Reported chemical compositions are listed in Table 1 and these highlight the high cleanness of the 0.5% chromium pipe with less than 20 ppm of sulfur content. As for the consumables, the nickel enrichment of the basic electrode and the higher carbon level, e.g. for the cellulosic electrode, should be noted. Whatever the technique, the use of 0.5% chromium consumables led to homogeneous chromium distribution throughout the weld bead and the HAZ equivalent to the base metal level.

Hardness distribution across the weld was homogeneous. Some discrepancies were observed but can be correlated to the welding trial parameters as shown in Table 2. An improvement to the welding procedure would overcome the detrimental hardness profile for H_2S resistance, however, this study focused on CO_2 behaviour only.

Mechanical properties of weldments and pipe body were determined by tensile tests on round specimens and by Charpy V-notch impact tests. The absorbed energy at –20°C was evaluated both in the centreline zone and at 2 mm from the fusion line. Results are displayed in Table 3. All the welds can easily meet the API 5L X65 grade with good HAZ toughness with impact energy values about 230 J. This level is reduced in the weld centre-line area for SMAW where it meets the usual criteria of 40 J. It is noticeable that GMAW guarantees a level higher than 150 J whatever the location in the weldment. Finally, guided bend tests were successfully performed according to API 1104 on both root and cap bend specimens; no cracks occurred.

2.2. Welded Large Diameter Line Pipe

Laboratory heats with no chromium (chromium < 0.01%) and 0.5% Cr were cast in 100 kg ingots. The chemical compositions of the heats are listed in Table 4. The ingots were rolled into 15 mm thick plate adopting a three-stage thermomechanical rolling schedule. The results of tensile tests on transverse specimens at room temperature are given in Table 4. Also included in the same table are the results of Charpy V-notch impact tests (transverse specimens) and BDWT (Battelle drop weight tear) tests. The 85% shear area transition temperatures are in a narrow range

Table 1. *Chemical analysis (10–3 wt%)*

		C	Si	Mn	P	S	Cu	Cr	Ni	Mo
Seamless pipe	Base metal	65	229	772	13	2	105	527	69	37
+ basic SMAW	Electrode	36	420	690	13	8	300	430	500	10
	Weld seam		359	783	15		288	412	430	9
	HAZ		291	599	15		251	407	364	14
+ cellulosic SMAW	Electrode	140	280	700	9	10	20	540	30	30
	Weld seam		271	956	10		35	518	60	20
	HAZ		222	765	10		66	543	61	28
+ GMAW	Wire	130	240	710	9	11	20	610	30	10
	Weld seam		526	1110	12		101	456	32	26

Table 2. *Hardness profile and welding parameters*

Table 3. *Mechanical properties*

HV 10 kg				Welding parameters				Seamless linepipe	Tensile test			Charpy impact tests –20°C (10 × 10 mm)			
o.d.	i.d.	HAZ		Heat input		Total passes	consum-ables dia. (mm)		YS (ksi)	UTS (ksi)	Failure location for welds	weld axis		2 mm fusion line	
		max.	min.	max kJ cm^{-1}	pass							av. (J)	min (J)	av. (J)	min (J)
203	191							Base metal	75	85		272	868		
201	170	160	152	30	3	4	2.5 & 3.15	Basic SMAW	67	80	HAZ	116	94	235	230
218	168	160	150	17.8	5	7	4 & 5	Cellulosic SMAW	70	84	Base metal	45	44	251	248
236	254	206	193	7.7	2	4	wire 1	GMAW	71	82	Base metal	165	142	243	240

Table 4. *Chemical and mechanical properties of plates for welded pipe*

Chemical analysis (× 10^{-3} wt%)						Plate	Tensile test			CVI	DWT
C	**Si**	**Mn**	**P**	**S**	**Cr**		**YS (ksi)**	**UTS (ksi)**	**E (%)**	**–20°C (J)**	**85% SA**
32	280	1320	2	1	10	0% Cr	81	85	24	317	–30°C
33	270	1000	2	1	430	0.5% Cr	77	86	26	305	–30°C

CVI = Charpy V-notch; DWT = drop weight tear test; SA = shear area.

of –25°C to –30°C. There is no effect of chromium content either on the absorbed energy or on the transition temperature. The high upper-shelf energy values (> 300 J) measured are indicative of good cleanness of the plates tested.

Simulated girth welds were produced and the results of the hardness measurements, the impact testing and the tensile tests are summarised in Table 5.

Guided-bend tests on root-bend specimens were also conducted. All the specimens could be bent through an angle of 180° without cracks.

3. Corrosion Test Results

3.1. Testing Procedure

The corrosion test programme for assessing the behaviour of weldments is summarised in Table 6. In particular, laboratory testing was conducted in order to reproduce the preferential weld corrosion experience by many operators in carbon steel systems both on large diameter welded pipes and seamless line pipes. A common test protocol was shared by MFI and CEV and is described below.

Machined coupons 50 × 10 × 5 mm were taken at inside-wall position (CEV) and mid-wall position of the plate (MFI). The position of the weld was parallel to the 10 mm edge in the specimen centre. Three specimens were used per test. The apparatus consisted of 2.5 L glass cylinders. The test solution contained 0.1% NaCl and was saturated with CO_2 (1 bar). The test temperatures were 20, 40, 70 and 90°C and the flow rate was 1 ms^{-1}. The test duration was 3 or 14 days.

3.2. Results and Discussion

As well as using the same test protocol, the two laboratories (CEV and MFI) shared a

Table 5. *Chemical and mechanical properties of welded plate*

HV10 kg					Welded		Tensile test		CVI–20°C	
HAZ up	WELD up	Base material	HAZ low	WELD low	plate	YS (ksi)	UTS (ksi)	Failure location	Weld axis Av. (J)	Min. (J)
180	232	193	165	176	0%Cr + SMAW	75	86	HAZ/BM	246	224
202	250	205	150	186	0%Cr + GMAW	76	88	HAZ/BM	*	*
175	225	185	169	188	0.5%Cr + cellulosic 1 SMAW	71	84	HAZ/BM	194	184
170	192	194	158	170	0.5%Cr + cellulosic 2 SMAW	70	83	WM/HAZ	*	*
*	*	*	*	*	0.5% Cr + basic SMAW	*	*	*	216	120
*	*	*	*	*	0.5%Cr + GMAW	*	*	*	145	136
*	*	*	*	*	0.5%Cr + SAW	*	*	*	249	220

common coupon, i.e. a 0% chromium steel, as a reference. The results shown in Fig. 1 highlight the similarity of the behaviour of the carbon steel whatever the test conditions and the test laboratory. Moreover, for each laboratory, the deviation within the 3 results from triplicate coupons is generally lower than 5%, as illustrated in Fig. 2. These comments emphasise the high reproducibility of the testing procedure. The average

Table 6. *Corrosion tests (Cr concentration in %)*

PIPE	STEEL	ELECTRODE & WELDING	CORROSION TESTS
X-60 SEAMLESS	0.5 Cr Q & T	0.5 Cr Cellulosic 0.5 Cr Basic 0.5 Cr Automatic	Dynamic test 40°C, 70°C, 90°C
X-60 WELDED	0.5 Cr TMCP	0.5 Cr Cellulosic 0.5 Cr Basic 0.5 Cr Automatic	Dynamic test 40°C, 70°C, 90°C
REFERENCE SEAMLESS	Cr < 0.05 Q & T	Cellulosic	Dynamic test 40°C, 70°C, 90°C
REFERENCE WELDED	Cr < 0.05 TMCP	Cellulosic	Dynamic test 40°C, 70°C, 90°C

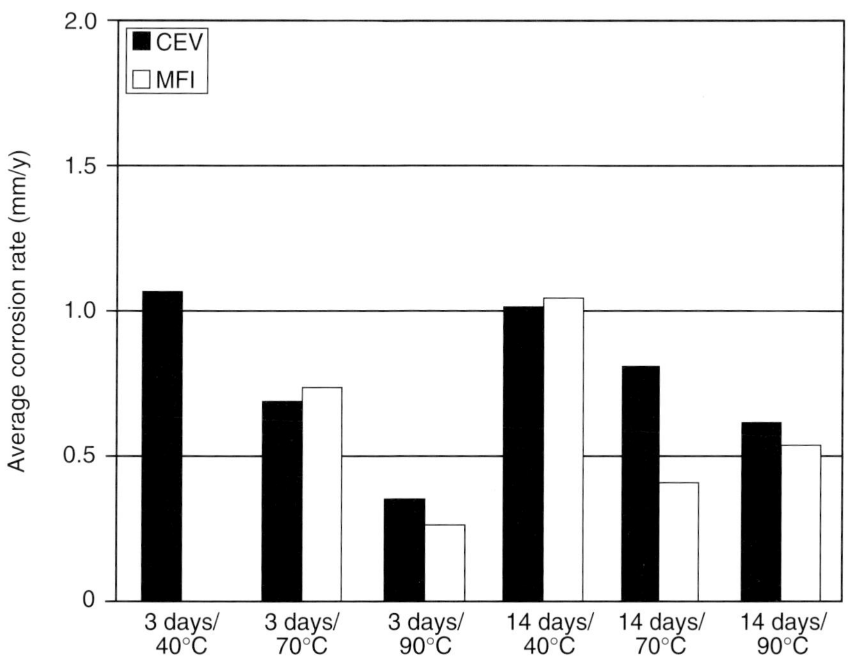

Fig. 1 *Reproducibility between CEV and MFI.*

corrosion rates of all specimens (for both seamless and welded pipes) are plotted in Fig. 3 and Fig. 4. The corrosion rates decrease with increasing temperature in the

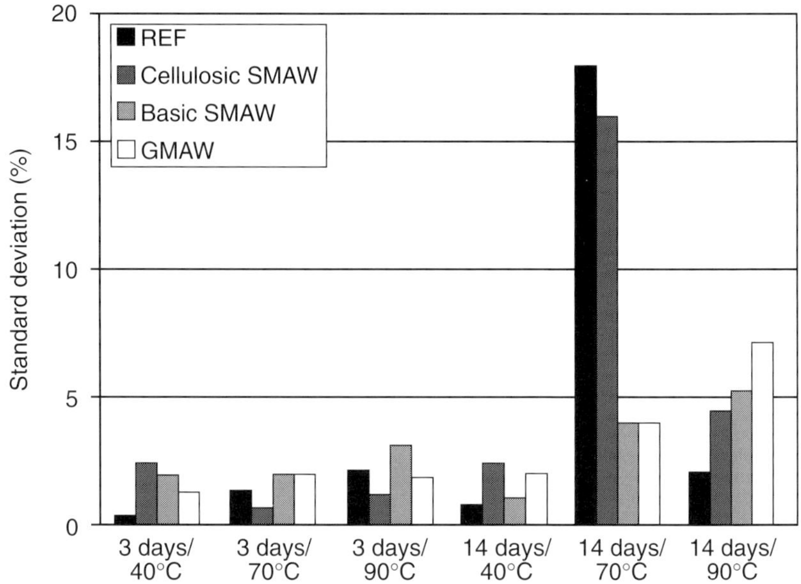

***Fig. 2** Reproducibility on triplicate coupons (CEV).*

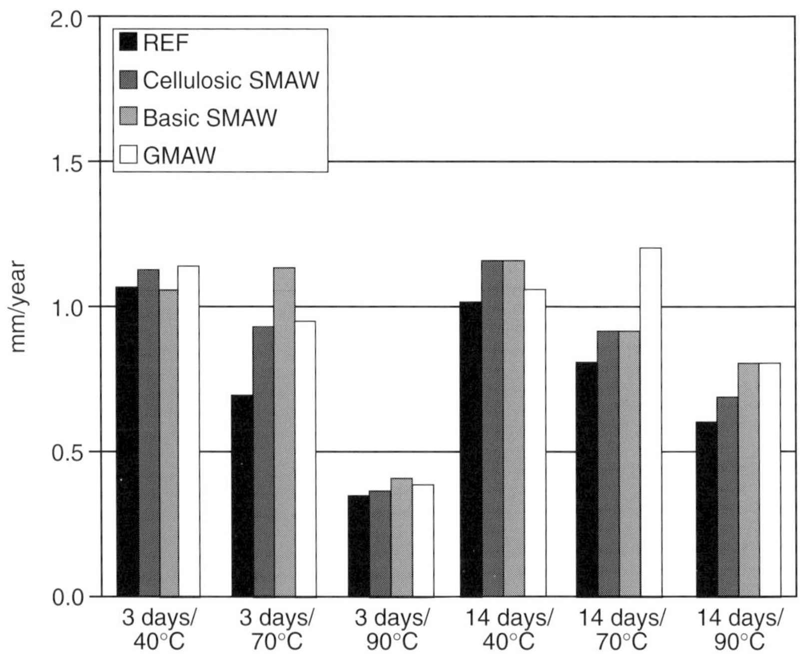

***Fig. 3** Welds on seamless pipe. Weight loss corrosion rate of X65 grade weldments in CO_2 bearing aqueous solutions.*

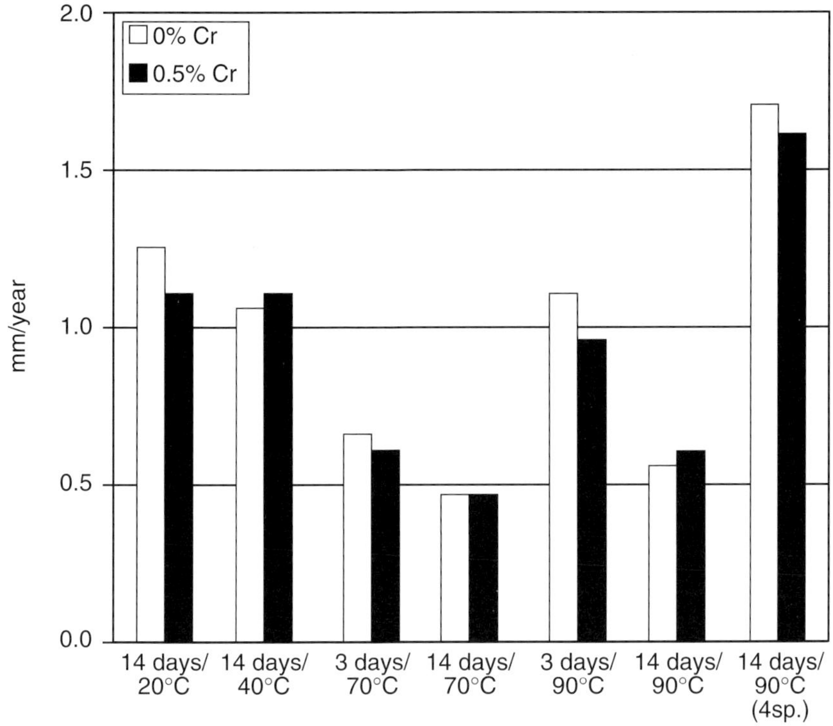

Fig. 3 *Welds on plates for welded pipe. Weight loss corrosion rate of X65 grade weldments in CO_2 bearing aqueous solutions.*

Table 7. *Iron content and renewals*

Test temperature	CEV [Fe] ppm	MFI [Fe] ppm	Renewals
20°C	**	170 380	After 1 day After 3 days
40°C	170–250 120	250 440	After 1 day After 3 days
70°C	80 80	140 140	After 1 day After 3 days
90°C	40 40	50 50	After 1 day After 3 days

range between 20 and 70°C. As can be seen from Table 7, there is an unexpected effect of the iron content in the test solution. Thus, at the lower temperatures, the iron content measured after a 3-day test was only twice that measured after one day. At temperatures of 70 and 90°C, the iron content was lower and independent of testing time. The same trends were observed with the acidity of the test solution. It appears that the corrosion rate is determined by the limited solubility of iron in the test solution. To verify this finding, a further test at 90°C was carried out using 4, instead of 12, specimens in each test vessel. The corrosion rates were markedly higher, see last column in Fig. 4, while the iron content and the pH values were the same as in the previous tests with 12 specimens. Also in this last test, there was no difference in the iron content between three-day and one-day testing times. Therefore, the measured corrosion rate still appears to be governed by the iron solubility in the test solution. The maximum possible corrosion rate at 90°C is likely to be higher than that measured in the present series of tests.

Despite the effects of testing variables on the level of corrosion rate, it can be concluded from the results that the corrosion rate is independent of the chromium content of the steel under the test conditions described above.

4. Conclusions

This study showed that:

- addition of 0.5 % Cr does not decrease the general corrosion rate (this confirms IFE results);
- no preferential corrosion was observed at weldments with the new consumables developed recently for welding of 0.5 % chromium;
- mechanical characteristics of welded samples are acceptable.

Based on the findings of this study and other available data, it is now possible to use this type of alloy for flowlines and trunklines with a reduced risk of localised corrosion.

5. Acknowledgements

Authors would like to thank F. Blanchard (Vallourec), J. P. Jansen (Europipe), R. Metz (Total) and J. Lacome (Serimer) for their help and support.

References

1. A. Ikeda, M. Ueda and S. Mukai, "CO_2 corrosion behaviour and mechanism of carbon steel and alloy steel". *Corrosion '83*, Paper No. 45, NACE International, Houston, Tx, 1983.
2. "Corrosion rest results of Cr bearing ERW and SML linepipes". Kawasaki Steel Corporation, 1991.
3. M. Kimura, Y. Saito and Y. Nakano, "Effects of alloying elements on corrosion resistance of high strength linepipe steel in wet CO_2 environment". *Corrosion '94*, Paper No. 18, NACE International, Houston, Tx, 1994.
4. "Technical document of 0.5 % Cr ERW pipe". NKK Corporation, 1994.
5. "Influence du Cr sur le comportement à la corrosion des line-pipes trempes-revenus en milieux CO_2-chlorures." CEV Rapport technique No.8819 C. September, 1993.
6. IFE Kjeller, *Sweet Corrosion* — V. Final report, 1997.

8
Effect of Environmental Factors and Microstructure on CO_2 Corrosion of Carbon and Cr-Bearing Steels

M. UEDA and H. TAKABE
Tubular Products Development Section, Wakayama Steel Works, Sumitomo Metal Industries, Ltd, 1850 Minato, Wakayama City, 640-8555, Japan

ABSTRACT
The effect of environmental factors and microstructure on CO_2 corrosion was investigated by using steels with Cr contents of 0, 2 and 13 mass%, and Steels J55 and N80 (API grade). Temperature, pH and H_2S contamination were studied as the environmental factors. The steels with 0 mass% (0Cr) and 2 mass% Cr(2Cr) showed a large increase in corrosion rate near a temperature, $T_{max.}$ with a maximum corrosion rate produced by the addition of 0.5% CH_3COOH to CO_2 environments. The steel with 13 mass% Cr(13Cr) did not show the $T_{max.}$ and its corrosion rate increased with increase of testing temperature with the addition of acetic acid. The H_2S contamination at 0.001 MPa suppressed the CO_2 corrosion at around the $T_{max.}$ due to FeS formation at the interface region between corrosion product and metal. Observations of the corrosion product formed on Steel J55 with a ferritic–pearlitic microstructure showed that the lamellar Fe_3C in the steel was retained while $FeCO_3$ was produced on the former ferrite phase. Therefore, the good resistance to localised corrosion of Steel J55 could be explained by an anchor effect of the lamellar remaining Fe_3C in the corrosion product.

1. Introduction

The CO_2 corrosion of carbon and low Cr steels known as 'sweet corrosion' has been one of the important problems in oil and gas industry because of both a high corrosion rate and severe localised corrosion [1]. The severity of the CO_2 corrosion depends particularly on temperature, CO_2 partial pressure, pH, H_2S contamination and material characteristics [2].

Thus, in the presence of carbon dioxide the temperature region where carbon steel shows general corrosion with a high corrosion rate below 60°C is referred to as Type 1, the region with highest susceptibility to severe corrosion occurs at around 100°C (Type 2), and corrosion resistance through the formation of a protective $FeCO_3$ film occurs above 150°C (Type 3). The corrosion rate of carbon steel increases with increasing CO_2 partial pressure. De Waard *et al.* have proposed a nomogram to estimate the corrosion rate from CO_2 partial pressure and temperature [3]. In particular, the presence of organic acids will lead to a decrease of pH and an increase in corrosion rate in CO_2 environments. Legezin *et al.* [4] have reported an increase of corrosion rate resulting from the addition of acetic acid and formic acid. However, the effect of organic acids on the three types of CO_2 corrosion has not been sufficiently investigated.

Low Cr steel with Cr contents below about 2 mass% has acceptable corrosion resistance to general and localised corrosion below 60°C, since low Cr steel has Cr-enriched corrosion products in CO_2 environments [2].

With regard to the effect of microstructure on the CO_2 corrosion of carbon steel, it is well known that carbon steel with a ferritic–pearlitic microstructure shows less localised corrosion at the temperatures below 80°C than those with a martensitic microstructure [4].

Therefore, in this paper, the effect of environmental factors and microstructure in CO_2 environments with and without acetic acid as the organic acid is investigated using carbon and Cr steels. The corrosion behaviour is discussed from the viewpoint of the thermodynamic stability and the morphology of corrosion products.

2. Experimental

2.1. Materials

The chemical compositions of materials used in this study are shown in Table 1. Materials melted in the laboratory and in mill were used. The steels melted in the laboratory, which were marked as 0Cr, 2Cr and 13Cr, were hot rolled and heat treated after melting, and had the normalised ferritic microstructure. On the other hand, the Steels J55 and N80 melted in mill were manufactured by the Mannesmann-mandrel process; Steel J55 with minimum yield strength of 379 MPa (55 ksi) had an as-rolled ferritic–pearlitic microstructure, and Steel N80 with minimum yield strength of 550 MPa (80 ksi) had a quenched and tempered martensitic microstructure.

2.2. Experimental Methods

Immersion tests were carried out in an autoclave with a stirrer. The inner wall of the autoclave was lined with titanium. Test specimens of 10 × 40 × 2 mm were used for pure iron (0Cr), the 2 and 13 mass% Cr steels. Specimens of 15 × 40 × 2 mm were used for Steels J55 and N80. These specimens were polished with silicon carbide No. 600 paper, rinsed with distilled water and then degreased in ethanol and acetone. The deaeration of the test solutions was performed by using a vacuum pump and bubbling of N_2 gas. The flow velocity at specimen surfaces was 2.5 m s^{-1} (500 rev/

Table 1. *Chemical composition of materials used for this study (mass%)*

Mark	C	Si	Mn	P	S	Ni	Cr	Mo
0 Cr	0.003	<0.01	0.05	<0.003	0.002	0.01	<0.01	0.02
2 Cr	0.001	0.004	0.03	0.002	0.004	0.014	2.15	0.02
13 Cr	0.0048	0.012	0.08	0.002	0.007	0.05	13.52	0.01
J55	0.51	0.20	0.98	0.024	0.014	–	0.05	–
N80	0.21	0.25	1.11	0.010	0.005	–	0.02	–

min). After the test, the specimens were removed from the specimen holder, visual observations were made, and the weight loss was measured after descaling in an aqueous ammonium citrate solution. X-ray diffraction, scanning electron microscope (SEM) and electron probe microanalysis (EPMA) methods were used to analyse the properties of corrosion products.

3. Results and Discussion

3.1. Effect of Environmental Factors on CO_2 Corrosion of 0Cr, 2Cr and 13Cr

The temperature $T_{max.}$ at which the corrosion rate showed a maximum value in the CO_2 environment as shown in Fig. 1 became higher as the Cr content in the steels increased. The $T_{max.}$ was 80°C in 0Cr, 100°C in 2Cr and 200°C in 13Cr steels with the corrosion rate of 13Cr at 200°C higher than that in 0Cr and 2Cr. It was shown earlier by Ikeda *et al.* that this behaviour related to the formation of $FeCO_3$ and of amorphous Cr-enriched oxide which is a corrosion product in CO_2 environments [2]. In the CO_2 environment with 0.5% CH_3COOH as shown in Fig. 2, the corrosion rate at the $T_{max.}$ was significantly increased by the addition of acetic acid. Thus, the corrosion rate at $T_{max.}$ in the CO_2 environment with 0.5% CH_3COOH was 5 times higher than that in the CO_2 environment. The temperature of $T_{max.}$ for 0 and 2Cr steels was the same in both environments, but the corrosion rate of 13Cr in the environment with acetic

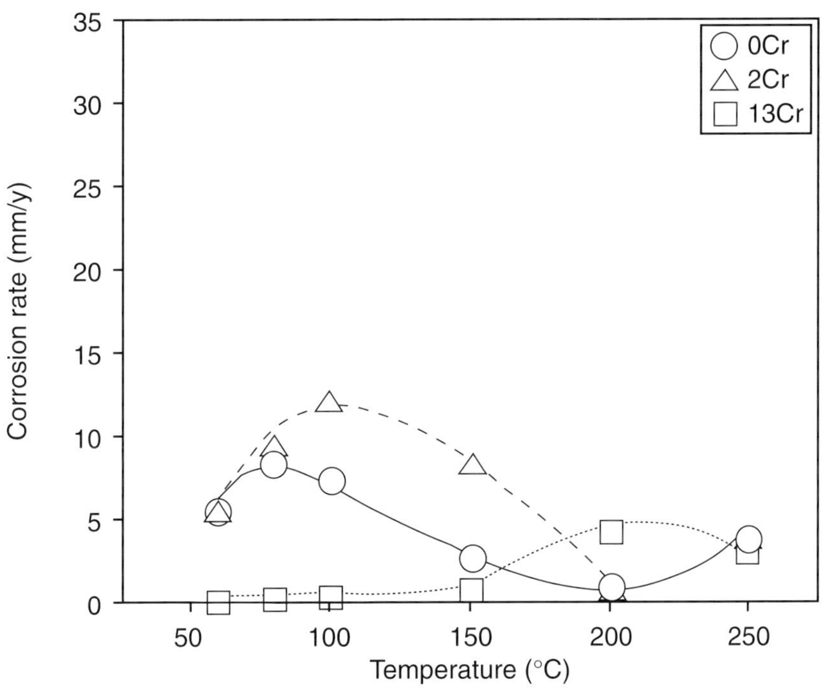

Fig. 1 *Effect of Cr content and temperature (3.0 MPa CO_2, 5% NaCl, 96h, 2.5 m s^{-1}).*

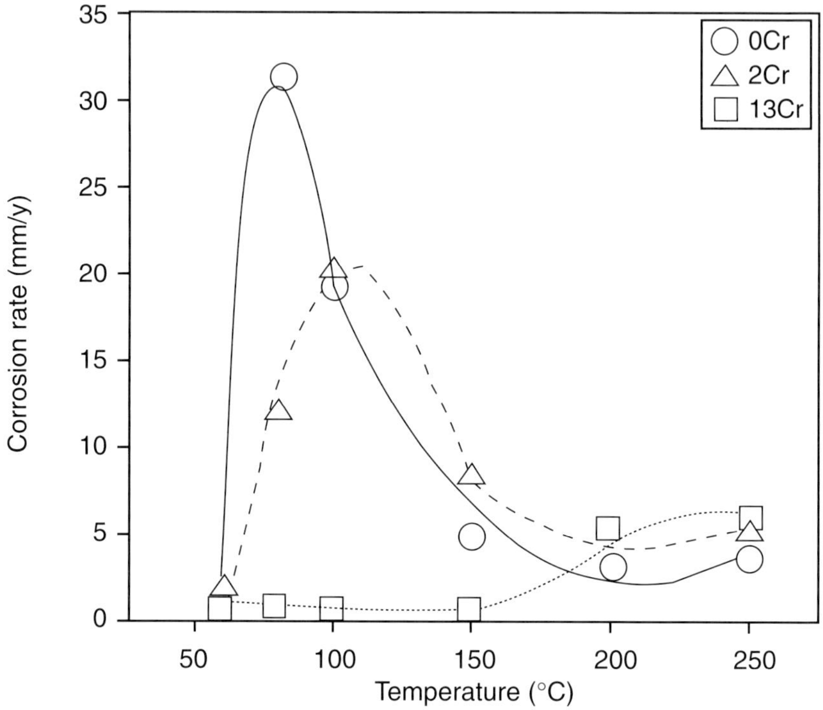

Fig. 2 *Effect of Cr content and temperature (3.0 MPa CO_2, 5%NaCl + 0.5% CH_3COOH, 96 h, 2.5 m s^{-1}).*

acid continued to increase at temperatures above the $T_{max.}$ obtained without acetic acid.

The results of X-ray analysis of corrosion products is shown in Table 2. Fe_3O_4 was detected together with $FeCO_3$ in the environment with acetic acid at temperatures above 150°C for 2Cr and at 250°C for 13Cr. $FeCO_3$ was detected at 60 to 250°C in the 0Cr material (pure iron) with acetic acid, and in 0Cr, 2Cr and 13Cr without acetic acid.

Table 2. *X-ray analysis results of corrosion products after test (CO_2: 5% NaCl, 3.0 MPa CO_2, 2.5 ms^{-1}, 96 h); (CO_2 + CH_3COOH: 5% NaCl + 0.5% CH_3COOH, 3.0 MPa CO_2, 2.5 ms^{-1}, 96 h)*

Condition	Mark	60°C	100°C	150°C		200°C		250°C	
		$FeCO_3$	$FeCO_3$	$FeCO_3$	Fe_3O_4	$FeCO_3$	Fe_3O_4	$FeCO_3$	Fe_3O_4
CO_2	0Cr			SS	–	S		S	
	2Cr			S	–	SS		MS	
	13Cr			M	–	M		MS	
CO_2 + CH_3COOH	0Cr	MW	S	MS	–	S	–	S	–
	2Cr	MW	MW	MS	W	MS	M	M	WW
	13Cr	–	MW	MW	–	WW	–	WW	WW

The Fe^{2+} ion concentration, $C(Fe^{2+})$ necessary for $FeCO_3$ formation on a steel surface in the initial corrosion stage was calculated based on a simple model by using a thermodynamic calculation method proposed by Ueda [5]. First, the pH in the CO_2 environment with and without 0.5% CH_3COOH was estimated by considering the following dissociation reaction, and by using Henry's constant and dissociation constants at elevated temperatures as described by Helgeson *et al.* [6].

Dissociation reactions

$$CO_2 + H_2O \rightarrow H_2CO_3(aq) \tag{1}$$

$$H_2CO_3(aq) \rightarrow H^+ + HCO_3^- \tag{2}$$

$$CH_3COOH \rightarrow CH_3COO^- + H^+ \tag{3}$$

Dissociation constants

$$[H_2CO_3(aq)] = K \times p_{CO_2} \tag{4}$$

$$\frac{[H^+][HCO_3^-]}{[H_2CO_3(aq)]} = K_1 \tag{5}$$

$$\frac{[H^+][CH_3COO^-]}{[CH_3COOH]} = K_2 \tag{6}$$

where [A] is concentration of A, mol L^{-1}, p_{CO_2} is partial pressure of CO_2, bar, K is Henry's constant, and K_1 and K_2 are dissociation constants. In the pH calculation, $[H^+] \doteq [HCO_3^-]$ was assumed in the CO_2 environment and $[H^+] \doteq [CH_3COO^-]$ was assumed in the CO_2 environment with 0.5% CH_3COOH. The result of the calculation is shown in Fig. 3. The pH at 25 to 300°C decreased by 0.3 due to the addition of 0.5% CH_3COOH at each temperature. The pH calculation result was in good agreement with the experimental value measured by Crolet *et al.* [7].

Next, Fe^{2+} ion concentration, $C(Fe^{2+})$ necessary for $FeCO_3$ formation on a steel surface in the initial corrosion stage was calculated by using the pH values shown in Fig. 3. The following $FeCO_3$ formation reaction was considered.

$$Fe^{2+} + H_2CO_3(aq) = FeCO_3 + 2H^+ \tag{7}$$

$$Fe^{2+} = \frac{[H^+]^2}{[Fe^{2+}][H_2CO_3(aq)]} \tag{8}$$

where K_3 is the equilibrium constant for reaction (7). The $C(Fe^{2+})$ in the CO_2 environment is given in eqn (9) and did not depend on p_{CO_2}.

$$[Fe^{2+}] = K_1/K_3 \tag{9}$$

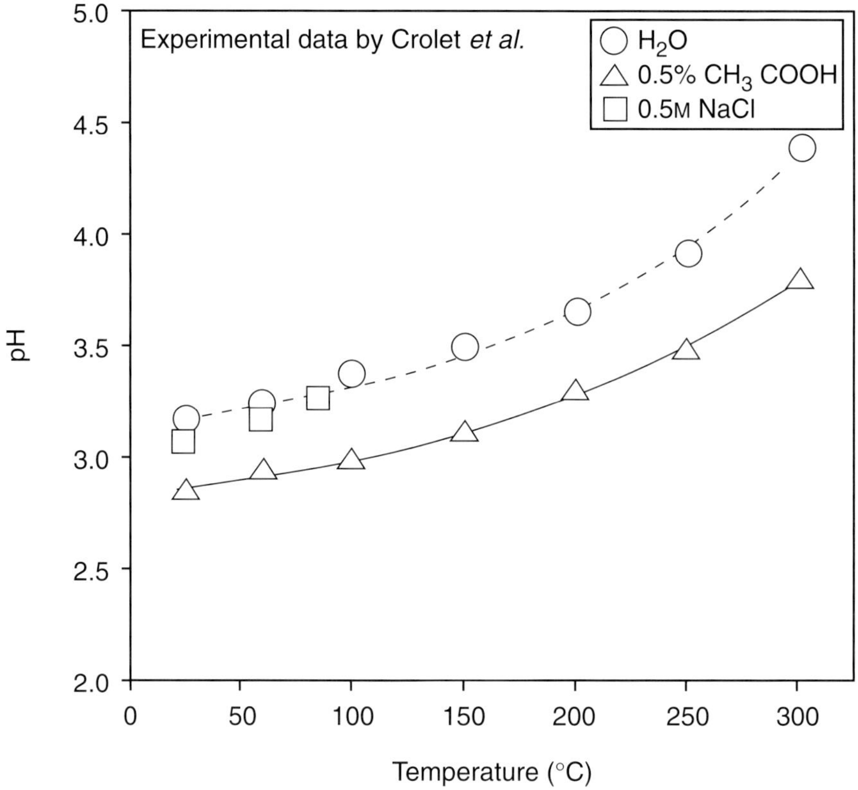

***Fig.* 3** *Calculated pH in the experimental conditions (3.0 MPa CO_2).*

The C(Fe^{2+}) in the CO_2 environment with 0.5% CH_3COOH is given in eqn (10) and this is a function of p_{CO_2}.

$$Fe^{2+} = \frac{[H^+]^2}{K_3 \cdot K \cdot p_{CO_2}} \tag{10}$$

The C(Fe^{2+}) calculated is shown in Fig. 4. The C(Fe^{2+}) decreased with the increase in temperature, and this anomalous behaviour is important in the understanding of CO_2 corrosion. Thus, $FeCO_3$ is easily produced as the temperature increases. The addition of 0.5% CH_3COOH, corresponding to a drop of 0.3 in pH, brought about 5, 7 and 20× increases in C(Fe^{2+}) at 100, 200 and 300°C respectively. The C(Fe^{2+}) based on the bulk $FeCO_3$ formation condition in the CO_2 environment without CH_3COOH, which was described by Ueda, is also shown in Fig. 4 [5]. In this test, the Fe^{2+} ion concentration measured after autoclave tests was between the C(Fe^{2+}) in the surface $FeCO_3$ formation model (eqn 9) and that in the bulk $FeCO_3$ formation model. Because the pH at the steel surface would become high due to the consumption of H^+ ion in corrosion reaction. If the corrosion rate begins to decrease as the Fe^{2+} ion concentration on the steel surface exceeds the solubility of $FeCO_3$, the $T_{max.}$ will shift to higher

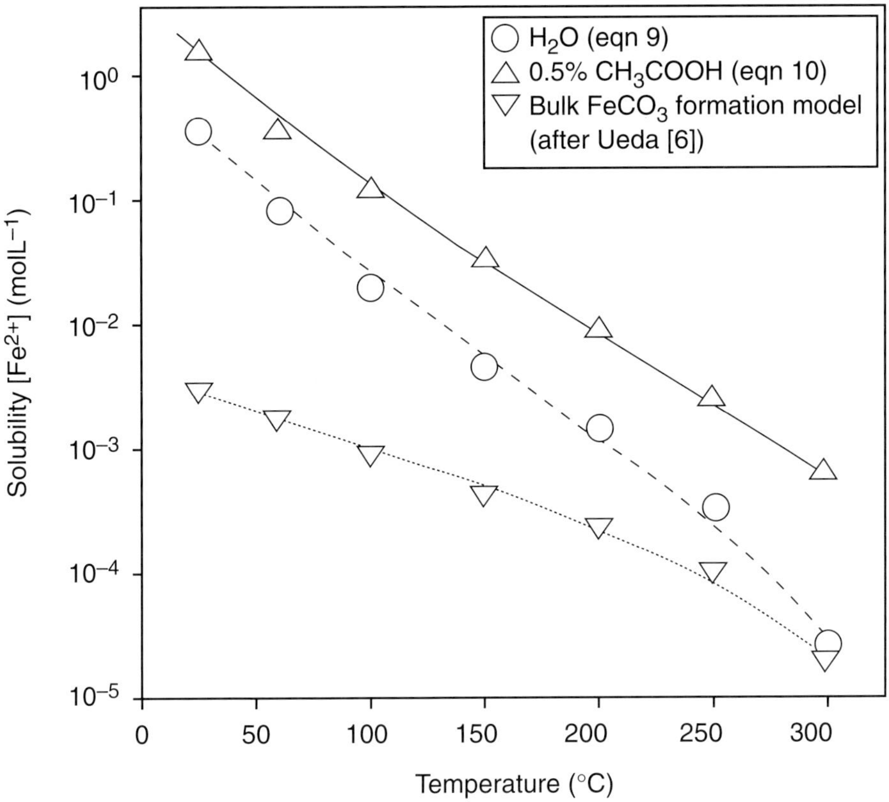

Fig. 4 *$FeCO_3$ formation condition based on thermodynamic calculation (3.0 MPa CO_2 at 25 °C).*

temperature following the addition of acetic acid. However, the experimental result in 0Cr and 2Cr showed the same $T_{max.}$ in both environments. Therefore, it is thought that the reaction rate of $FeCO_3$ formation plays an important part in the corrosion behaviour, and the addition of acetic acid accelerates the dissolution rate on the material surface with heterogeneous $FeCO_3$ film. The corrosion behaviour of 13Cr with acetic acid would not only relate to $FeCO_3$ formation on the steel surface, but also Fe_3O_4 and Cr oxide formations.

The effect of H_2S contamination (0.001 MPa) to CO_2 environments is shown in Fig. 4. The corrosion rate at around $T_{max.}$ was dramatically suppressed by the contamination of H_2S. The suppression effect was large for 0Cr. In EPMA analysis, the concentrated-S in a cross section of the surface film produced on Steel L80 (0.24C–0.48Cr) at 60°C was observed on the steel surface and C in the outer of that existed. These compounds were identified as FeS and $FeCO_3$, respectively.

So, the solubility of $FeCO_3$, FeS and $Cr(OH)_3$ was calculated at elevated temperatures in the solution saturating by 1 MPa CO_2 and 0.001 MPa H_2S. The following reactions were considered as dissolution reactions;

$$Fe^{2+} + H_2CO_3 \rightarrow FeCO_3 + 2H^+ \tag{11}$$

$$HFeO_2^- + H^+ + H_2CO_3 \rightarrow FeCO_3 + 2H_2O \tag{12}$$

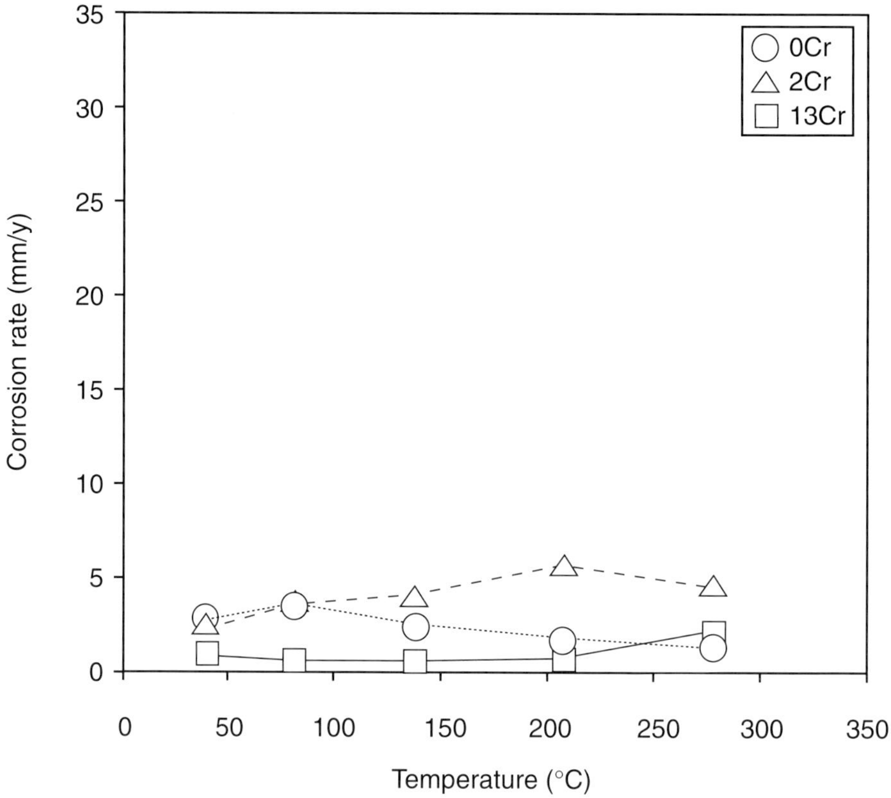

Fig. 5 *Effect of H_2S contamination (3.0 MPa CO_2 + 0.001 MPa H_2S, 5%NaCl, 96 h, 2.5 m s^{-1}).*

$$Fe^{2+} + H_2S \rightarrow FeS + 2H^+ \tag{13}$$

$$HFeO^{2-} + H^+ + H_2S \rightarrow FeS + 2H_2O \tag{14}$$

$$Cr^{3+} + 3H_2O \rightarrow Cr(OH)_3 + 3H^+ \tag{15}$$

$$CrO^{2-} + H_2O + H^+ \rightarrow Cr(OH)_3 \tag{16}$$

The effect of temperature on their solubility is shown in Fig. 6, respectively. Calculated pH in the solution containing 1 MPa CO_2 is 3.56 at 60°C and 3.71 at 100°C. The solubility at this pH becomes small in order of $FeCO_3$, FeS and $Cr(OH)_3$. Considering pH changes, the solubility of $FeCO_3$ and FeS becomes small together with the increasing of temperatures. FeS produces first on the steel surface, and then $FeCO_3$ also produces as corrosion product in CO_2 + H_2S environments at 60 and 100°C.

This FeS formation would be suppressing the localised corrosion caused by CO_2. From the low solubility of $Cr(OH)_3$, it could be explained that Cr oxide is the most stable corrosion product and Cr bearing steel is a corrosion resistant material.

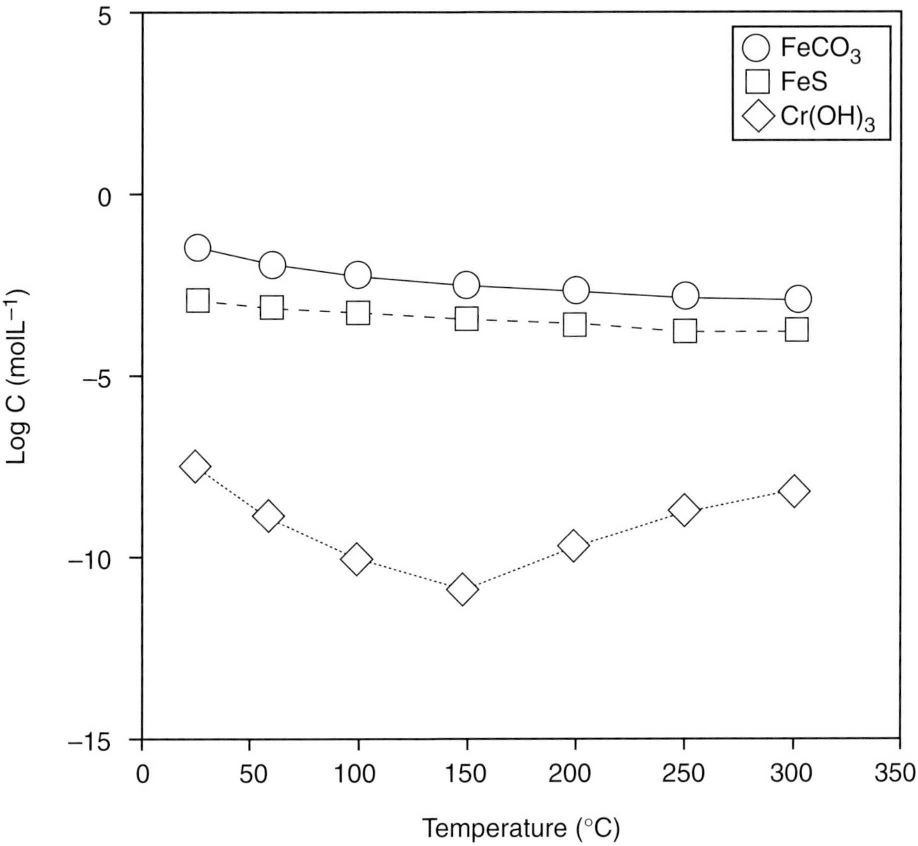

Fig. 6 *Effect of temperature on solubility of $FeCO_3$, FeS and $Cr(OH)_3$ (1 MPa CO_2 + 0.001 MPa H_2S).*

3.2. Effect of Microstructure and Corrosion Morphology on CO_2 Corrosion

The effect of microstructure on CO_2 corrosion was shown in Fig. 7. The corrosion rate was almost the same in Steel J55 and Steel N80. But general corrosion was observed in Steel J55 with ferritic–pearlitic microstructure, and localised corrosion was observed in Steel 80 with martensitic microstructure as well known. Homogeneous and adhesive corrosion product was observed by SEM in low magnification on the steel surface of Steel J55 as shown in Fig. 8. Therefore, in order to investigate formation behaviour of corrosion product at interface portion between corrosion product and metal, an etching was applied to corrosion product and metal, and SEM observation in high magnification was carried out. The result of SEM observations is shown in Fig. 9 with schematic illustration of corrosion product formation behaviour. It was found out clearly that the lamellar Fe_3C in the steel was left, and $FeCO_3$ formed on a former ferrite phase in the corrosion product after corrosion occurred. Therefore, the good resistance to localised corrosion of Steel J55 could be explained by the anchor effect by the lamellar Fe_3C remaining in corrosion product.

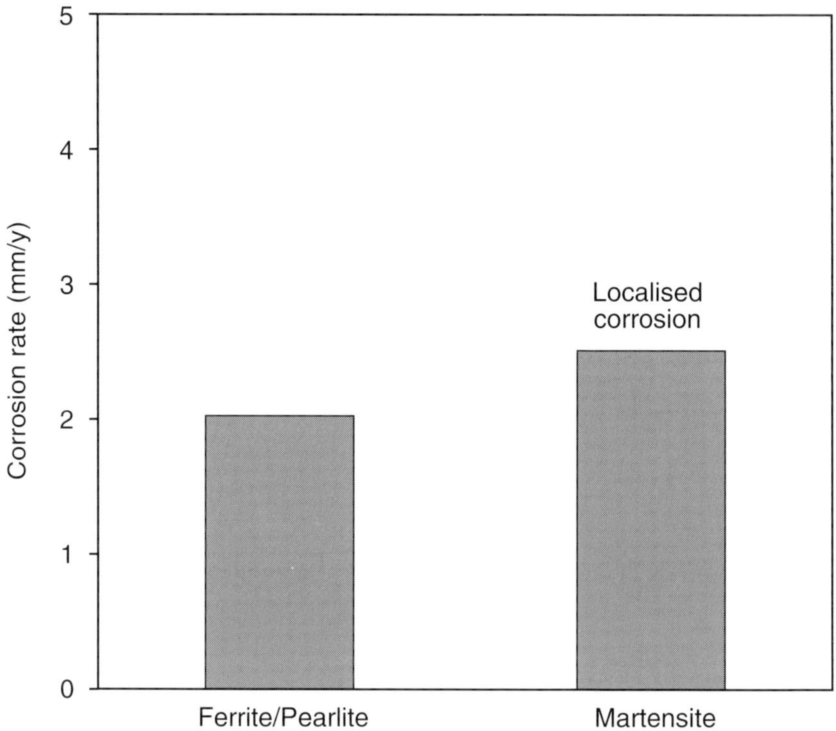

Fig. 7 *Effect of microstructure (80°C, 0.3 MPa CO_2, 70 000 ppm Cl^-, pH = 4.0, 720 h, 2.5 m s^{-1}).*

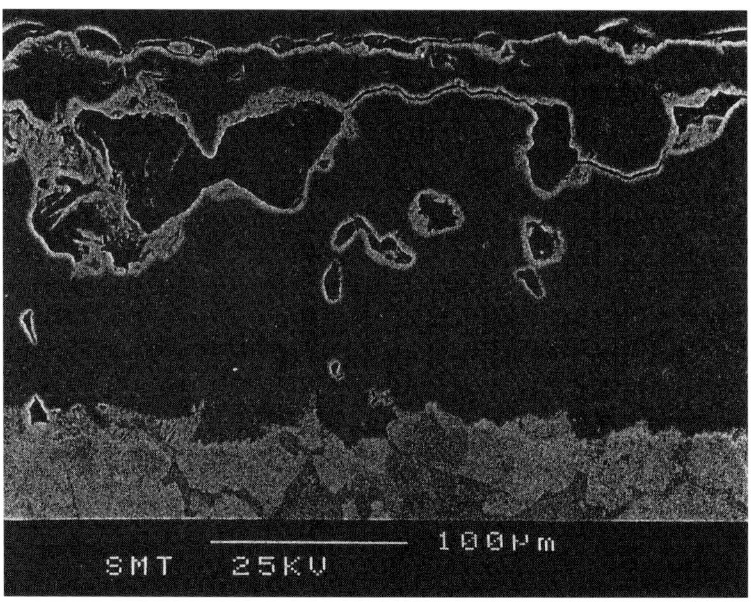

Fig. 8 *SEM observation result in cross-section of corrosion product.*

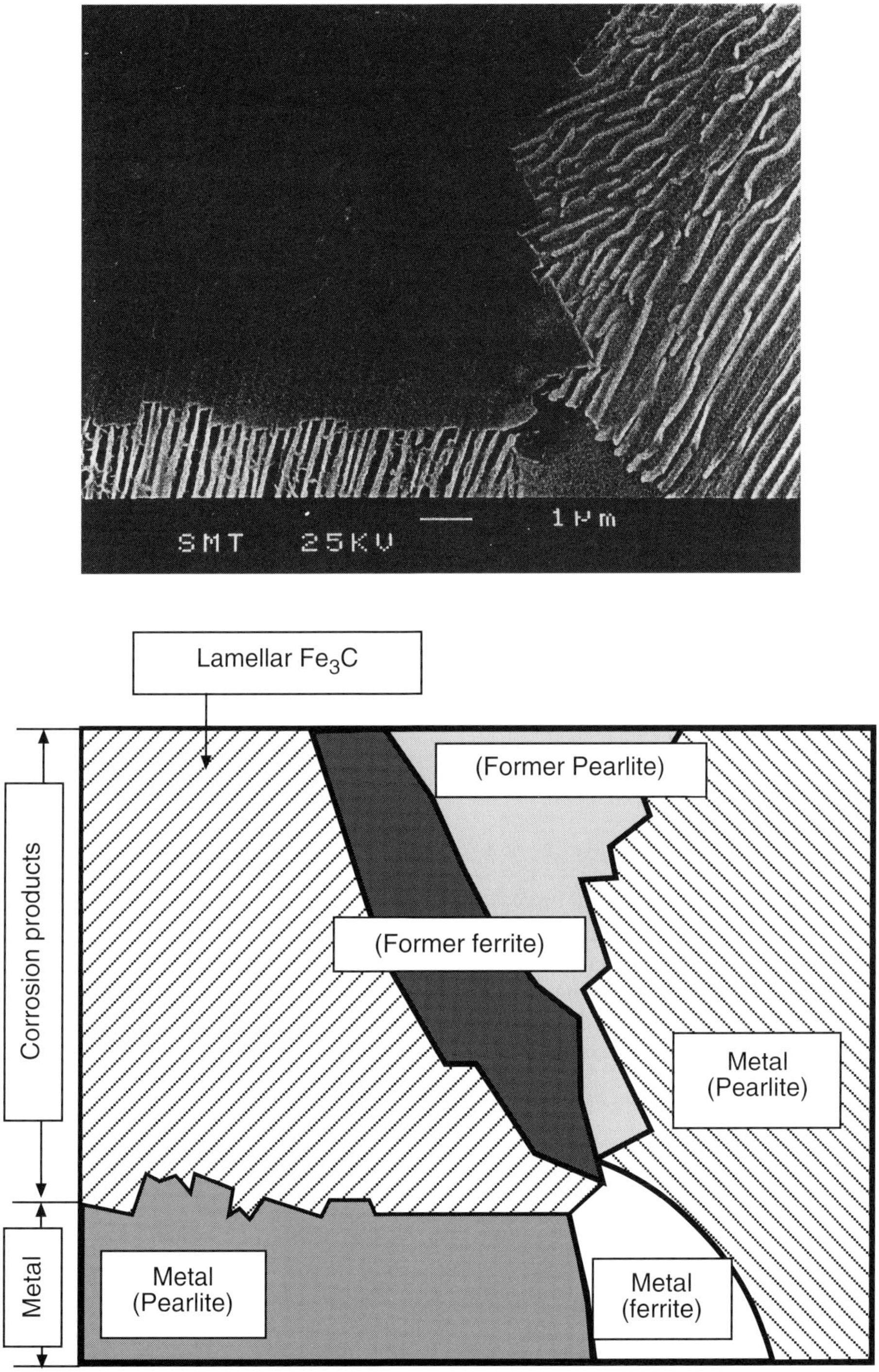

Fig. 9 *SEM observation result and its schematic illustration of corrosion product formation behaviour.*

4. Conclusions

1. Steels containing 0Cr and 2Cr showed a large increase of corrosion rate at around the temperature, $T_{max.}$ with the corrosion rates reaching a maximum by addition of acetic acid, although the value of the $T_{max.}$ did not change. A 13Cr steel did not show the $T_{max.}$ and its corrosion rate increased with increasing of testing temperature in the presence of acetic acid.

2. The H_2S contamination of 0.001 MPa to the CO_2 environment suppressed the CO_2 corrosion at around the $T_{max.}$ due to the FeS formation on the interface region between corrosion product and metal.

3. The good resistance of Steel J55 with ferritic–pearlitic microstructure to localised corrosion could be explained due to the anchor effect by the lamellar Fe_3C remaining in corrosion product.

5. Acknowledgements

The authors wish to thank Sumitomo Metal Industries, Ltd. for allowing publication of this research. The assistance and discussion of co-workers in laboratories are gratefully acknowledged.

References

1. P. A. Burke, *Advances in CO_2 Corrosion, Vol. 1.* NACE International, Houston, Tx, 1984, p.3.
2. A. Ikeda, M. Ueda and S. Mukai, *Advances in CO_2 Corrosion Vol. 1.* NACE International, Houston, Tx, 1984, p.39.
3. C. De Waard, U. Lotz and D. E. Milliams, *Corrosion '91.* NACE International, Houston, Tx, 1991, p.976.
4. N. E. Legezin and K. S. Zarembo, *Mater. Nauchn. Tech. Soveshch. Zashch. Korroz. Oborud. Neft. Gaz. Skvazhin Baku,* 1964, **1969**, 89.
5. G. Schmidt and B. N. Labus, *Corrosion '94,* Paper No. 37. Nace International, Houston, Tx, 1994.
6. M. Ueda, *Corros. Engng,* 1995, **44**, 159.
7. H. C. Helgeson, *J. Phys. Chem.*, 1967, **71**, 3121.
8. L. Cloret and M. R. Bonis, *Corrosion-NACE,* Vol. 39, No.1, 1983, p.39.

9
Effects of Chromium Contents of Low-alloyed Steel and of Dissolved Oxygen in Aqueous Solution on Carbon Dioxide Corrosion

K. NOSE, T. ISHITSUKA, H. ASAHI and H. TAMEHIRO
Nippon Steel Corporation, Steel Research Laboratories, 20-1 Shintomi, Futtsu-city, Chiba 293-8511, Japan

ABSTRACT
Effects of small additions of chromium in steel on carbon dioxide corrosion were studied under different levels of dissolved oxygen (DO) in aqueous solution. Rotating corrosion tests were carried out in 10mass% NaCl aqueous solution with relatively high and low DO levels. Flow rate, temperature, pH and partial pressure of CO_2 were 1 m s^{-1}, 80°C, 5.0 and 1 bar, respectively. The level of DO was controlled to lower than 50 ppb or over 100 ppb. Under DO conditions of 50 ppb or lower, while the corrosion rates of 0 to 1% Cr steel were relatively low, the corrosion rate of 0.5% Cr steel was slightly higher than that of Cr-free steel. When the DO exceeded 100 ppb, the corrosion rate decreased significantly with increase in Cr content. This tendency and the value of corrosion rate in this high DO level were similar to those reported previously.

1. Introduction

Carbon dioxide corrosion of carbon steel is one of the most aggressive corrosion situations. The corrosion rate is above 10 mm/y in typical simulation tests, for example, at a temperature of 80°C, under a CO_2 partial pressure of 0.1 MPa and a flow rate of 1 ms^{-1}. There are some previous reports on effects of small addition of Cr in steel. In particular, a 0.5% Cr addition was reported to reduce the corrosion rate of steel to half that of Cr-free steel [1,2]. On the other hand, there are some other reports, in which 0.5% Cr addition has no significant effect, especially on general corrosion [3]. Therefore, the effect of small additions of Cr and environmental conditions seems unclear.

In this paper, the effects of Cr additions on CO_2 corrosion were studied under different levels of dissolved oxygen (DO) in aqueous solution.

2. Experimental

2.1. Test Specimens

Three types of steel of typical ordinary line pipe steel but with different Cr contents of 0%, 0.6% and 1% were used. Their chemical compositions are listed in Table 1.

Table 1. *Chemical compositions of tested steels* (mass%)

Steel	C	Si	Mn	P	S	Cr	Al	Ti	Nb	N
A	0.04	0.22	1.06	0.011	0.040	–	0.011	0.013	0.036	0.030
B	0.04	0.23	1.06	0.011	0.039	0.59	0.007	0.012	0.038	0.029
C	0.04	0.23	1.06	0.011	0.039	1.00	0.004	0.015	0.040	0.028

They were melted in a vacuum furnace, cast into ingots and rolled into 15 mm thick steel plates using laboratory facilities. The microstructures of the plates are shown in Fig. 1. The steel plates were cut into 2 × 10 × 60 mm rectangular test specimens, of which the test surfaces were 1 mm below the original steel plate surfaces. The

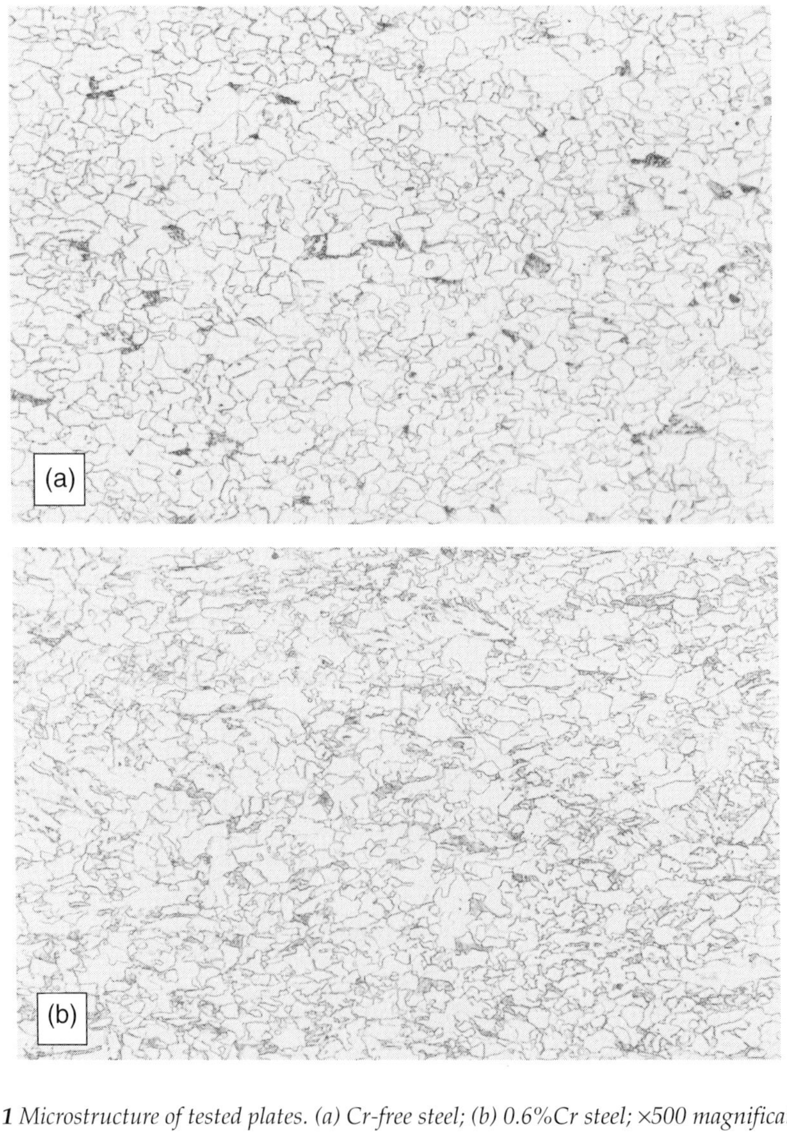

Fig. 1 Microstructure of tested plates. (a) Cr-free steel; (b) 0.6%Cr steel; ×500 magnification.

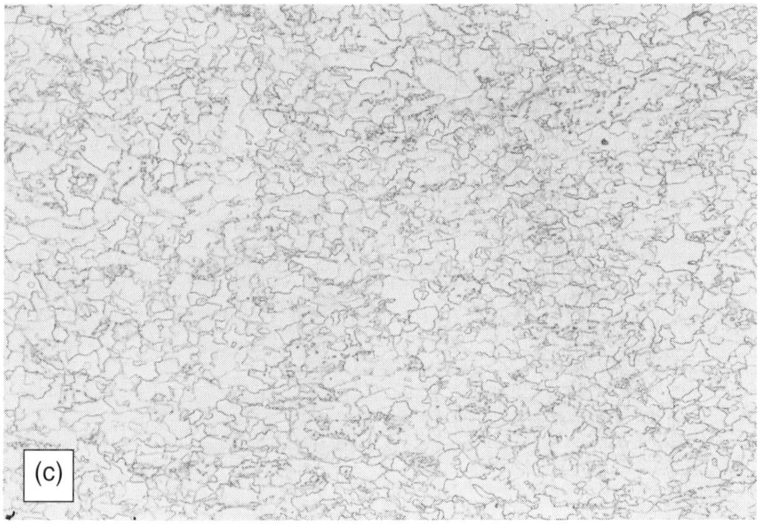

Fig. 1 (continued) (c) 1%Cr steel; ×500 magnification.

specimens were finished with dry emery paper (No.320), degreased with acetone, weighed, and masked with plastic tape except the test areas. Figure 2 shows the dimensions of a specimen.

2.2. Rotary Immersion Tester

A rotary immersion tester was used for the corrosion tests. Figure 3 shows a cross-section structure of the equipment which consists of a double-walled immersion cell and a rotating specimen holder. The rotating specimen holder is immersed into the test solution inside the cell. The double-wall of the cell constitutes a temperature-controlled bath. The tester was put inside a glove box to control the outside gas environment.

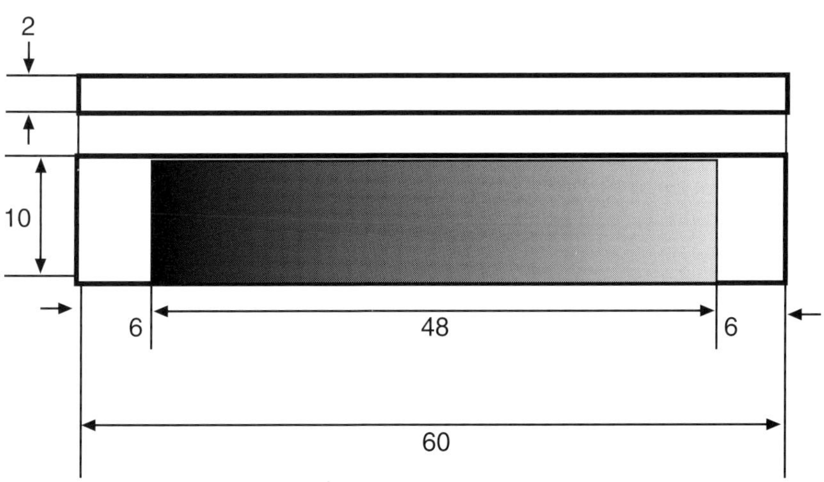

Fig. 2 Dimensions of test specimen (in mm; hatched area = test surface).

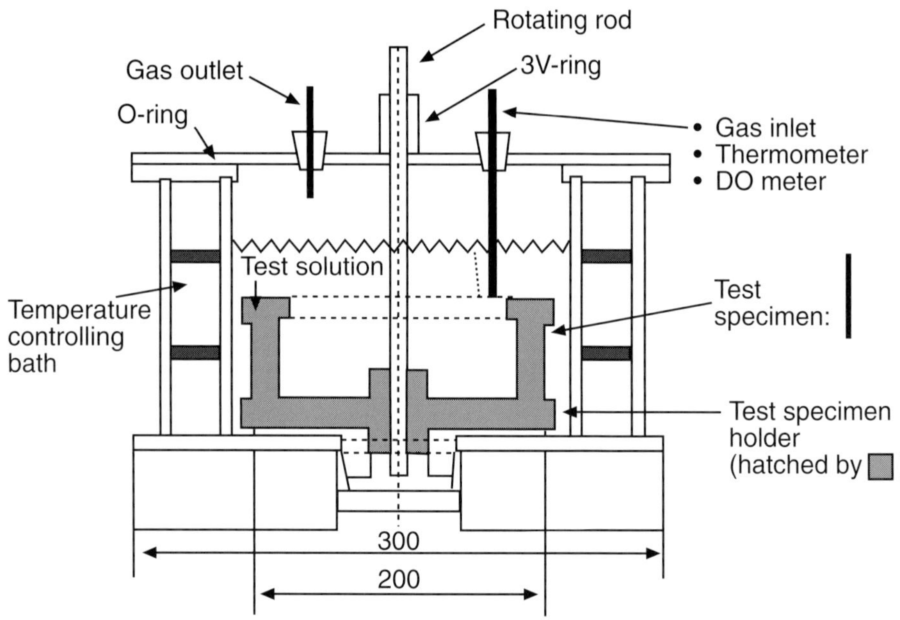

Fig. 3 Illustration of rotary immersion tester (dimensions in mm).

2.3. Test Method

A 10% NaCl aqueous solution was used as the test solution and the temperature was 80°C. The solution volume: test area ratio was 45:1. The test solution was heated in the test cell and N_2 bubbled for de-aeration. Carbon dioxide gas was bubbled into the solution during the test period. When the CO_2 gas was saturated at atmospheric pressure (0.1 MPa), the pH was adjusted to 5.0 by addition of NaOH. A relative velocity was established between the test cell and the test surface and was controlled at 1 m s^{-1}, referred to as the flow rate.

Three levels of DO were achieved by changing the oxygen (O_2) exposure method.

The test duration was set at 96 h. A 720 h test was carried out only for the test at the lowest DO level.

The test specimens were taken out of the tester immediately after the test period. Some of the specimens were washed by water, dried, and the corrosion products analysed by X-ray diffraction (XRD). The remainder of the specimens were pickled in 10wt% sulfuric acid solution containing inhibitor to remove the corrosion products, and then weighed to measure the corrosion weight loss.

3. Results

3.1. DO Variations During the Tests

Typical profiles of the variations of the DO value during the tests are plotted in Fig. 4. In each case, before the test specimens were immersed, the DO value was allowed to stabilise at an equilibrium value. The DO value increased at first up to the maximum

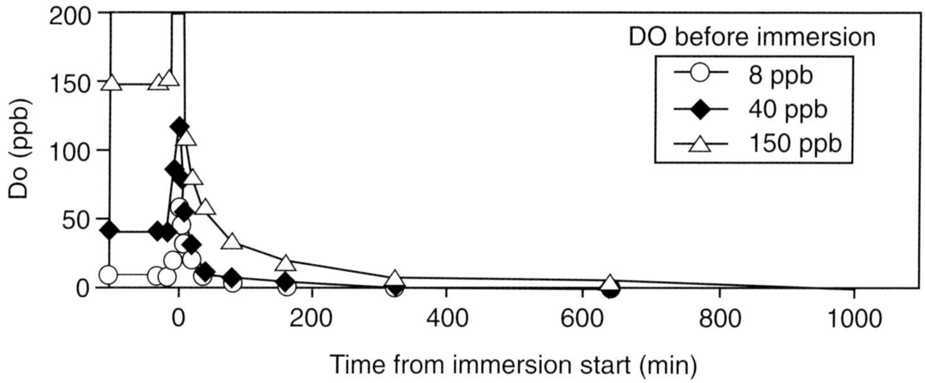

Fig. 4 DO value variation during test.

value because of the temporary opening of the test cell for immersion. The initial equilibrium values and the maximum values of DO are listed in Table 2. After the immersion, every DO value subsequently started to fall. The value continued to fall during the test and all DO values were below the measurable limit of 2 ppb at the end of the test. The rate of DO fall depended on the initial equilibrium value and therefore the DO values of the initial equilibrium values before immersion were referred to as the DO levels of the test conditions.

3.2. Effects of Cr addition and DO on the Corrosion Rates

The average corrosion rates in the three DO levels are plotted as a function of Cr content of the steel in Fig. 5.

The corrosion rates for high DO tests were higher than those for lower DO tests irrespective of the Cr content. However, there were different effects of Cr contents on corrosion rate which depended on the DO level.

At the highest DO level of 150 ppb, the corrosion rate obviously decreased with an increase in Cr content. The positive effect of Cr addition on the corrosion resistance was clear. The corrosion rate of about 10 mm/y for the Cr-free steel is close to the value reported previously [1].

However, at the lower DO levels of 8 or 40 ppb, the effect of Cr addition was not significant. The corrosion rates were almost constant irrespective of Cr contents, but there were small peaks of corrosion rates at 0.6% Cr at these lower DO levels. Even in the 720 h test as shown in Fig. 6, while the average corrosion rate was lower than that of the shorter duration test, the tendency for 0.6% Cr steel to corrode more than Cr-free steel remained.

Table 2. *Measured values of DO (ppb)*

DO level	DO before immersion (ppb)	Max. DO just after immersion (ppb)
Low	8	<50
Middle	40	120
High	150	500–800

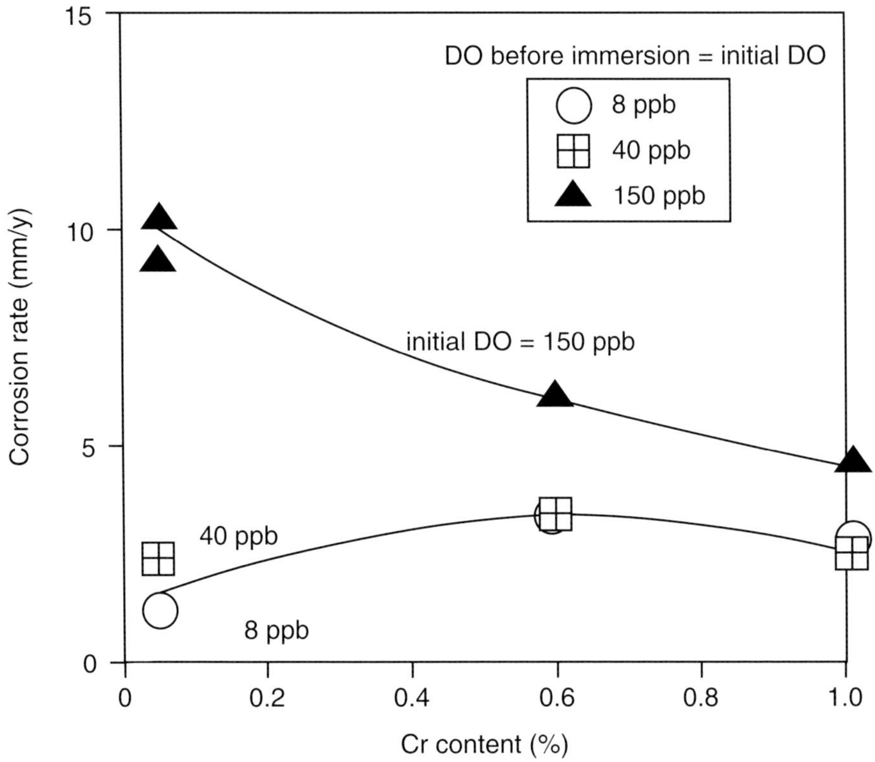

Fig. 5 *Effects of Cr content on corrosion rate in different DO environments (96 h).*

3.3. Analysis of the Corrosion Products

Typical XRD patterns obtained from the test surfaces are indicated in Fig. 7 and the results of the analyses are listed in Table 3. Almost all corrosion products were $FeCO_3$ in the low DO level regardless of Cr content. In contrast to this, $FeCO_3$ was not detected in corrosion products with the high DO level. $Fe_2(OH)_2CO_3$ was determined as a corrosion product for Cr-free steel and the broad XRD pattern for 0.6% Cr steel suggested formation of amorphous-like magnetite.

4. Discussion

4.1. Effect of Cr on Corrosion Products

The addition of Cr was effective in the relatively high DO environment where an oxide film was formed on Cr-containing steel. The Cr addition is thought to help the formation of the protective oxide film, or to make the oxide film more protective, while small amounts of Cr were not effective in the relatively low DO environment where the $FeCO_3$ film was formed regardless of Cr content. In particular, the 0.6% Cr addition seemed to make the $FeCO_3$ film less protective.

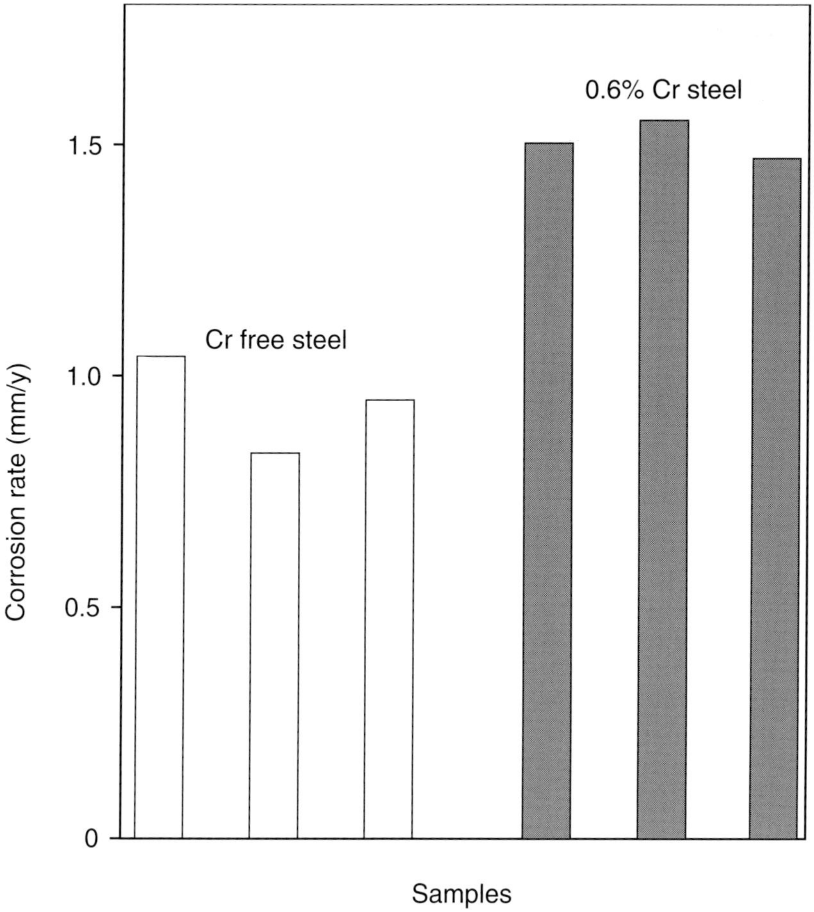

Fig. 6 *Result of immersion test in low DO environment (720 h).*

4.2. Measured DO Value

In this report, the initial equilibrium DO values before immersion were used to represent the DO levels. It should be noted that these initial DO values may be different from ordinary DO values as conventionally measured.

A DO value is a result of the balance of O_2 supply and consumption. At the initial equilibrium, O_2 supply in the test solution was balanced with O_2 consumption taking place only by CO_2 bubbling. When the test cell was opened temporarily for immersion of the specimens the DO value increased because the O_2 supply increased. After immersion, the DO value decreased due to O_2 consumption by corrosion of the specimens in addition to CO_2 bubbling. Since the corrosion rates in the CO_2 environment was relatively high, the O_2 consumption by the corrosion was larger than the O_2 supply even at the relatively high DO condition. In the results it was found that every DO value was below the measurable limit of 2 ppb at the end of the test. If the O_2 supply had been larger than the O_2 consumption, a definite DO value would have been measured. Therefore, the O_2 supply was not high enough to meet

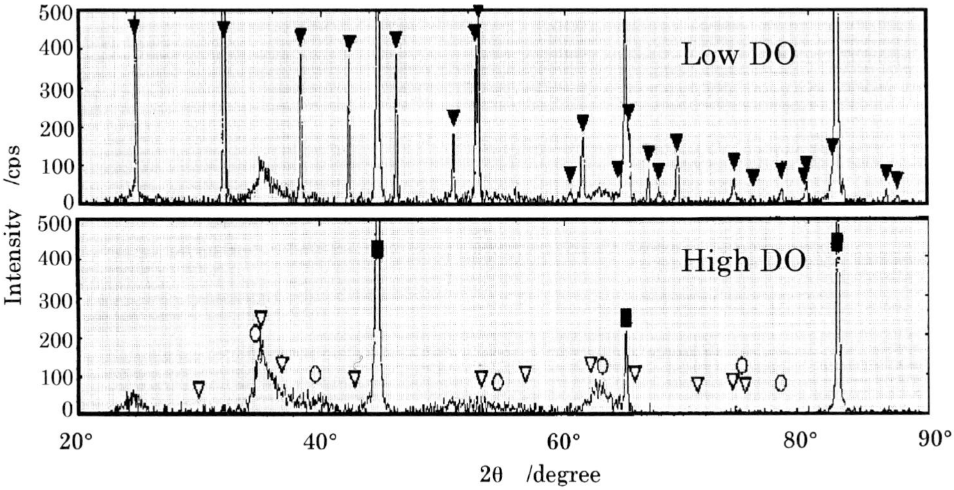

Fig. 7 X-ray diffraction patterns from test surface of 0.6% Cr-containing steel. Upper: low DO (initial DO = 40 ppb). Lower: high DO (initial DO = 150 ppb). (▼: $FeCO_3$, ∇: Fe_3O_4, ○: FeOOH, ■: Fe).

***Table 3**. Main products on test surface detected by XRD*

DO (ppb)	Cr free steel	0.6% Cr steel
40	$FeCO_3$	$FeCO_3$
150	$Fe_2(OH)_2CO_3$	(Fe_3O_4)*

* Not clear, amorphous like.

the O_2 consumption by corrosion even with the highest DO level used in this work.

Every DO value had the same very low value at the end of the test despite the fact that it should have been the same as before immersion of the specimens. Thus, a measured low DO does not necessarily mean a low O_2 supply since a low final DO will result even with a high initial DO if the O_2 consumption is higher than the supply. In another situation a high solution flow rate can cause a high O_2 supply if the solution contains low DO.

The reported corrosion rate is relatively close to the data of the high DO level test in this paper. So even if the measured DO value was low, the O_2 supply could be as high as for the high DO level test in this paper. If the O_2 supply to the steel surface is the same, the same phenomenon must occur.

5. Conclusions

At a relatively high DO level, such as 150 ppb, the corrosion rate obviously decreased with an increase in Cr content. However, at the lower DO levels, such as 8 and 40 ppb, the effect of Cr addition was not significant. This leads to the conclusion that the corrosion performance of Cr-containing steel and that of Cr-free steel depends on the DO level. When these types of steel are applied in the field, the DO levels have to be surveyed and considered. But if it is considered that the DO level changes over time, the Cr-containing steel should prove more corrosion resistant because its benefit in a high DO condition is much higher than in a low DO condition.

References

1. M . Kimura *et al., Corrosion '95*, Paper No. 60, NACE International, Houston, Tx, 1995.
2. H. Ishikawa *et al., 14th Int. Conf. on Offshore Mechanics and Arctic Engineering (OMAE)*, Copenhagen, Denmark, 1995.
3. R. Popperling *et al., EUROCORR '97*, Trondheim, Norway 1997, Publ. NTNU, Gløshavgen, N-7034, Trondheim, Norway, 1997, this volume p.84.

10
The Influence of Small Amounts of H_2S on CO_2 Corrosion of Iron and Carbon Steel

J. KVAREKVÅL*
Marintek, 3201 Sandefjord, Norway

ABSTRACT
Corrosion measurements on iron and AISI 1010 carbon steel were performed in a glass/titanium flow loop, containing a single-phase test fluid of 0.5M NaCl in distilled water, saturated with CO_2 or a CO_2 + H_2S gas mixture at a total pressure of 1 bar, and a temperature of 25°C. The flow rate was kept constant at 1.0 ms^{-1}. Corrosion rates were calculated from Linear Polarisation Resistance (LPR) and Tafel extrapolations. Scanning Electron Microscopy (SEM)/Energy Dispersive X-ray Analysis (EDX) investigations of surfaces and cross-sections were carried out for each test specimen. The effect of H_2S addition on CO_2 corrosion was detrimental, increasing the uniform corrosion rate by a factor of 4–6 compared to pure CO_2 solutions (from <1.5 mm/y to 5–6 mm/y). Cathodic Tafel slopes changed from activation control (0.12 V/decade) to diffusion control (1.0 V/decade) when H_2S was added, while the anodic slopes remained unaffected (0.08 V/decade). Results show that H_2S has a catalytic effect on corrosion in low pH–CO_2 environments. This could be caused by a proposed mechanism, in which H_2S accelerates hydrogen evolution by acting as a catalyst. Also, sulfide inclusions in corrosion products could have a similar effect. In pure CO_2 solutions, corrosion seems to be controlled by charge-transfer at room temperature. Introducing H_2S to the solution can increase the cathodic reduction rate until diffusion of proton donors becomes rate-determining.

1. Introduction

Internal corrosion of oil-producing equipment by carbon dioxide has been thoroughly investigated by many workers [1]. Materials selection for installations handling well fluids with CO_2 and H_2S is often dictated by the H_2S concentration alone, according to accepted standards [2,3]. For short-life wells, this may lead to over-specification and unnecessarily high construction costs. Normal carbon steel could be a candidate for pipelines carrying an H_2S-containing water phase. Also, in the oil fields off the Norwegian shores, the H_2S level is less than or within the sweet/sour corrosion transition region [4]. Motivated by these facts, research on the topic has been carried out at the University of Oslo [5]. The work presented in this paper is part of a more extensive study sponsored by Norsk Hydro asa.

---

*Present address: Institute for Energy Technology, P.O.Box, 40, 2027 Kjeller, Norway.

2. Experimental

Experiments were run in a flow loop with test sections designed for electrochemical measurements. The electrolyte reservoir consisted of a 20 L glass vessel with a Quickfit lid. Operations such as temperature and pressure control, gas purging and pH measurements were conducted in this part of the apparatus. The loop was equipped with a 1.1 kW magnetic drive gear pump with frequency inverter. Tubing, valves and connections were made from grade 2 Titanium. The test section was a 40 cm long, horizontal channel with a flush-mounted test specimen. The exposed area was approx. 0.5 cm^2, and the ratio between the brine volume and steel surface area was at least 40 L cm^{-2}. Platinum foil counter electrodes with a total area of 15 cm^2 per section were arranged along the walls close to the test specimen. A reference electrode of the saturated calomel type was connected to the test section by a salt bridge.

The test solution was doubly distilled water with 3% NaCl saturated with CO_2 or a $CO_2 + H_2S$ gas mixture. Most of the test specimens were cut from AISI 1010 carbon steel sheet. In order to investigate the effects of low-alloy elements in the steel (Ni, Cr, Cu), Goodfellow super pure iron electrodes were also prepared. The surface was polished with 1200 grit SiC paper, then rinsed with distilled water and methanol. One specimen at a time was introduced to the deaerated test section and, along with the counter and reference electrodes, connected to a Gamry PC3/CMS 100 potentiostat. Flow velocity through the test section was 1.0 ms^{-1}. Linear Polarisation Resistance (LPR) measurements were performed at intervals of 15–30 min. Total exposure times varied from 24 to 300 h. Potentiodynamic sweeps were run at the beginning and at the end of each exposure, and when test parameters were altered. Corrosion rates and Tafel slopes were calculated from these curves. Also, the Tafel slopes were used in the Stern–Geary equation for conversion of polarisation resistance into corrosion current.

After exposure, the steel specimen was removed from the test section, dried and stored in a desiccator. The corroded steel surfaces were examined by scanning electron microscopy (SEM), and energy dispassive X-ray analysis (EDX) was used to determine the composition of the corrosion film.

3. Results

Figure 1 shows the pH dependence of steady-state corrosion rates on carbon steel at room temperature. The total pressure was 1 bar, the H_2S partial pressures were 0.1 and 5 mbar with the balance CO_2. The corrosion rates were calculated from Tafel extrapolations after a minimum of 24 h of exposure. In pure CO_2 solutions with pH 4, steady-state corrosion readings were in the range 0.4–1.5 mm/y (see Fig. 2). However, only the highest value is cited in Fig. 1. Corrosion rates in CO_2/H_2S environments with pH 4 were 5–6 mm/y, which is a significant increase compared to the exposures without H_2S. At higher pH, corrosion rates were all less than 1 mm/y, and the effect of H_2S is ambiguous.

Using the LPR technique, the corrosion was monitored during exposure. Figure 2 illustrates the aggravated corrosion behaviour resulting from H_2S introduction to CO_2-saturated brine. Also, the H_2S effect appears to be independent of the duration

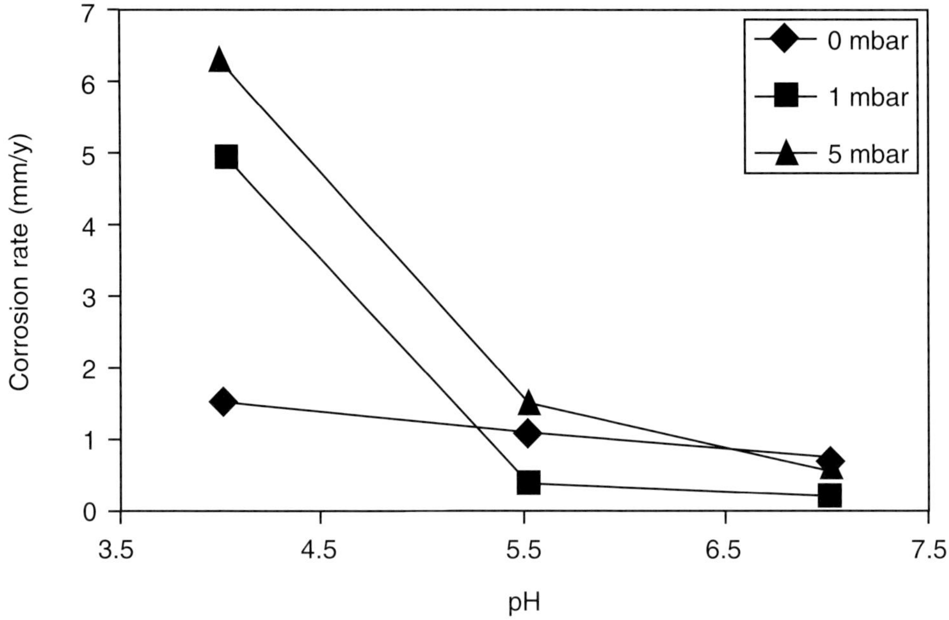

***Fig. 1** Steady-state corrosion rates vs pH. Temp. 25°C, 1 bar CO_2, flow velocity 1.0 ms^{-1}. H_2S partial pressure; see legend.*

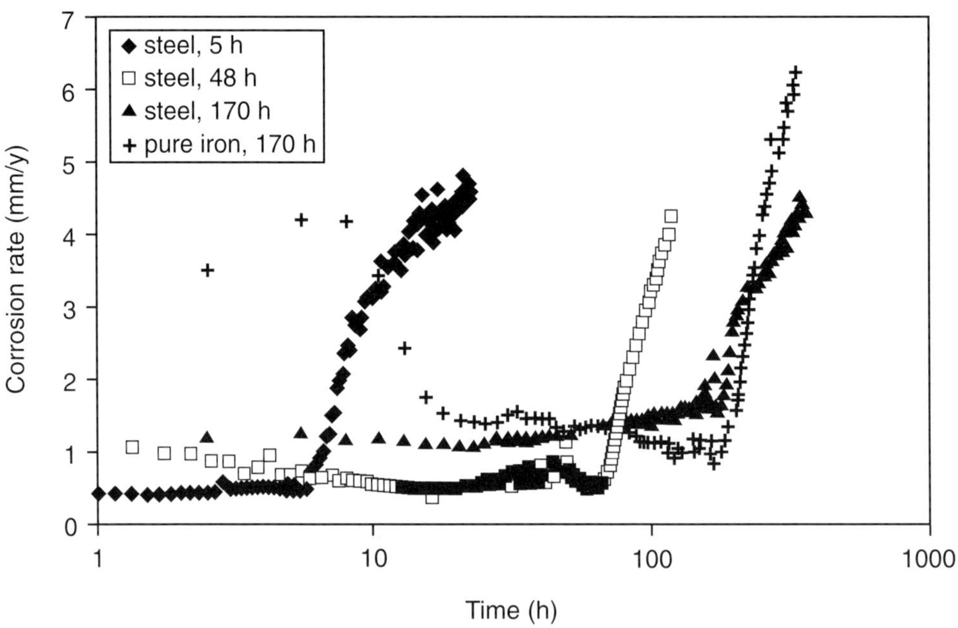

***Fig. 2** Corrosion rate monitoring of carbon steel and super pure iron. Temp. 25°C, 1 bar CO_2, flow velocity 1.0 ms^{-1}. After initial exposure to pure CO_2 (see legend), 1 mbar H_2S was added. Anodic/cathodic Tafel slopes used for LPR conversion (V/decade): 0.08/0.12 (CO_2 exposure); 0.08/1.0 (CO_2/H_2S exposure).*

of exposure prior to H_2S addition. Super pure iron specimens behaved even worse than steel in the CO_2/H_2S environment, and corrosion rates higher than 4 mm/y were observed during exposure in pure CO_2 solutions.

Potentiodynamic sweeps were performed on the steel specimens with and without H_2S, as shown in Fig. 3. The shift from activation-controlled cathodic reactions with CO_2 to diffusion control with CO_2/H_2S is evident, as the cathodic Tafel slope changes from 0.12 to 1 V/decade respectively. The displacement of the corrosion potential caused by H_2S addition was 25 mV in the positive direction. For all tests at pH 4, the Tafel slopes for anodic iron dissolution were 40–60 mV/decade initially and 80 mV/decade or more after some hours of exposure.

Results from EDX surface analysis of carbon steel and pure iron corroded in CO_2/H_2S solutions are given in Fig 4. As expected, peaks for sulfur (at 2.3 keV) and iron (at 6.3 and 7.1 keV) are found in both spectra. For carbon steel, relatively high peaks are also seen for nickel (at 7.5 and 8.3 keV), indicating a Ni concentration of more than 10% in the corrosion film. This is remarkable, since the nickel content of the steel used is very low (> 0.1%).

4. Discussion

The H_2S-promoted increase in corrosion rate is almost one order of magnitude. Similar results have been reported by others [6]. The dominating corrosion mechanism in CO_2/H_2S solutions is probably somewhat different from that in pure carbonic acid. As established from polarisation sweeps, H_2S accelerates the cathodic corrosion processes, i.e. hydrogen evolution from protons and proton-donating species.

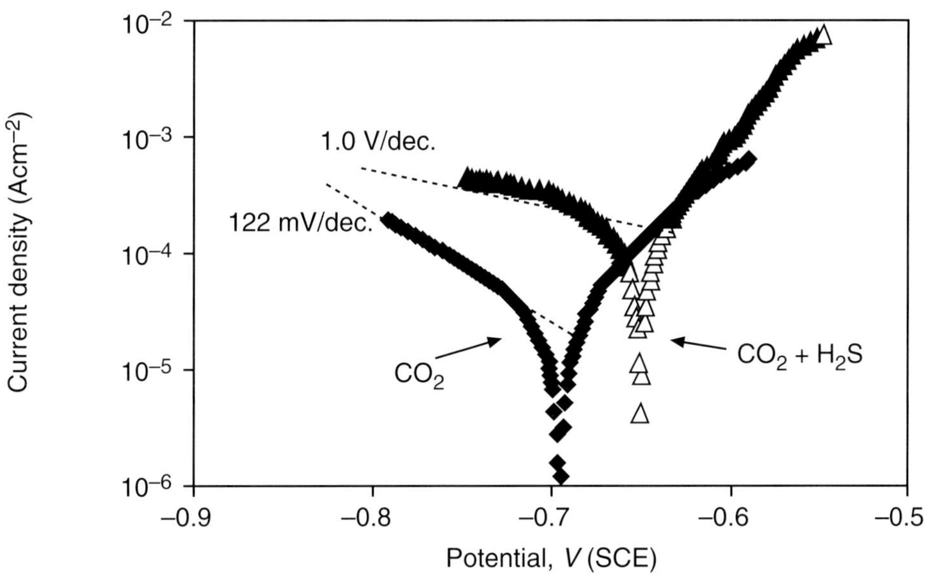

***Fig.* 3** *Potentiodynamic sweeps in CO_2 solutions with/without H_2S. Temp. 25°C. p_{CO_2} 1 bar. p_{H_2S} 1 mbar. pH 4. Flow velocity 1.0 ms^{-1}.*

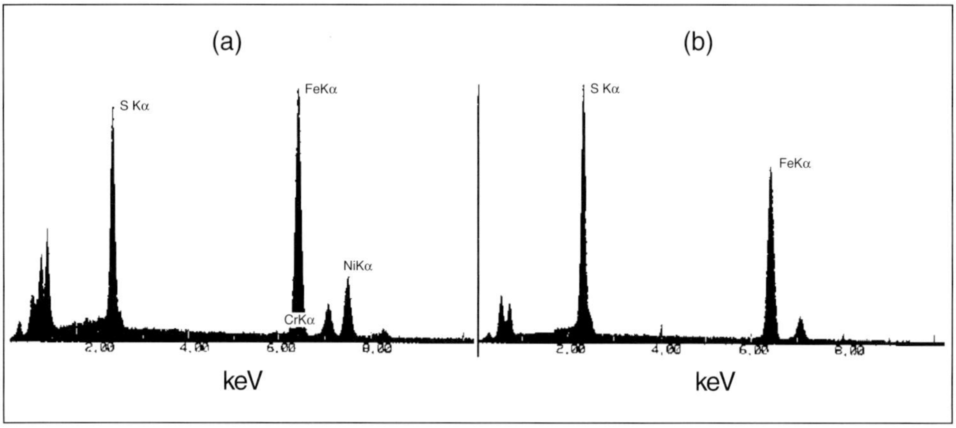

Fig. 4 *EDX spectra of corrosion films formed in CO_2/H_2S solutions. Temp. 25°C. p_{CO_2} 1 bar. p_{H_2S} 1 mbar. pH 4. Flow velocity 1.0 ms^{-1}. (a) Carbon steel. (b) Super pure iron.*

It is conceivable that porous precipitates of conductive iron sulfides promote corrosion by acting as a high-area cathode to the steel. Although this could be the case at pH 4, the presence of iron sulfides formed at higher pH did not involve higher corrosion rates (Fig. 1).

One should also pay attention to the possibly harmful effect of non-ferrous sulfide inclusions in the corrosion film. Research has show that the hydrogen evolution exchange reaction on $Ni–S_X$ substrates is up to 200 times faster than on pure nickel [7]. In the steel used here, nickel is present only as a contaminant. However, it has been shown that the concentrations of low-alloy elements in the corrosion film can be more than 100 times higher than in the steel matrix [8]. On pure iron corroded in CO_2/H_2S solutions, iron sulfide is observed to be the predominant corrosion product, as confirmed by others [9].

Earlier research has shown that H_2S reacts rapidly with hydrated electrons [10]. Based on these results, a mechanism has been suggested for H_2S-promoted proton discharge [11]. Molecules of H_2S adsorb on the metal surface and react with free electrons from the metal (eqn.1). These intermediates act as bridge-forming ligands with adsorbed protons (eqn. 2). Finally, hydrogen is liberated on disintegration of the complexes (eqn. 3).

$$H_2S_{ad} + e^- \rightarrow H_2S^-{}_{ad} \quad (1)$$

$$H_2S^-{}_{ad} + H^+{}_{ad} \rightarrow H_2S...H_{ad} \quad (2)$$

$$H_2S...H_{ad} \rightarrow H_2S_{ad} + H_{ad} \quad (3)$$

Consequently, H_2S can increase the hydrogen evolution rate without taking part in the net reaction. Since corrosion behaviour deteriorated after H_2S addition on both carbon steel and pure iron, it appears to be unaffected by the presence of non-ferrous compounds in the corrosion products. Thus, the catalytic mechanism given by eqns (1–3) seems to be a more probable cause of the increased corrosion rates.

The anodic polarisation behaviour of steel does not change significantly with the addition of hydrogen sulfide. Anodic Tafel slopes of 80 mV/decade are not traditional values, but are consistent with recent work on iron dissolution in CO_2 solutions [12].

5. Conclusions

Corrosion measurements of carbon steel and pure iron in the presence of CO_2/H_2S were performed in a flow loop. With only CO_2 present, the pH dependence of the steady state corrosion rate was small, and readings from 0.4 to 1.5 mm/y were found at pH 4. In non-buffered solutions (pH 4), the presence of H_2S accelerated the corrosion rate by a factor 4–6 compared to pure CO_2 solutions. However, at pH 5.5–7, no such effect of H_2S was seen.

In pure CO_2 solutions, corrosion seems to be controlled by charge-transfer at room temperature. Even at low concentrations, H_2S can accelerate the cathodic reaction rates, and the overall corrosion rate can be governed by diffusion of proton donors.

The increased corrosion rates may have been caused by sulfides of low-alloy elements concentrated in the corrosion film, or by H_2S acting as a catalyst for hydrogen evolution. Experimental results support both alternatives, but the effect of the latter is the stronger.

References

1. EFC Publication No. 13, *A Working Party Report on Predicting CO_2 Corrosion in the Oil and Gas Industry*. The Institute of Materials, London, UK, 1995.
2. NACE Standard MR 0175: "Sulphide Stress Cracking Resistant Metallic Materials for Oilfield Equipment".
3. NORSOK Standard M-DP-001: "Materials Selection".
4. EFC Publication No. 16, *Guidelines for Material Requirements for Carbon & Low Alloy Steels in H_2S-Containing Environments*. The Institute of Materials, London, UK, 1995.
5. K. Videm and J. Kvarekval, *Corrosion*, 1995, **51**, (4), 260–269.
6. A. Ikeda, M. Ueda and S. Mukai, "Influence of Environmental Factors on Corrosion in CO_2 Source Well", *Advances in CO_2 Corrosion*, Vol. 2, NACE International, Houston, Tx, 1985.
7. I. Paseka, *Electrochim. Acta*, 1993, 38, (16), 2449–2454.
8. A. Dugstad, L. Lunde and K. Videm, *Corrosion '91*, Paper No. 473. NACE International, Houston, Tx, 1991.
9. J. B. Sardisco, W. B. Wright and E. C. Greco, *Corrosion*, 1963, **19**, 354–359.
10. E. J. Hart and M. Anbar, *The Hydrated Electron*. John Wiley & Sons, New York, 1970.
11. A. Kawashima, K. Hashimoto and S. Shimodaira, *Corrosion*, 1976, **32**, (8), 321–331.
12. S. Nesic, N. Thevenot, J.-L. Crolet and D. M. Drazic, *Corrosion '96*, Paper No. 3. NACE International, Houston, Tx, 1996.

11
Hydrogen-Related Stress Corrosion Cracking in Line Pipe Steel

L. V. NIELSEN
Department of Manufacturing Engineering, Materials Technology,
Technical University of Denmark, Denmark

ABSTRACT
A correlation between hydrogen concentration (C_0) and the critical stress intensity factor for propagation of hydrogen-related cracks (K_{ISCC}) has been established by fracture mechanics testing of compact tension, CT, specimens for the heat affected zone (HAZ) of an API 5L X70 pipeline steel. The correlation can be expressed as

$$K_{ISCC} = 89 \exp(-0.09C_0).$$

In turn, the hydrogen concentrations resulting from cathodic protection (CP) of steel exposed in a clay sediment and a peat bog have been assessed using a recently developed hydrogen sensor based on the Arup type. A comparison of the fracture mechanics test results and the hydrogen concentrations assessed by field measurements in the clay and in the peat bog indicates that no critical hydrogen concentrations can be generated by CP in these sediments.

1. Introduction

Assessments of the susceptibility of line pipe steel to hydrogen-related stress corrosion cracking are relevant when considering cathodic protection criteria for buried pipelines for oil and gas transmission. The combined effect of cathodic protection and an anaerobic sediment with high activity of sulfate-reducing bacteria (SRB) may introduce a critical concentration of hydrogen into the pipe as a result of enhanced hydrogen evolution reaction kinetics and a poisoning of the hydrogen recombination brought about by the hydrogen sulfide formed during the SRB metabolism.

The aims of the present study were to establish the threshold stress intensity factor (K_{ISCC}) as a function of lattice hydrogen concentration (C_0) for high strength line pipe steel and compare this with field measurements of the hydrogen uptake in cathodically protected steel exposed to various sediments .

The establishment of the K_{ISCC} vs C_0 relationship was made by fracture mechanics testing using pre-cracked compact tension CT specimens in slow strain rate (SSR) tests. The specimens were cathodically charged and the hydrogen concentration assessed by the electrochemical permeation technique. The field measurements were performed using an electrochemical probe especially designed for evaluation of soil corrosivity in general.

2. Experimental

2.1. Establishment of K_{ISCC} vs C_0

The CT-specimens were cut from the pipe with dimensions according to the ASTM standard E-399, although the thickness was determined by the thickness of the pipe wall. The resulting dimensions were typically 50 × 6 mm ($W \times B$; W = distance from load line to bottom of specimen). In this paper, results are reported for an API 5L X70 steel with the chemical composition and mechanical properties given in Tables 1 and 2. Pre-cracks were made using a hydraulic fatigue testing machine and were located in the heat affected zone (HAZ) with maximum hardness 4 mm from the weld centreline (found by microhardness measurements Fig. 1). A 50 kN tensile machine was used for the SSR testing. The tensile machine was equipped with a 7 L electrochemical cell for control of charging conditions, purging gas, and temperature. All tests were made in an acetic acid/sodium acetate buffer system (pH = 6) at 10°C and the CT specimens galvanostatically charged by 2 $mAcm^{-2}$ cathodic current. A wide range of hydrogen permeation levels was established in the different tests by adding different recombination poisons (none, As_2O_3-saturation, and purging with 1000 ppm H_2S in N_2). Prior to establishing a constant slow strain rate of 7 μmh^{-1}, the CT specimen was left for 24 h with a load of 1 kN in order to saturate the specimen with hydrogen. The crack growth was followed by compliance measurements vs time [1]. The actual crack length a was calculated from the fracture mechanics relations between crack length and compliance, starting with the initial crack length. The critical stress intensity factor K_{ISCC} for crack propagation was assessed from established K vs log da/dt curves as the K-value at which da/dt exceeded 10^{-10} ms^{-1}.

In each experiment, a steel membrane was mounted in the wall of the electrochemical test cell for the purpose of studying the hydrogen permeation. Throughout the experiment, the membrane was kept at the same charging current density as the CT-specimen, and the resulting hydrogen permeation was determined by the electrochemical technique [2]. The resulting hydrogen concentration was established by

$$C_0 \text{ (ppm by weight)} = (M_H \cdot i_{p,ss} \cdot L)/(F \cdot D \cdot \delta) \quad (1)$$

where M_H is the hydrogen molar mass (1 g $mol.^{-1}$), $i_{p,ss}$ is the steady state hydrogen permeation current density (μAcm^{-2}), L is the membrane thickness (0.1 cm), F is

Table 1. Chemical composition of the pipeline steel according to certificate

C	S	Mn	P	S	Cr	Ni	Mo
0.071	0.493	1.591	0.019	0.004	0.029	0.018	0

Table 2. Mechanical properties of the pipeline steel according to specification and certificate

	YS (MPa)	TS (MPa)	Elong. (%)	Hardn. (HV)
Specification	480 (min.)	600–750	20 (min.)	220 (max.)
Certificate	530	706	24.7	212

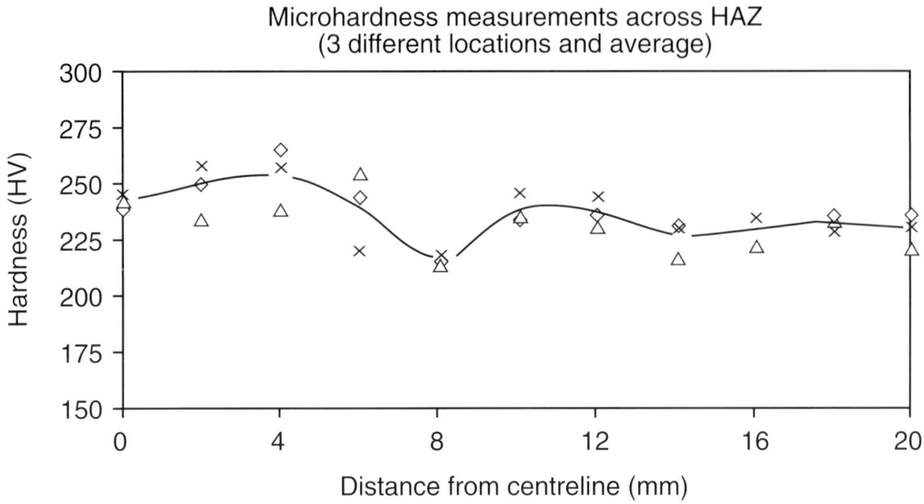

Fig. 1 *Microhardness measurements across HAZ of the pipeline steel.*

Faraday's constant, D is the diffusion coefficient determined from the rising permeation transient ($cm^2\ s^{-1}$), and δ is the density of iron (7.79 $g\ cm^{-3}$).

2.2 Hydrogen Permeation Measurements in the Field

Assessments of the hydrogen uptake in cathodically protected steel exposed in natural sediments have been made using a sealed electrochemical Devanathan cell mounted in a probe designed for evaluation of soil corrosivity in general (Fig. 2). The soil hydrogen permeation electrode (SHPE) is a modification of the Arup-type cells [3,4]. A membrane in the form of a cylinder is built into the general purpose probe, and a metal–oxide cathode placed inside the cell in an alkaline electrolyte acts as the counter electrode for reoxidation of the hydrogen diffusing across the membrane. The hydrogen flux is measured as the potential drop across a resistor placed at between the membrane and the metal–oxide cathode. The active site of the general purpose probe is concentrated within 20 cm. During measurements, the active site of the probe was placed in a depth of approximately 1 m (corresponding a typical depth of a buried pipeline) in the sediment under investigation. The membrane was cathodically charged firstly by connecting it directly to the cathodic protection system of the pipe and secondly by using a sacrificial anode (Mg) connected to the membrane via a resistor. Different charging conditions were established by varying this resistor. The charging conditions (potential and current) were recorded on a data-logger over a period of 3–4 days per charging condition. The IR-free membrane potentials were obtained at the end of each charging period by interrupting the charging current for a short time and reading the potential immediately after the interruption (the 'on-off'-technique).

Two different types of sediment have been investigated so far: a heavy clay with no significant SRB-activity, and a peat bog with modest SRB-activity. Sediments with severe sulfate-reduction are under investigation.

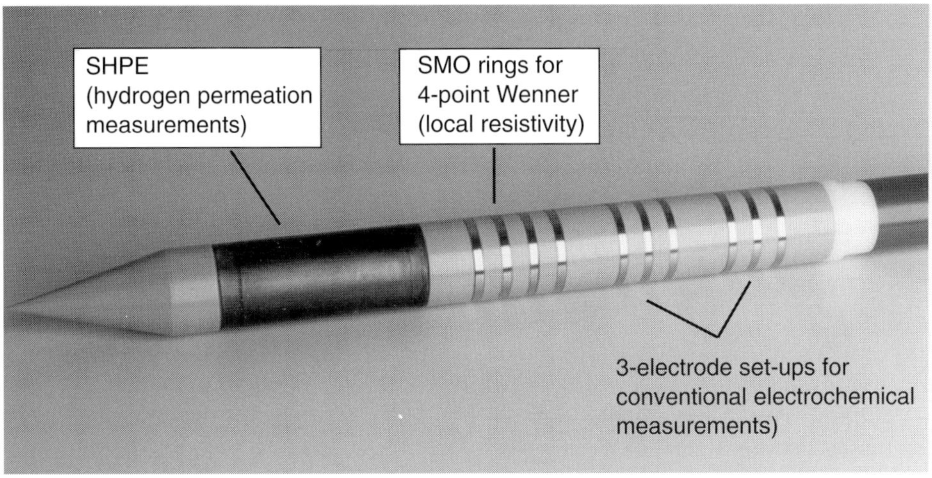

***Fig.* 2** *Active site of the soil corrosion probe. Electrodes are concentrated within 20 cm.*

3. Results and Discussion

3.1. K_{ISCC} vs C_0 Established in the HAZ of the API 5L X70 Pipeline Steel

The correlations between the observed critical stress intensity factor for crack propagation K_{ISCC} and the steady state hydrogen permeation current density $i_{p,ss}$ for the three hydrogen charging conditions are presented in Table 3. The resulting hydrogen concentrations calculated according to eqn (1) are given as well. It has been suggested in the literature [5] that the correlation between hydrogen concentration and K_{ISCC} can be described by the general formula:

$$K_{ISCC} = \alpha \exp(-\beta C_0) \tag{2}$$

where α and β are material-related constants. Using this equation with the data presented in Table 3, it can be shown that the observations fit quite well into the expression:

$$K_{ISCC} = 89\exp(-0.09C_0) \tag{3}$$

***Table* 3.** *Correlation between* $i_{p,SS}$, K_{ISCC}, *and* C_0 *found in fracture mechanics tests of the X70 pipeline steel. Diffusion coefficient* $D = 5.5 \times 10^{-7}$ $cm^2\ s^{-1}$ *calculated from the rising permeation transient*

Recombination poison	$i_{p,SS}$ (μA cm^{-2})	K_{ISCC} (MPa m$^{1/2}$)	C_0 (ppm)
None	1.5	87.0	0.35
As_2O_3	10.0	69.5	2.40
1000 ppm H_2S	15.0	65.0	3.65

3.2. C_0 Resulting from Hydrogen-Absorption by Cathodically Charged Steel Exposed in Natural Sediments

The results from field measurements of the hydrogen absorption by cathodically charged steel exposed in the clay sediment and the peat bog are presented in Table 4 in terms of E_{off} (IR compensated potential), i_c (charging current density), $i_{p,ss}$ (steady state hydrogen permeation current density), and C_0 (resulting hydrogen concentration calculated from eqn (1)). The potentials reflect a slight overprotection since, according to common practice, the protection potential of the pipe should not be more cathodic than –1100 to –1150 mV vs the saturated $Cu/CuSO_4$ reference electrode. The current consumption in the peat bog seems to be more than twice the consumption in the clay sediment, whereas the hydrogen absorption and resulting hydrogen concentration tends to be a factor 2–3 higher in the peat bog.

A comparison between the observed K_{ISCC} vs C_0 (eqn (3)), found by fracture mechanics testing in the HAZ of the API 5L X70 steel, and the field measurements of the hydrogen uptake can be made by adapting a commonly used fracture mechanics approach. The stress intensity factor build up at semicircular flaws (e.g. resulting from weld defects) when the pipe is stressed by the amount σ (resulting from the operating pressure) can be described by [6]:

$$K = 1.12\sigma\sqrt{(\pi a/Q)} \tag{4}$$

where a is the size of the semicircular defect and Q is a flaw shape parameter, which for a semicircular defect is given by (YS = yield strength):

$$Q = \sqrt{(\pi^2/4 - 0.212(\sigma/\sigma_{YS})^2)} \tag{5}$$

Calculations based on the above equations are presented in Fig. 3, showing the stress

Table 4. *E_{off} (measured vs saturated $Cu/CuSO_4$), i_c, $i_{p,ss}$, and C_0, found for the two test sites. Diffusion coefficient $D = 1.13 \times 10^{-7}$ $cm^2\ s^{-1}$*

Heavy clay E_{off} (mV)	i_c (μA cm^{-2})	$i_{p,ss}$ (μA cm^{-2})	C_0 (ppm by weight)
–1090	0.8	0.04	0.05
–1160	3.8	0.12	0.14
–1180	12	0.18	0.22
–1220	19	0.19	0.22
Peat bog E_{off} (mV)	I_c (μA cm^{-2})	$i_{p,ss}$ (μA cm^{-2})	C_0 (ppm by weight)
–1112	1.5	0.21	0.24
–1125	7.5	0.29	0.34
–1180	34	0.56	0.65
–1200	46	0.57	0.67

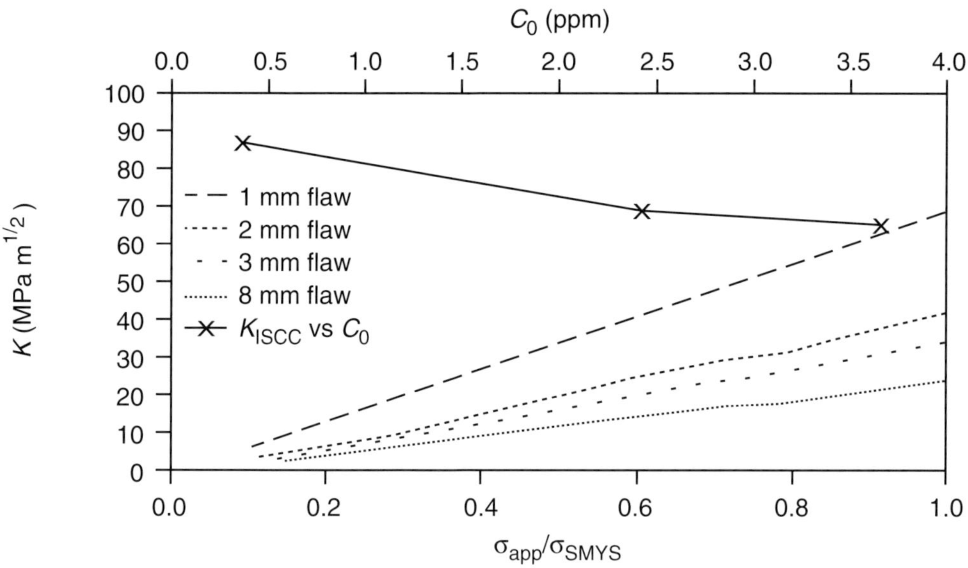

Fig. 3 Stress intensity factors at flaws of size, 1, 2, 3 and 8 mm as a function of the stress (fraction of SMYS) resulting from pipeline operating pressure. The correlation between K_{ISCC} *and* C_0 *is shown as well.*

intensity factor build up at flaws of sizes, 1, 2, 3 and 8 mm. It may be suggested that flaws of 1 mm size are probably present, flaws of 2 mm size may be present, whereas flaws of 3 mm size are seldom present in the pipe. Flaws of 8 mm size have been included since these represent a leak in the pipe. The established correlation between K_{ISCC} and C_0 is also given in Fig. 3 and indicates that even the highest hydrogen concentration cannot provoke crack propagation at a 2 mm flaw. It is further indicated that a hydrogen concentration of 3.5–4.0 ppm in combination with yielding is necessary if a leak should result in a catastrophic bursting of the pipe. It is noted that common practice allows pipelines to be operated at pressures corresponding to a maximum of 72% of the specified minimum yield strength (SMYS), although, local yielding may occur. It should be pointed out that, the field observations made so far have not reflected hydrogen concentrations higher than 0.7 ppm, but severe sulfate-reducing conditions have yet to be investigated. Earlier findings by the Force Institute in such environments indicate that the cathodic current consumption approximately doubles when severe sulfate-reduction is present, whereas the hydrogen uptake increases by a factor of 10 or more [7]. On the present basis not even these conditions would lead to crack propagation at a 3 mm flaw present in the HAZ of the API 5L X70 steel investigated here.

4. Conclusions

1. A correlation between C_0 and K_{ISCC} has been established for the HAZ of an API 5L X70 pipeline steel by fracture mechanics testing using pre-cracked CT-specimens: $K_{ISCC} = 89\exp(-0.09C_0)$.

2. The hydrogen concentrations resulting from cathodic protection of steel exposed in a clay sediment and in a peat bog have been assessed using a recently developed hydrogen sensor. A comparison of the fracture mechanics test results and the hydrogen concentrations assessed by field measurements in the clay and in the peat bog indicates that no critical hydrogen concentrations are generated by CP in these sediments.

3. Further work includes assessments of the hydrogen concentrations resulting from CP of steel exposed under severe sulfate-reducing conditions.

5. Acknowledgements

The studies were carried out under the Danish Energy Research Programme (EFP-95) supported by the Danish Ministry of Environment and Energy. The National Oil & Gas Company of Denmark (C. Juhl), Greater Copenhagen National Gas (B. Baumgarten), and The Force Institute (N. K. Bruun) are thanked for support and participation in the project. The Soil Hydrogen Permeation Electrode was designed and manufactured in co-operation with the Force Institute (T. S. Nielsen and N. K. Bruun). Final reports are available [8,9].

References

1. A. Saxena and S. J. Hudak Jr, "Review and extension of compliance information for common crack growth specimens", *Int. J. Fract.*, 1978, **14**, (5), 453–468.
2. M. A. V. Devanathan and Z. Stachurski, "The adsorption and diffusion of electrolytic hydrogen in palladium", *Proc. Roy. Soc*, 1964, **270A**, 90–102.
3. H. Arup, "Measuring hydrogen activity with a sealed Devanathan cell", *9th Scand. Corros. Congr.*, Copenhagen, 1983.
4. C. Christensen, H. Arup, P. Press and P. B. Mortensen, "Measurement of effluent hydrogen in pressure vessel and linepipe steel", in *Hydrogen Transport and Cracking in Steel* (A. Turnbull Ed.), pp 161–171. The Institute of Materials, London, 1995.
5. P. J. Kilgallon and M. J. Robinson, "Effect of marine biofilms on hydrogen absorption by cathodically protected steel", *UK Corrosion and EUROCORR '94*, pp. 71–84. Inst. of Corrosion, Leighton Buzzard, UK, 1994.
6. D. Broek, *Elementary Engineering Fracture Mechanics*. Martinus Nijhoff Publishers, Boston, USA, 1983.
7. E. Maahn, C. Christensen and H. Arup, "Environmental effects on slow straining of marine pipelines", The Danish Corrosion Centre (now Force Institute), 1987.
8. L. V. Nielsen, "Assessment of soil corrosovity using an electrochemical soil corrosion probe", ENS 1313/95. The Technical University of Demark, 1998.
9. L. V. Nielsen, "Evaluation of hydrogen-assisted stress corrosion cracking in cathodically protected high-pressure gas transmission lines", ENS 1313/95. The Technical University of Denmark, 1998.

12
The Role of Microstructure in Sulfide Stress Cracking Resistance of Thermomechanically Processed High Strength Low Alloy Steels

A. GINGELL and X. GARAT
IRSID, 57283 Maizieres les Metz, France

ABSTRACT
The sulfide stress cracking (SSC) behaviour of a high strength low alloy (HSLA) line pipe steel has been investigated to determine the role of microstructure, or, more precisely, the type of second phase. The SSC resistance, evaluated using uniaxial tensile tests at 1 bar H_2S, exceeded 80% of the actual yield stress for all the given treatments. Nevertheless, it was found that the least SSC resistance was associated with the presence of certain pearlitic structures, either isolated or accompanied by bainite. Microstructures containing isolated islands of martensite displayed excellent SSC resistance. The relative behaviour of the different microstructures is discussed in terms of the observed modes of cracking.

1. Introduction

The ongoing need to reduce the costs associated with line pipe projects has led to the development of higher strength grades in order to reduce wall thickness. This strengthening is unattainable by alloying additions as service and commercial conditions are prohibitive. Therefore thermomechanical controlled processing (TMCP), comprising controlled rolling and accelerated cooling, has been developed as a method for producing high strength microalloyed steels [1]. These processes are of great metallurgical interest as they require careful control of the processing parameters in order to attain the necessary steel microstructure and mechanical properties.

The HSLA steels such as the API 5L X60 line pipe steels are insufficiently alloyed to produce microstructures other than ferritic matrix at the cooling rates attained in the plate mill. However, the TMCP parameters such as final rolling temperature, accelerated cooling rate and final cooling temperature can be controlled so as to develop different second phase microstructures, ranging from pearlite for air cooling to martensite for rapid cooling to low temperature. The development of multi-phase microstructures is important for the attainment of certain mechanical properties, but can be detrimental for resistance to sulfide stress cracking (SSC), a type of embrittlement specific to H_2S-containing environments. Carbon rich phases such as pearlite, bainite or martensite can be particularly susceptible to this mode of hydrogen embrittlement [2].

This study considers the SSC resistance of a API 5L X60 line pipe steel subjected to laboratory thermomechanical treatments to generate various second phase microstructures. The microstructures are compared in terms of their SSC threshold stress values and in terms of the damage modes observed by microscopy. The role of microstructure has been examined using a second line pipe steel differing from the first only in its nickel content. This alloying element lowers the austenite–ferrite transformation temperature (Ar_3) and therefore leads to a refinement of the microstructure. The stabilisation of austenite to lower temperatures also enhances the formation of martensite at lower cooling rates.

2. Experimental Procedure

Two API 5L X60 line pipe steels were provided in the form of 300 mm thick blooms. The compositions of the two steels, designated D and G, are given in Table 1. The steels, with very low carbon, sulfur (<0.001%) and phosphorus (<0.01%) are alloyed with manganese and microalloyed with niobium and vanadium. Steel G differs from steel D in its nickel content.

Blanks with dimensions 250 × 70 × 70 mm^3 were machined from the quarter-thickness of each bloom and rolled at the experimental hot rolling mill at IRSID. Thermocouples were placed in drilled holes at mid-thickness of each blank to control the hot rolling and cooling processes. The blanks were homogenised and rolled from an initial thickness of 70 mm to a final thickness of 12 mm. After rolling the specimens were cooled either in air or by water-quenching, the cooling rate being varied by the addition of polymeric liquids to the water-bath.

Microstructural characterisation of the microstructures generated by the thermomechanical treatments was conducted using optical microscopy, scanning electron microscopy and transmission electron microscopy on carbon replicas. Image analysis was used to determine the fraction of second phase present with the constituent phases being differentiated by the use of various chemical agents. Although nital was used as an etching agent for standard optical microscopy, image analysis required the use of picral to isolate bainite and the use of Villela's reagent to differentiate martensite and bainite constituents.

Sulfide stress cracking resistance was determined using standard smooth round-bar tensile specimens described in the NACE standard TM0177-90 method A [3]. These specimens were machined in the transverse direction of each billet. The tests were conducted following the above NACE standard using proof rings to maintain a constant stress. Initial solution pH was approximately 2.7 and the H_2S pressure was maintained at 1 bar. Threshold was taken as the stress value at which there was no failure up to 720 h given as a percentage of the actual yield stress.

After testing, specimens were sectioned normal to the fracture surface and examined in the scanning electron microscope for crack initiation sites. Crack counting

Table 1. *Steel compositions (10^{-3} wt%)*

Steel	C	Mn	Si	P	S	Al	Ni	Nb	V
D	52	1360	283	4	0.3	44	—	38	55
G	42	1410	215	8	0.7	46	215	36	65

was also carried out at a magnification of ×2000 on the area covered by three scans across the specimen thickness immediately below the fracture surface.

3. Results

All the thermomechanical treatments developed ferritic matrices. However, the variation of cooling rates from 1 Ks^{-1} to 40 Ks^{-1} generated second phases ranging from pearlite to martensite (Table 2). The percentage of second phase is increased by faster cooling rates as the volume fraction of transformed ferrite decreases. The dual effect of nickel in enhancing martensite at the expense of bainite and in reducing second phase grain size is clearly demonstrated.

The mechanical properties and the SSC threshold stress of the different microstructures are given in Table 3 along with the property requirements for X60 line pipe steel. All the microstructures tested display excellent SSC resistance (≥ 90% YS). Nevertheless, there are small differences depending on the second phase microstructure, notably a trend in SSC resistance increasing in the order pearlite < bainite < martensite. The addition of nickel does not change this order but does modify the tensile properties of the steels. The yield stresses of the ferrite–pearlite

Table 2. *Second phase microstructures as a function of cooling rate*

Treatment	Cooling rate (Ks^{-1})	% pearlite	% bainite	% martensite	Island size (μm)
D3	1.5	5	—	—	6
D4	15	—	6	—	3
D8	40	—	—	15	7.5
G3	1.5	4	—	—	1
G4	15	—	2	2	1.2
G8	40	—	—	16	1.4

Table 3. *Mechanical properties as a function of microstructure*

Treatment	Second phase	YS (MPa)	YS/TS	σ_{th} (%YS)
D3	5% P	405	0.84	90
D4	6% B	479	0.88	95
D8	15% M	428	0.63	97.5
G3	4% P	437	0.86	92.5
G4	2% B & 2% M	418	0.66	95
G8	16% M	458	0.66	97.5
Grade X60	—	≥ 413	—	≥ 72

P = pearlite B = bainite M = martensite.

and ferrite–martensite microstructures are increased by the addition of Ni, probably as a result of solid solution strengthening. The ferrite grain size does not differ appreciably between the two steels so the strengthening effect does not appear to be due to grain refinement. At the intermediate cooling rate the Ni-containing steel has a lower yield stress, presumably as the volume fraction of second phase is reduced.

Metallographic observation of the specimens after SSC failure distinguished two modes of microcracking. The first mode, identified as SSC, describes microcracking transverse to the tensile axis. This mode of cracking is almost exclusively confined to the second phases, (Fig. 1). The second mode describes cracking in the rolling plane and hence parallel to the tensile axis of the specimen. This mode has been described as hydrogen-induced cracking (HIC) and occurs on a larger scale than SSC (Fig. 2). The combination of both modes can lead to a SOHIC (Stress-Oriented Hydrogen-Induced Cracking) contribution to the overall mode of failure.

The observation of SSC and cracking behaviour linked to the second phase microstructure has led to the simulation of TMCP to develop a range of second phases including various mixes of pearlite + bainite and bainite + martensite. The SSC resistance, as well as the defect density close to the fracture surface, are given in (Fig. 3) as a function of second phase microstructure. There is no direct correlation between the crack density and SSC resistance. In general, the SSC threshold increases as the second phases pass from pearlite through bainite to martensite. Hydrogen induced cracking cracking is observed in microstructures containing pearlite or bainite but is practically absent in those containing martensite. The density of SSC cracks at failure varies considerably between the microstructures and does not appear to be directly related either to the microstructure or the SSC threshold stress.

The variation of SSC threshold stress (expressed in terms of the ratio σ_{th}/YS) with

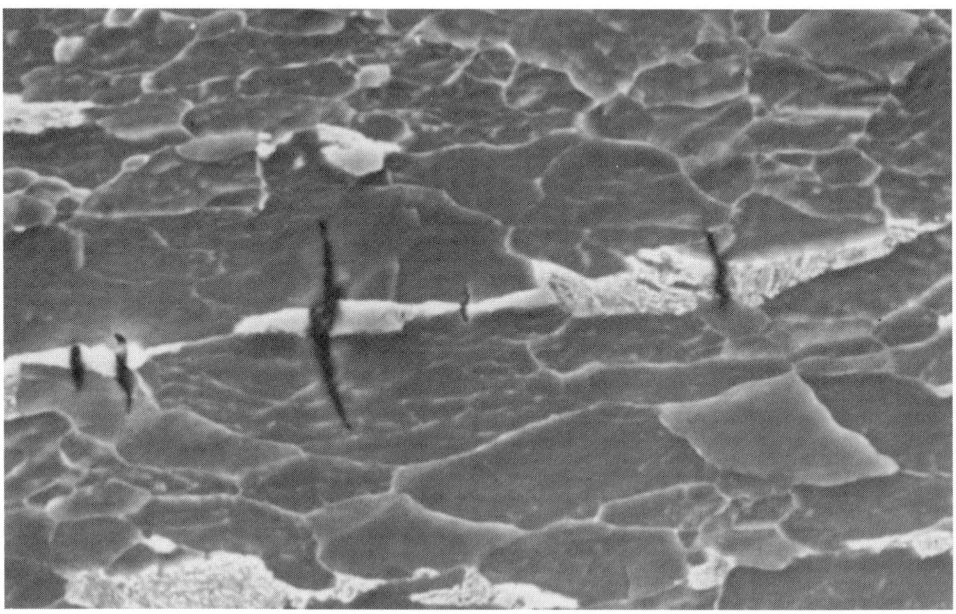

Fig. 1 *SSC defects in second phase, perpendicular to tensile axis (×2500).*

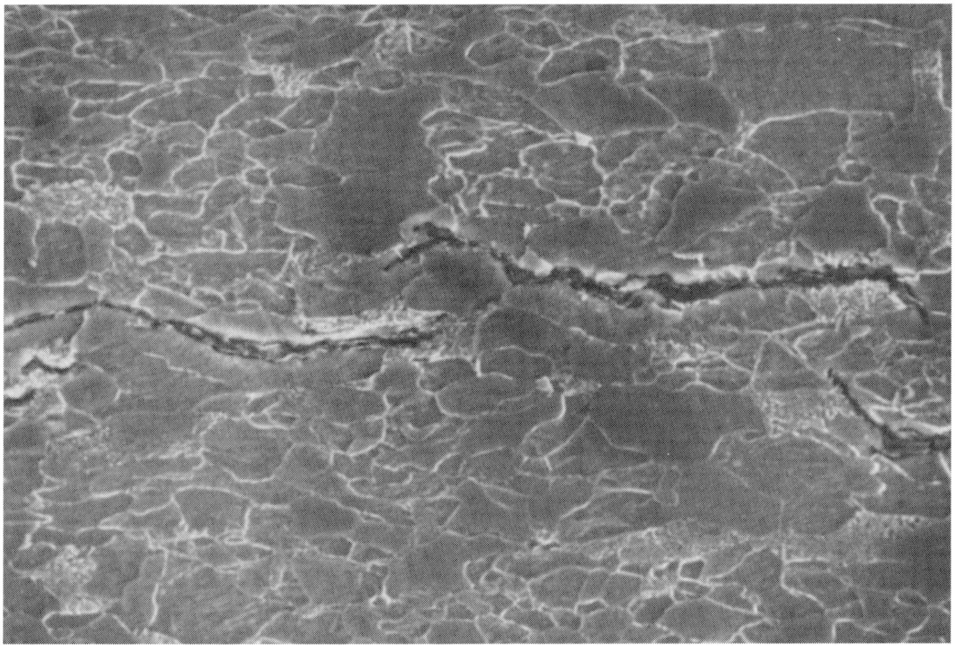

Fig. 2 *HIC-type defect parallel to the tensile axis (×1250).*

yield stress is given in Fig. 4 for both steel D and steel G. In general there is a slight trend for the threshold stress ratio to decrease with increasing yield stress for both steels. Comparing the data for the two steels, there does not appear to be a marked influence of Ni content on the SSC threshold stress.

4. Discussion

The SSC behaviour of API 5L X60 line pipe steels has been extensively studied with respect to their heat-affected zone (HAZ) microstructures after welding [4–10]. At the same time, much work has concentrated on the possibly harmful role of nickel additions on the SSC resistance [7,8,11,12]. Little work, however, has been published on the role of base metal microstructure and its second phases. In doing so, this study reveals interesting differences with the conventional literature on the effects of certain microstructural phases. It is well documented that banded microstructures have poor SSC and HIC resistance [4,6]. The use of TMCP to disperse pearlite colonies improves HIC resistance [13] and reduces the incidence of SOHIC [10]. The generation of mixed ferrite–bainite microstructures by TMCP leads to an improvement in SSC resistance over ferrite–pearlite microstructures [4,9,14]. Martensitic structures display varying SSC behaviour depending on the degree of tempering and the yield stress [11,13]. Local hard HAZ microstructures containing martensite–austenite constituents have been associated with crack initiation during SSC failure of weld regions [6,10].

The line pipe steels used in this study are characterised by their extremely low S content and by the homogeneity of their microstructures and low degree of

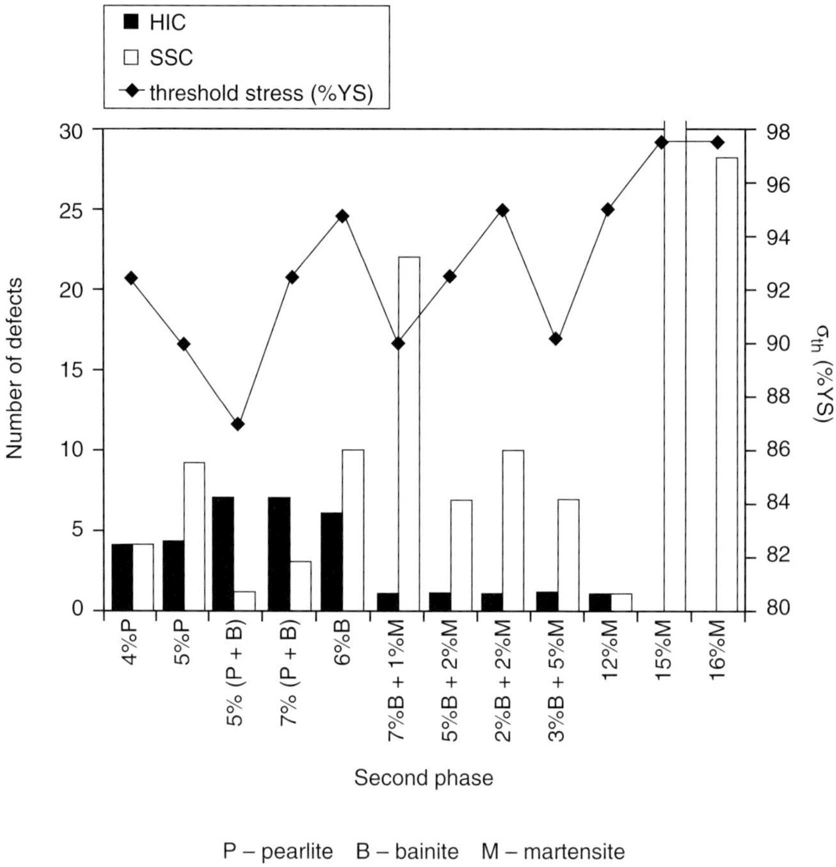

Fig. 3 *Crack density and SSC threshold stress as a function of second phase microstructure.*

segregation. It is therefore not surprising that they exhibit excellent SSC resistance. The differences between the SSC behaviour of the various treatments can therefore be attributed to the diverse features of the second phase microstructure. Pearlite colonies tend to favour the development of HIC-type cracks, leading to SOHIC failure. Martensite islands, on the other hand, act as SSC initiation sites at the complete expense of HIC-type defects. These SSC defects tend to be on a finer scale and are less likely to propagate. Thus, the main difference between the second phases lies in their propensity towards either SSC or HIC-type defects.

Additions of Ni have been reported to increase sensitivity to SSC of API 5L X60 and X65 line pipe steels [8,11]. Nevertheless, this study shows that the addition of 0.2% Ni has little direct influence on the SSC resistance of such steel grades. Although the precise role of Ni is difficult to determine, since the comparison of similar microstructures with identical strength levels both with and without Ni is difficult, there is evidence that it increases the hydrogen permeation rate [7] and that it accelerates crack initiation through pitting corrosion [12]. It is also reported that Ni favours the formation of fresh martensite which can lead to a degradation of SSC performance. This latter point is interesting with respect to the microstructures

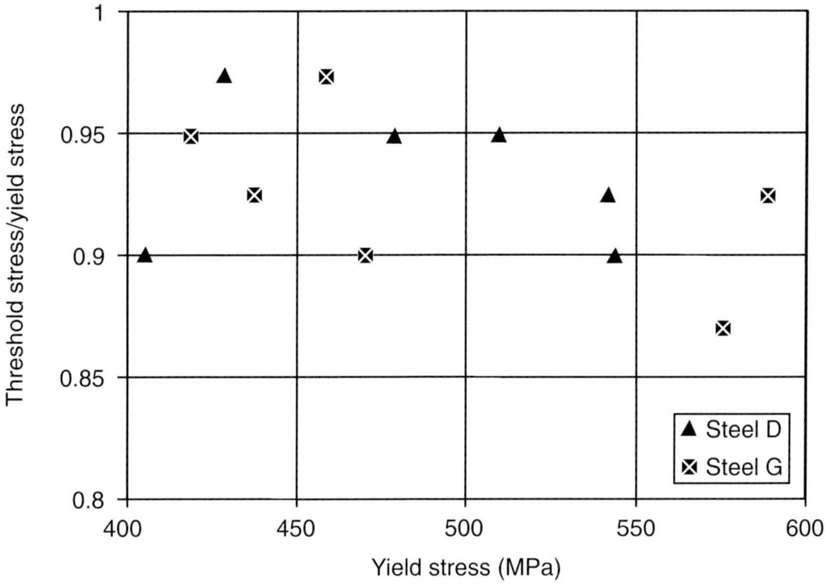

***Fig. 4** Dependence of threshold stress ratio on yield stress.*

developed by TMCP. The martensitic second phases generated in this study are to a certain degree auto-tempered in that each island contains a population of fine carbides. The hydrogen trapping capacity of such islands is effectively increased by this carbide precipitation. Crack initiation in tempered martensite has been reported by Kimura *et al.* [4], notably without further crack propagation.

5. Conclusions

1. Two low carbon API 5L X60 line pipe steels have undergone simulated TMCP treatments which have generated various ferritic microstructures containing second phases ranging from pearlite to martensite. These microstructures exhibit excellent SSC resistance.

2. Differences in SSC behaviour have been related to the second phase microstructure generated by TMCP.

3. The relatively poor SSC resistance of certain microstructures containing pearlite or bainite is related to the presence of HIC-type defects, leading to a SOHIC mode of failure.

4. HIC-type defects are entirely absent from microstructures containing martensite. Such microstructures can undergo extensive SSC microcracking.

6. Acknowledgements

This work was carried out as part of the European Coal and Steel Council (ECSC) project No. 7210-KB/323, with the support of DH-GTS, Dillingen, Germany. The financial aid of the ECSC is gratefully acknowledged.

References

1. A. Streisselberger, J. Bauer, W. Schütz and V. Schwinn, *Rev. Métall.*, 1995, January, 109–119.
2. J. C. Charbonnier, H. Margot-Marette, F. Moussy, D. Bridoux and C. Perdrix, *Rev. Métall.*, 1988 January, 91–105.
3. NACE Standard TM0177-90, "Laboratory testing of metals for resistance to sulfide stress cracking in H_2S environments", NACE International, Houston, Tx, USA, 1990.
4. M. Kimura *et al.*, *Corrosion*, 1989, **45**. (4), 340–346.
5. S. Endo, *CAMP-ISIJ*, (Current Advances in Materials and Processes – Iron and Steel Institute of Japan), 1993, **6**, (3), 632–635.
6. T. Kushida, *CAMP-ISIJ*, 1993, **6**, (3), 636–639.
7. A. Takahashi, *CAMP-ISIJ*, 1993, **6**, (3), 644–647.
8. H. Ishikawa, *CAMP-ISIJ*, 1993, **6**, (3), 648–651.
9. H. Asahi, *CAMP-ISIJ*, 1993, **6**, (3), 652-655.
10. F. Kawabata, *CAMP-ISIJ*, 1993, **6**. (3), 656–659.
11. C-M. Liao and H-Y. Liou, *China Steel Techn. Report*, 1990, (4), 88–95.
12. T. Kaneko and A. Ikeda, *Trans. ISIJ*, 1988, **28**, (7), 575–577.
13. H-Y. Liou, S-C. Wang and R-I. Hsieh, *Corrosion '94*, Paper No. 224. NACE International, Houston, Tx, USA, 1994.
14. C. Bao-Iturbe and S. Gutierrez de Saiz-Solabarria, *Rev. Métal.*, 1993, **29**, (1), 3–12.

13
The Effect of Microstructure on the K_{ISSC} of Low Carbon Low Alloy Steels

G. ECHANIZ, C. MORALES and T. PÉREZ
Center for Industrial Research (FUDETEC), Siderca, Simini 250, Campana (2804) Buenos Aires, Argentina

ABSTRACT
The effect of austenite grain size and microstructure on the crack rate propagation in double-cantilever beam specimens was evaluated using an ultrasonic technique. The materials had approximately the same yield strength and different microstructures. Additionally, their sulfide stress cracking resistance (SSC) was determined using the Double Cantilever Beam Test (Method D NACE TM0177-90). The materials that possessed a large austenitic grain or those that showed some percentage of a type of upper bainite (a structure with laths of approximately 1 μm wide and large carbides at the border of the laths) had significantly lower SSC resistance. The negative effect of a large grain size and of a wider lath structure was also evident on comparing the crack growth rate.

1. Introduction

Sulfide Stress Cracking (SSC) is one of the major problems in the oil and gas industry and imposes restrictions on the selection of the materials for Oil Country Tubular Goods (OCTG). Increasing development of deep sour wells requires High Strength Low Alloy (HSLA) steels with good SSC resistance. The chemical composition and the thermo-mechanical treatment are the main variables available to reach these objectives.

Most of the pipes used for sour service are made of quenched and tempered low alloy low carbon steels, such as AISI Type 4130 or similar, modified by microalloy additions (mainly B, Nb and Ti). The important role of microstructure in determining the SSC resistance of HSLA steels is known [1–3]. Tempered martensite is generally considered to be the best microstructure for sour service [3–5]. Several authors have proposed the recrystallisation of quenched martensite as a way of improving the resistance to SSC [1,3]. It is generally accepted that the SSC susceptibility increases as the austenite grain size increases.

In the present work the effect of microstructure and the influence of austenite grain size on SSC resistance were studied. The SSC susceptibility was evaluated using the Double Cantilever Beam (DCB) Test (NACE TM0177-90, Method D [6]). Additionally, an ultrasonic technique, previously described [7], was used to determine crack propagation rates in DCB specimens.

2. Experimental

2.1 Tested Materials

The chemical compositions of the tested steels are listed in Table 1. The C content was between 0.22 and 0.28%, the Cr content about 1% and the Mo ranged from 0.45 to 0.66%. Steels J1, S92, T11-7 and T11-8 were Nb microalloyed.

The materials were quenched and tempered to obtain similar mechanical properties (Table 2); this is an important point since SSC susceptibility is a function of the material strength. Two austenitising temperatures were used to change the austenite grain size.

2.2. Test Method

The microstructural characterisation of the steels was carried out using optical and electron microscopy techniques.

Double Cantilever Beam tests were performed according to NACE TM0177-90.

An ultrasonic technique previously described [7] was used to follow crack growth as a function of the applied stress intensity factor in DCB specimens. Materials S1 and S92 were tested according to this technique.

Table 1. *Chemical composition (%) of the tested materials*

Material	C	Mn	S	P	Cr	Mo	Nb	Ti	B
S1	0.24	0.54	0.003	0.010	0.94	0.51	0.004	0.043	0.0013
SC	0.23	0.46	0.002	0.012	0.90	0.50	0.004	0.003	0.0001
T11-7	0.26	0.48	0.002	0.011	0.93	0.66	0.032	0.004	0.0001
T11-8	0.22	0.51	0.001	0.009	0.91	0.63	0.027	0.031	0.0010
J1	0.28	0.48	0.006	0.011	0.93	0.46	0.030	0.012	0.0010
S92	0.25	0.43	0.005	0.007	0.90	0.45	0.025	0.028	0.0014

Table 2. *Mechanical properties of the tested materials*

Material	YS		UTS		Hardness
	ksi	(MPa)	ksi	(MPa)	RC
S1	96.0	(661.9)	111.7	(770.1)	22.0
SC	95.5	(658.4)	111.6	(769.5)	22.5
T11-7	96.0	(661.9)	114.5	(789.4)	21.4
T11-8	101.4	(699.1)	117.0	(806.7)	23.8
J1	95.4	(657.8)	110.8	(763.9)	22.0
S92	99.7	(687.4)	112.0	(772.2)	21.6

3. Results and Discussion

The prior austenite grain (PAG) size of the S1, SC, J1 and S92 materials ranged from 18 to 32 μm. Materials T11-7 and T11-8 were austenitised at higher temperatures to obtain grain size as large as 60–76 μm (Table 3).

The microstructure of the materials S1, SC was mainly tempered martensite. A small percentage of a structure with wider laths and large carbides located mainly at the borders of the laths and in some cases inside them, was also observed (Fig. 1).

Materials T11-7 and T11-8 presented a uniform tempered martensite microstructure; a few packets of laths were observed in a large proportion of the austenite grains.

The microstructure of materials J1 and S92 was a mixture of tempered martensite and a 'ferrite-carbide' structure; more detailed analysis is being carried out to characterise this structure (Figs 2 and 3). A fine and homogeneous distribution of NbCN was observed by transmission electron microscopy.

The average values of K_{ISSC} obtained for the tested materials are shown in Table 4.

These results show that K_{ISSC} was influenced by the austenitic grain size and by the presence of structure with wider laths. The materials with these characteristics exhibited significantly poorer SSC resistance. Previous work conducted with HSLA steels having austenite grain size between 10 and 30 μm showed no dependence of this variable on SSC susceptibility as determined by DCB testing [8].

Table 3. *Prior austenitic grain size of the tested materials*

Material	S1	SC	T11-7	T11-8	J1	S92
PAG size (μm)	32	20	60	76	18	18

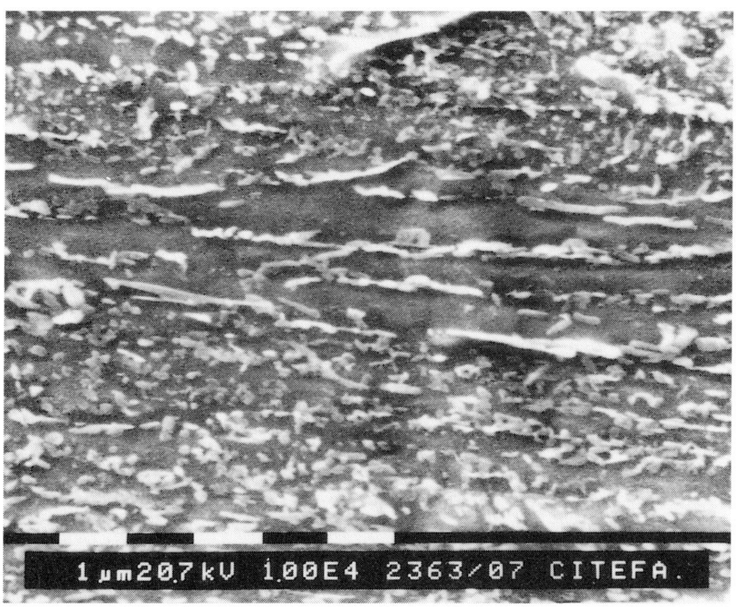

Fig. 1 *Material SC.*

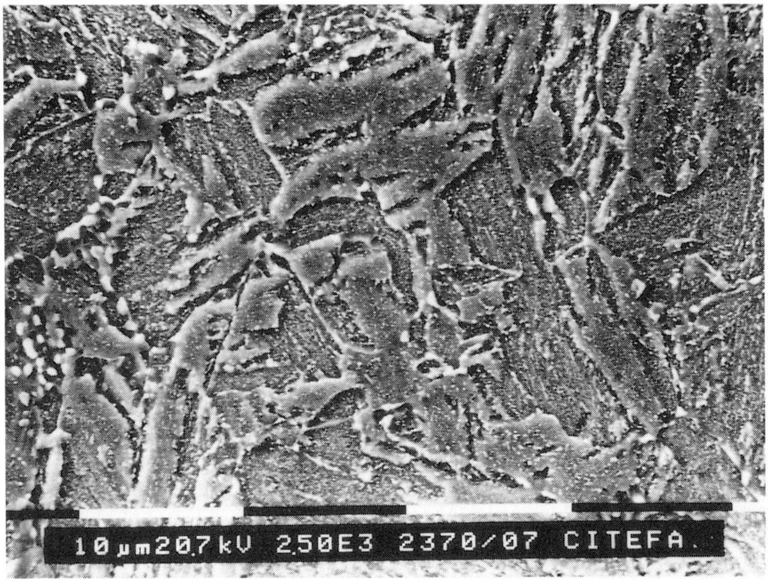

Fig. 2 Material S92.

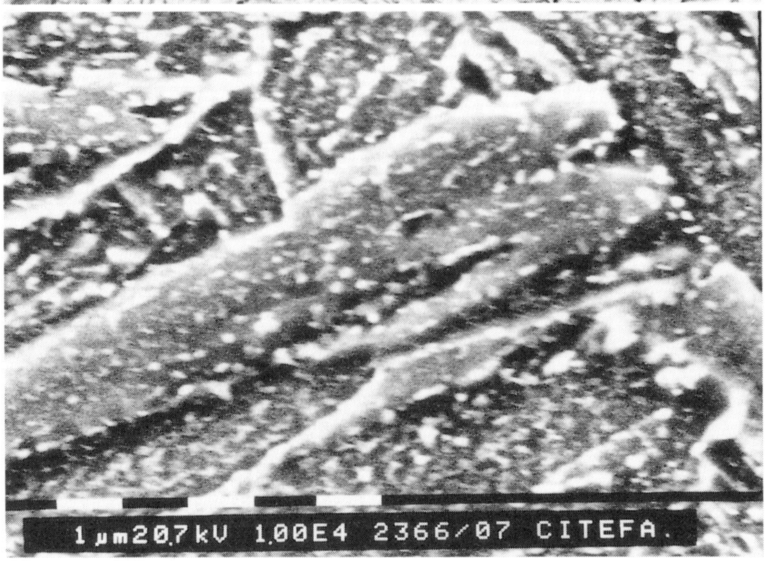

Fig. 3 Material S92.

Table 4. *DCB test results*

Material	K_{ISSC} (average)	
	ksi in$^{0.5}$	(MPa m$^{0.5}$)
S1	27.8	(30.5)
SC	28.4	(31.2)
T11-7	28.2	(31.0)
T11-8	22.5	(24.7)
J1	39.9	(43.8)
S92	39.1	(43.0)

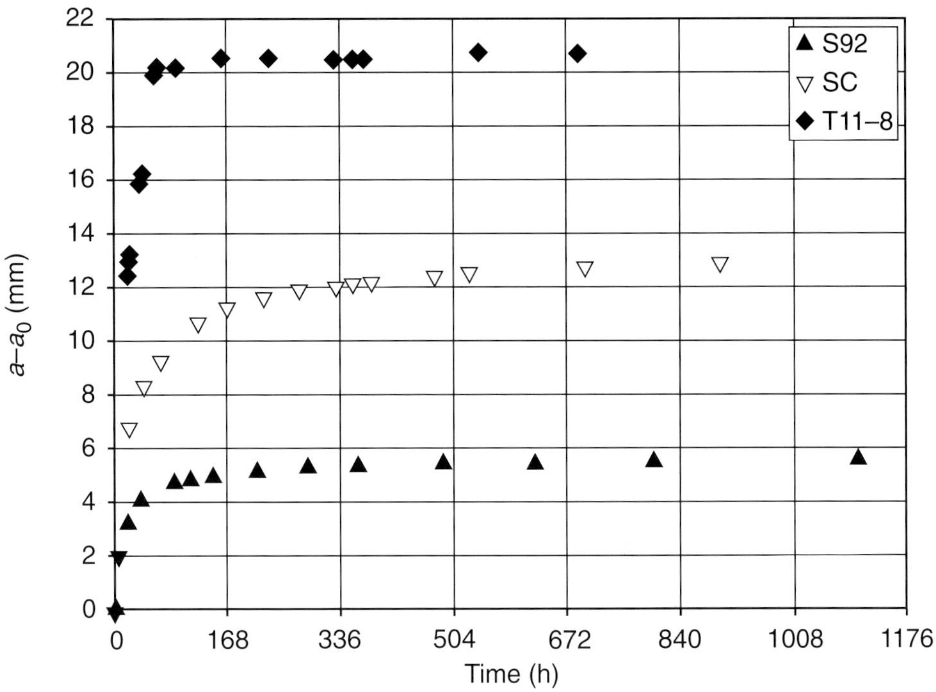

Fig. 4 *Crack growth as a function of time.*

Crack propagation tests in DCB specimens were conducted following a previously described technique [7]. The results obtained for materials T11-8, SC and S92 are shown in Fig. 4 (above). The initial crack propagation rate is markedly affected by the austenitic grain size and to a lesser extent by the microstructure.

Conclusions

1. The microstructure has a strong influence on SSC resistance when this is evaluated using the Double Cantilever Beam test (NACE TM0177-90). Large grain size (> 30 μm) and the presence of wide martensite laths with large carbides are detrimental.

2. The crack propagation rate in DCB specimens is also influenced by the microstructure. In this case, larger grain size appears to be more detrimental than the microstructure in the initial crack propagation rate.

References

1. E. Anelli, L. Cariboni and A. Mascanzoni, *Processing, Microstructure and Properties of HSLA Steels*, A de Ardo. Jr, ed. The Minerals, Metals and Material Society 1988, p.477.

2. J. L. Albarrán, L. Martinez and H. F. López, *Corrosion '95*, Paper No. 154. NACE International, Houston, Tx, USA, 1995.
3. M. Watkins and R. Ayer, *Corrosion '95*, Paper No. 50. NACE International, Houston, Tx, USA, 1995.
4. A. Ikeda and M. Kowaka, *Chem. Econ. Eng. Rev.*, 1978, **10**, May/June, 12–22.
5. R. Garber, T. Wada, F. Fletcher and T. Cox, *J. Mater. Energy Systems*, 1985, (9), 91–103.
6. NACE Standard TM-0177-90, Standard Test Method, *Laboratory Testing of Metals for Resistance to Sulfide Stress Cracking in H_2S Environments*, 1990. NACE International, Houston, Tx, USA.
7. C. R. Morales, T. E. Pérez and G. L. Fitzsimons, *Corrosion '97*, Paper No. 52. NACE International, Houston, Tx, USA, 1997.
8. C. R. Morales, Thesis of Master in Science and Technology of Materials, Inst. of Technology, (UNGSM-CNEA), Argentina, 1996.

14
Coiled Tubing and Pipe for CO_2 and H_2S Service

R. P. BADRAK
Precision Tube Technology, Inc., P.O. Box 24746, Houston, Tx 77229, USA

ABSTRACT
Coiled tubing materials of construction and their applications are discussed in relation to service in oil and gas environments. The resistance of these materials of construction, including high strength low alloy and ferritic stainless steels, to corrosion in CO_2 and H_2S environments is discussed. The resistance to sulfide stress cracking (SSC) of coiled tubing up to the 110 ksi (758 MPa) strength level and test results measuring the resistance to hydrogen induced cracking (HIC) of coiled line pipe are presented. Stainless steel coiled tubing and coiled tubing lined with thermoplastics are presented as materials suitable for corrosive CO_2 service. The problems associated with welding of stainless steel coiled tubing are presented and discussed.

1. Introduction

Coiled tubing is different from jointed tubing or welded pipe in that the product is made continuously from flat strip or sheet using a roll forming operation and a continuous longitudinal seam weld. Coiled tubing has been used for such diverse activities as drilling, well logging, gas lift, acidising/fracturing, injection, production strings (including velocity strings), and sand control. Coiled tubing has also been used more recently (in the last 3–4 years) for flowline and pipeline applications.

Corrosion problems in oil and gas production can be categorised as corrosion problems associated with existing applications and corrosion problems which may be addressed through new applications of coiled tubing and pipe.

For the first type of corrosion problems, estimates of 25% [1] and 37% [2] have been put forward as the percentage of coiled tubing failures that were attributable to corrosion between 1996 and 1997. From a historical perspective, the percentage of failures attributable to corrosion has been rising [3]. The cause for the rise is associated with a combination of a decrease in problems arising from manufacturing defects, an increase in the operators ability to estimate normal service life (failures would have previously been attributed to other causes) and increasingly severe applications.

Some corrosion related failures have occurred as the result of low cycle fatigue cracks that have initiated at the base of corrosion pits. This type of corrosion, generally caused by acidising environments or corrosive produced fluids, has been known to reduce dramatically the useable life of coiled tubing. For example, we have observed fatigue failures at 10–20% of the life predicted through fatigue modelling programmes.

The second type of corrosion problem is only now being addressed through the use of coiled tubing. As the trend for more corrosion related problems and applications increases the needs for corrosion resistant coiled tubing also increase. The types of corrosion resistant materials available have varied but will develop to meet the needs.

2. Materials for Coiled Tubing

The materials used to make coiled tubing correspond to the class of high strength low alloy (HSLA) steels and are sometimes referred to as micro alloyed steels. Several process routes are possible although the most prevalent is hot working with a warm finish temperature, seam welding with a seam normalisation step and full tube body stress relieving.

Downhole grades of coiled tubing typically contain about 0.5% chromium and low carbon contents (less than 0.15%) with selected additions of minor and microalloying elements that are used to achieve desired properties. Pipeline grades of coiled tubing have had more variability in composition but still fit the category of low carbon and low total alloy content; the carbon equivalents are generally below 0.4.

There have been some limited applications for coiled tubing made from corrosion resistant alloys (CRAs); tubing for these applications has been mostly made by fabricating lengths of alloy tube together and spooling the consolidated product.

As the trend for more corrosion related problems and applications increases, it follows that the needs for corrosion resistant coiled tubing will also increase. Table 1 lists some coiled tubing materials with the year in which they were first produced.

No evidence of production or use of corrosion resistant coiled tubing was found until the 1990s. As can be seen from Table 1, the history to date (mid-1998) with corrosion resistant materials has been very limited.

Despite the cost, most of the work in the coiled tubing industry prior to 1997 in CRA materials has been in titanium. The advantages or impetus behind the

Table 1. *Types, applications and dates of first introduction of coiled tubing*

Material	Applications	1st Date
Carbon steel	All	1962
HSLA steel	All	1980
Q & T steel	Limited due to welding & life	1991
Titanium Grade 12	Concentric injection	1992
Titanium Grade 18	Production string between packers	1996
316L	Oxygen injection	1997
12Cr–Ni stainless steel	In progress	1998
Composite pipe	Test strings	1998

development has been superior corrosion resistance coupled with lightweight and high strength. It is these features that have prompted the feasibility studies into the use of titanium coiled tubing for North Sea operations [4].

The use of weldable stainless steels such as the ferritic, ferritic–martensitic dual phase, and ferritic–austenitic duplex families of stainless steels in coiled tubing is in its infancy. Of these alloys, the first to meet application demands will probably be a ferritic or ferritic–martensitic stainless steel that has corrosion resistance of the 12% chromium family. The alloy will be readily weldable and will find a place in acidising operations and completion applications in corrosive CO_2-containing fields such as those that exist in the North Sea.

Several production trials with a ferritic stainless steel manufactured in up to 350 m (1150 ft) lengths have been completed and fatigue, strength and corrosion tests conducted [5]. The yield strength levels of this stainless have been measured to be in the 414 MPa (60 ksi)–517 MPa (75 ksi) range. The fatigue resistance of 1.5 and 2 in. o.d. coiled tubing was measured and the results were not markedly different from those expected from a HSLA coiled tubing of equal strength levels. The primary failure mode for coiled tubing for downhole or servicing applications is low cycle fatigue. This is the reason why low cycle fatigue tests and performance are the benchmarks from which the field performance is predicted. Initial low cycle fatigue tests were conducted and results indicate that the material will become a viable option for those applications requiring corrosion resistance similar to that of 12% chromium steel.

This 12% chromium coiled tubing (continuously milled product) is expected to be introduced into the marketplace this year. The next generation material in the

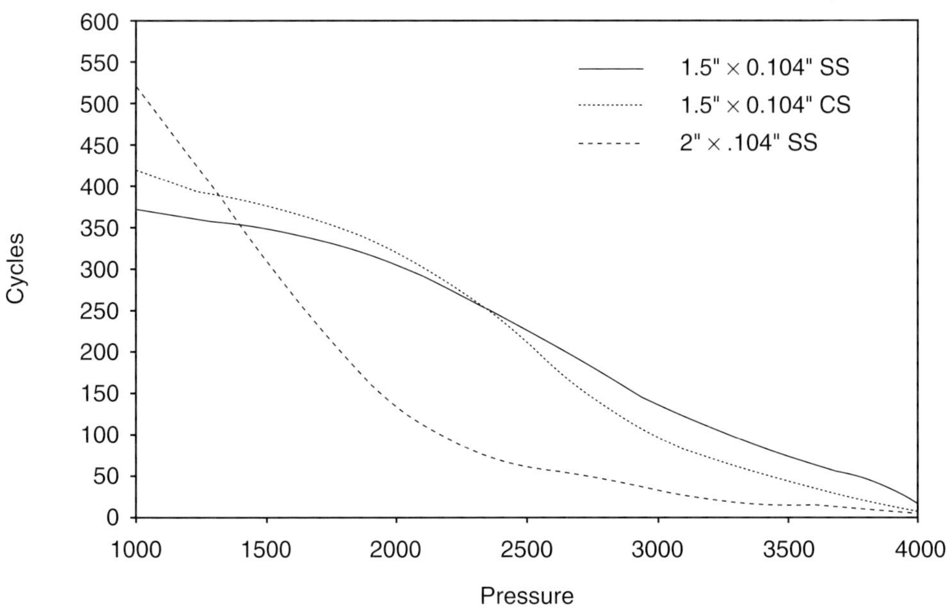

***Fig. 1** Low cycle fatigue life comparison between high strength low alloy steel (CS) and ferritic stainless steel (SS). Data based on 415 MPa (60 ksi) yield strength materials.*

corrosion resistant alloy family is believed to be a super 13–15% chromium material with corrosion resistance approaching that of a 316 stainless steel.

3. Welding of Stainless Coiled Tubing

Though we have had success in manufacturing high quality defect free stainless steel coiled tubing, the process of its development has not been without its share of problems and learning curves. The primary difficulty that we have experienced has been fusion boundary oxides that result from the high temperature welding operation.

The same feature that renders the stainless steel 'stainless', i.e. the great affinity for oxygen that encourages the formation of protective chromium oxides on the surface, also results in a much greater tendency to form oxides on the fusion boundary during the high temperature welding operation. However, unlike typical silicon manganese oxides that form during HSLA steel welding, these oxides are much more tenacious and they are not easily pressed or squeezed out of the weld seam during the high frequency induction (HFI) welding process. This affinity for oxygen increases with increasing corrosion resistance

Two basic techniques (or a combination thereof) have successfully been used with the HFI welding process to prevent or minimise the formation of these oxides. These are inert gas shrouds over the welding process and proprietary coatings. These techniques work by preventing the oxygen from reaching the heated fusion surface during the welding operation. Tests performed using both of these mitigation techniques have indicated that, though successful, there is some risk with respect to the quality level in the weld seam. Though this risk or a lower quality level may be suitable for some applications such as hang-off type production strings including velocity strings, pipeline products would require higher quality levels. Also, typical workover applications could suffer reduced product life.

The other welding processes that were examined for the manufacture of these corrosion resistant alloys were automated gas tungsten arc welding (GTAW) and laser welding (LW). The GTAW process was found to exhibit quality levels that were similar to the inert gas shrouded HFI processed stainless tubing. The highest quality levels were observed using the laser welding process. Plans have been made to evaluate further the laser welding process for coiled tubing pipeline type applications.

4. Stress Corrosion Cracking Resistance

Sulfide stress corrosion cracking (SSC) is a pervasive cracking mechanism in the oil & gas industry when the produced environment contains hydrogen sulfide. Coiled tubing made with carefully selected composition and processing variables exhibits excellent resistance to SSC cracking [6,7]. Tests have been conducted on 70, 80 and 90 ksi (483, 552 and 621 MPa) minimum yield strength grades in both transverse and longitudinal directions in NACE TM0177 [8] tests with no cracking or failures detected at 100% of the actual yield strength.

Table 2. *Longitudinal SSC tests on a new 110 ksi coiled tubing grade*

Sample description	Test	Applied stress	% of Y.S.	Results
Longitudinal tube section	4-point bent beam	99 ksi (683 MPa)	90%	Failed
Longitudinal tube section	4-point bent beam	88 ksi (607 MPa)	80%	Failed
Longitudinal tube section	4-point bent beam	77 ksi (531 MPa)	70%	No cracks
Longitudinal tube section	4-point bent beam	66 ksi (455 MPa)	60%	No cracks

Longitudinal SSC tests have been conducted on a new 110 ksi (758 MPa) coiled tubing grade using 4 point bent beam specimens. The test environment was the standard NACE TM0177 composed of 1 atm H_2S, 5% sodium chloride, 0.5% acetic acid in deionised water with an initial pH of about 2.7. The test results are presented in Table 2.

With the next highest strength grade, 90 ksi (621 MPa), hardnesses are in compliance with NACE MR0175 [9] and tensile strengths below 115 ksi (793 MPa). This grade has been marketed since mid-1996. Two types of SSC tests have been conducted on this grade, i.e. the 4-point bent beam tests and the C-ring tests in accordance with NACE TM0177 Method C (but using i.d. tensioning) and used the standard NACE environment described above. The results of these tests (all tests duplicated) are presented in Table 3.

Though all three of the coiled tubing grades with minimum yield strengths below 100 ksi (690 MPa) exhibited excellent resistance to SSC, we believe that there is an advantage in increased resistance to SSC in using lower strength grades. An example would be the use of an 80 ksi (552 MPa) instead of a 90 ksi (621 MPa) or higher grade of coiled tubing.

Table 3. *SSC tests on the 90 ksi coiled tubing grade*

Sample description	Test	Applied stress	Actual Y.S.	Results
Longitudinal tube section	4-point bent beam	91.5 ksi (631 MPa)	91.5 ksi (631 MPa)	No cracks
Longitudinal tube section	4-point bent beam	90 ksi (621 MPa)	91.5 ksi (631 MPa)	No cracks
Non-welded C-ring	C-ring	90 ksi (621 MPa)	91.5 ksi (631 MPa)	No cracks
Non-welded C-ring	C-ring	81 ksi (558 MPa)	91.5 ksi (631 MPa)	No cracks
Welded C-ring	C-ring	90 ksi (621 MPa)	91.5 ksi (631 MPa)	No cracks
Welded C-ring	C-ring	81 ksi (558 MPa)	91.5 ksi (631 MPa)	No cracks

Furthermore, we consider that, though we do not see a lowering in SSC resistance with increasing yield strength in coiled tubing grades with minimum specified yield strengths below 90 ksi (621 MPa), there is a difference in resistance that is manifested in the range of large plastic strains, and that the practical end result would be a lower life expectancy with the 90 ksi (621 MPa) and higher strength grades. We have some field evidence that suggests that the low cycle fatigue resistance of HSLA coiled tubing decreases in the presence of hydrogen sulfide-containing environments and it is probable that this decrease becomes more pronounced as the strength level increases.

5. Hydrogen Induced Cracking Resistance

For pipeline applications, another effect of diffused hydrogen called hydrogen induced cracking (HIC) must be considered. This form of cracking may occur with no applied external stress. Products made from flat steel products such as hot rolled plates, strip steels, and tubes (rolled or roll-formed) have been known to be susceptible to this cracking phenomenon.

Tests conducted have confirmed that conventional coiled tubing products generally exhibit fair to poor resistance to this type of cracking. However, tests have revealed that conventional seamless products may also exhibit poor performance and coiled tubing can be designed to exhibit good resistance. Table 4 illustrates the degree of HIC that can be observed in seamless and seamed tubulars when care is not exercised. These results were obtained from randomly selected samples from two pipe sections and examined after 96 h exposure to hydrogen sulfide in accordance with NACE TM0284 [10]. The only difference was that the test environment was the more severe environment required for testing to NACE TM0177. The CSR, CLR and CTR listed in Table 4 are the crack sensitivity ratio (CSR), crack length ratio (CLR) and crack thickness ratio (CTR) respectively and were calculated by the following formulae:

Table 4. *Hydrogen induced cracking tests (NACE TM 0184) with conventional seamless and coiled tubing*

Description	*CSR %	†CLR%	§CTR%
Seamless API 5L X52	2.38%	11.83%	20.12%
Seamless API 5L X52	5.99%	17.12%	35.00%
Seamless API 5L X52	1.31%	8.31%	15.75%
X52 type coiled tubing	6.97%	36.49%	44.01%
X52 type coiled tubing	7.53%	34.26%	44.09%
X52 type coiled tubing	16.31%	67.02%	47.16%

*CSR = crack sensitivity ratio;

†CLR = crack length ratio;

§CTR = crack thickness ratio.

$$CSR = 100\% \times (\Sigma\, (a \times b))/(W \times T)$$

$$CLR = 100\% \times (\Sigma\, a)/W$$

$$CSR = 100\% \times ((\Sigma\, b)/T$$

a = crack length;
b = crack thickness;
W = section width;
T= test specimen thickness.

The steels used to make coiled tubing can be selected to avoid centre segregation that has been found to be the primary cause of the susceptibility to HIC type cracking. For example, a production quantity of an X70 type, 70 ksi (483 MPa) minimum yield strength, coiled tubing was tested from a selected heat of steel with the results given in Table 5.

These test results demonstrate that the HIC resistance can be excellent with coiled line pipe or tubing products. The good and bad results presented here demonstrate that care must be taken to ensure good resistance regardless of the product form (seamless, seamed, or coiled tubulars).

6. Conclusions

The coiled tubing products continue to evolve to meet market demands. Corrosion resistant alloys, coated pipe, lined pipe, and HIC resistant coiled pipe are either recent innovations or in the initial phase of product development.

References

1. J. A. Campbell, SPE 36339, "Coiled Tubing Management System: A Review After Two Years in Operation", *1st SPE/ICoTA North American Coiled Tubing Roundtable*, World Oil, Conroe, Tx, February, 1996.
2. R. K. Stanley, "Failures in Coiled Tubing", *5th International Conference on Coiled Tubing and Well Intervention*, World Oil, Houston, Tx, February, 1997.
3. L. Smith and J. Misselbrook, SPE 38413, "An Overview of Inspection Technology and Criteria for Inspecting Coiled Tubing", *2nd SPE/ICoTA North American Coiled Tubing Roundtable*, World Oil, Montgomery, Tx, April, 1997.
4. M. Christie and B. Gaven, SPE 38416, "Titanium as Alternative to Conventional Coiled Tubing: A North Sea Case Study", *2nd SPE/ICoTA North American Coiled Tubing Roundtable*, World Oil, Montgomery, Tx, April, 1997.
5. R. P. Badrak, "Coiled Tubing Innovations For Corrosive Service", *EUROCORR '97*, European Federation for Corrosion, Trondheim, Norway, September 1997. This volume p. 149.
6. M. S. Cayard and R. D. Kane, SPE 38410, "Serviceability of Coiled Tubing for Sour Oil and Gas Wells", *2nd SPE/ICoTA North American Coiled Tubing Roundtable*, World Oil, Montgomery, Tx, April, 1997.
7. R. P. Badrak, "High Strength Materials for Coiled Tubing", *6th Int. Conf. on Coiled Tubing Technologies*, 27–29 October, 1997, Houston, Tx.

8. NACE Standard TM0177, 'Test Method — Laboratory Testing of Metals for Resistance to Sulfide Stress Cracking in H_2S Environments'. NACE International, Houston, Tx.
9. NACE Standard MR0175, 'Standard Material Requirements — Sulfide Stress Cracking Resistant Metallic Materials for Oilfield Equipment'. NACE International, Houston, Tx.
10. NACE Standard TM 0284, 'Test Method — Evaluation of Pipeline and Pressure Vessel Steels for Resistance to Hydrogen-Induced-Cracking'. NACE International, Houston, Tx, 1996.
11. J. G. Haimbl and J. Geiser, "Polyethylene Coatings in Europe: a Look Back on 30 Years of Experience', Prevention of Pipeline Corrosion Conference, Houston, Tx, 31 October–2 November, 1995 (Gulf Coast Publishing).

15
Coiled Tubing Innovations for Corrosive Service

R. P. BADRAK
Precision Tube Technology, Inc., P.O. Box 24746, Houston, Tx 77229, USA

ABSTRACT
A broad view of coiled tubing operations and environmental considerations is presented. This overview is introduced using a historical perspective of where we have been, where we are and what we may be able to expect from this emerging technology. The specific topics of the use of polymeric materials, corrosion resistant alloys, and composite materials to mitigate corrosion are addressed. Field applications of stainless steel coiled tubing, outside diameter three layer coatings and polymeric liners are presented.

1. Introduction and Overview

Coiled tubing has been used successfully for over 30 years in well servicing and workover applications. These applications have been increasing in recent years as operators and end users discover new ways to reduce operating costs and perform certain tasks which had been previously impossible or difficult to perform. Coiled tubing has expanded into such diverse activities as drilling, well logging, gas lift, acidising/fracturing, injection, production strings (including velocity strings), and sand control. Coiled tubing has also been used more recently (last 2–3 years) for flowline and pipeline applications.

The drop in oil prices in the mid 1980s spurred on development in areas, including coiled tubing, that would permit more economic development and production of oil and gas. This development was evident in the active programmes in operating and end user companies. These new applications for coiled tubing have resulted from the combination of better coiled tubing equipment, better operating procedures and the advances made in the materials and quality.

The advances in coiled tubing diameter, length, strength level, quality, and corrosion resistance have in many instances been the impetus behind new successful applications. These coiled tubing advances have been accelerating in recent years; for example, the continuous tubing length between butt welds has increased from 15 m (50 ft) lengths in the mid-1960s to 450 m (1500 ft) lengths in 1980 and over 7300 m (24 000 ft) lengths in 1990. The increases in coiled tubing diameter and strength level as a function of year available are presented in Fig. 1.

From Fig. 1 it can be seen that strength levels up to 100 ksi are available and that tubing sizes up to almost 7 in. (actually 6 $\frac{5}{8}$ in o.d. from Precision Tube Technology) are available today.

In the 1996–1997 time period, estimates of 25% [1] and 37% [2] have been put forward as the percentage of failures that were attributable to corrosion. From a and chemical resistance, buckling, surface wear/damage and field repairability.

The Hydril program utilises a combination of pultrusion and filament winding [7] whereas Fiberspar utilises a braided pultruded product [8]. Both JIPs are looking at similar liners and resins. Hydril is using carbon and glass filaments for their products whereas Fiberspar uses carbon, glass and aramid filaments. The Fiberspar JIP is more advanced than the Hydril programme in that tubing samples have been tested by Conoco and Halliburton and the first actual field trial of a 38 mm (1.5 in.) o.d. was scheduled for June 1997 [8]. Hydril has scheduled a manufacturing date of October, 1997 for a 1500 m (5000 ft) test run of a 38 mm (1.5 in.) o.d. fibre–epoxy resin composite [7].

The composites will ultimately fill a niche market. However, the immediate problems that will need to be addressed are (i) composite tube handling/surface damage, (ii) temperature limitations of approximately 100°C (212°F), (iii) elevated temperature aging/creep, and (iv) cost.

4. Lined Coiled Tubing

Thomas [5] has described five techniques for lining pipe; these are sliplining, modified sliplining, fold and formed, spiral wound, and deformed and formed. Two techniques have been examined for lining coiled tubing: sliplining and pipe reduction techniques (modified sliplining). Medium density polyethylene has been successfully lined and spooled into coiled tubing by Precision Tubing Technology. The current limit for continuous lengths of lined coiled tubing without joints is in the 600–1000 m range depending upon pipe diameter.

The following are current development subjects in lined coiled tubing: (i) connection technology, (ii) polyamide 11 and polyvinylidene difluoride (PVDF) type thermoplastic materials for linings, (iii) increasing the length of lined coiled tubing between joints and (iv) co-extruded thermoplastics.

For most corrosion resistant applications, where polyethylene is not suitable because of its temperature and permeability limitations, it is believed that polyamide 11 will become an economical alternative material. With polyamide 11, the primary failure mode will be hydrolysis. A published relation between hydrolysis rate and temperature for polyamide 11 is presented in Fig. 2 [9].

From Fig. 2 it becomes apparent that the polyamide 11 can only be used at temperatures lower than 100°C. It has been estimated that the life would be expected to be about 9000 days (24.7 years) for this material at 80°C [9] .

5. Externally Coated Coiled Tubing

Though a variety of external coatings have been used as discussed previously, this investigator believes that the three layer (epoxy plus polyolefins) will be the external coating of choice for corrosion resistance, damage protection and cost. In Continental Europe, two and three layer polyethylene coating systems have been successfully applied to pipelines for over 30 years [10]. These coatings have resulted in very low current requirements with the cathodic protection used for buried pipelines.

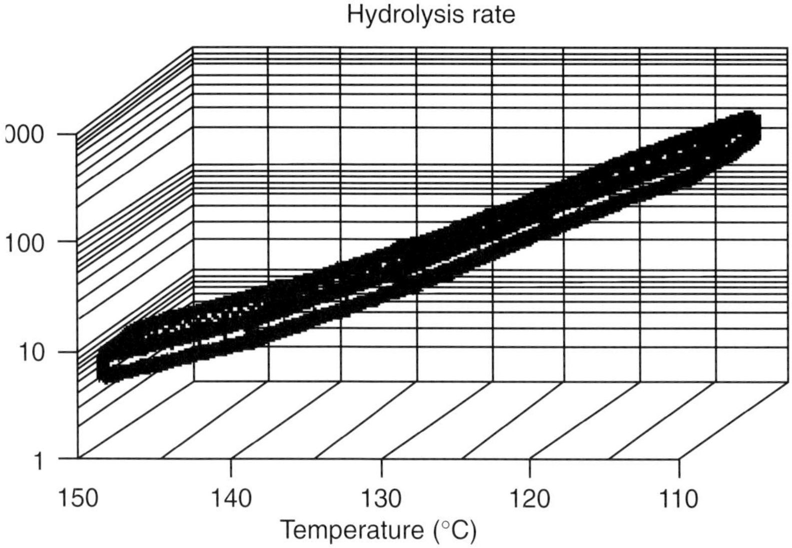

Fig. 2. Hydrolysis rate as function of temperature (after Mason [9]).

The three layer coating comprises a fusion bonded epoxy primer, adhesive intermediate layer, and polyethylene or polypropylene top layer. The epoxy primer or initial coat provides corrosion resistance with excellent adhesion to the steel pipe and tube. The adhesive layer is necessary because the polyolefin does not exhibit good adhesion directly onto the epoxy or the steel. The polyolefin layer provides additional resistance to corrosion and is the layer that can withstand mechanical damage.

One coiled tubing manufacturer has the internal capability to apply fusion bonded epoxy and co-extrude copolymer (or adhesive layer) and polyolefins directly onto coiled tubing during the manufacturing process in both North America and Europe. Up to January 1999, this company has shipped over 300 000 m of coiled tubing coated by the three layer process.

6. Conclusions

From the state-of-the-art of coiled tubing for corrosion resistant applications, the following conclusions have been made:

1. Pipeline applications will accelerate and the primary external corrosion control will be three-layer (epoxy plus polyolefins).

2. Internal liners will be used for some downhole shallow injection/gas lift operations.

3. Internal liners will see extensive use in flowlines and small diameter pipelines for production, transportation and refinery applications.

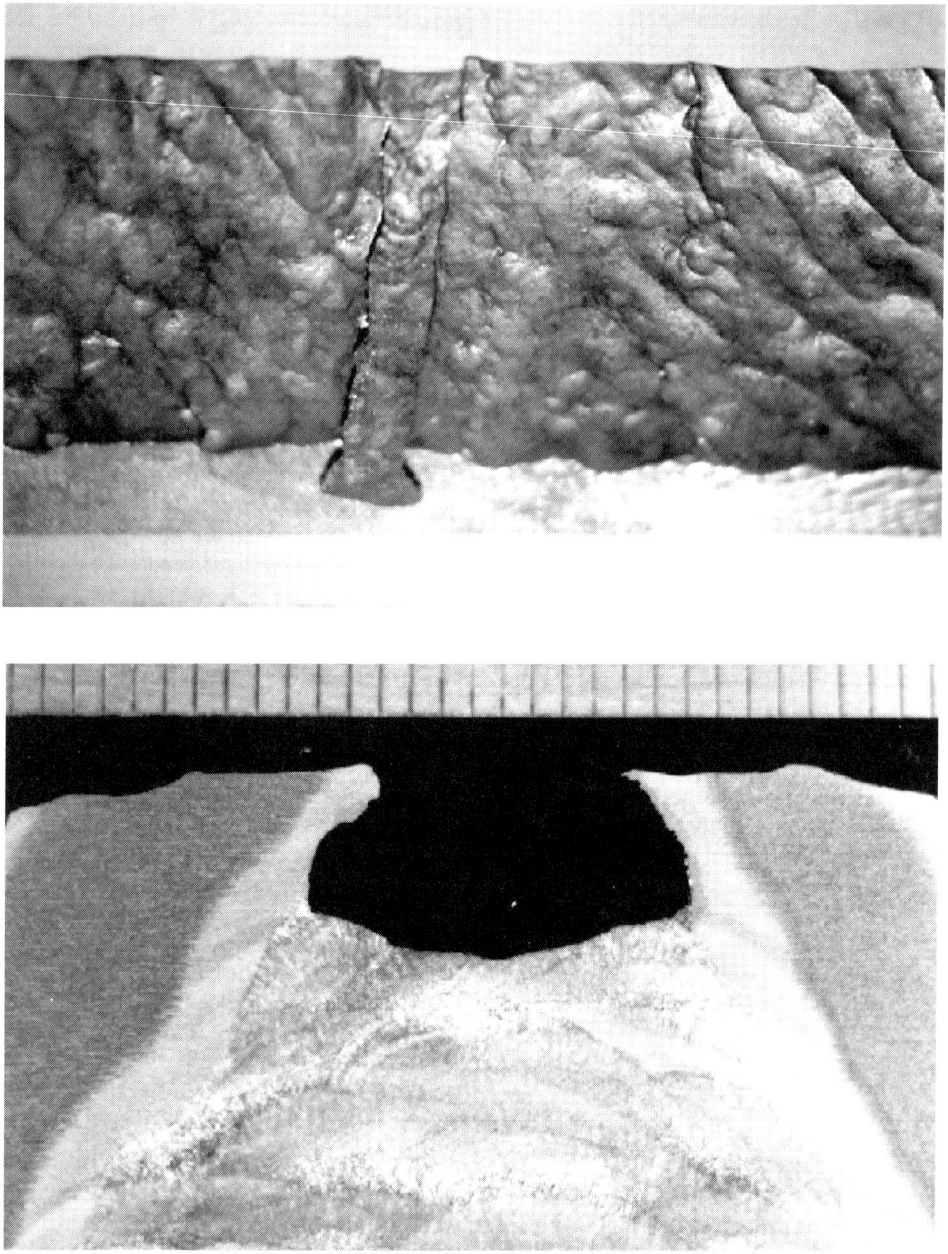

***Fig. 1** Example of preferential weld corrosion in a CO_2 environment.* *(Resolution of original photographs supplied at 72 dpi.)*

affected zone (HAZ) may be mitigated by modifying the alloy composition of the parent to a more noble potential and by control of the microstructure through specification of the thermal cycle. The typical morphologies of preferential weldment corrosion, depicted in Fig. 2, are influenced by several factors. These include the

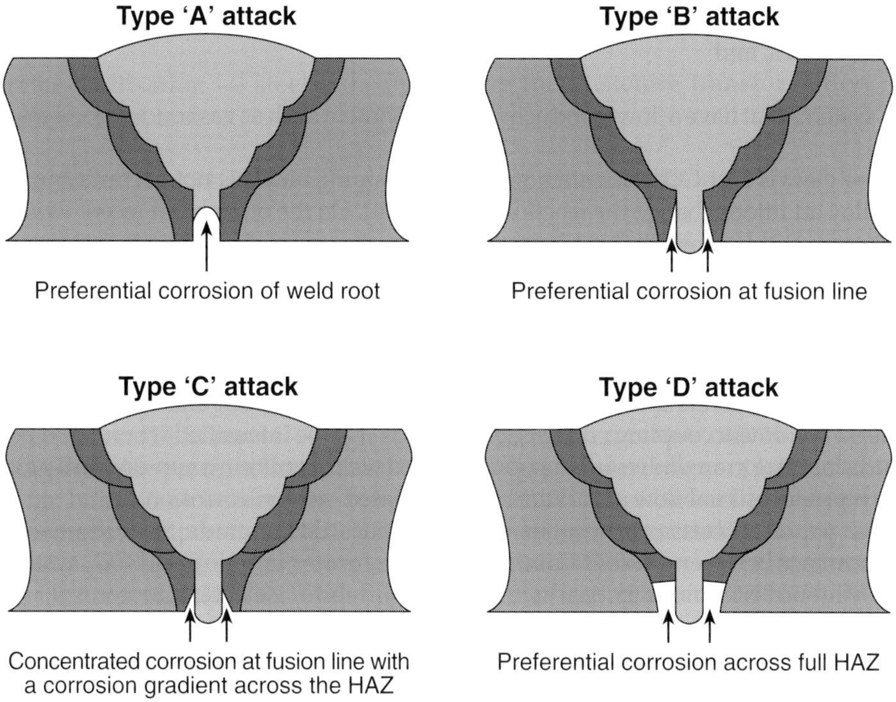

Fig. 2 *Typical morphologies of preferential weldment corrosion.*

environment, the flow conditions, the parent steel composition, the weld composition and the weld procedure; changes in any one can cause a significant difference in corrosion behaviour.

Early weld corrosion failures due to galvanic effects were overcome by addition of Cu, Cr, Ni and Mo to the weld consumable to produce a cathodic (noble) weld. However, by the 1960s problems of HAZ corrosion and fusion line attack, such as 'tramline' corrosion in sulfuric acid, became more common. Weld procedures were therefore further developed that not only optimise the strength requirements and provide requisite hardness levels to prevent post-weld cracking, but also introduce suitable compositions and structures appropriate for the corrosive environment.

This approach was also relatively successful for the control of weldment corrosion in sea water injection systems during the late 1980s and early 1990s. For example, a joint industry sponsored project [10] conducted by CAPCIS and The Welding Institute during the late 1980s, with sea water tests under flowing conditions on a range of consumables and weld processes, concluded that a weld deposit with about Ni 0.6%–Cu 0.4% or Ni 1% should be satisfactory, with low silicon being preferred.

The use of Ni-containing welds has also been adopted by some operators in sweet production environments. Although this has generally led to satisfactory performance, high corrosion rates have been reported in some systems [11]. Laboratory tests have also identified a range of performance with different inhibitor packages, some of which even accelerated preferential corrosion on the weld or HAZ [12]. An evaluation of such problems [13] indicates that two major operational environments of concern can be identified:

4. Measurement Procedures

Early testing of weld corrosion was by weight loss or surface profile measurements from long term exposure to natural/artificial sea water or CO_2-containing brines. The practical exposure results were subsequently confirmed by rapid assessment electrochemical techniques. These include coupling to an anode to maintain a fixed potential, on the assumption that corrodability is proportional to anodic dissolution rate [24], potentiodynamic polarisation to predict corrosion currents for galvanically coupled mixed metal systems [25] and ZRA current data from sectioned welds [26,27].

One of the most successful, and now widely used, techniques combines electrode probes and specialised electrochemical instrumentation based on coupled zero resistance ammeters [28–30]. This approach, proposed by Eden and used to assess flow induced galvanic corrosion [31], was first employed to assess weld corrosion in 1984 [32] and has subsequently been developed for laboratory testing and on site monitoring.

Samples of appropriate weldments are first etched to reveal the weld metal and HAZ regions (Fig. 4). Individual components are then sectioned from the parent plates/weldment. Test electrodes are prepared by spot welding electrical connections to the back of each segment and encapsulating them, at equal spacing, using inert mounting compound into electrode holders. Once mounted into the test system or insert probe, each segment of the weldment is coupled through a zero resistance ammeter. This allows galvanic coupling currents and corrosion rates on individual sections of the weldment to be monitored, either freely corroding in the test solution, Fig. 5 (a) and (b) or polarised by a conventional potentiostat arrangement. Typically, coupling currents and corrosion rates on individual parts of the weld are logged throughout the tests. At the end of a test, depth of attack measurements are compared with electrochemical data. Normal procedures for deaeration (< 20 ppb O_2) and pH control are observed in the test loop or rotating cylinder apparatus employed to

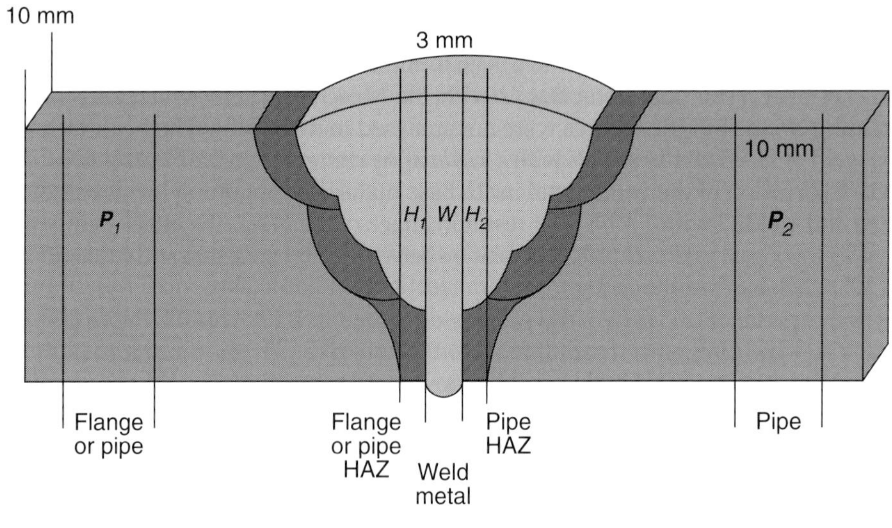

Fig. 4 Weldment sections used as electrode elements.

simulate line flow conditions. The same electrochemical monitoring procedure has been used to assess the effect of biocides and inhibitors in the field.

Weld corrosion studies have been successfully employed to assess the risk of preferential weld corrosion in simulated service environments, to evaluate the effects of flow and low conductivity and to assess the effectiveness of inhibitors. If the weld metal or HAZ is found to be anodic to the parent pipe, this indicates possible susceptibility to preferential corrosion. However, a cathodic weld metal is not necessarily an indicator of a 'safe' metallurgy/environmental combination.

The effects of weldment microstructure on iron carbonate scale retention and the influence of inhibition on corrosion rates is similar to those found by Schmitt *et al.* [17] under flow induced conditions. Their conclusions are that intrinsic stresses within the growing scale give rise to fracture and loss of adhesion; a ferritic-pearlitic substrate being more able to resist than a martensitic–bainitic. The lack of pearlite influences the performance of some inhibitor packages and may account for the observation of high corrosion on some weldments. Although there are techniques available to evaluate the risk of preferential attack in specific environments, failures observed in the field illustrate the danger of attempting to assume that a 'safe' weldment in one environment can be applied under different conditions. An important factor influencing weld corrosion behaviour is the environment conductivity; this is illustrated by the test results described below.

Initial tests in a formation water environment showed that preferential corrosion of weld or HAZ segments did not occur, although the corrosion rate of the parent pipe 1 was significantly lower than that of parent 2 probably because of its higher Mo content (Fig. 5 (a) and (b)). However, tests on the same weldment in a 'dilute' brine (10% concentration of the formation water composition) gave a different result Fig. 5 (c). In this case, the weld was again cathodic but preferential corrosion of the weld metal was observed [33]. This indicated that in the 'dilute', lower conductivity environment, there was a significantly greater risk of preferential corrosion than in the tests with conducting formation water. It was not thought that these differences in behaviour were related to differences in scaling tendency. These tests indicated potential problems if a future tie-in results in dilution of the formation water with condensed water.

5. Improving the Capability to Predict Preferential Corrosion

Preliminary modelling studies demonstrated the effect of environment conductivity [34] and this approach has been further developed in order to improve the capability to extrapolate data and assess various environments. Prediction of the galvanic current 'throwing power' at a weld is based around the ability of the parent plate to provide a protection current, as illustrated in the Evans diagram (Fig. 6), which shows interactions of electrode kinetics and solution conductivity.

When the weld and parent plate are uncoupled and in a conductive electrolyte they will adopt corrosion potentials, E_{w1} and E_{p1}, separated by the potential difference ΔE_1. The corrosion current on the weld, i_{w1}, is greater than that on the plate, i_{p1}. (Note that the current density would be even higher on the weld but relative areas would be difficult to represent on the diagram.) Coupling the two electrodes together is

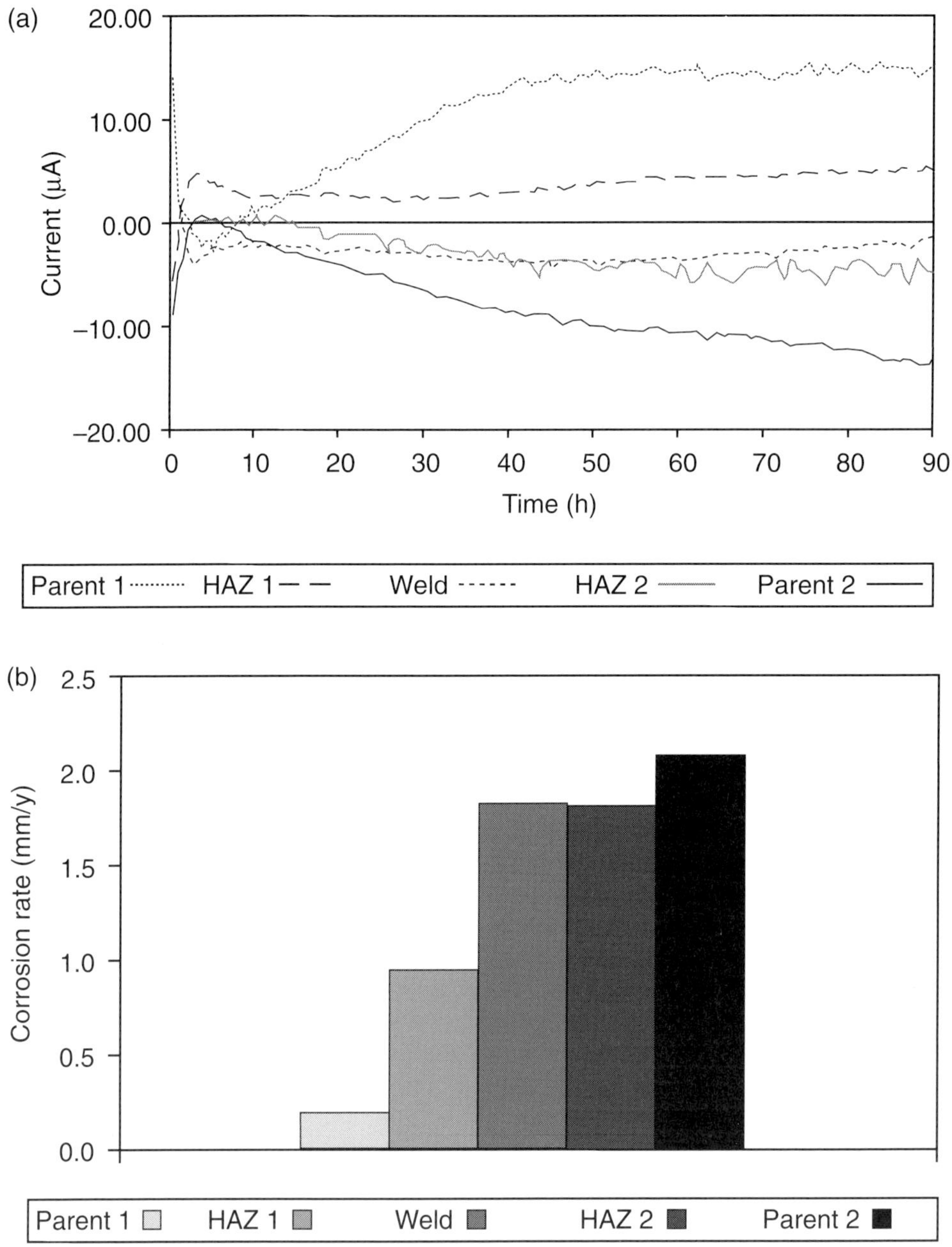

Fig. 5 (a) Current (zero resistance ammeter) vs time in formation water, 1 bar CO_2, 71°C, and (b) corrosion rates for these materials.

essentially the same as a sacrificial anode cathodic protection system, the current from the anodic plate increases and the corrosion rate of the cathodic weld decreases (the current densities should now be somewhat similar).

In a 'low' conductivity electrolyte, the IR effect will polarise the anodic and cathodic reactions on both uncoupled electrodes (dashed lines and intersections w_2 and p_2).

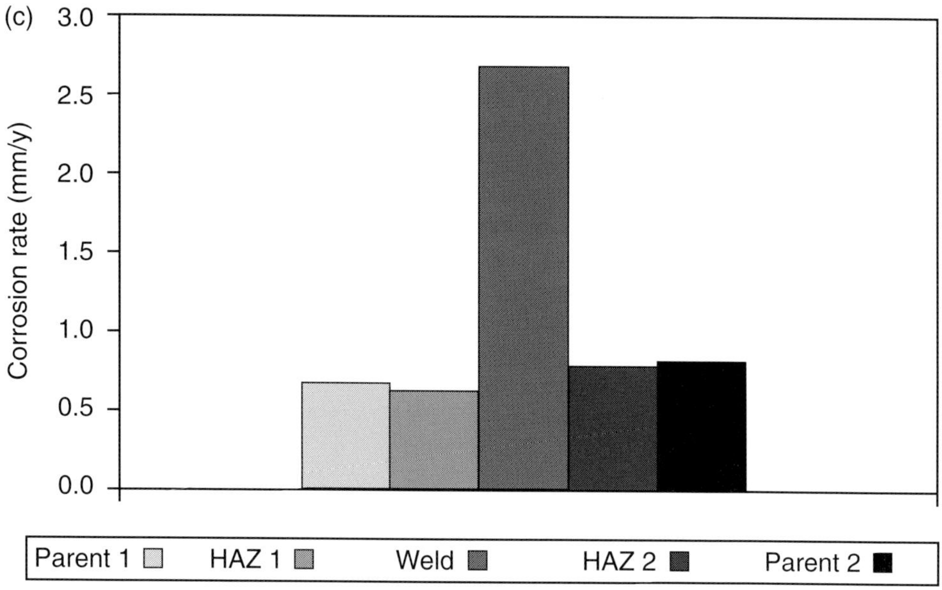

Fig. 5 *(c) Corrosion rates , in 'dilute' formation water, 1 bar CO_2, 71°C.*

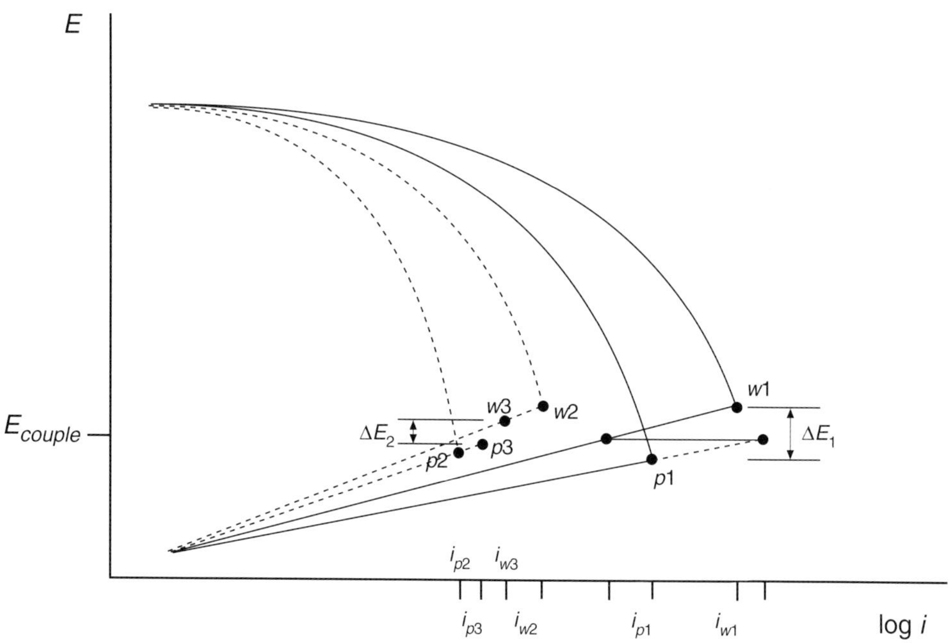

Fig. 6 *Schematic Evans diagram depicting corrosion of a Ni-containing weld.*

Coupling the electrodes will also mean a further *IR* effect on the galvanic current, thus although the electrode potentials will move towards each other, there will still be a potential difference, ΔE_2, between w_3 and p_3. The weld remains cathodic but still corrodes at a higher rate than the plate. The worst case is when the solution resistance is sufficiently high to maintain the potential difference, $E = IR$, at a value equivalent to ΔE_2. Note the logarithmic current scale which can mean in practice that apparent minor differences on the axis equate to a change of several mm/y in corrosion rate. The cathodic protection offered by the parent plate will depend on factors such as the voltage difference between the weld and parent, the relative Tafel slopes of the electrodes, the specific resistivity of the corrosive medium and the configuration of the weld and parent plate (a thin condensed liquid film or a fully flooded pipe).

These situations are analogous to cathodic protection current and potential distribution within a crevice as initially modelled by Schwenk [35], but here the electrolyte layer on the pipe is the same as the width of a crevice and the depth of the crevice is the distance from the interface at the weld to plate. This analysis provides the potential distribution along the weld metal surface, i.e. from the plate interface, as a function of the current density distribution for a given solution resistivity.

The model [36] then requires a valid value of the maximum current density that can be provided by the parent plate at the interface. This will again be determined by the electrolyte resistance but also by the Faradaic resistance (or polarisation resistance) of the anode plate along its surface away from the weld interface. The analysis is achieved by means of an equivalent resistance network or matrix. The number of resistance components or sequences included in the computational network will depend on the relative values of the polarisation and solution resistances and currents involved. The analysis, involves an iterative recalculation process achieved by a computational determinant method to obtain the total protection current that is provided by the plate area adjacent to the weld. A similar potential distribution is then employed to determine the corrosion rate of the weld using the anodic Tafel value.

Corrosion rate data of 0.75 to 5.5 mm/y, obtained from de Waard–Milliams predictions [37] over a typical CO_2 partial pressure range, i.e. 0.5 to 10 p_{CO_2} (bar), were used to make a preliminary assessment in the model to predict the effect of environment conductivity on weld corrosion. It was further assumed that the specific resistance of the corrosive media was between 10–10 000 $\Omega\,cm^{-2}$. A typical distribution across a weld for a potential difference of 40 mV between plate and weld is shown in Fig. 7. The low conductivity electrolyte is unable to pass a protective current whereas in a conductive brine and especially in a filled pipe, the galvanic coupling could mitigate the corrosion of the weld.

6. Conclusions

Although general conclusions on the causes and preventative measures must be made with caution in such a complex subject area, some practical information is summarised as:

- Nickel-containing weldments provide required strength but have a high risk of preferential corrosion of the weld metal in CO_2 systems with low conductivity environments as found in wet gas pipelines and condensate lines/pipework.

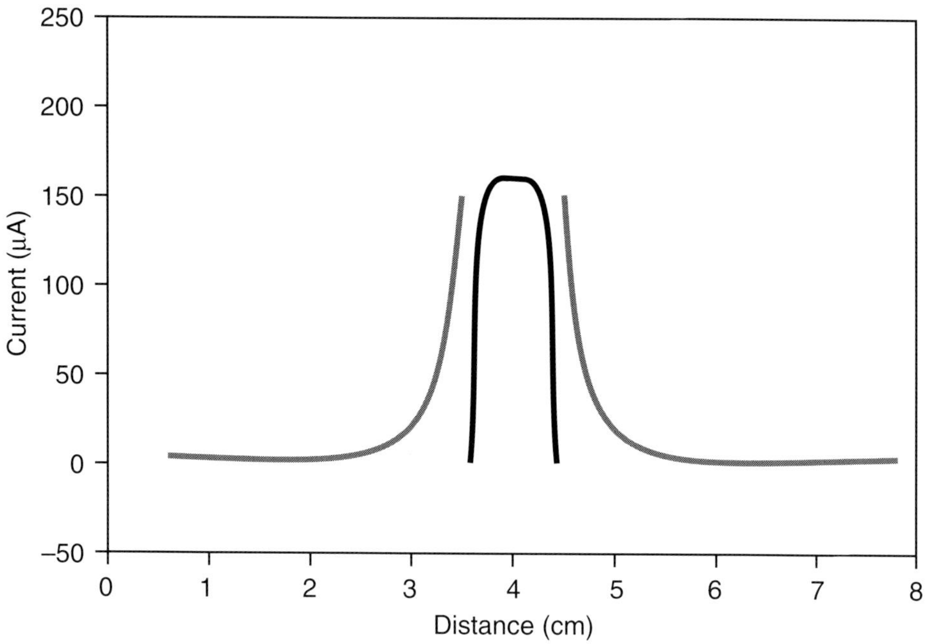

Fig. 7 *Corrosion prediction of a Ni-containing weld in a low conductivity environment.*

- There is also a risk of preferential weld metal attack in more conductive environments under conditions which favour formation of semi-protective corrosion product scales, e.g. 70–80°C, pH > 5. High flow rates appear to exacerbate the problem and this may be due to poor adhesion of iron carbonate scales formed on specific weldment microstructures.

- The risk of preferential corrosion of the HAZ can be reduced by using 1% Ni consumables with unalloyed parent pipe and welding with high heat input (> 1.0 kJ mm^{-2}) and high preheat temperatures (180°C). With these restrictions, nickel welds can be used successfully in high conductivity environments at temperatures below 60°C provided high local turbulence does not occur.

- In tests to evaluate preferential weld corrosion behaviour, it is recommended that flow conditions are simulated. It should also be noted that a cathodic weld does not necessarily indicate a safe weld metallurgy/environment combination.

- Inhibition can be used to control even the most severe preferential weld metal corrosion but specific packages must be first assessed under appropriate simulated field conditions as some inhibitors do not protect all weldment components. This is a particular problem at high fluid to wall shear stress levels.

- Experience from site and laboratory data shows that preferential weld corrosion behaviour is very specific to the environment conditions, it is therefore not advisable to extrapolate results that may be relevant in other systems to a particular application.

- The potential mode of failure that can result from weld corrosion, a local circumferential groove, highlights the need to predict environments of concern and to identify suitable inhibitors that are most effective at protecting against preferential corrosion.

- The capability to extrapolate data could be improved by the further development and evaluation of a model to understand the influence of environment conductivity on preferential corrosion. The basis for such a model has been described.

References

1. G. A. Nelson, *Corrosion*, 1958, **14**, (3), 415.
2. E. Rasane, and K. Relander, *Scand J. Met.*, 1951, **7**, 11.
3. J. N. Banbury and J. C. Rowlands, *Brit. Weld. J.*, 1962, **9** (8), 476.
4. G. Bailey and A. N. Rothwell, Weld corrosion in oil and gas production, *Conference on Downhole Corrosion*, Oyez, London, 1984.
5. A. Baumel, *Werks. Korros.*, 1972, **23** (7), July.
6. H. Lajain, *Werks. Korros.*, 1972, **23**, 5375–45.
7. R. Bhavsar, *Corrosion '87*, Paper No. 229, NACE International, Houston, Tx, 1987.
8. Z. A. Foroulis, and H. H. Uhlig, *J. Electrochem. Soc.*, 1964, **111**, 522–528.
9. A. Takamura *et al.*, *Corrosion '72*, Paper No. 466, NACE International, Houston, Tx, 1972.
10 "Preferential Corrosion of Carbon Steel Weldments in Oil and Gas Production Systems". Joint Industry Project, CAPCIS/TWI, 5563/28/89 (1989).
11. CAPCIS Report HJPKHN, January (1989).
12. J. W. Palmer, J. L. Dawson, C. M. Fowler and G. Dicken, Challenges for corrosion inhibition. *Corrpipe '96 Conf.*, Aberdeen, Pub, IIR, 1996.
13. CAPCIS Report HJPKEN, Janaury 1989.
14. A. N. Rothwell, PhD. Thesis, University of Manchester, 1992.
15. A. D. Krijgsman, private communication from Shell report. *'Metallurgical Aspects of Preferential Weld Attack of Carbon Steels in* CO_2 *Environments'*, June 1996.
16. MR0175, Pub. NACE, Houston, Tx.
17. G. Schmitt, T. Gudde, and E. Strobel-Effertz, *Corrosion '96*, Paper No. 9, NACE International, Houston, Tx, 1996.
18. G. Schmitt and B. Labus, *Corrosion '94*, Paper No. 37, NACE International, Houston, Tx, 1994.
19. A. Dugstad, L. Lunde, Paper No. 473, *Corrosion '91*, Pub. NACE International, Houston, Tx, 1991.
20. K. Denpo and H. Ogawa, *Corros. Sci.*, 1992, **20**, 285–288 [Corrosion & Protection Centre 20th Anniversary Conference, UMIST.]
21. M. W. Joosten and G. Payne, Preferential corrosion of steel in CO_2 containing environments, *Corrosion '88*, Paper No. 211, NACE International, Houston, Tx, 1988.
22. M. W. Joosten, J. Kolts, P. G. Humble, J. H. Kiefer, and J. Marlow, Aspects of selective weld and HAZ attack in CO_2 containing production environments, *Corrosion '96*, Paper No. 79, NACE International, Houston, Tx, 1996.

23. S. Olsen, B. Sundfaer, and J. Enerhaug, Weld corrosion in C-Steel Pipelines in CO_2 Environments — Comparison Between Field and Laboratory Data, *Corrosion '97*, Paper No. 43, NACE International, Houston, Tx, 1997.
24. R. Raty, V. O. Valanti *et al.*, *Acta Polytech. Scand.*, Series No. 78, 1968.
25. H. P. Hack and J. R. Scully, *Corrosion*, 1986, **42**, 2.
26. C. Kato *et al.*, *Tetsu-To-Hagane*, 1977, **63**, (1), 130–138.
27. K. G. Compton, and J. A. Turley, *Galvanic & Pitting Corrosion — Field and Laboratory Studies*, ASTM STP 576, 56–68, 1976, ASTM, West Conshohocken, Pa.
28. J. J. Dawson, J. S. Grill, A. A. Al-Zanki and R. C. Woolam, DECHEMA Monograph, **101**, 235, Pub. VCH, 1986.
29. J. W. Palmer, J. L. Dawson, T. Ulrich, and A. N. Rothwell, *Corrosion '93*, Paper No. 119, NACE International, Houston, Tx, 1993.
30. A. N. Rothwell, J. L. Dawson, D. A. Eden and J. W. Palmer, Interpreting electrochemical impedance spectra from segmented electrodes. *Electrochemical Impedance: Analysis and Interpretation*. ASTM STP 1188, 237–251, 1993, ASTM, West Conshohocken, Pa.
31. R. C. Woollam, Msc Dissertation, UMIST, 1983.
32. A. N. Rothwell, Msc Dissertation UMIST, 1984.
33. CAPCIS Report CGDJGB Dec. 1996.
34. J. W. Palmer, J. L. Dawson, H. Al-Ajwad, S. Turgoose and R. Cottis, Developments in corrosion inhibitor evaluation techniques, *8th Middle East Corrosion Conf.*, Bahrain, May, 1988.
35. W. Schwenk, *Corros. Sci.*, 1983, **23**, (8), 871–886.
36. P. J. Moreland, J. C. Dawson and J. W. Palmer, to be presented at the *Int. Corr. Congr.*, Cape Town, Sept. 1999.
37. C. de Waard, U. Lotz and A. Dugstad, *Corrosion '95*, Paper No. 128, NACE International, Houston, Tx, 1995.

17
Experimental Simulation of Multiphase Flow for Assessing System Corrosivity

S. SRINIVASAN and R. D. KANE
InterCorr International, Inc. (formerly CLI International, Inc.), 14503 Bammel-N, Suite 300, Houston, Tx, USA

ABSTRACT
Multiphase systems present an imposing challenge from a standpoint of corrosion evaluation and prediction because of the need to capture synergistically the roles of important environmental, flow and metallurgical variables and underlying mechanisms of corrosion. This paper focuses on a methodology to model fluid flow in order to assess wall shear stress in multiphase systems. Methods of determining wall shear stress for differing flow regimes are discussed. An experimental setup for multiphase flow and shear stress simulation is also covered.

1. Introduction

Carbon dioxide/hydrogen sulfide corrosion in oil and gas production environments represents one of the most important areas of corrosion research because of the critical need to assess corrosion severity in oil and gas production operations. The question of assessing corrosion in multiphase systems requires, in addition to accurate modelling of the electrochemical effects of different corrosion species, an effective method to model the contribution of the flowing medium to corrosion acceleration or mitigation.

Numerous predictive models have been developed and reported [1–5]. However, many of these models are very conservative or focus only on a narrow range of parametric effects, thereby limiting their scope of applicability. The background to the current research effort has been extensively described elsewhere [6–8]. The Predict™ model [6–8] was developed in order to address the following concerns:

- Predicting corrosion severity by using commonly available parameters.
- Using existing laboratory/field data and theoretical models for realistic assessments of corrosivity and corrosion rates.
- Developing a computational/experimental approach that integrates both numerical ('laboratory trends') and *heuristic* (field data and experience) information and knowledge about corrosivity prediction.

This paper describes further research undertaken in critical areas of corrosion prediction for multiphase CO_2/H_2S systems. The work described here and the data obtained will form the basis for improving and augmenting the existing corrosivity prediction model.

In this paper, a methodology for assessing and experimentally simulating fluid flow induced corrosion in multiphase systems is described. This description forms an integral part of ongoing research and test system development that focuses on the following tasks in corrosion prediction:

1. Simulation/modelling of flow parameters and regimes to assess flow-related effects in multiphase systems.

2. Characterisation of H_2S corrosion and scaling in multi-phase CO_2/H_2S systems (scale nucleation, morphology and stability as a function of temperature, pH and acid gas content).

3. Evaluation of the role of hydrocarbon chemistry and its persistence in promoting/inhibiting corrosion (assessment of conditions for creation of an oil wet as opposed to a water wet metal surface).

4 Examination of the role of steel metallurgy and composition in defining corrosion resistance.

Task 1 forms the focus of further discussion in the paper; tasks 2, 3 and 4 have been described elsewhere [9].

2. Flow Effects in Multiphase Systems: Background

Corrosion in multiphase systems stems from the electrochemical interaction of the corrosive species (CO_2, H_2S, etc. dissolved in an aqueous medium) with steel leading to metallic oxidation and subsequent metal loss. This situation is significantly exacerbated through flow-related effects in multiphase systems where high wall shear stresses developed as a consequence of different flow regimes can lead to accelerated corrosion. Experimental determination of corrosion as a consequence of fluid flow involves

(i) determination of the prevalent flow regime in terms of gas/liquid interaction and

(ii) characterisation of the fluid flow in terms of the wall shear stress and/or a mass transfer coefficient as well as dimensionless parameters such as Reynold's number, Sherwood Number, etc.

Mass transfer and corrosion are inextricably linked in multiphase systems. Mass transfer relates to the transport of various corrosive species from the bulk fluid medium to the metal surface. Because of the electrochemical reactions progressing as a consequence of corrosion, there exists a concentration gradient from the bulk to the pipe wall. This gradient promotes movement of corrosive species from the bulk to the pipe wall. Since fluid flow has a significant impact on the rate of mass transfer, and since it has been shown that the corrosion rate is directly proportional to the turbulent kinetic energy of the flowing medium [10], modelling the flow in terms of the wall shear stress has shown this factor to be directly proportional to the potential corrosivity of the medium.

Several researchers have reported the importance of wall shear stress in influencing corrosion in different flow regimes and have developed empirical/ numerical relationships to determine corrosion as a function of wall shear stress developed in a flowing system [11,12]. Efird *et al.* [13] performed experiments with three different systems including a horizontal pipe flow system, a jet impingement set-up, and a rotating cylinder electrode apparatus. In all three cases, they established a correlation between corrosion rate and wall shear stress as follows:

$$R_{COR} = a\tau_w^{\ b}$$

where R_{COR} = corrosion rate in mm/year, τ_w = wall shear stress in Pa, and a and b are constants.

Jepson *et al.* [11,12] have also extensively reported on the relationship between wall shear stress and corrosion rates in multiphase systems in slug flow.

Any efforts to characterise corrosion in flowing systems require modelling mass transfer and momentum transfer effects. The mass transfer relates to the movement of corrosive species between the pipe wall and the bulk flowing medium, and wall shear stress represents momentum transfer effects as a result of the physical forces applied by a turbulent flowing medium on the pipe material surface. Since mass transfer and momentum transfer effects are interdependent, a change in wall shear stress will affect mass transfer and vice versa. Accurate characterisation of wall shear stress can, therefore, be used to assess the effect of fluid flow on corrosion.

This paper considers a generic approach to wall shear stress determination using generally available hydrodynamic data in multiphase systems and an experimental set-up to re-create field conditions in simulated multiphase flow. The detailed approach is outlined in the following steps:

1. Use fluid flow data and flow maps to determine the appropriate flow regime for a specific service environment.

2. Determine pressure drop in the system due to different components, e.g. pressure drop due to friction, pressure drop due to elevation changes and pressure drop due to acceleration changes for different flowing systems and corresponding flow regimes.

3. Calculate wall shear stress as a function of the pressure drop or other hydrodynamic or dimensionless fluid flow parameters.

4. Simulate wall shear stress in the laboratory under identical process conditions (brine, hydrocarbons H_2S, CO_2 etc.) to obtain accurate characterisation of fluid flow and corrosion effects from the field to the laboratory.

5. Use corrosion data generated in laboratory evaluation to assess potential for field corrosion.

3. Modelling Horizontal and Vertical Flow in Multiphase Systems

Flow regimes in multiphase flow develop due to the different velocities that the different phases exhibit. In a three phase gas–oil–water system, velocities of the three phases define the flow regime that is dominant. Flow regimes in horizontal flow differ from those in vertical flow, but the parameters that define the flow regimes are the same in both cases. They can be estimated as a function of superficial velocities in the liquid and gas phases.

Horizontal flow may be divided into five major flow regimes as

- bubble flow,
- stratified flow,
- slug flow,
- annular flow, and
- mist flow. (Mist flow is also referred to as drop or spray or dispersed flow [14].)

Figure 1 shows the different regimes in horizontal flow.

Bubble flow occurs when bubbles of gas move along the upper part of the pipe at approximately the same velocity as the liquid. Here, the gas is present as small bubbles randomly distributed, and their diameters also vary randomly.

Stratified flow occurs when the liquid flows along the bottom of the pipe and the gas flows over a smooth liquid–gas interface.

Slug flow occurs when waves of liquid are picked up by a more rapidly moving gas to form a frothy slug, which passes through the pipe at a much greater velocity than the average liquid velocity. The gas phase is more pronounced than in bubble flow. Although the liquid phase is still continuous the gas bubbles coalesce to form stable bubbles with almost the diameter of the pipe. Slugs of liquid separate these large bubbles. The velocity of the bubbles is greater than that of the liquid and can be predicted in relation to the liquid velocity. The liquid surrounding the gas bubble travels at different velocities compared to the rest of the liquid phase. These variations result in varying wall friction losses and also in liquid hold-up. Both the gas and liquid phases have significant effect on the pressure gradient.

Annular flow occurs when the liquid forms a film around the inside wall of the pipe and the gas flows at high velocities in the centre of the pipe.

Mist flow is the regime in which most or nearly all of the liquid is entrained as spray by the gas. This is also called dispersed flow/spray flow and may result in droplet impingement on the metal surface, causing significant erosion damage. Here, the gas phase is continuous and the bulk of the liquid is entrained in the gas.

Different liquid and gas superficial velocities and distribution properties determine the flow regime. A flow map may be used to identify the flow regime from calculated values of liquid and gas superficial velocities. Figure 2 shows a typical flow map, proposed by Taitel and Dukler [14].

3.1. Vertical Flow

Flow regimes in vertical flow are affected, in addition to all the other factors that affect horizontal flow, by the force of gravity. The flow regimes observed in vertical flow are also very similar to those in horizontal flow. The primary flow regimes in vertical flow conditions are bubble, slug, annular and mist/dispersed/spray/drop flow.

3.2. Determination of Pressure-drop and Wall Shear Stress in Horizontal Flow

Numerous researchers have reported on methods of determining flow regime transitions. pressure-drops and the concomitant wall shear stress for different flow regimes [15–18] for flowing systems. In general, for systems in horizontal flow, wall shear stress is given as

$$\tau_w = \Delta P\ D/(4\ \Delta L)$$

where ΔP = pressure drop, D = pipe diameter and ΔL = segment length.

Determination of pressure drop for a two-phase (gas/liquid) flowing system in horizontal flow involves the following steps:

1. Determination of volumetric liquid fraction in the line.

2. Calculation of two-phase friction factor.

3. Determination of two-phase Reynold's number and the corresponding liquid hold up factor for the volumetric liquid fraction.

4. Determination of friction pressure drop as a function of two phase friction factor and other physical characteristics of the system (such as velocity, density, diameter, etc.).

3.3. Calculation of Pressure Drop Shear Stress in Vertical Flow

As detailed above, an accurate determination of pressure drop in vertical flow may be obtained using factors that account for liquid hold-up and also consider flow regimes in vertical flow. Three suggested methods, widely referenced in literature for these calculations are those proposed by Hagedorn and Brown [17], Duns and

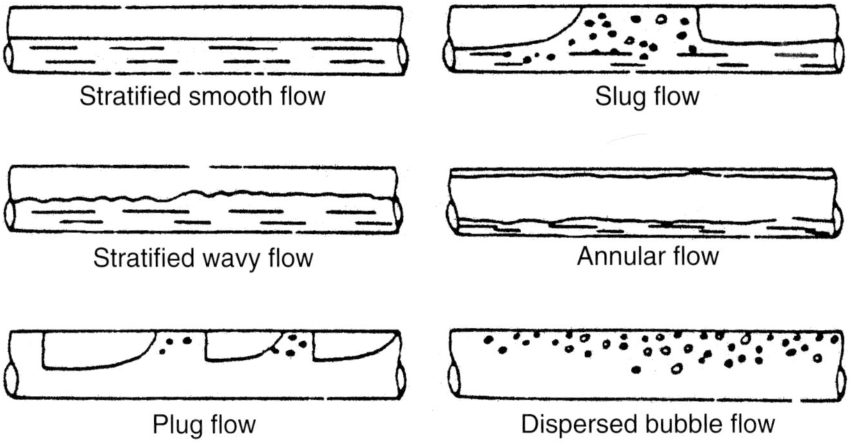

Fig. 1 *Sketch of flow regimes in horizontal flow.*

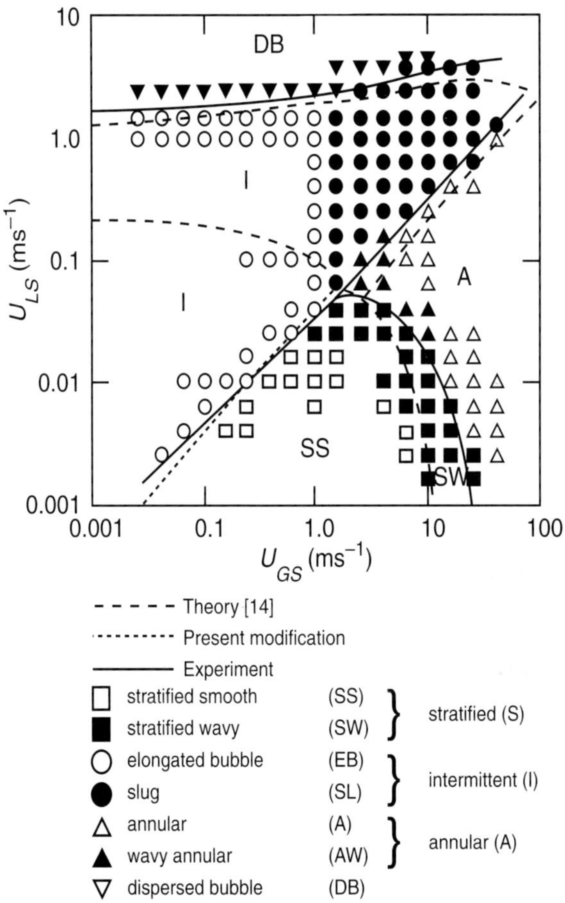

Fig. 2 *Flow map proposed by Taitel and Dukler [14].*

Ros [18], and Griffith and Wallis [19]. The Griffith–Wallis method was modified by Orkiszewski [20] to consider four different flow regimes. These are bubble flow, slug flow, annular-slug transition flow and annular-mist flow.

The identification of a flow regime may be achieved by the boundaries defined by Griffith and Wallis between bubble and slug flow and Duns and Ross between all other flow regimes. The different limits are:

Limits	Flow regime
$q_g/q_t < (L)_B$	Bubble
$q_g/q_t > (L)_B$, $v_{gD} < (L)_g$	Slug
$(L)_M > v_g > (L)_g$	Annular
$v_{gD} > (L)_M$	Mist

where v_{gD} = dimensionless gas velocity= $q_g\ (\rho_l/g\sigma)^{1/2}$;
$(L)_B$ = bubble slug boundary (dimensionless) = 1.071–0.2218 ($v_t 2/d_h$) with the limit $(L)_B >/0.13$;
$(L)_s$, slug transition boundary (dimensionless) = 50 + 36 $v_{gD} q_l/q_g$;
$(L)_M$, transition mist boundary (dimensionless) = 75 + 84 $(v_g q_l/q_g)^{0.75}$;
q = volumetric flow rate, subscript g = gas, subscript L = liquid, subscript t = total.

Once the flow regime is identified in terms of these parameters, and other required variables such as reduced temperature and pressure, viscosity and compressibility are calculated and then pressure drop and wall shear stress for different flow regimes can be determined.

The methodology described above can be used to determine wall shear stress for both horizontal and vertical flowing systems in a generic manner. However, how can this determination be useful in the simulation of field corrosion in the laboratory? Given a specified corrosive medium operating under specific field conditions, the following procedure will facilitate accurate simulation of multiphase flow in the laboratory:

1. Determine appropriate flow regimes, pressure drops and wall shear stress as detailed above.

2 Create a flowing medium with an electrochemical (in terms of corrosive constituents present in the aqueous medium) identical to that in the field. This involves aspects of brine composition and appropriate partial pressures of H_2S and CO_2.

3. Simulate, through a flowing medium, wall shear stress levels identical to those determined to be relevant to the field conditions.

4. Obtain corrosion data and determine corrosion rates that are valid for field conditions.

By creating the aqueous medium with the exact make-up of corrosive constituents (for simulation of mass transfer effects) and the wall shear stress (for simulation of momentum transfer effects), the above method provides a convenient basis to re-create and assess multiphase flowing systems.

The above procedure requires an experimental set-up that can facilitate the engineering of an aqueous medium which can simulate multiphase conditions in terms of the dissolved gases (H_2S, CO_2) as well as other corrosive agents. The system should also be capable of variable flow rates and utilisation of differing specimen configurations. Such an experimental set-up is described in the next section.

4. Experimental Set-up

The flow loop used in this research effort is a proprietary Multiphase Autoclave Pipeline Simulator (MAPS) system. A schematic of the MAPS system is given in Fig. 3. It consists of a five litre Alloy C-276 lined autoclave and provides for flow simulation with an Alloy C-276 external loop and pump system. The flow loop includes three types of test coupons:

- The flow through electrode (FTE) coupon involves a tubular geometry and allows for the corrosive environment to flow through the i.d. of the specimen to study flow-related effects and different flow regimes.

- The impingement electrode (IE) coupon consists of a ring of test material embedded into a non-conductive mount. This permits tests to be conducted under constant rheology and conditions of liquid phase impingement.

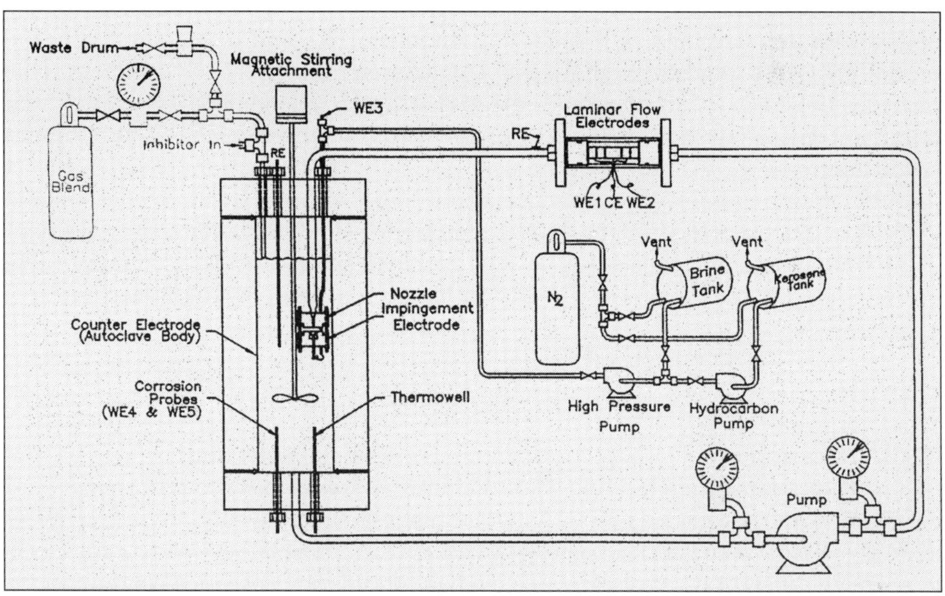

***Fig. 3** Multiphase autoclave pipeflow simulator (schematic).*

- The static zone electrode (SZE) coupon facilitates experimental assessment of mass loss data and electrochemical corrosion monitoring in an autoclave reservoir (conditions of low levels of fluid velocity).

The FTE, IE and SZE coupons are monitored using linear polarisation resistance technique as described in ASTM G59 [21]. An external high pressure Ag/AgCl reference electrode is used in combination with the C-276 system as the counter electrode for making potential measurements Electrochemical impedance spectroscopy (EIS) is used to examine fundamental properties of corrosion film formation. The test duration may vary up to four days depending on the severity of the environment.

Upon completion of the tests, the FTE and IE coupons are evaluated to determine both scale weight and thickness as well as average mass loss rate and maximum localised corrosion rate according to ASTM G1 [22] and G46 [23]. For environments with low water cuts, the electrochemical monitoring set up can be replaced by electrochemical resistance measurements, ASTM G59.

5. Discussion

The methodology described in this paper facilitates laboratory simulation of conditions essential for assessment of both mass and momentum transfer effects in multiphase flowing environments. However, the true benefit is in the ability of the methodology detailed to reduce the complexity normally associated with simulation of multiphase flow. The described test system allows for a simple, cost- effective experimental apparatus can simulate multiphase flow more easily than that typically required for simulation systems operating with all three phases.

The simplification results from the use of properly constituted two phase environments (water and oil) containing dissolved H_2S and CO_2 saturated at the partial pressures identical to those that characterise the service environment. This provides the basis for simulation of the mass transfer phenomena associated with dominant corrosion reactions (either CO_2 or H_2S). The same two phase environment (with dissolved gases) may then be used to re-create momentum transfer effects through attainment of wall shear stress conditions produced by three phase media in the field. Such a procedure results in accurate corrosion simulation and assessment in the laboratory of three phase (oil/water/gas) field environments, and is applicable even for specific flow regimes such as droplet impingement or slug flow.

6. Conclusions

Predicting corrosivity of multiphase flowing systems is a complex and challenging task requiring a clear understanding and characterisation of both mass transfer and momentum transfer-related effects. A methodology to simulate multiphase flow using hydrodynamic parameters has been described. An experimental set-up to conduct tests to re-create different types of corrosive media and differing specimen configurations has been developed and is being used to generate corrosion assessment data.

References

1. C. S Fang *et al.*, "Computer model of a gas condensate well containing carbon dioxide", *Corrosion '89*, Paper No. 465, NACE, Houston, Tx, 1989.
2. C. De Waard and U. Lotz, "Prediction of CO_2 corrosion of carbon steel", *Corrosion '93*, Paper No. 69, NACE International, Houston, Tx, 1993.
3. C. De Waard and D.E. Milliams, "Carbonic acid corrosion of steel", *Corrosion*, 1975, **31**, 5, 177.
4. E. Dayalan *et al.*, "Modelling CO_2 corrosion of carbon steels in pipe flow", *Corrosion '95*, Paper No. 118, NACE International, Houston, Tx, 1995.
5. L. H. Gatzky and R. H. Hausler, "A novel correlation of tubing corrosion rates and gas production rates", *Advances in CO_2 Corrosion*, Vol. 1, pp. 87. Published by NACE, Houston, Tx, 1984.
6. S. Srinivasan and R. D. Kane "Prediction of Corrosivity of CO_2/H_2S Production Environments", *Corrosion '96*, Paper No. 11, NACE International, Houston, Tx, 1996.
7. S. Srinivasan *et al.*, Prediction of corrosivity of multiphase CO_2/H_2S systems. *Proc. EUROCORR '97* (extended abstract), Trondheim, Published NTNU, Glushaugen, N-7034, Trondheim, Norway 1997.
8. S. Srinivasan, "Role of critical factors in predicting corrosivity of multiphase CO_2/H_2S systems". *Corrosion '98*, NACE International, Houston, Tx, 1998.
9. S. Srinivasan and R. D. Kane, "Prediction and assessment of corrosivity in multiphase CO_2/H_2S systems", CLI Joint Industry Project description, May 1997.
10. R. Zhang *et al.*, "Development of a mechanistic model for predicting corrosion rate in multiphase oil/gas/water flows", *Corrosion '97*, Paper No. 601, NACE International, Houston, Tx, 1997.
11. W. P. Jepson *et al.*, "Predictive model for sweet corrosion in horizontal multiphase slug flows", *Corrosion '96*, Paper No. 19, NACE International, Houston, Tx, 1996.
12. W. P. Jepson *et al.*, "Model for sweet corrosion in horizontal multiphase slug flow", *Corrosion '97*, Paper No. 11, NACE International, Houston, Tx, 1997.
13. K. D. Efird *et al.*, "Experimental correlation of steel corrosion in pipe flow with jet impingement and rotating cylinder laboratory tests", *Corrosion '93*, Paper No. 81, NACE International, Houston, Tx, 1993.
14. Y. Taitel and A. E. Dukler, "A model for predicting flow regime transitions in horizontal and near-horizontal gas liquid flows", *AIChE J.*, **22**, 47.
15. J. C. B. Lopes and A. E. Dukler, "Droplet dynamics in vertical gas liquid annular flow", *AIChE J.*, **33**, 1013.
16. Flanigan Orin *et al.*, "Effect of uphill flow on pressure drop in design of two phase gathering systems", *Oil Gas J.*, 1958, 132.
17. A. R Hagedorn and K E Brown, "Experimental study of pressure gradients occuring during continuous two-phase flow", *J. Pet. Tech.* No. 1, 1965, April, 475-484.
18. H. Duns, Jr and N. C. J. Ros, "Vertical flow of gas and liquid mixtures in wells". *Proc. 6th World Petroleum Congr.*, Section 2, Paper No. 22, Frankfurt, Germany, 1963.
19. P. Griffith and G. B. Wallis, "Two phase slug flow", *J. Heat Transfer Trans., ASME*, 1961, August, 307.
20. J. Orkiszewski, "Predicting two phase pressure drops in vertical pipes", *J. Petrol. Technol.*, 1967, June.
21. ASTM G59 Practice for Conducting Potentiodynamic Polarization Resistance Measurements. American Society for Testing and Materials, West Conshohocken, Pa.
22. ASTM G1-90 (1994) Preparing Cleaning and Evaluating Corrosion Test Specimens.
23. ASTM G46-94 Guide for Examination and Evaluation of Pitting Corrosion.

18
Investigation of Premature Failure of a Well Fluid Pipeline in an Indian Offshore Installation

A. K. SAMANT, V. K. SHARMA, S.THOMAS, P. F. ANTO and S. K. SINGH
Institute of Engineering & Ocean Technology, Oil and Natural Gas Corporation Ltd, Panvel, New Mumbai-410221, India

ABSTRACT
Premature failure of one of the well fluid pipelines was observed in an Indian offshore installation. Detailed investigations were carried out by evaluating various operating parameters and also by laboratory analysis. Various studies indicated low flow velocity, non-pigging of the pipeline and presence of sulfate reducing bacteria to be responsible for localised under-deposit corrosion which led to the leakage of the pipeline.

1. Introduction

A 6.5 km long well fluid pipeline, which was commissioned in an Indian offshore installation in May 1994 to transport well fluid from the well platform to the process platform, failed due to leakage $2^1/_2$ years after commissioning. The diameter of the pipe line was 16 in. with a wall thickness of 22.2 mm. As per the design specifications, the pipe line material was API 5L X60. Details of the pipeline are given in Table 1.

Table 1. *Details of pipe line*

Design details		Production details	
Type	Well fluid	Oil	954 BOPD*
Diameter	16″ (406 mm)	Water	3102 BWPD**
Wall thickness	0.875″ (22.2 mm)	Gas	0.0436 MM^3***
Design pressure	141.0 kg cm^{-2}	Water cut	75%
Design temperature	97.0°C	Operating pressure well/process side	11.0 kg cm^{-2}/ 7.4 kg cm^{-2}
Material grade	API 5L X60	Temperature well/process side	70°C/41°C
Concrete coating thickness	30 mm	CO_2 in associated gas	2.31%
As laid length	6514.28 metres	H_2S in associated gas	2 ppm
Number of joints	524	–	–

* Barrels of oil per day. ** Barrels of water per day. *** Million cubic metres.

2. Leakage History

The first leak in the pipeline was noticed by operators on 29th January 1997. During a leak detection survey and repair job, three holes with sizes 5 mm, 5 mm and 10 mm respectively were noticed at the 6 o'clock position at a distance of 168 m from the process platform towards the well platform. These holes were located in a 45 mm length of the pipe (leak location-1) as shown in Fig. 1. A detailed cathodic protection (CP) survey and a wall thickness survey were carried out. The wall thickness survey showed thickness reduction only at two places, i.e. 3.5 mm near the first hole and 11.8 mm near the third hole. At other places the thickness was between 20.8 mm to 23.4 mm at the 6 o' clock position in a 920 mm segment of pipeline.

Later on, during pressure testing, after repair of the first leak by clamps, another heavy oil leak was observed about 10 m from the previous leak location. During a survey two other holes with sizes 13 mm and 10 mm diameter (Fig. 1, leak location-2) at a distance of about 192 m from the process platform towards the well platform side were found. The distance between the two holes was 300 mm. The minimum wall thickness on both sides of the holes was 21.8 mm and the maximum was 23.6 mm.

The cathodic protection potential values at both the locations were found to be between –920 mV to –935 mV and no external damage was observed by divers.

3. Laboratory Investigation

For laboratory investigation various operating data and well fluid test data related to this pipe line were analysed. Since the leaked pipeline sample was not available for failure analysis, produced water samples and associated gas samples were collected from the well and process platforms for detailed laboratory investigation.The following studies were carried out.

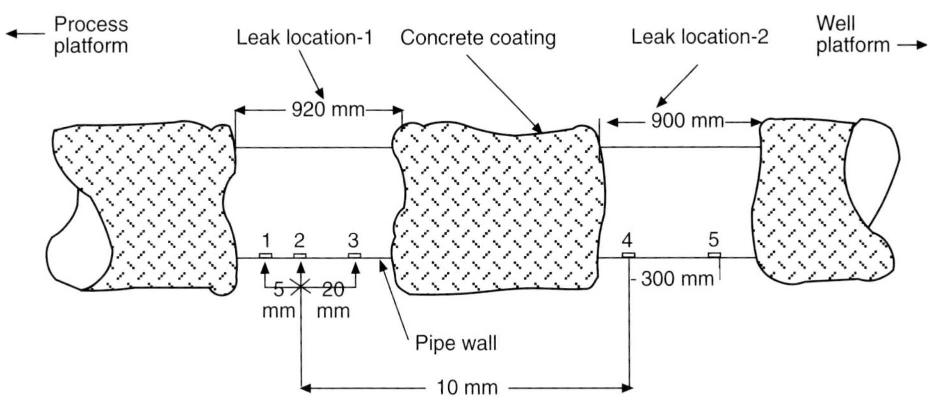

Fig. 1 Diagrammatic representation of section of pipeline showing leak location.

3.4. Simulation Corrosion Studies

Simulation gravimetric corrosion studies in the produced water samples from the well platform were carried out in the presence of a carbon dioxide (2.5%) and H_2S (10 ppm) environment at 70°C in a high pressure high temperature autoclave. For this purpose, small metal coupons of API 5L X60 material were taken. After grinding and polishing, the area and weight of the coupons were measured. The coupons were then exposed to the produced water samples under the above conditions after purging the solution with oxygen-free nitrogen to remove dissolved oxygen. The rate of corrosion was calculated after measuring the weight loss. The result showed a rate of corrosion of 122 mpy (3.04 mm/year) in the water sample .

3.5. Study of Flow Pattern

Different types of flow pattern exist in well fluid pipelines. Flow pattern configuration provides the information regarding stratification of liquid phase. In order to have a realistic flow pattern, software based on the standard pressure drop correlation was used to predict the velocity of different phases in the pipeline and also the flow pattern. The analysis indicated that most of the time flow pattern in the pipeline was transitional in nature. Flow velocities of the various phases were as follows:

Liquid velocity	0.06 m s^{-1} (0.2 ft s^{-1})
Gas velocity	0.52 m s^{-1} (1.7 ft s^{-1})
Total velocity	0.58 m s^{-1} (1.9 ft s^{-1})

Since the total velocity of the fluid was less than 1.0 m s^{-1} and the water cut was also about 75%, water will wet the pipeline even in the transition flow condition, especially at low lying areas where the chances of water accumulation is greater. Therefore, aggressive localised corrosion has to be expected in these stagnant areas.

3.6. Study of Bathymetric Profile

The bathymetric profile of the pipeline was also examined in detail. The profile showed a low lying area in and around the 1.5 km location of the pipeline from the process side towards the well platform and so the probability of water stagnation and deposit accumulation in this segment is expected to be very high. This segment is, therefore, considered highly susceptible to under-deposit localised corrosion. The reported leaks were also in this segment.

3.7. Pigging Details

A review of platform records indicated that the pigging programme was not regular and systematic. After commissioning in May 1994, this pipeline was pigged only twice — the first pigging was carried out in June 1995 and the second during November 1995. Later on, pigging was not carried out in this line until leaks were reported. The absence of a systemic pigging practice, despite the presence of a highly corrosive fluid, low fluid velocity and high water cut in the pipeline, was one of the probable reasons for the accumulation of deposits, localised microbial activity and ineffectiveness of the bactericides, which together ultimately resulted in the pipeline leaks.

3.8. Corrosion Monitoring Practice

For corrosion monitoring only the iron count technique was being used in this pipeline as there was no provision for installation of electrical resistance probes or any other probes or coupons. The iron count data from the platform were also analysed. The reported values showed a corrosion rate of less than 1.0 mpy for most of the time, which is very low. This indicated that other factors such as accumulation of deposits and activity of bacteria in the stagnant areas of the pipeline might have contributed to enhance extremely localised corrosion in this pipeline besides the corrosion due to carbon dioxide.

4. Discussion

Corrosion normally occurs in pipelines whenever water wets the pipe wall. If the amount of water is more than 30%, the pipe can corrode at almost any location depending upon its corrosivity. If continuous hydrocarbon wetting is occurring, then the corrosion risk will be extremely low. Therefore the oil/water ratio and emulsion tendency/stability is the key factor. For many crude oils, the presence of a water cut of approximately 35% will lead to water becoming the continuous phase for a fully mixed oil/water system to the extent that corrosion then becomes a continuous potential risk. Similarly, if the gas/oil ratio (GOR) is > 5000 then continuous water wetting can be expected. The pipeline under investigation fulfilled both these criteria and therefore water wetting was expected.

Water can stratify in pipelines if the line velocity is less than that required to sweep the oil through the line. Low velocity lines essentially act as long thin separators and can accumulate huge water volumes at a steady state. The velocity at which water is entrained in flowing oil depends upon the relative densities of the oil and water, the interfacial tension between the oil and water, and the pipe diameter. Entrainment velocity is typically in the 0.3–3.0 ft s^{-1} (0.09–0.9 ms^{-1}) range, and is greater in larger pipe sizes [1]. In the present case, the liquid velocity was 0.2 ft s^{-1} (0.061 ms^{-1}) and the diameter of pipe was also large, i. e. 16 in. Further, a bathymetric profile of this line also indicated a low area near the segment with leaks and so the probability of accumulation of water existed in this segment of pipeline.

Once water accumulated in low lying areas, the pipe wall might have become water wet and corroded severely because of the presence of corrosive water and dissolved carbon dioxide gas present in the line. Since bacteria were already reported in this pipeline, their activity might have aggravated the corrosion rate.

4.1. Role of Bacterial Corrosion

Microbiological corrosion has prove to be a major problem in pipelines [2,3]. Bacteria can enter a line during the hydrotest and remain viable for long periods. Further, most lines receive fluids from bacterially-contaminated production systems, which continuously feed bacteria-contaminated fluids into the pipeline. It was also not known whether pigging was carried out after hydrotesting of this pipeline.

Low flow or stagnant conditions such as stratified water are ideal for bacterial growth, and can lead to serious bacterial corrosion problems. Bacteria frequently cause corrosion at rates far greater than estimated on the basis of the CO_2 or H_2S contents. Bacterial corrosion treatment usually requires a combined pigging and batch chemical treating program or sustained bactericide treatment [4]. As mentioned earlier, this pipeline was contaminated with aerobic bacteria and sulfate-reducing bacteria. In view of the extremely high corrosion rate, the possibility of the existence of anaerobic acid-producing bacteria under deposits cannot be ruled out. Anaerobic acid producing bacteria (APB) produce organic acids like acetic acid, butyric acid, propionic acid as by-products of their metabolic activities. Studies carried out by Pope *et al.* [5] showed a level of acids of up to several thousand ppm at corrosion sites. The impact of acidic metabolites is intensified when they are trapped at the biofilm/metal interface. These acids produced by bacteria lower the pH to a value of 2.00 [6] and accelerate the corrosion rate. Since the failed pipeline section and corrosion products were not made available, the presence of these bacteria could not be identified. Nevertheless, as sulfate-reducing bacteria were already reported in the produced water and H_2S was also present, the possibility of corrosion due to bacterial metabolites cannot be ruled out.

Furthermore, an increase in the concentration of H_2S from 2 to more than 15 ppm, a few days after batching of bactericides, indicated that the bactericide treatment programme was ineffective and this was perhaps due to non-implementation of a frequent and systematic pigging programme.

4.2. Effect of CO_2 and H_2S Corrosion

Hydrogen sulfide and CO_2 are acidic gases. They readily dissolve in water to produce weak acids. Both contribute to increased acidity (i.e. decreased pH) with increasing partial pressure and, in theory, could increase the corrosion rate of steel. In the present pipeline, the partial pressure of CO_2 was around 3.5 psi. whereas that of H_2S was very low. Therefore, the ratio of CO_2 to H_2S partial pressure was more than 200. Under this condition, carbon dioxide corrosion will dominate [6]. Further, the CO_2 corrosion rate nomogram [7] predicts carbon dioxide corrosion at 3.5 psi partial pressure to be around 2.0 mm/year. After back calculation, the average rate of corrosion, which led to the leakage from the 22.2 mm thick pipeline in $2^1/_2$ years, came out to be 7.2 mm/year. On taking the total rate of corrosion into account, the contribution due to carbon dioxide comes out to be about 30%.

However, the line leaked within $2^1/_2$ years through a loss of 22.2 mm metal wall thickness. This clearly indicates that besides carbon dioxide other acidic corrodents — most probably produced by microbial activity — might have become active in the line and aggravated the corrosion rate which ultimately resulted in the formation of holes in the pipeline.

5. Conclusions

Various studies confirmed that leakage in a pipeline was the result of extremely localised internal corrosion which led to the formation of holes. The fluid flowing through the pipeline was highly corrosive due to presence of dissolved carbon dioxide and hydrogen sulfide gas. Localised corrosion of the pipeline was accelerated by under deposit microbial activity.The low flow velocity and non-pigging of the pipeline further aggravated the corrosion rate which ultimately resulted in the leakage of the line.

6. Acknowledgement

The authors are thankful to the Executive Director, MRBC and Director (Tech.) for giving permission to publish this paper.

References

1. J. S. Smart *et al.*, *Pipeline Gas Ind.*, 1995, March, 29.
2. A. K. Samant *et al.*, *Proc. ISOPE-93 Conf.*, Singapore, 1993.
3. A. K. Samant *et al.*, *Corrosion '98*, Paper No. 280, NACE International, Houston, Tx, 1998.
4. J. S. Smart *et al.*, *Corros. Prev. Control*, 1991, Feb., 5.
5. D. H. Pope *et al.*, *Corrosion '88*, Paper No. 79, NACE, Houston, Tx, 1988.
6. B. Little *et al.*, *Corrosion '87*, Paper No. 370, NACE, Houston, Tx, 1987.
7. R. D. Kane, *Petrol. Eng. Int.*, 1995, Oct, 53.

19
Corrosion Resistance of Thermomechanical Control Process (TMCP) Steels for Cargo Oil Tanks of Very Large Crude Oil Carriers (VLCC)

H. MIYUKI, A.USAMI*, K. MASAMURA†, Y. YAMANE§ and Y. KOBAYASHI¶

Corporate Research and Development Laboratories, Sumitomo Metal Industries Ltd, 1-8, Fuso-cho, Amagasaki, Hyogo, 660-0891, Japan.

*Steel Research Laboratories, Technical Development Bureau, Nippon Steel Corporation, 20-1 Shintomi,Futtsu, Chiba, 293, Japan

†Keihan Material Center, NKK Corporation, 1-1 Minami Wataridacho, Kawasakiku, Kawasaki, Kanagawa, 210-0855, Japan

§Technical Research Laboratories, Kawasaki Steel Corporation, Kawasakidori, 1-chome, Mizusima,Kurasiki,Okayama, 712, Japan

¶Plate Technology Section, Plate & Wire Rod Department, Iron & Steel Division, Kobe Steel Ltd, l, Kanagawacho,Kakogawa, Hyogo 075-01, Japan

ABSTRACT

In order to compare the corrosion resistance of TMCP (Thermomechanical Control Process) steels with that of conventional mild steels and CR (Controlled Rolled) steels in upper deck environments of double hull VLCC (Very Large Crude Oil Carrier) cargo oil tanks, various measurements and corrosion tests were conducted. These included measurements of polarisation curves and cyclic immersion tests carried out in sea water. Furthermore, exposure tests were conducted in simulated wet inert gas environments, which contained 13%CO_2, 5%O_2 and a small amount of SO_2. Results obtained all revealed that the three types of steels have almost the same corrosion resistance in sea water and wet inert gas environments, and only general corrosion was observed. Corrosion rates increased with increasing temperature and O_2 and SO_2 contents in the inert gas. The temperature effect was particularly significant.

1. Introduction

Recently vapour space corrosion of upper deck steel plates in double hull VLCC (Very Large Crude Oil Carrier) cargo oil tanks has been the subject of particular attention because of the severity of the corrosion. The steel plates are exposed to corrosive environments with high temperature, high humidity and inert gas containing corrosive components such as carbon dioxide, oxygen and sulfur dioxide. Thermomechanical Control Process (TMCP) steel plates are used in addition to conventional mild steels (MS) and Controlled Rolled (CR) steel plates. The production processes are different for these three types of steels. Although little difference in chemical compositions of the steels can be observed, there are differences in microstructure. The effect of the manufacturing procedure of these steels on their corrosion resistance has not so far been studied.

In the present work, electrochemical measurements and various corrosion tests were conducted to investigate the corrosion resistance of TMCP steels, mild steels and conventional CR steels and to elucidate the effect of environmental variables on the corrosion behavior in simulated vapor space environments of double hull VLCC cargo oil tanks.

2. Experimental

2.1. Specimen Materials

Table 1 shows the chemical composition of the materials used. Series A were used for the corrosion tests in sea water environments. Series B were used for the tests in simulated wet inert gas environments. Typical examples of the microstructures are shown in Fig. 1. Lamellar structures of ferrite and pearlite were observed for both TMCP and CR steels.

2.2. Test Conditions

2.2.1. Basic corrosion tests

In order to clarify the corrosion behavior of the steels, measurements of polarisation curves were carried out under the conditions shown in Table 2 as well as cyclic immersion tests in sea water environments. Table 3 shows the test conditions. Long-term corrosion tests were conducted in a sea water tank. Figure 2 shows the test apparatus of volume 2 m^3. Corrosion tests were continued for 3 months under the condition shown in Table 4.

2.2.2. Vapour space corrosion test under wet inert gas environments

Upper deck plates are exposed in service to wet inert exhaust gases from the ship's engine; these gases contain corrosive substances such as CO_2, O_2 and SO_2. Corrosion tests were therefore carried out with a simulated inert gas under the conditions shown in Fig. 3.

After the corrosion tests, the corrosion products were qualitatively and quantitatively analysed by X-ray diffraction method, using KCl as an internal standard [1].

Table 1. *Chemical composition of the materials used (mass%)*

	Steel	C	Si	Mn	P	S	Cu	Ni	Cr	Mo	V	Nb	Ti	Al
A-1	MS	0.17	0.15	0.50	0.011	0.003	0.01	0.02	0.02	<0.01	<0.01	<0.01	<0.01	0.01
A-2	TMCP	0.14	0.18	1.04	0.014	0.003	0.01	0.02	0.02	0.01	<0.01	<0.01	0.01	0.03
A-3	CR	0.16	0.31	1.24	0.016	0.002	0.01	0.02	0.03	0.01	<0.01	<0.01	0.01	0.03
B-1	MS	0.14	0.18	0.69	0.017	0.006	0.02	0.03	0.04	0.01	<0.01	<0.01	<0.01	0.01
B-2	TMCP	0.14	0.16	1.02	0.010	0.003	0.02	0.02	0.03	0.01	<0.01	<0.01	<0.01	0.02
B-3	CR	0.16	0.40	1.30	0.017	0.007	0.01	0.02	0.04	0.01	<0.01	0.01	0.01	0.03

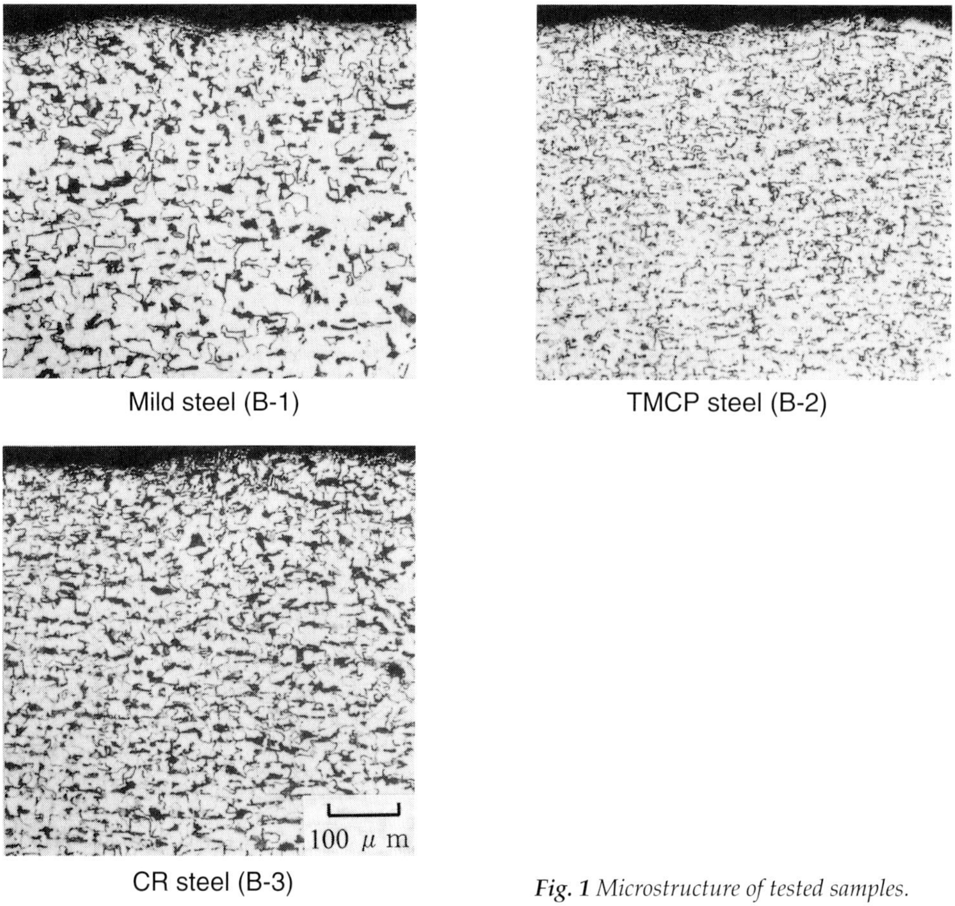

***Fig. 1** Microstructure of tested samples.*

***Table 2.** Conditions of polarisation measurement*

Item	Condition
Solution	Synthetic sea water (ASTM-D1141-52)
Temperature	50°C
Potential Sweep rate	20 mV min^{-1}
	1 Anodic polarisation (deaerated) 2 Cathodic polarisation (air saturated)

***Table 3.** Cyclic immersion test conditions*

Item	Condition
Solution	Synthetic sea water (ASTM-D1141-52)
Cycle	Wet (1 h) ↔ Dry (3 h)
Temperature	30°C, 60°C
Duration	15 and 30 days

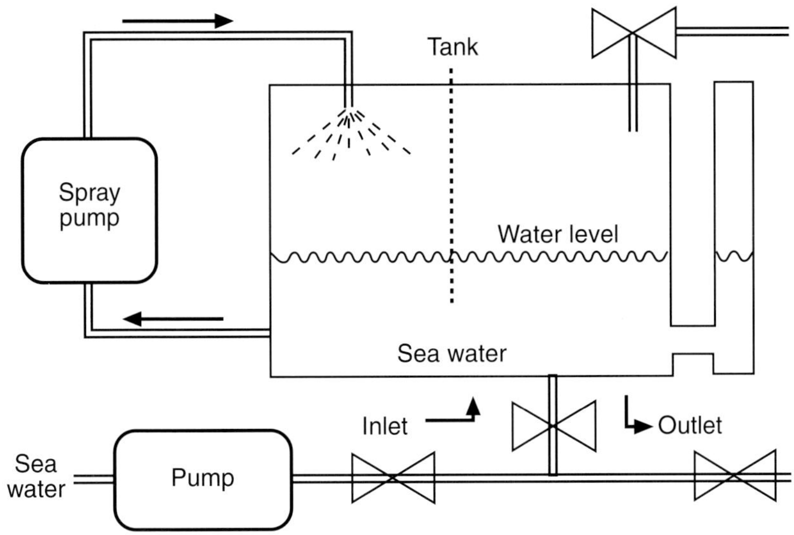

***Fig.* 2** *Corrosion test apparatus using a sea water tank.*

Table 4. *Conditions of corrosion test in a sea water tank*

Item	**Condition**
Solution	Natural sea water
Position	1 Vapour phase 2 Liquid phase
Temperature	Ambient temperature
Test duration	3 months
Specimen	150 × 70 × 5 (mm)
Sea water spray	Six times a day for 15 min

3. Results and Discussion

3.1. Basic Corrosion Tests

The anodic and cathodic polarisation curves of mild, TMCP and CR steels in synthetic sea water are shown in Fig. 4. The three kinds of steels exhibit almost the same polarisation behaviour in both the anodic and the cathodic curves. Clearly, the shape of the cathodic polarisation curves is not determined by the nature of the steel since the cathodic current is controlled by the diffusion of oxygen to the steel surface. The cyclic immersion test results are shown in Fig. 5. The three steels show the same corrosion rates at 30 and 60°C. The TMCP steel has almost the same corrosion resistance as other steels in the sea water environment. Figure 6 shows the corrosion test results from the sea water tank, and again there was little difference in corrosion rates among the three steels.

Composition of inert gas

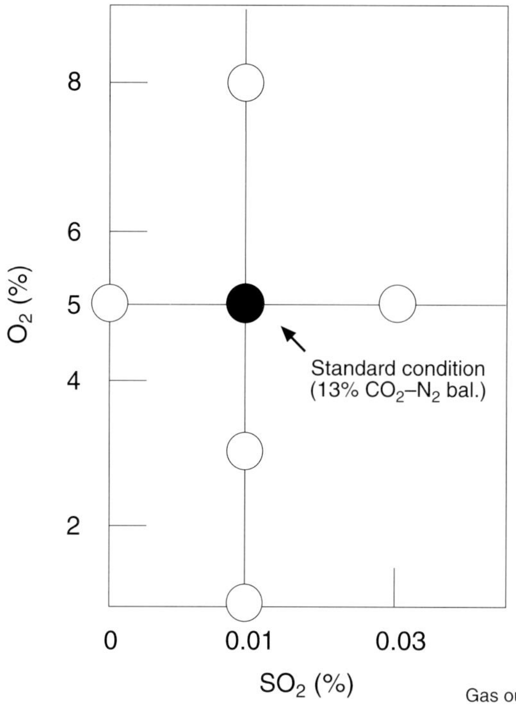

Test apparatus

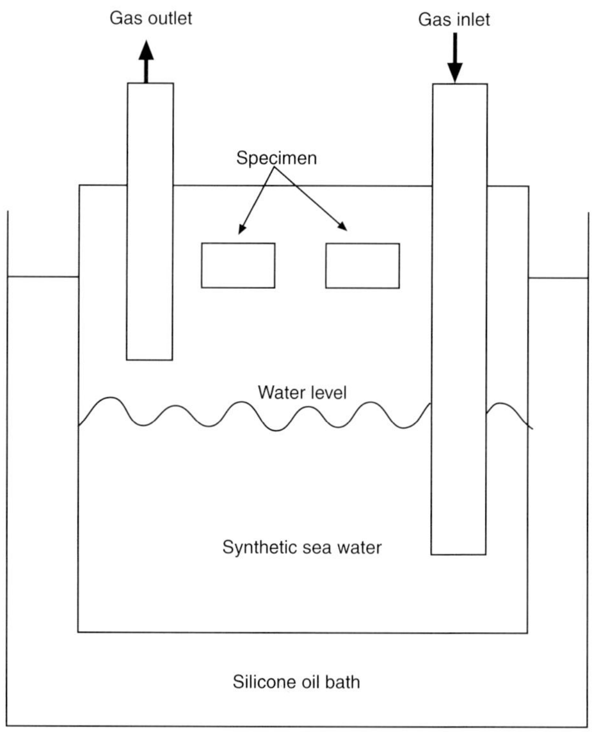

***Fig. 3** Corrosion test conditions in simulated upper deck environments and test apparatus* (below)*. Temperature: 30–60 ℃, test duration: 60 days.*

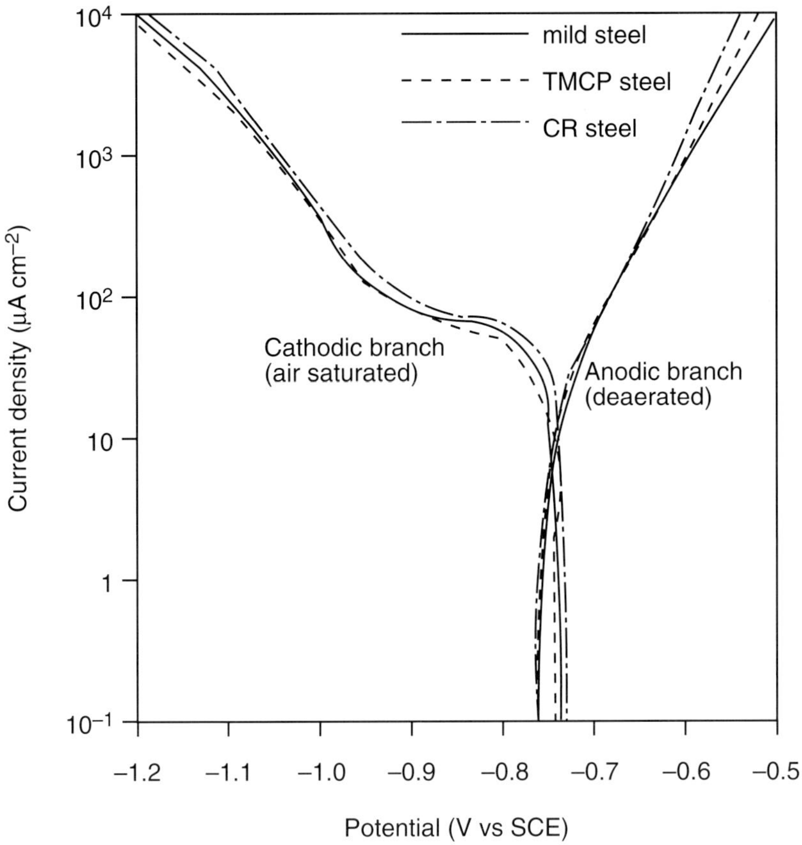

Fig.4 *Polarisation curves of mild, TMCP and CR steels in synthetic sea water.*

3.2. Vapour Space Corrosion Tests

Figure 7 shows the corrosion test results under simulated vapour space environments for 60 days. In general, the kind of steel had little effect on corrosion, though corrosion was determined by environmental factors such as inert gas composition and temperature. Corrosion rates of steels increased with increasing temperature and with increase in O_2 and SO_2 concentrations in the inert gas. The temperature effect was especially significant.

Table 5 shows the means and standard deviations of corrosion rates of all the steels tested. The average corrosion rate of mild, TMCP and CR steels is close to unity and the standard deviation is very low, which suggests that three types of steels have almost the same corrosion resistance.

Table 6 shows the corrosion deposit constituents as determined by X-ray diffraction measurements. Corrosion products on the steels are mainly α-FeOOH and Fe_3O_4 as crystalline substances. The presence of a little γ-FeOOH can also be detected. There are only small differences among the constituents of the corrosion products from the three steels.

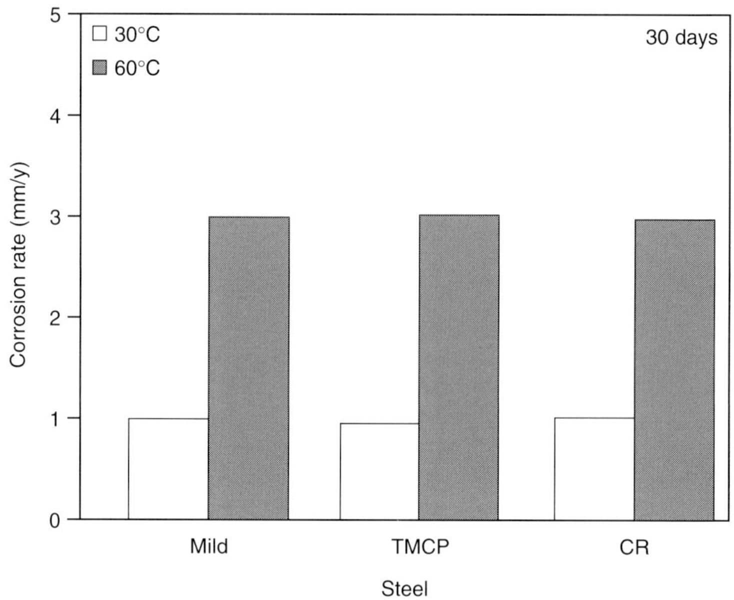

Fig. 5 *Cyclic immersion test results under synthetic sea water environment.*

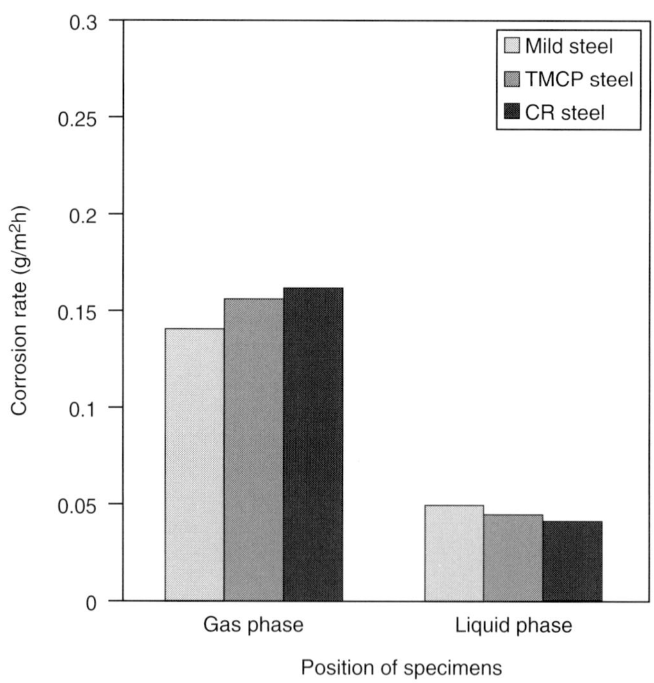

Fig. 6 *Comparison of corrosion rate among mild steels, TMCP steels and CR steels in sea water tank test for 90 days.*

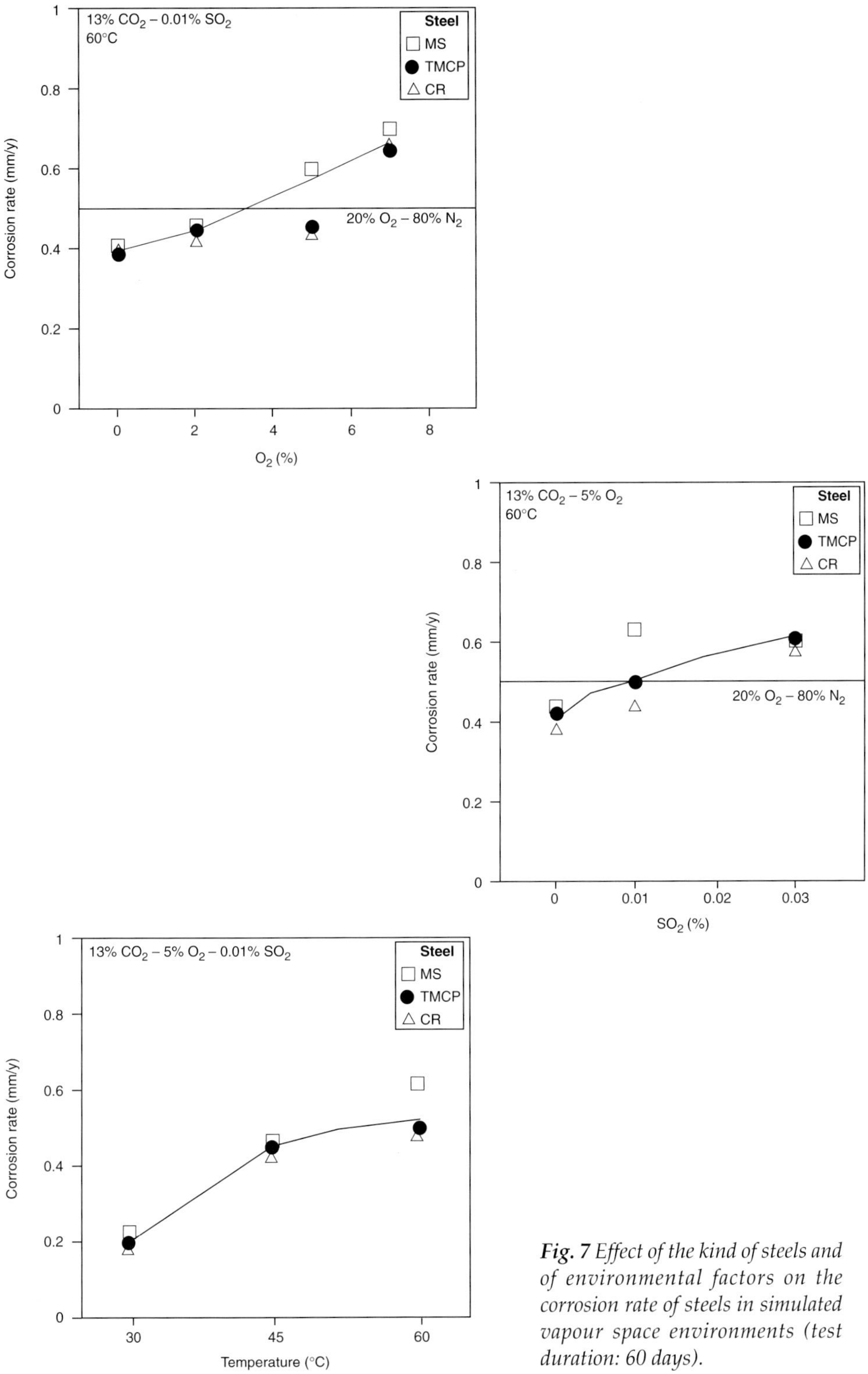

Fig. 7 Effect of the kind of steels and of environmental factors on the corrosion rate of steels in simulated vapour space environments (test duration: 60 days).

Table 5. *Mean and standard deviation of corrosion rate of tested steels*

	Mild steel	**TMCP steel**	**CR steel**
Average mm/y	0.99	0.99	1.02
Standard deviation	0.07	0.06	0.06

Table 6. *Analytical results of corrosion products by X-ray diffraction method. (Standard Condition, mass%)*

Material	**α-FeOOH**	**β-FeOOH**	**γ-FeOOH**	**Fe_3O_4**	**NaCl**	**Others**
TMCP steel	26.9	0	0.6	20.4	a little	52.1
Mild steel	58.3	0	3.3	25.7	a little	12.7
CR steel	26.1	0	4.0	9.4	a little	60.5

The effects of temperature [2] and O_2 concentration [3] on the corrosion rate of steels is well known. However, little is known on the effect of SO_2.

It seems that SO_2 reacts with water to form H_2SO_3, which dissociates to form HSO_3^-. The HSO_3^- ion plays a role as an oxidant in addition to O_2 and accelerates corrosion. As a result, elemental sulfur can be formed. Sulfur is also able to be formed by the oxidation of sour constituents (for example, H_2S in crude oil) by O_2.

Thus,

$$SO_2 + H_2O = H_2SO_3$$

$$H_2SO_3 = H^+ + HSO_3^-$$

$$\left.\begin{aligned} HSO_3^- + 5H^+ + 4e &\rightarrow S + 3H_2O \\ O_2 + 2H_2O + 4e &\rightarrow 4OH^- \end{aligned}\right\} \text{Cathodic reaction}$$

$$2H_2S + O_2 \rightarrow 2S + H_2O$$

4. Conclusion

The effects of manufacturing procedures for steels and of environmental factors on the vapour space corrosion of steels for VLCC cargo oil tanks were investigated. The results obtained are as follows:

1. It was found that Thermomechanical Control Process, controlled rolling and mild steels showed no differences in corrosion resistance either in basic corrosion tests in sea water environments or in tests in simulated inert gas environments.

2. Corrosion of all steels was accelerated in vapour space environments containing corrosive substances such as CO_2, O_2 and SO_2.

References

1. T. Misawa, K. Asami, K. Hashimoto and S. Shimodaira, *Corros.Sci.*, 1974, **14**, 279.
2. F. Speller, *Corrosion, Causes and Prevention*, p.168, McGraw-Hill, New York, 1951.
3. H. Uhlig, D. Triadis and M. Stern, *J. Electrochem. Soc.*, 1955, **102**, 57.

20
Effect of Applied Potential on Cracking of Low-alloyed Pipeline Steel in Low pH Soil Environment

M. TOUZET, N. LOPEZ and M. PUIGGALI
Laboratoire de Mécanique Physique, Université Bordeaux 1, 33405 Talence cedex, France

ABSTRACT

Transgranular corrosion cracking has been observed on carbon steel pipelines in the NS4 near-neutral pH simulated soil environment with CO_2 bubbling. Previous research had shown an effect of hydrogen in this case. The sensitivity to cracking was evaluated by slow strain rate tests with various levels of cathodic polarisation. Polished material and as-received surface material were tested. The results showed that mechanical resistance was reduced by a cathodic polarisation applied at the beginning of the test; this effect increases on as-received surface material and becomes more important when the cathodic polarisation was applied in the plastic strain range. Observations of crack distribution confirmed the development of the mechanical behaviour.

These results showed the effect of hydrogen to be particularly important when it was localised at the tip of pits (initiated at the corrosion potential) or on surface heterogeneities.

1. Introduction

Underground pipelines are protected from the effects of the environment by coatings and cathodic polarisation. Coating degradation and disbondment [1] are often associated with a modification of the cathodic protection as result of changes of soil conductivity and can induce local conditions favourable to crack initiation. Various kinds of cracking can occur on carbon steel pipelines in areas of coating disbondment depending on the composition of the environment and the pH value [2,3]. The first, which has been recognised for a long time, is that of intergranular cracking in an environment consisting of a highly concentrated solution of carbonate and bicarbonate ions with a pH 9–11 [4]. The second, discovered more recently, is transgranular cracking observed in more dilute solutions at pH 6–7. In this case, it appears that hydrogen plays an important role. An increase in stress corrosion cracking sensitivity has been induced by cathodic polarisation [5,6]. Mao *et al.* [7] explained that cracking can be attributed to hydrogen enhanced dissolution near the free corrosion potential and to hydrogen induced cracking below a certain applied potential.

The aim of this study was to evaluate the effect of cathodic polarisation on the sensitivity to cracking for various surface conditions. Polished material and as-received surfaces were tested by slow strain rate tests in the NS4* simulated soil environment with 5% CO_2–95% N_2 bubbling.

2. Experimental Procedure

Samples were prepared from segments of an API 5L X60 carbon steel pipeline of diameter 32.6 mm and wall thickness 7.2 mm. Chemical composition of this steel (wt%) is presented in Table 1. Metallographic examinations showed (Fig. 1):

- from the external surface to the half thickness: a homogeneous distribution of small grains of ferrite and pearlite (Fig. 1a),
- from the half thickness to the internal surface: elongated alternating grains of ferrite and pearlite (Fig. 1b).
- non-metallic inclusions: manganese sulfide oriented longitudinally and titanium carbonitrides.

In the literature, tests have been proposed with various specimens cut from pipes to study stress corrosion cracking (SCC) phenomena. These include:

- symmetrical four-point loading tests: these produce a constant tensile stress only on the outer surface and so have been used for initiation studies,
- parallel-sided gauge length specimens cut in the longitudinal direction of the pipeline: these allowed the original surface to be maintained but the orientation of cracks and inclusions was perpendicular to the actual direction in service and so mechanical properties were different in the longitudinal and transverse directions,
- cylindrical tensile specimens machined from pipelines: these allowed slow strain rate tests or fatigue tests to be conducted, but in this case the outer surface was not maintained and the microstructure could be different, as explained above,
- fatigue pre-cracked specimens: these were used to study the propagation stage of cracks but this situation was probably only representative of real conditions when cracks were deep.

Table 1. *Chemical composition of the API 5L X60 pipeline steel (wt%)*

C	Mn	Si	Cr	Ni	Mo	S	P	Cu	Ti	Al	Fe
0.21	1.17	0.23	0.12	0.19	<0.05	0.032	0.022	0.17	0.04	0.028	Bal.

*A test electrolyte designated NS4 by Trans Canada Pipe Lines containing KCl 0.122, $NaHCO_3$ 0.483, anhydrous $CaCl_2$ 0.093, $MgSO_4$ $7H_2O$ 0.131 gL^{-1}.

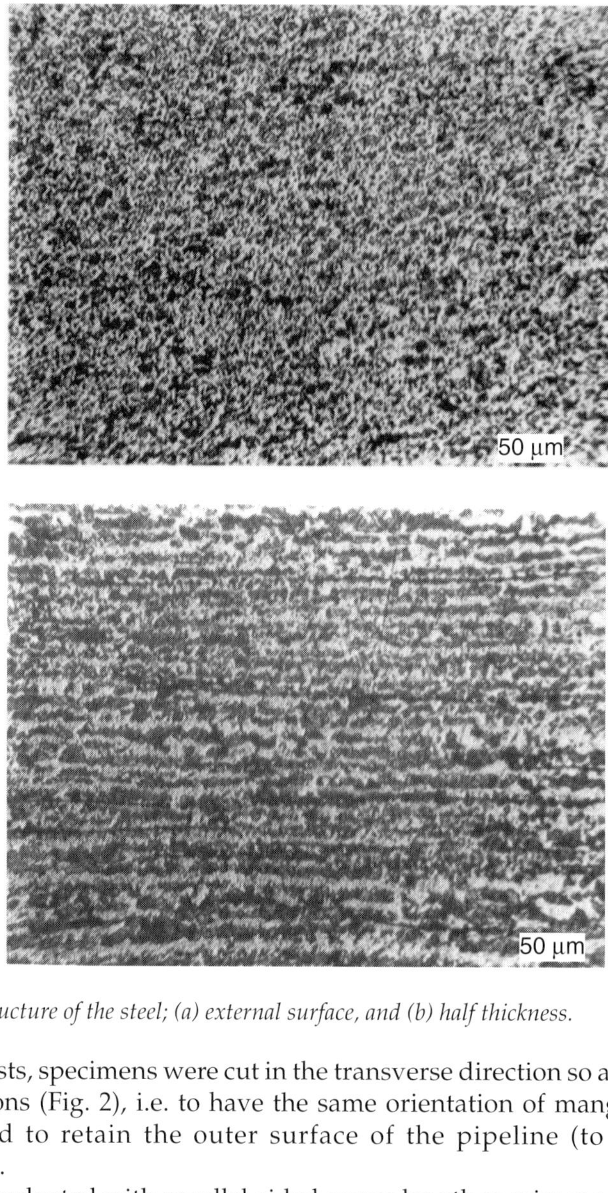

Fig. 1 Microstructure of the steel; (a) external surface, and (b) half thickness.

So, for our tests, specimens were cut in the transverse direction so as to be close to service conditions (Fig. 2), i.e. to have the same orientation of manganese sulfide and cracks and to retain the outer surface of the pipeline (to preserve the microstructure).

Tests were conducted with parallel-sided gauge length specimens 20 × 8 × 4 mm. To obtain a constant tensile stress in all the specimens during the tests, the mechanical system was designed so that the loading axis was applied in the centre of the gauge length section after the straightening stage (in the first part of the test) of the curved specimen (Fig. 2). Electrical resistance strain gauges were attached to the gauge length (three on the internal surface and three on the external surface of the specimen). The best position of the initial loading axis was chosen from results of strain measurements.

To simulate pressure fluctuations in service, fatigue tests are usually used. Such accelerated tests do not allow a comparison with initiation and propagation kinetics

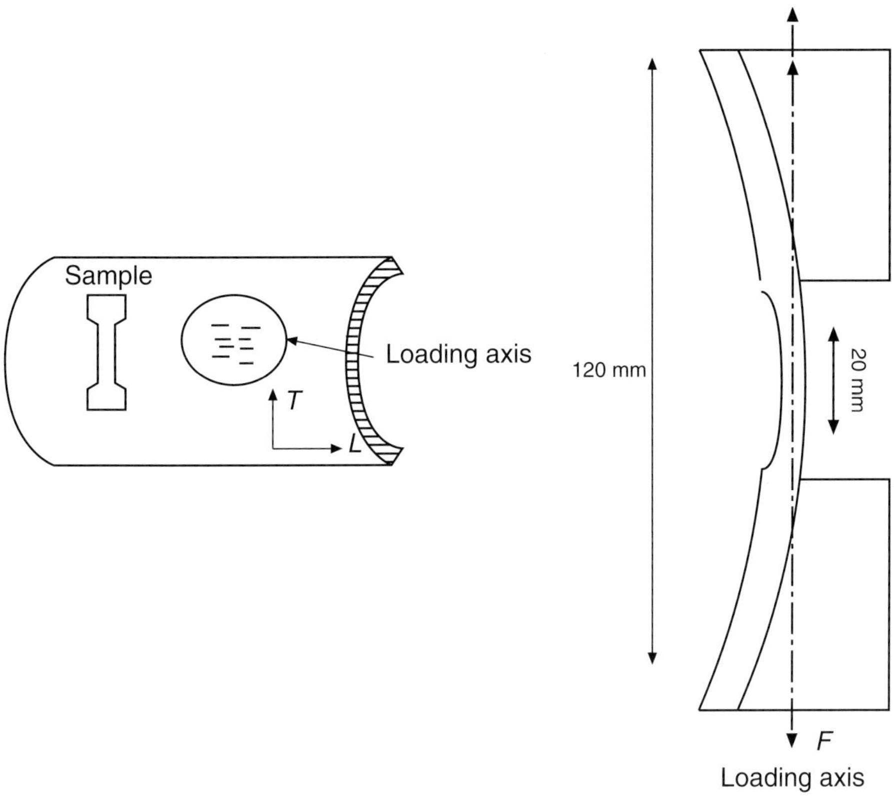

Fig. 2 *Position of the specimen relative to the orientation of cracks and inclusions as observed on pipe and specimen.*

of cracks in service. In this study, Slow Strain Rate Tests (SSRT) were chosen to study susceptibility to SCC with respect to surface conditions and cathodic polarisation. Slow strain rate tests with a strain rate of 4×10^{-7} s^{-1} were performed up to rupture in NS4 solution saturated with 5% CO_2/balance N_2 (pH = 6.7) at room temperature. These conditions simulated the dilute ground water in presence of carbon dioxide that has been associated with transgranular stress corrosion cracking of pipelines. The tests were conducted on polished and non-polished samples at various applied potentials.

Scanning electron microscopy observations were made to observe rupture surfaces. Longitudinal sectioning of each tested specimen allowed the density and depth of cracks to be determined.

All potentials are given with respect to the saturated KCl calomel electrode (SCE).

3. Experimental Results and Discussion

3.1. Fractographic Observations

Figure 3 shows micrographs of the rupture surface obtained after tests conducted up to rupture. Transgranular cracks were observed at the corrosion potential or when cathodic polarisation was applied. This transgranular form is often detected in service and associated with pH values of 6.5 resulting from the presence of carbon dioxide.

3.2. Cathodic Polarisation on Polished Samples

Figure 4 presents the stress-strain behaviour of polished samples in air and in the NS4 corrosive environment at the open circuit corrosion potential (OCP) E_{corr} = –780 mV and at –950, –1500 and –2000 mV. Rupture elongation and ultimate tensile strength were nearly the same for the specimens tested in air and at E_{corr}. At cathodic potentials, the mechanical behaviour was clearly modified.

The distribution of cracks for tests conducted at E_{corr} (OCP), –1500 and –2000 mV are also given in Fig. 4. Whatever the potential conditions, the density and the depth

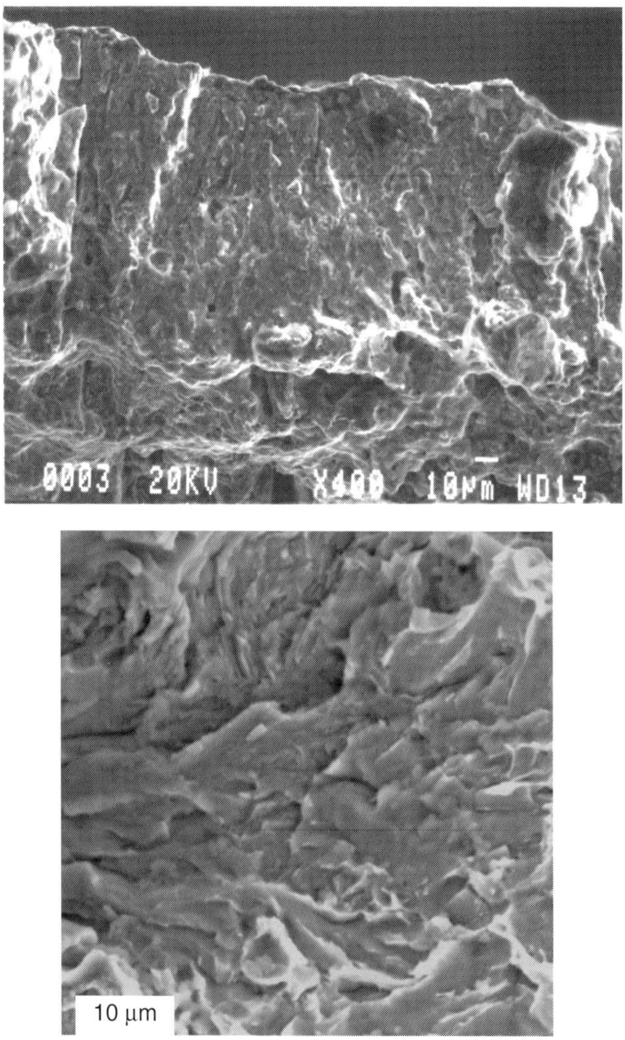

Fig. 3 Transgranular crack observed on rupture surface.

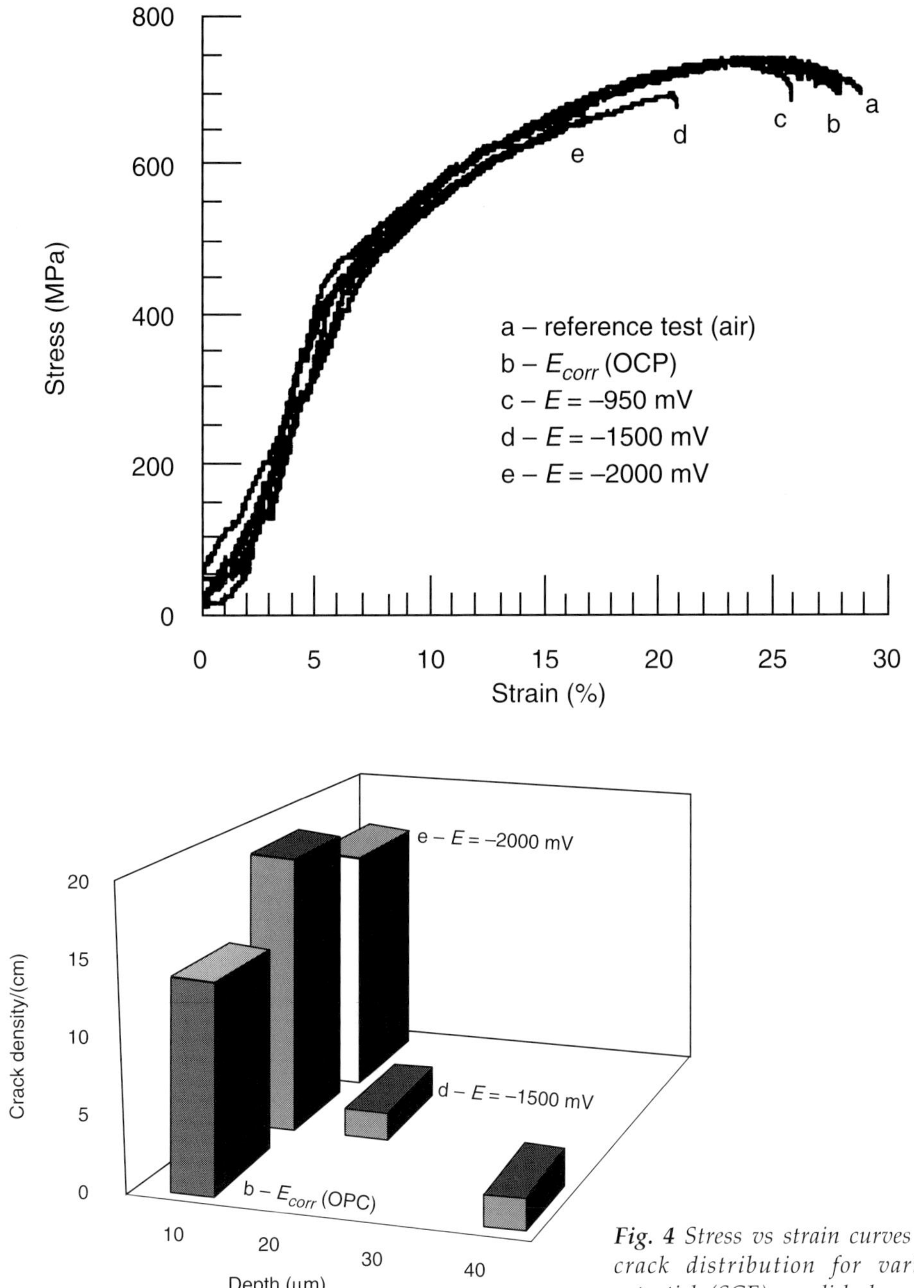

***Fig. 4** Stress vs strain curves and crack distribution for various potentials (SCE) on polished samples.*

of cracks remained the same, the cracks being very short (mainly smaller than 10 μm). So, the modification of the mechanical behaviour of the material was not due to the presence of microcracks since these were not deep enough, but is probably brought about by a uniform diffusion of atomic hydrogen into the metal leading to a decrease of ductility.

3.3. Surface Conditions

Figure 5 presents the stress–strain behaviour of polished and non-polished specimens in the corrosive environment under open circuit potential conditions of –780 mV for the polished and –580 mV for the non-polished. The distribution of cracks is also shown. The ductility was slightly modified for the non-polished tested material. On the other hand, four times as many cracks were observed on the as-received surface sample. In both cases, cracks were short (smaller than 10 μm).

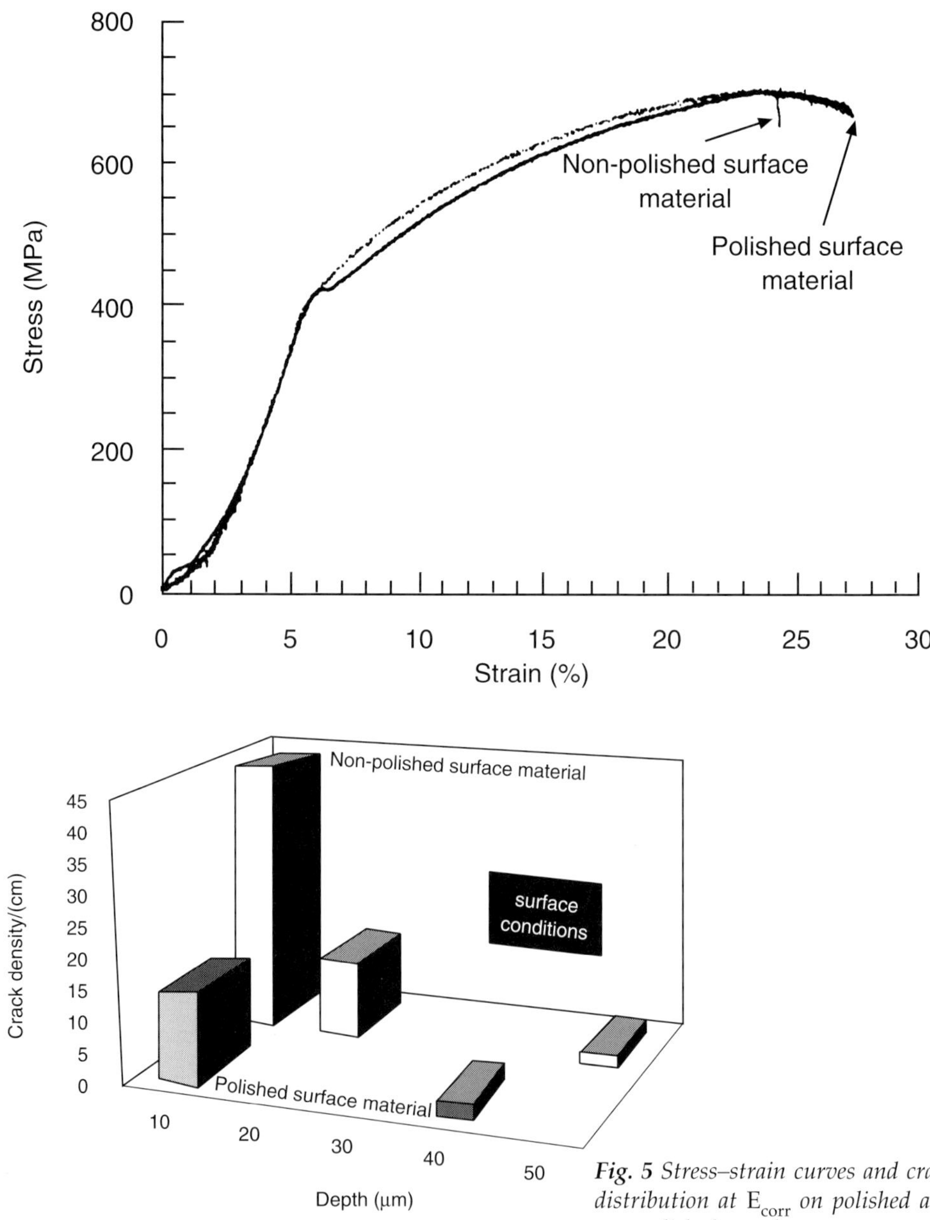

Fig. 5 Stress–strain curves and crack distribution at E_{corr} on polished and non-polished samples.

Figure 6 shows defects observed on a longitudinal section of two specimens (polished and as-received surfaces). Initiation of damage was very different in the two cases. The geometry of pits obtained on the as-received surface was comparable to that of pits from which cracks often initiated in service.

Initiation of cracks was significant when a non-polished sample was used, but propagation was not accelerated at the corrosion potential.

3.4. Cathodic Polarisation Applied in the Plastic Range

Figure 7 presents the stress–strain behaviour on non-polished surface samples in the NS4 corrosive environment at E_{corr} (OCP), at –1500 mV and for a test started at E_{corr} with the steel then cathodically polarised to –1500 mV in the plastic region. The distribution of cracks for each test is also presented. Mechanical behaviour was significantly modified between E_{corr} and both the other tests. A decrease in ductility

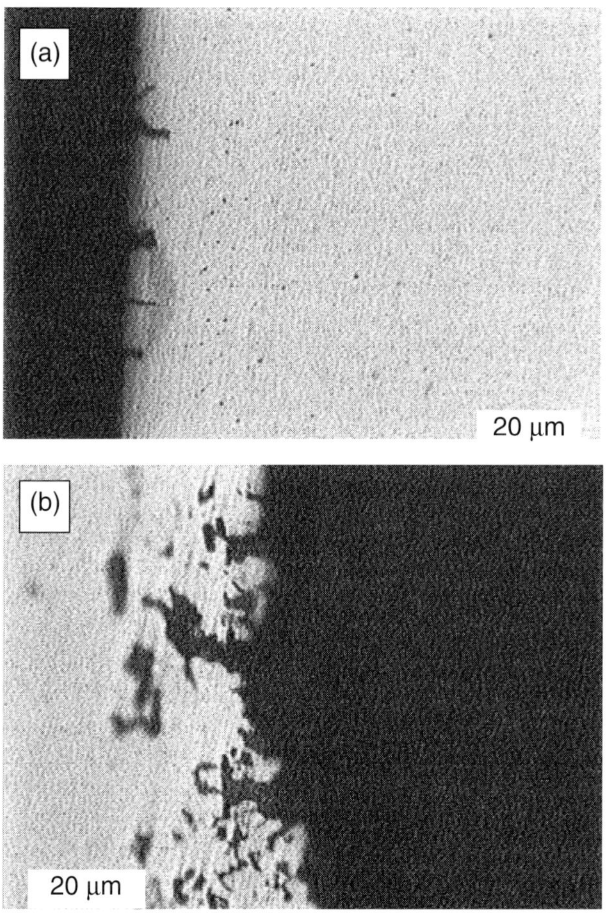

***Fig. 6** Microscopic observations of longitudinal sections of specimens after a test conducted at the corrosion potential; (a) polished specimen, and (b) as-received surface specimen.*

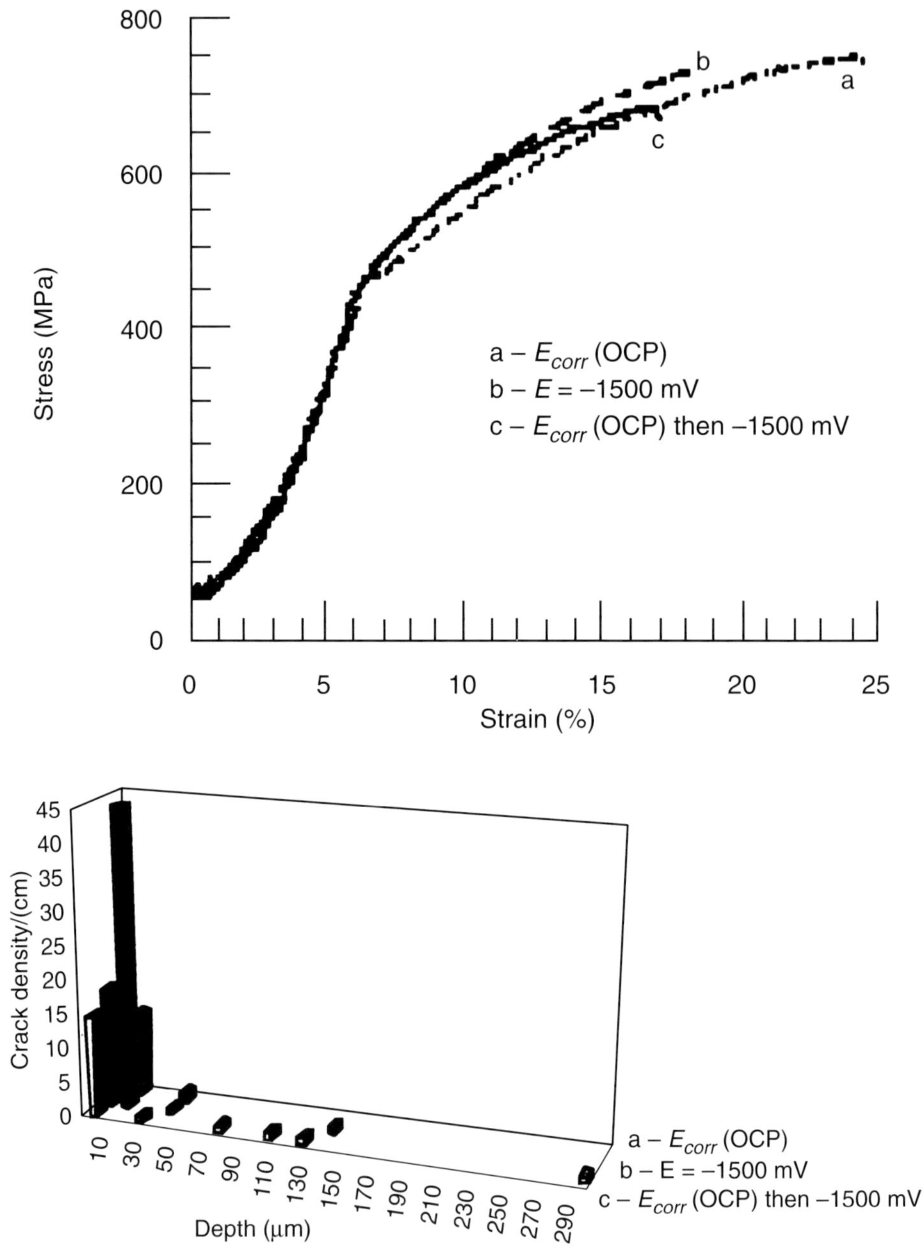

Fig. 7 *Stress–strain curves and crack distribution at* E_{corr}, *–1500 mV and for a cathodic polarisation at –1500 mV applied in the plastic range on non-polished samples.*

was observed when cathodic polarisation was carried out in the plastic region (test c) compared with test (b). The distribution of cracks was very different for each test, thus with:

- test (a): small cracks or pits,
- test (b): fewer cracks than at E_{corr} but one is deep (150 µm),
- test (c): same density of cracks as at –1500 mV but four long cracks were found.

Figure 8 shows a deep crack (150 µm) obtained in a test started at E_{corr} and then cathodically polarised at –1500 mV in the plastic region (test c).

These results were compared to studies carried out by Magnin [8] on aluminium alloys and austenitic stainless steels in which cathodic polarisation applied in the plastic region decreased the mechanical resistance because initiation of pits in the first part of the test (at E_{corr}) promoted the localised entrance of hydrogen into the material when the cathodic polarisation was applied.

3.5. Sensitivity of the Steel to Corrosion-Deformation Interactions

To compare all the results obtained in this study, a criterion was chosen to quantify the sensitivity of the steel for various conditions. This criterion $S(E)$ was defined by the relation:

$$S(E) = 1 - \frac{E_{SCC}}{E_{ref}}$$

where E_{SCC} and E_{ref} were areas under stress–strain curves for tests conducted in the NS4 corrosive environment and in air respectively. Figure 9 shows the change of $S(E)$ with potential for the various cases presented above.

Those results clearly show:

(i) an increasing sensitivity of the tested material to the environment whatever the surface conditions when the cathodic polarisation is increased,

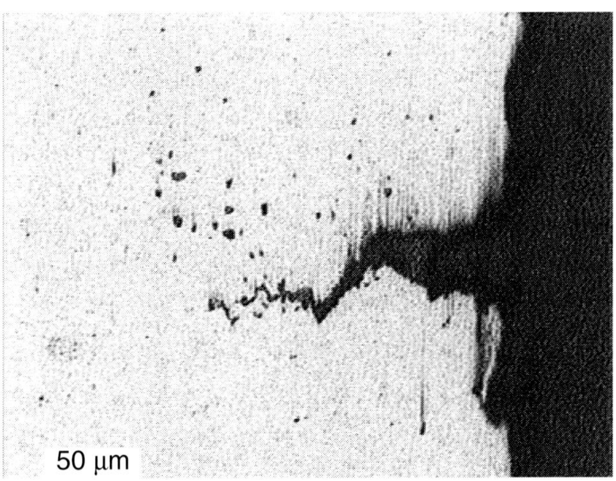

Fig. 8 Microscopic observation of longitudinal section of specimen after test (c).

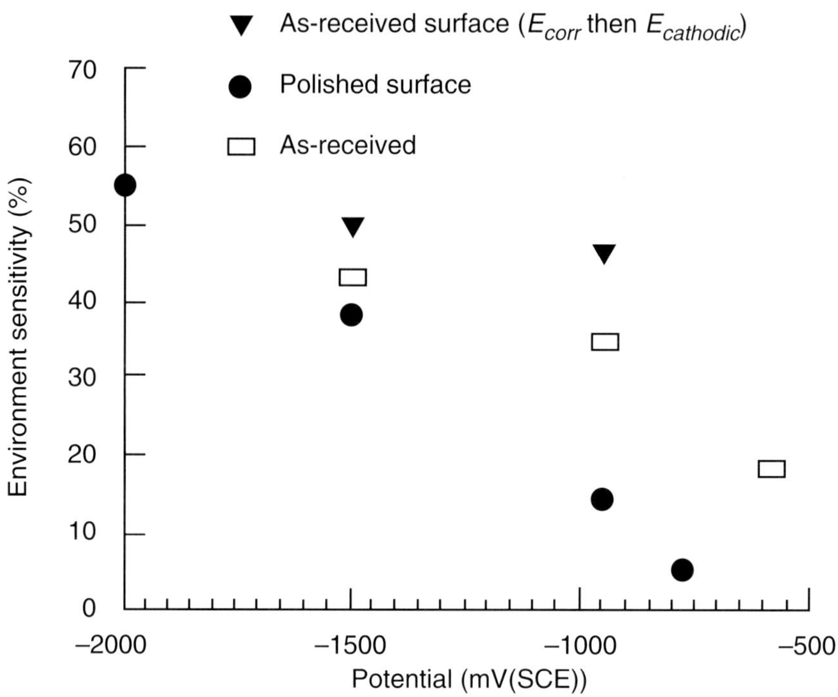

Fig. 9 *Influence of surface conditions and applied potential on the sensitivity to damage.*

(ii) for a fixed potential, sensitivity of the material increased when the as-received surface was used: localised entrance of hydrogen could explain this phenomenon,

(iii) this effect of localisation of damage was more important when cathodic polarisation was applied in the plastic range.

4. Conclusion

This study has confirmed a hydrogen effect in cracking observed on pipeline steel in near neutral pH soil environment. The hydrogen effect appears to be enhanced when a localisation of the corrosion occurs. Heterogeneities of the pipe outer surface (for example, surface defects, coating residues) probably induce these phenomena. The results obtained with cathodic polarisation imposed in the range of plastic strain after pit initiation at the corrosion potential could be representative of real conditions. Depending on the season and on the soil humidity — which can lead to variations in soil conductivity — the cathodic protection can be more or less effective. These conditions could promote pit initiation and then local penetration of hydrogen into the material. As shown in this study, these situations can be the most damaging to steels.

5. Acknowledgements

This work was conducted at the University of Bordeaux 1 and supported by Trapil France.

References

1. M. Wilmott, B. Erno and T. Jack, *International Pipeline Conf.*, Vol.1, 399–408, ASME, 1998.
2. E. A. Charles and R. N. Parkins, *Corrosion*, 1995, **51**, 518–526.
3. R. N. Parkins and S. Zhou, *Corros. Sci.*, 1997, **39**, 159–173.
4. A. K. Pilkey, S. B. Lambert and A. Plumtree, *Corrosion*, 1995, **51**, 91–97.
5. R. B. Rebak, Z. Xia, Z. Sklarska-Smialowska, *Corrosion*, 1996, **52**, 396–405.
6. E. Sinigaglia and M. Cabrini, *Corrosion '97*, Paper No. 202, NACE International, Houston, Tx 1997.
7. S. X. Mao, J. L. Luo, B. Gu, W. Yu, *International Pipeline Conf.*, Vol. I, 485–492, ASME, 1998.
8. T. Magnin, *Advances in Corrosion-Deformation Interactions*, Mat. Sci. Forum, Vol. 202, TransTech Publications, 1996.

21
Finding Optimum Positions for Field Signature Method (FSM) Corrosion Monitoring of Oil and Gas Pipelines*

P. O. GARTLAND
CorrOcean asa, Teglgaarden, N-7005, Trondheim, Norway

ABSTRACT
Internal corrosion protection and monitoring are important factors in the design and management of pipelines. The non-intrusive FSM (Field Signature Method) technology for internal corrosion monitoring offers the possibility of monitoring the corrosion rate on the pipe wall at any position along the pipeline. The present paper presents a method for finding optimum positions for corrosion monitoring. Examples of case studies on pipelines carrying wet and dried gas are presented.

1. Introduction

Some years ago CorrOcean introduced the non-intrusive FSM (Field Signature Method) technology for monitoring internal corrosion, erosion and cracking in pipelines and process piping [1]. The Field Signature Method monitors corrosion directly at the pipe wall of a selected section of the pipeline, with a high sensitivity (1/1000 of the wall thickness for general corrosion). The life of the instrumented pipeline section is identical to the life of the pipeline, and accordingly no regular maintenance and repair of the pipeline installed equipment is required. Field Signature Monitoring is therefore ideal for monitoring locations where access is difficult, e.g. at sections of underground and subsea pipelines.

Figure 1 shows a typical arrangement and instrumentation for FSM monitoring of a pipe section. The FSM method is based on feeding an electrical direct current through the selected section of the pipeline to be monitored and at the same time measuring the potential differences in the pipe wall between pairs of electrodes in the longitudinal direction. When the signals from the electrode pairs are compared to the signals obtained when the monitoring started, the changes will reflect the reduction in the wall thickness. The electrodes and the current feed connections are all welded to the pipe wall on the outside. The

*The greater part of this paper was presented at the NACE *Corrosion '98* conference and NACE International in the copyright holder.

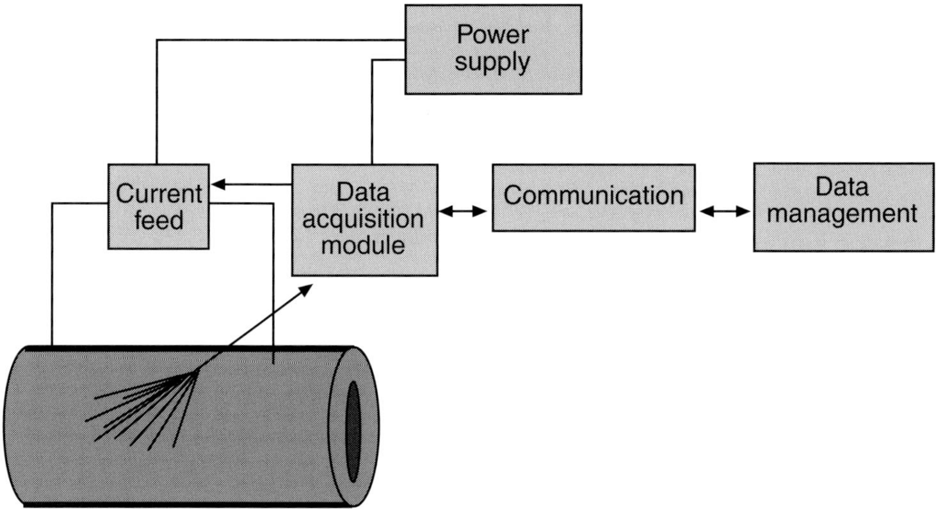

Fig. 1 *Schematics of the FSM principle and instrumentation.*

measured signals will, however, give information about corrosion and corrosion rates at the inside of the pipe.

In order to utilise the high positional flexibility of the FSM technology there has been an obvious need for a method that can pick out optimum locations. An optimum location in this respect is a location that satisfies the requirement of monitoring a representative high corrosion rate combined with the requirement of minimising the costs associated with installation and signal transmission. In this paper we outline the ideas and the methods invoked to establish CORPOS, the working name of the program that has the scope of predicting the corrosion rate versus position along any pipeline carrying corrosive hydrocarbons.

2. Strategy

The strategy has been as follows:

- Collect information about CO_2 corrosion mechanism and models describing the mechanism.
- Identify corrosion related parameters that vary with the position along the pipeline.
- Establish methods of calculating the corrosion related parameters along the pipeline.
- Select an appropriate CO_2 corrosion model.
- Establish a positional corrosion model (CORPOS) by taking all the positional

dependent corrosion related parameters as input to the chosen corrosion model.

3. Identification of Corrosion Related Parameters

The following parameters are considered as important for the variation of the corrosion rate along a pipeline: fluid flow, presence and nature of a water phase, temperature, pressure and pH. In fluid flow both the flow regime and the flow rate are of importance. The water may be present as produced water, condensed water or glycol/water phase in dried gas systems. The temperature has a direct influence on the corrosion rate and an indirect influence through the water phase (water condensation, water percentage in glycol), the flow conditions and the films formed on the surface. The pressure has a direct influence through the CO_2 partial pressure, and an indirect influence through the water condensation rate and the flow conditions. The pH has a direct influence on the corrosion rate.

4. Methods of Calculating the Corrosion Related Parameters

4.1. Flow

Estimation of flow characteristics in two and three phase flow is today routinely simulated by the use of computer based simulation programs like OLGA [2]. If it is felt that the flow characteristics are crucial for the results we normally recommend to run an OLGA simulation of the problem. For problems where the flow is of secondary importance a simpler flow model can be used to estimate the flow data. The model is based on the equations relevant for two phase flow in nearly horizontal pipelines, as originally presented by Taitel and Duckler [3].

4.2. The Water Phase

In a three phase system of gas, oil and water, the water may mix with the oil to form a dispersed phase or it may exist as a separate phase at the bottom of the pipe. The corrosivity will normally be higher in the latter case. In CORPOS we use the method of Wicks and Fraser [4] to calculate the probability of water drop out.

In systems with negligible amounts of produced water the liquid water phase may arise from water condensing from the vapour phase. The water condensation rates are calculated from thermodynamic equilibria.

Thermodynamic equilibria also form the basis for calculating the water content in the glycol carried into dried gas transportation pipelines.

4.3. Temperature

Temperature profiles along a pipeline are obtained as a result of heat transfer calculations, based upon the internal heat capacity of the pipe fluid, the average transportation rate and the heat transfer resistance of the pipe wall, the coating and the pipeline surroundings at the seabed.

4.4. Pressure

Pressure profiles are obtained as a result of the flow calculations, i.e. by the use of programs like OLGA.

Temperature and pressure profiles are usually available from the pipeline operator at a very early stage in the development project, and it has up to now not been necessary to run such calculations within a CORPOS study.

4.5. pH

The pH is a complex function of several parameters, like the CO_2 partial pressure, the temperature and the content of ions in the water phase. The pH can be calculated from the ion concentrations using equations describing the chemical equilibria involved, but simpler relations between the pH and the Fe^{2+} content are also used when appropriate. Since the pH itself is a parameter of the corrosion model the pH profile and the corrosion rate have to be calculated in a forward stepping approach down the pipeline.

5. Selection of a CO_2 Corrosion Model

As the basic CO_2 corrosion model we have chosen to use the Shell model as it has been presented in its latest revision of 1995, jointly by Shell and the Norwegian Institute of Energy Technology (IFE). Following de Waard, Lotz and Dugstad [5] the overall corrosion rate, V_{corr}, can be written as:

$$\frac{1}{V_{corr}} = \frac{1}{V_r} + \frac{1}{V_m} \tag{1}$$

where V_r is the highest possible corrosion rate determined by the reaction rate, i.e. when the mass transfer rate V_m is infinitely large.

The best fit equation describing V_r in dependence of the environmental parameters is:

$$\log\,(V_r) = 6.23 - \frac{1119}{t+273} + 0.0013t + 0.41\log(ap_{CO_2}) - 0.34pH_{act} \tag{2}$$

For V_m a best fit equation is given :

$$V_m = 2.45\frac{U^{0.8}}{d^{0.2}}ap_{CO_2} \tag{3}$$

The parameters entering the equations are the temperature t, the partial pressure p_{CO_2} the actual pH_{act}, the water velocity U, the hydraulic diameter d and the fugacity coefficient a.

In the presence of glycol in the water phase it has been shown that the corrosion rate decreases with increasing glycol concentration. This has been introduced as a multiplication factor F_{glyc}, to be multiplied with the overall corrosion rate [6]:

$$\log\,(F_{glyc}) = 1.6\log(W\%) - 2 \tag{4}$$

6. Establishing a Positional Corrosion Model (CORPOS)

With all the parameters defined and calculated as outlined above, the effect on the corrosion rate as a function of the pipe position, $CR(x)$, may be obtained as indicated by the general flowchart as shown in Fig. 2. In this process, CORPOS will invoke different sub-programs for different systems like gas/oil/water, gas/condensate, wet gas and dried gas.

7. Examples of Application

7.1. Dried Gas

For a pipeline carrying dried gas two corrosive modes have been considered. The first one is corrosion due to a glycol/water phase. Glycol may enter the pipeline as 'carry over' from the glycol contactors. This glycol absorbs a few percent water from the gas. The corrosion rate is generally very low, and lowest near the inlet of the pipe, as shown in Fig. 3. This is due to the lowest water content being where the

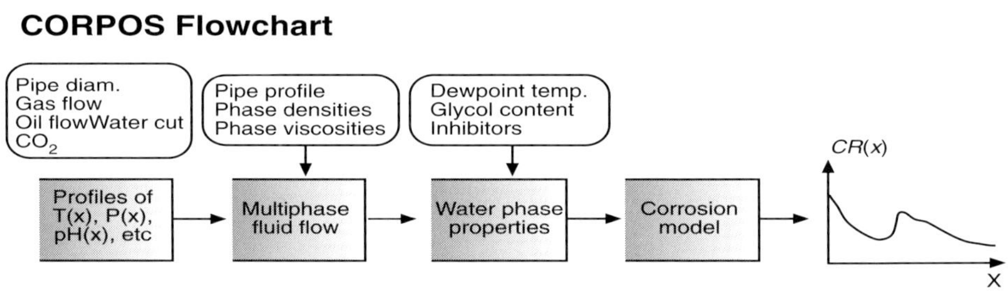

Fig. 2 Flowchart indicating the process of calculating the corrosion rate CR(x).

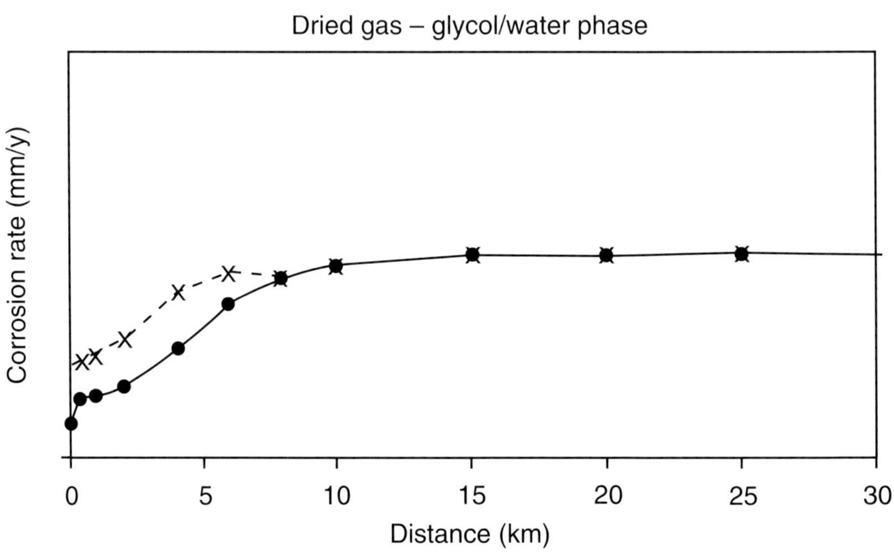

Fig. 3 CR-*profile in a pipeline carrying dried gas, at two levels of glycol input.*

temperature is highest. The second mode of corrosion is due to increased dewpoint temperature of the gas. This may cause some corrosion towards the cold end of the pipe, as shown in Fig. 4. For both corrosion modes the optimum FSM location is towards the coldest end of the pipe.

7.2. Wet Gas

A flowline carrying wet gas will have a corrosion profile with a maximum corrosion rate at the inlet. This is caused by two factors: the temperature and the pH. The corrosion rate is highest where the temperature is highest and the pH lowest, i.e. at the inlet. With the use of pH stabilising chemicals the pH effect is much reduced, but the corrosion will still have a maximum at the pipe inlet. The result of a wet gas case study with two wells coupled to the same flowline is shown in Fig. 5. In this case relevant positions for FSM are immediately downstream of the wells, with W1 as the primary optimum position.

8. Conclusions

- A mathematical model describing the internal corrosivity in oil and gas pipelines as a function of the position has been established.
- The model has been based on available empirical corrosion models, combined with methods for calculating variations in the multiphase fluid flow, temperature, pressure, pH and the water phase conditions.

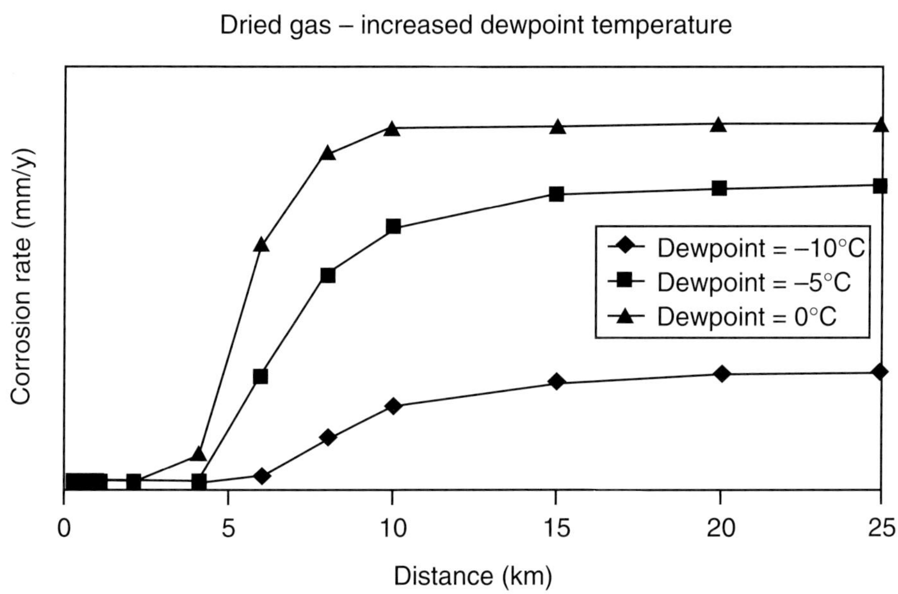

Fig. 4 CR-*profile in a pipeline carrying dried gas, at three levels of increased dewpoint temperature.*

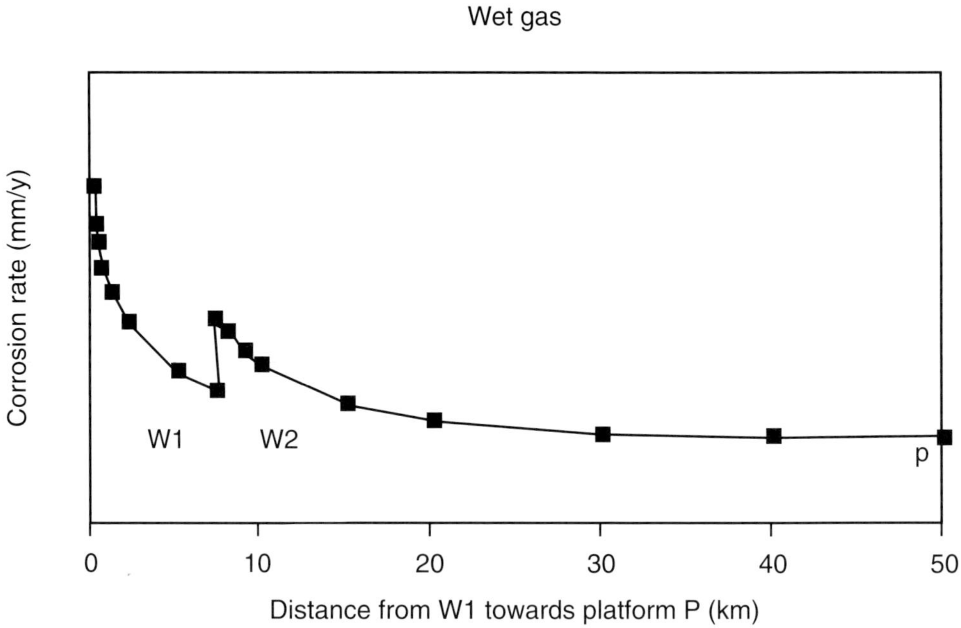

Fig. 5 CR-*profile in a pipeline carrying wet gas from wells W1 and W2.*

- The purpose of the model is to assist in the determination of the optimum locations for corrosion monitoring on a given pipeline.

- Application of the model to gas pipelines shows that the position of the maximum corrosivity will depend strongly on the conditions of the gas in the pipeline.

References

1. R. D. Strømmen, H. Horn and K. R. Wold, *Corrosion '92*, Paper No. 7, NACE International, Houston, Tx, 1992.
2. P. Fuchs and S. Nuland, "Three Phase Modelling is a Must", *Multiphase Transportation III*, Arranged by Norwegian Petroleum Society, Røros, September, 1992.
3.Y. Taitel and A. E. Dukler, "A model for predicting flow regime transitions in horizontal and near horizontal gas-liquid flow", *AIChE J.*, 1976, **22**, (1).
4. M. Wicks and J. P. Fraser, "Entrainment of water by flow oil", *Mater. Perform.*, 1975, (5), 9–12.
5. C. De Waard, U. Lotz and A. Dugstad, "Influence of liquid flow velocity on CO_2 corrosion: A semi-empirical model", *Corrosion '95*, Paper No. 128, NACE International, Houston, Tx, 1995.
6. C. De Waard, U. Lotz and D. E. Milliams, "Predictive model for CO_2 corrosion engineering in wet natural gas pipelines", *Corrosion*, 1991, **47**, (12), 1991.

Part 3

Martensitic Stainless Steels

22
Corrosion Resistance of Weldable Modified 13Cr Stainless Steel for CO_2 Applications

H. TAKABE, H. AMAYA, H. HIRATA and M. UEDA
Wakayama Steel Works, Sumitomo Metal Industries Ltd, 1850 Minato, Wakayama City, 640-8555, Japan

ABSTRACT
The effect of alloying elements on the localised corrosion resistance of weldable 13Cr stainless steels was investigated in CO_2 environments at elevated temperature. Chromium and molybdenum were effective elements to reduce the corrosion rate and localised corrosion tendency. To obtain martensitic single phase and high toughness at the heat affected zone in the non-post weld heat treated (PWHT) condition, Ni was added and the C content decreased. From these test results, weldable modified 13Cr stainless steel with ultra low C, Ni 5, Cr 13, Mo 0.7 and Ti 0.05 mass% has been developed as an improved corrosion resistant steel for CO_2 environments as compared to conventional 13Cr steel (API 13Cr). The weldable modified 13Cr stainless steel pipe, which was welded by Gas Tungsten Arc Welding (GTAW) using super duplex stainless steel (DP3WT) welding material, showed good toughness of more than 230 J cm^{-2} at –30°C in weld metal, fusion line and HAZ in the non-PWHT condition. The corrosion rate and localised corrosion resistance of the welded joint were as those of a super 13Cr stainless steel which had been developed for CO_2 + H_2S application.

1. Introduction

Weldable super 13Cr stainless steel with a chemical composition with ultra low C, Cr 12, Ni 6, Mo 2.5 and Ti of 0.07 mass% had been developed for CO_2 environments containing a small amount of H_2S [1]. This weldable super 13Cr stainless steel has higher resistance to localised corrosion at elevated temperature and to sulfide stress cracking (SSC) at room temperature than API 13Cr OCTG [2]. However, the requirements for the optimum material for each field condition have been increased recently in order to reduce the cost of well developments. Therefore, a new weldable modified 13Cr martensitic stainless steel with resistance to localised corrosion at elevated temperatures has been sought. A resistance to SSC the same as that of API 13Cr stainless steel (hereinafter this steel is referred to as conventional 13Cr stainless steel) was also required for the new steel.

In this paper, the alloy design of a weldable modified 13Cr martensitic stainless steel to be used with no post weld heat treatment (PWHT) is investigated with attention given to the resistance to localised corrosion at elevated temperature.

2. Experimental

2.1. Materials

Materials with chemical compositions given in Table 1 were used for investigating alloy design. A range of graded materials — hot rolled, quenched in water and tempered — were manufactured in the laboratory. The yield strength of these steels covered the range of grades from 550 MPa (80 ksi) to 757 MPa (110 ksi) specified minimum yield strength. Table 2 shows the chemical compositions and tensile properties of the modified 13Cr developed stainless steel, super 13Cr stainless steel OCTG with Mo 2 mass% and a conventional 13Cr stainless steel. These materials were manufactured in the mill and were quenched and tempered.

2.2. Experimental Methods

General and localised corrosion resistance were evaluated by weight loss and observation of coupons after 96 and 336 h exposure to the hostile environment in autoclaves with titanium liners. Figures 1 and 2 show the schematic illustration of the autoclave and the test specimens used to evaluate mechanical properties and corrosion performance of the welded joint. These specimens were isolated electrically from the autoclave. De-aeration was carried out by repeated alternate vacuum treatment and N_2 gas bubbling. Then, CO_2 gas was charged to achieve the experimental pressure at room temperature.

The analysis of the surface film formed on the materials in the CO_2 environment at elevated temperature was carried out by SIMS (Secondary Ion Mass Spectrometry). The susceptibility to SSC in the welded joint of the weldable modified 13Cr stainless steel was evaluated by the constant load test method with an applied stress of 90% actual yield strength of the base metal and the cyclic slow strain rate test (SSRT) method in the CO_2 environment containing a small amount of H_2S. In the cyclic SSRT, the tensile and compression test were repeated 15 times between 0.95 and 0.60

Table 1. *Chemical composition range of materials used for alloy design (mass%)*

C	Si	Mn	Cr	Mo	Ni
0.001/0.05	0.10/0.50	0.1/1.0	9.0/25.0	0/3.0	0/6.0

Table 2. *Chemical compositions and yield strength of 13Cr martensitic stainless steels*

Chemical compositions (mass%)							YS	Remarks
C	Si	Mn	Ni	Cr	Mo	Ti	(MPa)	
0.007	0.26	0.67	4.59	12.77	0.60	0.055	574/786	Modified
0.010	0.19	0.72	5.26	12.92	0.77	0.079	591/793	13Cr
0.007	0.23	0.35	6.05	12.20	2.01	0.074	591	Super 13Cr
0.21	0.46	0.52	0.12	12.84	0.02	0.001	632	Conventional 13Cr

YS = yield strength.

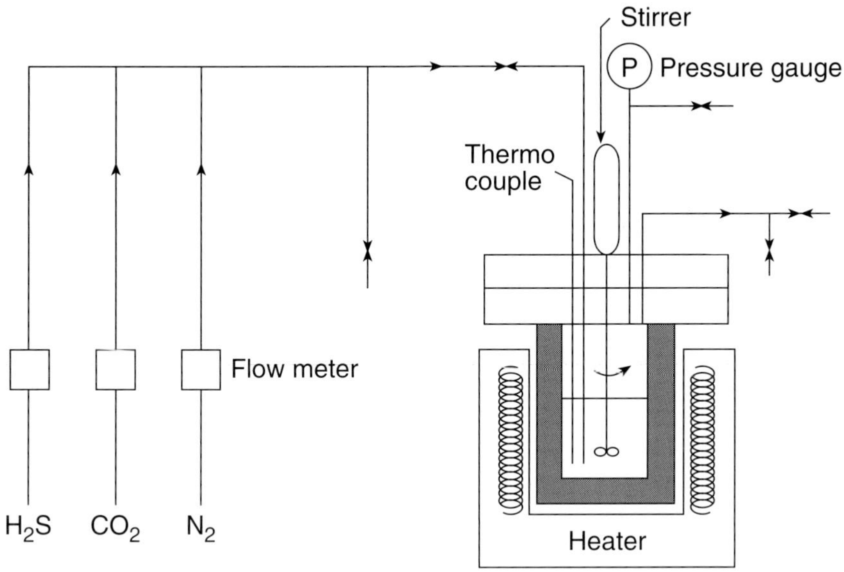

Fig. 1 *Schematic illustration of autoclave tester.*

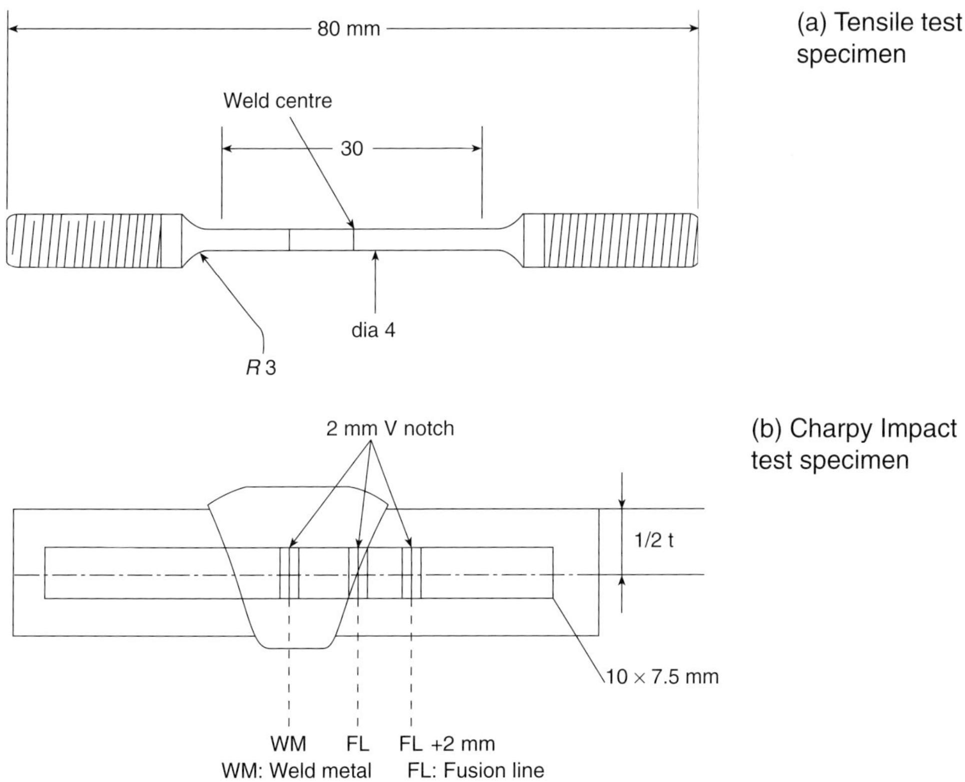

Fig. 2 *Schematic illustrations (shown above and overleaf) of the test specimens used to evaluate mechanical properties and corrosion performance of welded joints.*

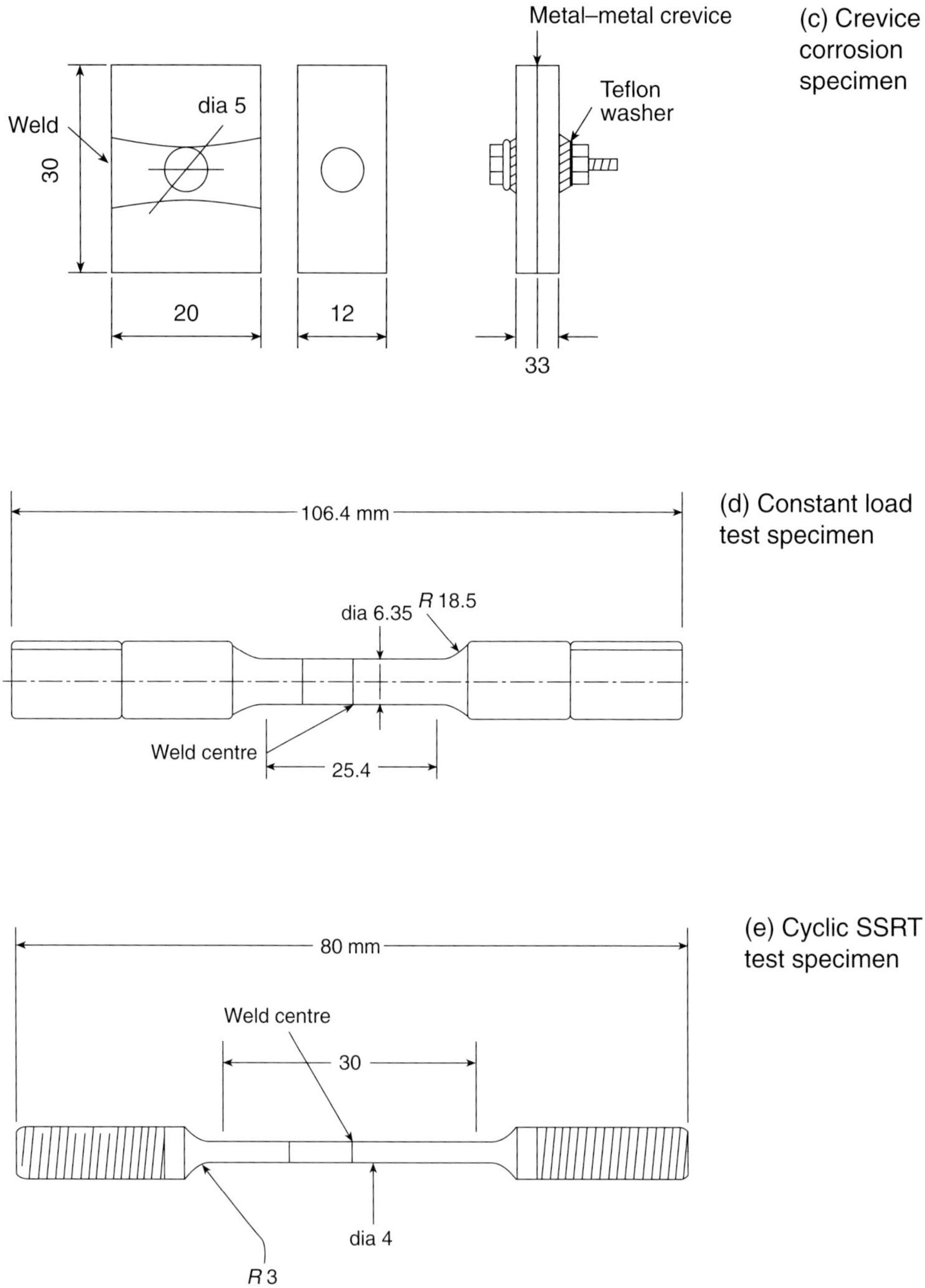

Fig. 2 *(Continued) Schematic illustration of the test specimens used to evaluate mechanical properties and corrosion performance of welded joints.*

of the actual yield strength of the base metal at a strain rate of $4 \times 10^{-6}\ s^{-1}$ in the hostile environment.

3. Results and Discussion

3.1. Alloy design

The corrosion rate of Cr-bearing steels decreased with increase in Cr content in the CO_2 environments at 150°C as shown in Fig. 3. The susceptibility to localised corrosion is shown in Fig. 4. A small amount of Mo was also effective in reducing the corrosion rate at 175°C, and no pitting was observed in the steels with more than Mo 0.25 mass%.

Furthermore, a low C content and a martensitic single phase were required to obtain good toughness and a lower increase in hardness in the heat affected zone (HAZ). The martensitic, single phase region, when austenitised at 1050°C, is shown in Fig. 5, and further demonstrated by the microstructural observations of the steels containing Cr 12–19, Ni 2–14 and Mo 0–3 mass%.

Based on these test results, the chemical composition region of the modified 13Cr stainless steel is given in Fig. 6 with the Ni content fixed at 5 mass%. Titanium was added to the new steel to prevent microstructural coarsening as in the weldable super 13Cr stainless steel. Ultra low C (0.01 mass%), 13Cr–5Ni–0.7Mo–Ti was the selected chemical composition for the newly developed modified 13Cr steel.

The good general and localised corrosion resistance of the new steel was also confirmed from analysis by SIMS of the surface film formed in the hostile environment. Figure 7 shows the distributions of elements in the surface film formed in the CO_2 environment at 150°C as compared with before immersion testing. After the immersion test, a Cr-enriched surface layer of more than 30 nm thickness was observed. It is considered that this Cr enrichment in the surface film was responsible for the good corrosion resistance of the new steel.

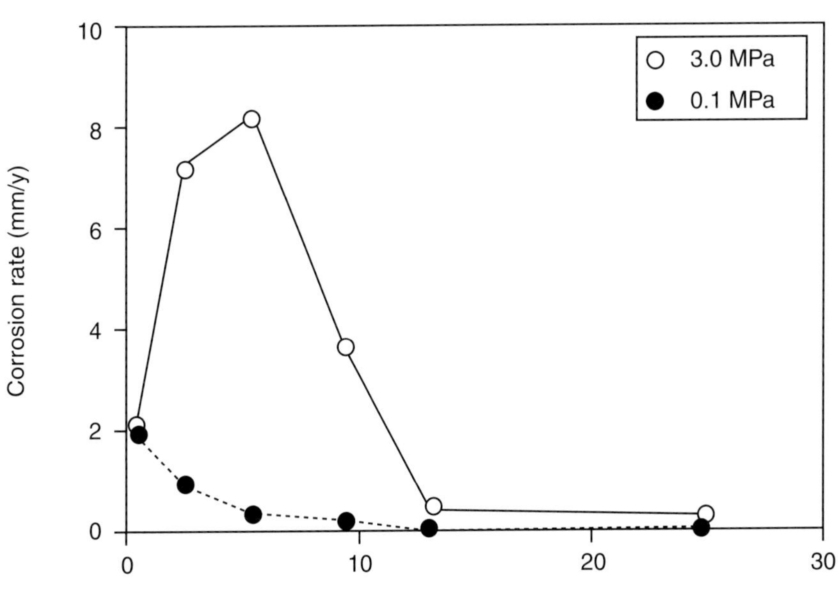

Fig. 3 Effect of Cr content and CO_2 partial pressure on corrosion rate (5%NaCl, 150°C, 96 h).

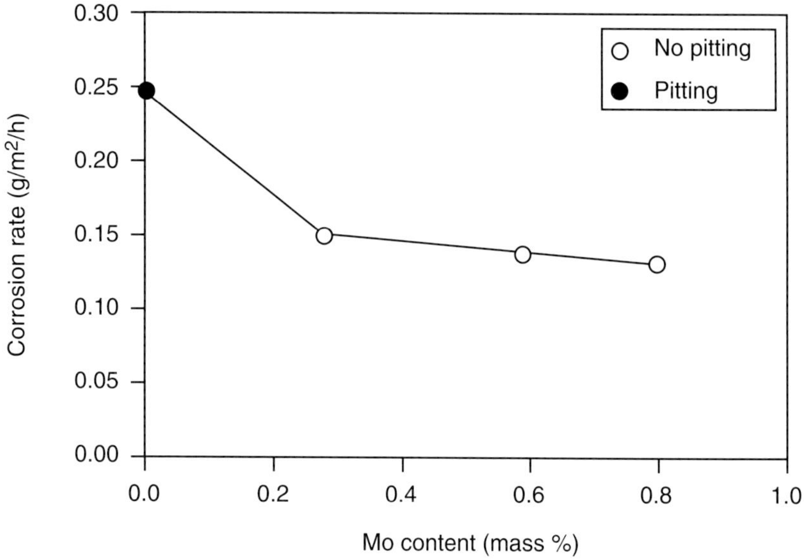

Fig. 4 *Effect of Mo content on corrosion rate and localised corrosion (25%NaCl, 3.0 MPa CO_2, 175°C, 336 h).*

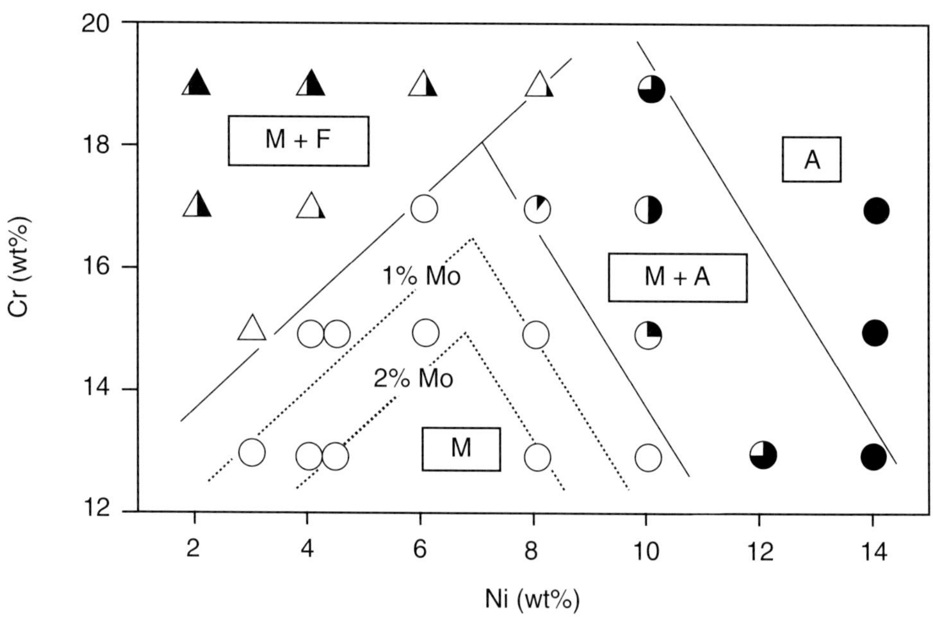

Fig. 5 *Effect of Mo content on martensitic single-phase range (0% Δ: Ferrite content in martensitic phase 100%▲, 0%○: Austenite content in martensitic phase 100%●).*

3.2. Corrosion Performance of the Welded Joint

The corrosion behaviour of the new weldable modified 13Cr stainless steel with a minimum yield strength of 550 MPa described in Table 2 was investigated. The new

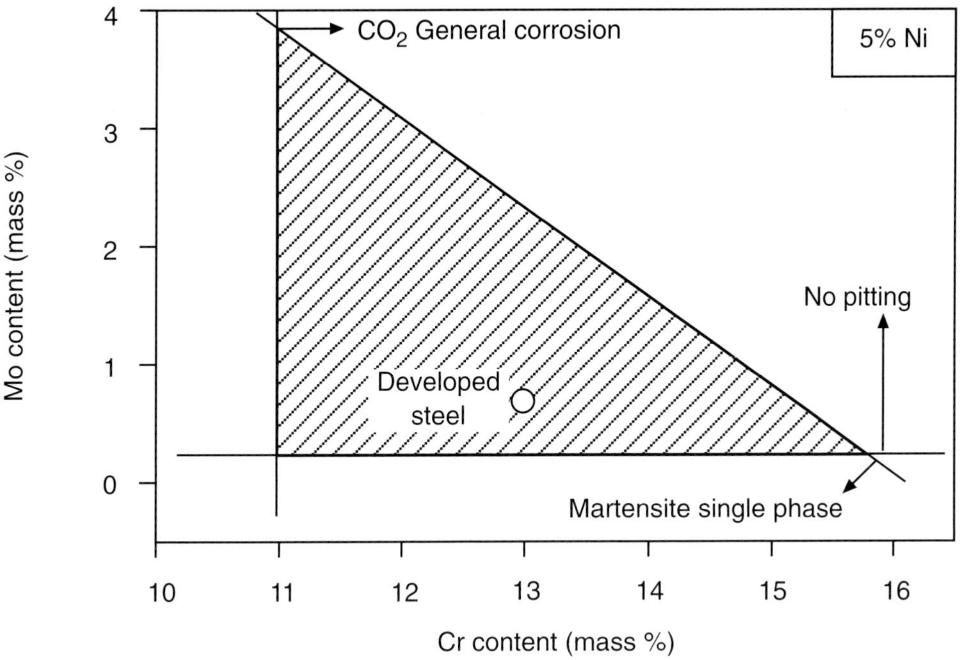

Fig. 6 *Recommended Cr and Mo contents of the new weldable modified 13Cr martensitic stainless steel for CO_2 environments.*

steel was melted and manufactured in the mill as seamless pipe of 88.9 mm o.d. × 9.52 mm w.t. The welded joint was produced by the gas tungsten arc welding gas tungsten are welding (GTAW) process using super duplex stainless steel, DP3WT (Cr25–Ni9–Mo3–W2–N0.3) as welding material. The welding conditions are shown in Table 3: PWHT was not conducted.

No noticeable coarsening was observed in the HAZ due to the Ti addition of 0.05% mass. The tensile test results of the welded joint showed that the tensile strength was more than 620 MPa and the fracture portion in the base metal was as shown in Table 4. The impact value at –30°C was more than 230 J cm^{-2} in the weld metal, fusion line and HAZ (2 mm from the fusion line) as shown in Table 5. In a side bend test of the welded joint, no cracking was observed on the surface of the bent specimen.

The hardness distribution in the welded joint 1 mm in from the inner surface is

Table 3. *Welding condition*

Pass	Welding current (A)	Welding voltage (V)	Welding speed (cm/min)	Welding sequence
1	110	13	10	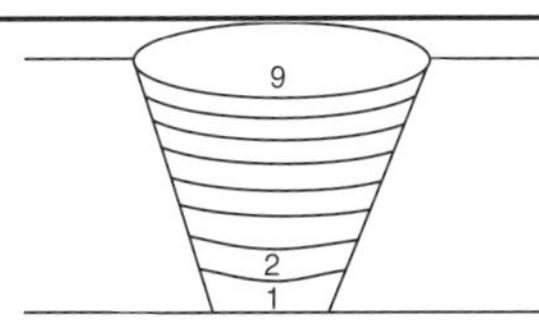
2	130	13	10	
3–9	160	14	10	

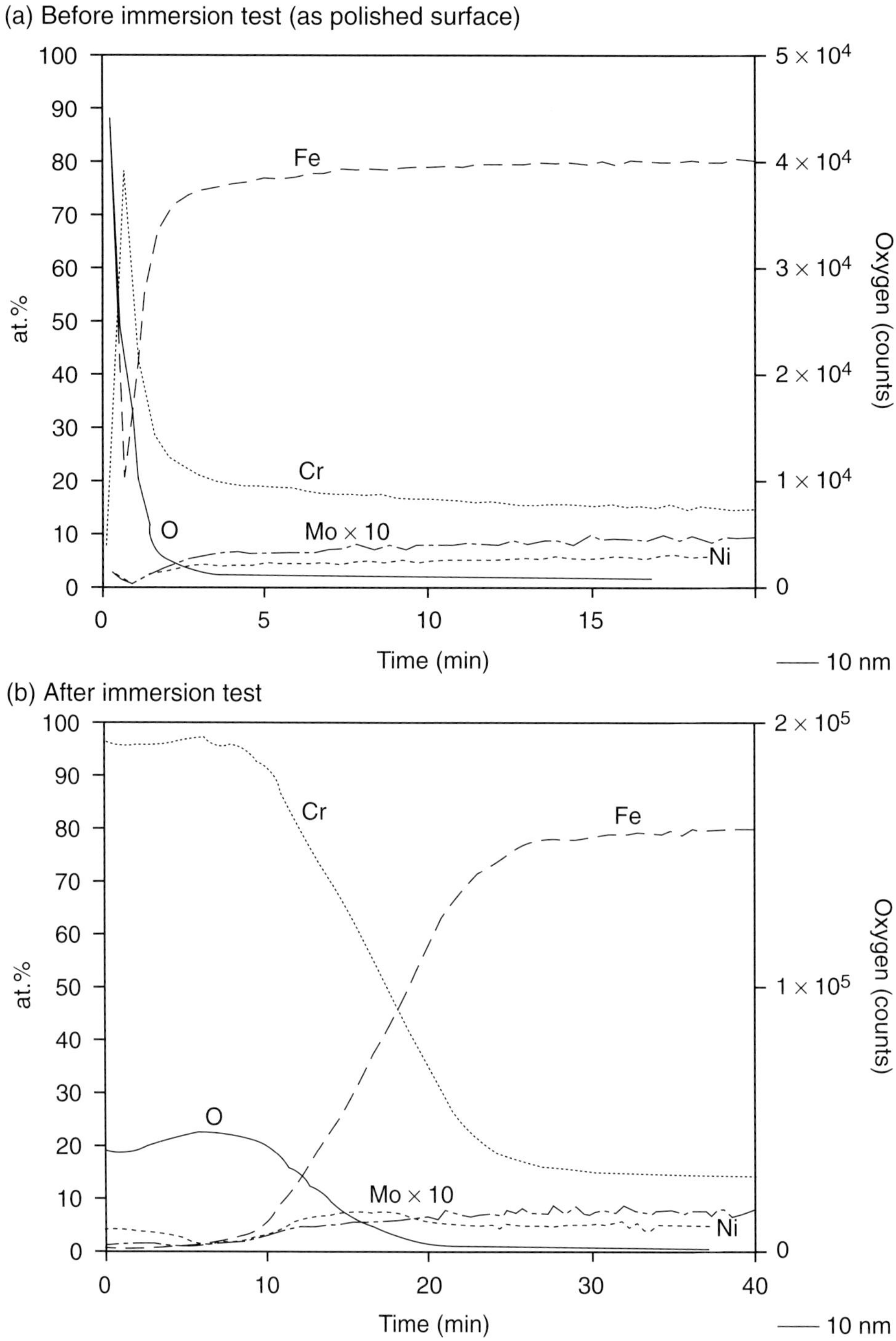

Fig. 7 *Element distribution in surface film by SIMS analysis (14 gL^{-1} Cl$^-$, 2.7 MPa CO_2, 150°C, 48 h) (oxygen (counts): arbitrary count units in SIMS analysis).*

shown in Fig. 8. No increase of hardness in the HAZ, fusion line and weld metal was observed because of the ultra-low C content in the steel and the use of DP3WT as welding material.

The resistance to localised corrosion of the welded joint was evaluated by using crevice corrosion test specimens with metal-to-metal as crevice former. One of the crevice specimens had weld metal in the centre, and the other was taken from the base metal. No crevice corrosion was observed (see Fig. 9) even in the CO_2

Table 4. *Tensile properties of welded joint*

Tensile strength (MPa)	Fracture portion
763	Base metal
766	Base metal
X80 spec.: Min. 620MPa	

Table 5. *Results of welded joints Charpy impact test (J cm^{-2})*

Notch portion	Weld metal	Fusion line	Fusion line +2 mm
0°C	293	330	300
	275	338	310
	295	320	300
–30°C	253	323	298
	233	315	298
	263	328	300

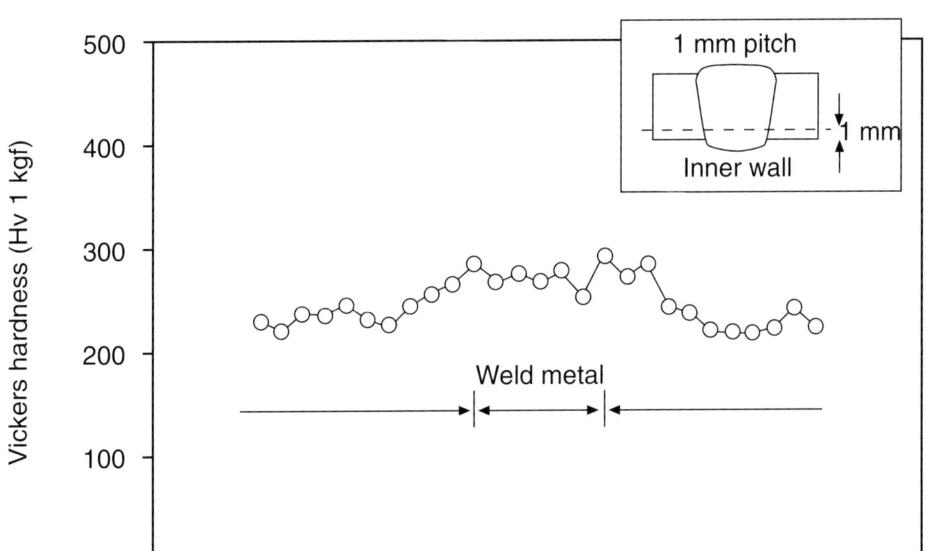

Fig. 8 *Hardness distribution in weldment.*

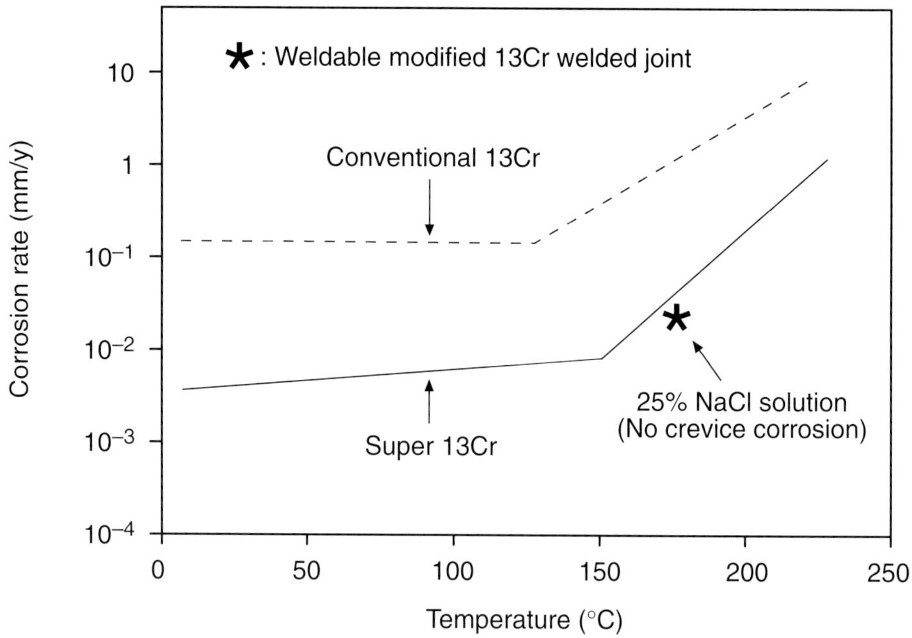

***Fig. 9** Effect of high Cl^- ion concentration on CO_2 corrosion of the new weldable modified 13Cr steel (3.0 MPa CO_2, 5%NaCl, duration 336 h).*

environment with 25% NaCl at 175°C. Also, the corrosion rate of the welded joint was the same as that of the super 13Cr stainless steel base metal OCTG with Mo 2 mass%. Sulfide stress cracking resistance was also evaluated by the constant load method and cyclic SSRT using specimens with weld metal in the centre of the gauge length as shown in Figs 2 (d) and (e). In constant load tests, the test condition was 0.0008 MPa H_2S (CO_2 balance), 68 000 ppm Cl^-, pH 4.5, applied stress of 90% actual yield strength of base metal, and a test duration of 720 h. No cracking in the welded joint was observed under these conditions. Table 6 shows the test condition for the cyclic SSRT. Figure 10 shows the cyclic SSRT test and constant load test results. The domain, where welded joints of the modified 13Cr steel did not suffer SSC, corresponds to Domain 1, designated as Non-Sour Domain by Kermani *et al.* [3].

***Table 6.** Test conditions of corrosion resistance in CO_2 environment with a small amount of H_2S*

Test method	Temp. (°C)	H_2S (MPa)	CO_2 (MPa)	Solution	pH	Applied stress
Constant load	25	0.0008	Bal.	68 000 ppm Cl^-	4.5	90% actual YS of base metal
Cyclic SSRT	25	0.0003 /0.01	Bal.	10% NaCl	3 /4.5	15 cycles from 60% to 95% actual YS of base metal

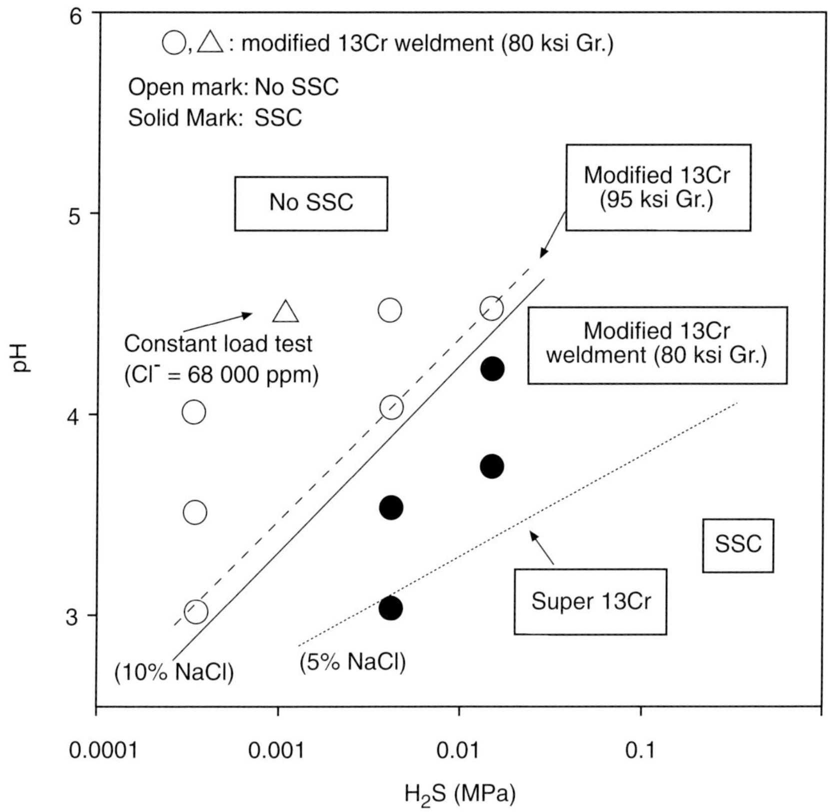

***Fig. 10** Effect of H_2S partial pressure and pH on SCC susceptibility of modified 13Cr stainless steel (cyclic SSRT test results).*

4. Conclusions

1. A weldable modified 13Cr stainless steel with ultra low C, Ni 5, Cr 13, Mo 0.7 and Ti addition of 0.05 mass% has been developed, as an improved corrosion resistant steel for CO_2 environments as compared with conventional 13Cr steel.

2. The weldable modified 13Cr stainless steel pipe, which was welded by GTAW using super duplex stainless steel (DP3WT) welding material, showed good toughness more than 230 J cm^{-2} at –30°C in the weld metal, fusion line and HAZ in the non-PWHT condition.

3. The corrosion rate and localised corrosion resistance of the welded joint were as those of super 13Cr stainless steel OCTG containing Mo 2 mass% at 175°C in a CO_2 environment with 25% NaCl.

5. Acknowledgements

The authors wish to thank Sumitomo Metal Industries, Ltd. for allowing publication of this research. The assistance and discussion of co-workers in laboratories are gratefully acknowledged.

References

1. M. Ueda, H. Amaya, K. Kondo, K. Ogawa, T. Mori and T. Kobayashi, Weldable Super 13Cr stainless steel for CO_2 environment with a little amount of H_2S, *EUROCORR' 96*, Session VII, Corrosion in Oil and Gas Production, 1996. Extended Abstracts. CEFRACOR International, Soc. Chim. Ind., Paris 1996.
2. M. Ueda, H. Amaya, K. Kondo, K. Ogawa and T. Mori, *Corrosion '96*, Paper No. 58, NACE International, Houston, Tx, 1996.
3. M. B. Kermani, D. Harrop, M. L. R. Truchon and J. L. Crolet, *Corrosion '91*, Paper No. 21, NACE International, Houston, Tx, 1991.

23
Corrosion Performance of Weldable 12% Chromium Stainless Steel Seamless Line Pipes

Y. MIYATA, M. KIMURA, T. TOYOOKA, Y. NAKANO and F. MURASE
Technical Research Laboratories, Kawasaki Steel Corporation, 1-1 Kawasaki-cho, Handa, Aichi 475-8611, Japan

ABSTRACT

The corrosion performance of two types of weldable martensitic stainless steel seamless pipe was studied: the 11Cr steel pipe (C0.01, Cr11, Ni1.5, Cu0.5, N0.01) for CO_2 environments and the 12Cr steel pipe (C0.01, Cr12, Ni5, Mo2, N0.01) for environments containing CO_2 and a small amount of H_2S.

The 11Cr steel had better CO_2 corrosion resistance than the API 5CT L80 13Cr OCTG steel. This is attributed to the fact that the two steels had the same amount of dissolved Cr and because the 11Cr steel also contained Ni, both elements being effective in promoting resistance to CO_2 corrosion.

Pitting which leads to sulfide stress corrosion (SSC) occurs more easily in the heat affected zone (HAZ) of welds when stress is applied because the passive film on the HAZ breaks down more easily than that on the base metal; the quenched martensite of the HAZ elongates more than that on the base metal when stress is applied.

The cathodic protection of 11Cr and 12Cr steel pipes in sea water does not cause hydrogen embrittlement, since hydrogen hardly enters these steels.

1.Introduction

As oil and gas resources have continued to be exhausted over recent years, most of the fields that remain untapped are located in frontier regions in deep water, offshore or in the Arctic region. Therefore, further technical innovations and cost saving in development and operation in the field are required. Moreover, the materials with higher corrosion resistance that are used for Oil Country Tubular Goods (OCTG) or pipelines are in strong demand as the number of more corrosive fields increases.

The oil and gas transported in flowlines and in-field gathering lines are not dehydrated or treated chemically to remove corrosive substances. Carbon dioxide corrosion and sulfide stress cracking (SSC) are major problems for these pipelines. They have been protected generally from corrosion by injecting inhibitor into the pipeline [1] or by using corrosion resistant materials such as duplex stainless steel [2]. However, inhibitor injection presents a number of problems, such as high operating costs in offshore pipe lines, variable effectiveness at high temperatures, and environmental pollution caused by leaks of inhibitor. On the other hand, although duplex stainless steel has excellent corrosion resistance, its cost is high.

Two types of martensitic stainless steel seamless pipe with good weldability and corrosion resistance have been developed to provide a more economical material to replace conventional flowline pipes using either carbon steel with inhibitor injection or duplex stainless steel [3]. One is an 11Cr steel pipe for CO_2 environments which has higher corrosion resistance than the conventional API 5CT L80 13Cr OCTG, and the other is a 12Cr steel pipe for environments containing CO_2 and a small amount of H_2S with good SSC resistance. This paper describes the development of these steel pipes and discusses their corrosion performance.

2. Development of 11Cr and 12Cr Steel Pipes

Martensitic stainless steels show effective CO_2 corrosion resistance and are comparatively low in cost. However, they have rarely been used in pipelines because of difficulties in welding. Since cracking often occurs due to hardening caused by martensitic transformation in the heat affected zone (HAZ) and hydrogen dissolved in the weld metal during welding, preheating and post weld heat treatment (PWHT) are required [4]. The newly developed martensitic stainless steel seamless pipes avoid weld cracking by reducing the C and N contents to about 0.01%.

The 11Cr steel pipe for CO_2 environments has been developed with the aim of providing CO_2 corrosion resistance as high as that of the API 13Cr OCTG steel. From the results of investigations on CO_2 corrosion of 11Cr and 12Cr steels with various chemical compositions it has been found that the corrosion rate could be assessed by a CO_2 corrosion index (CCI) defined by Cr–10C+2Ni, as shown in Fig. 1. It has also been confirmed that the addition of Cu improves the pitting resistance of these steels.

Sulfide stress cracking is the most important danger for steel used in H_2S-containing environments. The SSC in martensitic stainless steel initiates from pitting in conditions where a passive film is formed on the steel surface. Accordingly, SSC resistance can be improved by increasing the pitting resistance of the steel, and therefore molybdenum, which improves the pitting resistance of stainless steel [5], was added to the 12Cr steel for use in environments contaning CO_2 and a small amount of H_2S.

Hot workability is an important factor in determining the surface quality of seamless pipe and it can deteriorate drastically if delta-ferrite is present during hot rolling [6]. It can be improved by the addition of austenite-forming elements such as Ni, Mn and Cu and the reduction of ferrite-forming elements such as Cr and Si.

From this research, the compositions of two kinds of steel were developed: C0.01–Cr11–Ni1.5–Cu0.5–N0.01 for CO_2 environments and C0.01–Cr12–Ni5–Mo2–N0.01 for environments containing CO_2 and a small amount of H_2S, as shown in Table 1.

Table 1. *Nominal chemical compositions of steels (mass%)*

	C	Si	Mn	Cr	Ni	Cu	Mo	N
11Cr line pipe	0.01	0.2	1.2	11	1.5	0.5	–	0.01
12Cr line pipe	0.01	0.2	0.4	12	5.0	–	2.0	0.01
13Cr OCTG	0.20	0.2	0.4	13	–	–	–	0.02

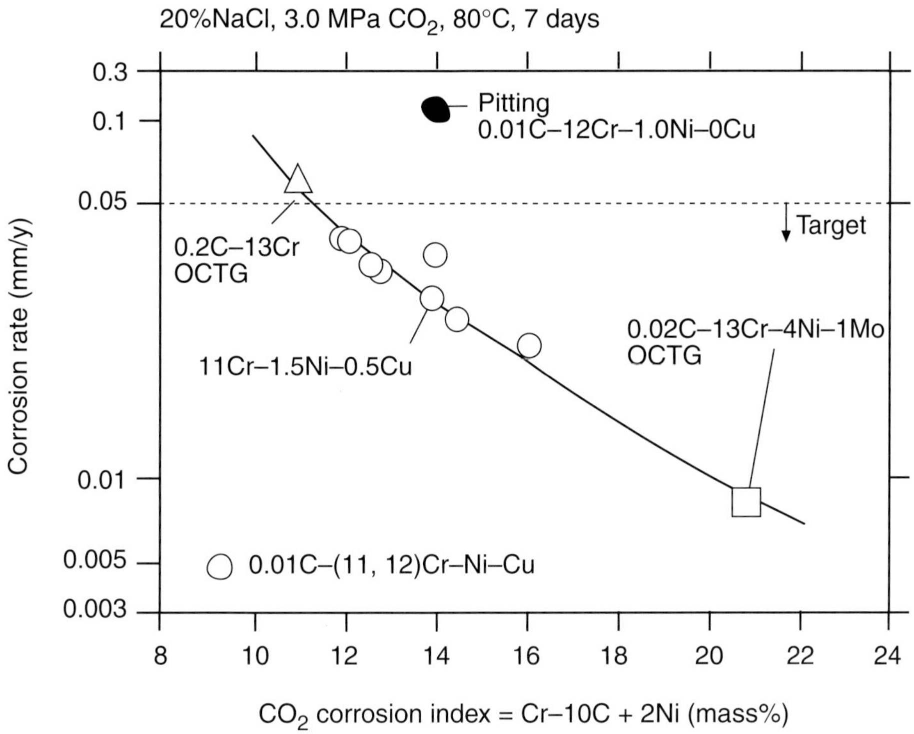

Fig. 1 *Relation between CO_2 corrosion index and corrosion rate.*

Quenching and tempering of the base materials provides yield strengths corresponding to X80 grades. The steel pipes proved to have good weldability for girth welding using a duplex stainless steel welding consumable. No cracking occurred in the welds although welding was conducted without preheating or post weld heat treatment. The results of tensile and side bend tests for welded joints of 11Cr and 12Cr steel pipes are shown in Table 2. The results of Charpy test for welded joints are shown in Figs 2 and 3. Welded joints of the steel pipes gave superior impact

Table 2. *Mechanical properties of girth welded joints of 11Cr and 12Cr steel pipes*

Material	Welded joint Base metal	YS (MPa)	TS (MPa)	El (%)	Fracture position	Side bend test
11Cr line pipe	Base metal	593	726	22	–	–
	GTAW	–	717	24	BM	No crack
	GTAW + GMAW	–	748	25	BM	No crack
12Cr line pipe	Base metal	634	827	34	–	–
	GTAW	–	845	30	BM	No crack
	GTAW + GMAW	–	856	30	BM	No crack

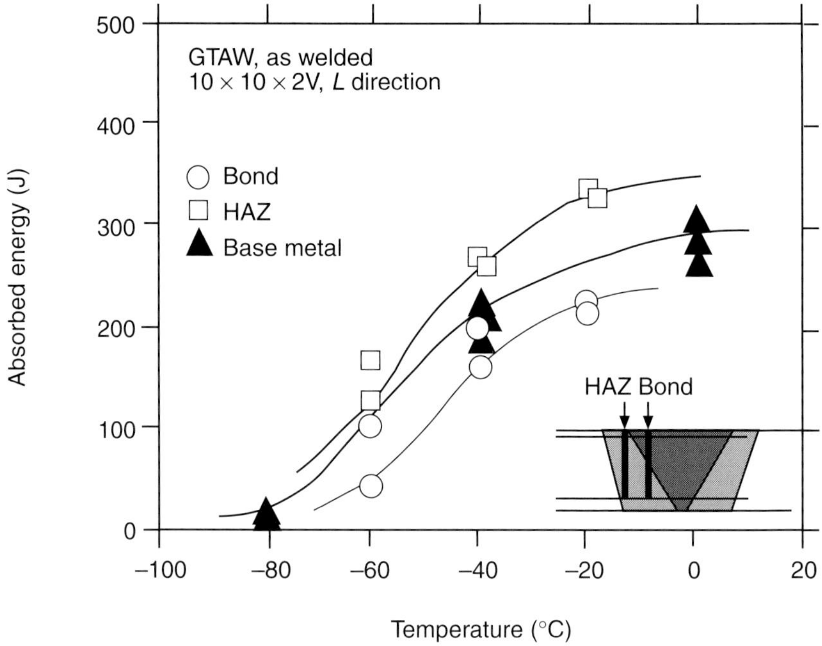

Fig. 2 *Charpy test results of the 11Cr steel pipe welded joint.*

energies, which were more than 100 J at –40°C for the 11Cr steel pipe and more than 200 J at –80°C for the 12Cr steel pipe.

3. CO_2 Corrosion Resistance

Table 3 shows the CO_2 corrosion test conditions and results for the 11Cr and 12Cr steel pipes. The 11Cr steel pipe gave CO_2 corrosion resistance as good as the API 13Cr OCTG and can be used in CO_2 environments below approximately 100°C. The 12Cr steel pipe showed higher resistance to CO_2 corrosion than the 11Cr steel pipe as a result of higher Cr and Ni contents. The 12Cr steel pipe can be used even at 6.4 MPa of CO_2 at 150°C. No preferential corrosion was observed in the weld portion, although the welded joints were made with a duplex stainless steel welding consumable. It is considered that the passive film, which is stable in this environment, prevents galvanic corrosion at the welded joint.

The CO_2 corrosion resistance of stainless steel is affected by the stability of the passive film of Cr oxides on the steel surface. The Cr dissolved in the matrix contributes to the formation of the passive film, while that precipitated as carbides does not [7]. The Cr effective for CO_2 corrosion resistance should, therefore, be considered in terms of the dissolved Cr. Accordingly, the effective Cr contents of the 11Cr and API 13Cr OCTG steels were estimated by analysing precipitated Cr contents. The results are summarised in Table 4. The precipitated Cr content is approximately ten times as much as the total C content. It indicates that the coefficient in the CCI

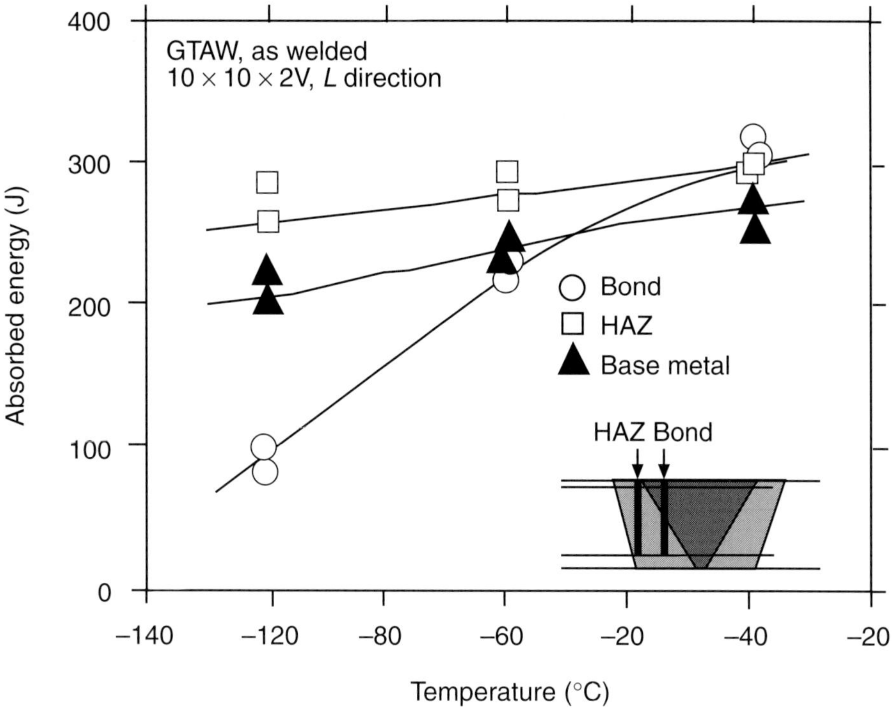

Fig. 3 *Charpy test results of the 12Cr steel pipe welded joint.*

Table 3. *CO_2 corrosion test results of the steels for line pipes. Duration: 168 h*

Material		**NaCl (%)**	**CO_2 (MPa)**	**Temp. (°C)**	**Corrosion rate (mm/y)**
11Cr line pipe	Base metal	10	3.0	80	0.033
	GTAW				0.030
	GTAW + GMAW				0.029
13Cr OCG	Base metal				0.059
11Cr line pipe	Base metal	3.5	1.0	100	0.104
13Cr OCTG	Base metal				0.112
12Cr line pipe	Base metal	20	6.4	150	0.088
	GMAW				0.085

obtained empirically is appropriate for estimating Cr precipitation. The precipitated Cr content of the 11Cr steel is less than that of the API 13Cr steel since the C content of the 11Cr steel has been reduced to about 0.01%. Consequently, the dissolved Cr content of the 11Cr steel is as great as that of the API 13Cr steel, although the total Cr content of the 11Cr steel is lower. The 11Cr steel also contained Ni which is effective

***Table 4.** States of Cr and C in 11Cr and 13Cr steels*

	Total Cr	Total C	Precipitated Cr	Precipitated C	Dissolved Cr	Dissolved C
11Cr line pipe	11.06	0.013	0.17	0.010	10.89	0.003
13Cr OCTG	13.16	0.20	2.1	0.16	11.06	0.04

for CO_2 corrosion resistance. As a result, the 11Cr steel had better CO_2 corrosion resistance than the API 13Cr OCTG steel.

4. SSC Resistance

Figure 4 shows SSC test results for a welded joint of the 12Cr steel pipe in the CO_2 and low H_2S environments. The welded joint was made with a duplex stainless steel welding consumable without PWHT. The SSC test was conducted using a constant load SSC test method. Mixed solutions of 5 or 10% NaCl and 0.5% acetic acid with various pH values were used as the test solutions to simulate a formation water. The applied stress was 567 MPa which corresponds to 90% of the actual yield strength of the base metal. Sulfide stress cracking occurred in the HAZ under the conditions of pH 3.5 and 0.004 MPa H_2S. Pitting occurred in the HAZ at pH 4.0 and with 0.007

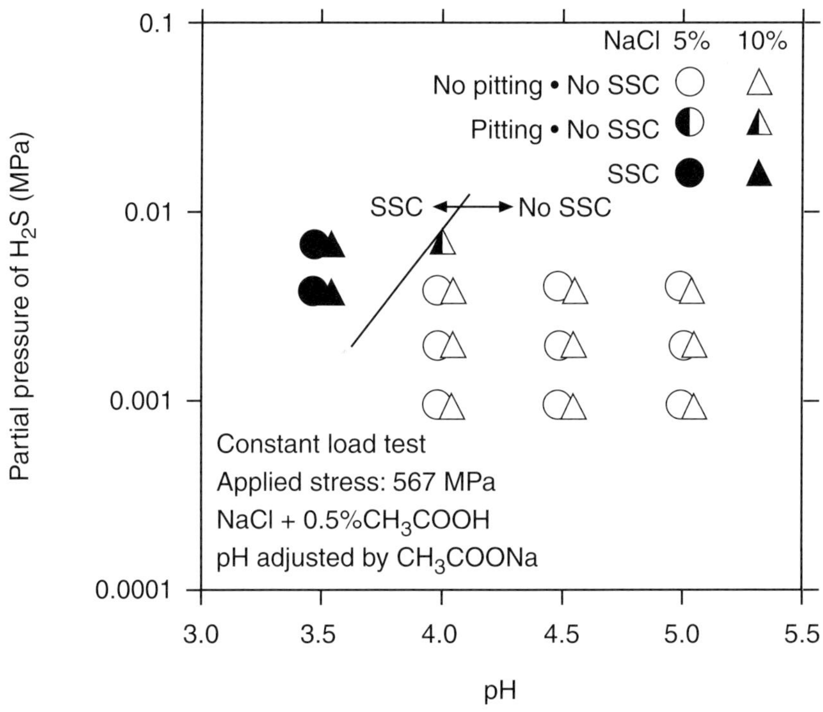

***Fig. 4** SSC test results of the 12Cr steel pipe welded joint.*

MPa H_2S, although it did not lead to SSC. The border line for the occurence of SSC appears to lie between pH 3.5 and 4.0, and with about 0.005 MPa of H_2S.

Sulfide stress cracking occurred in the HAZ under the severe conditions of a high H_2S partial pressure and low pH. It initiated from the bottom of pits, as shown in Fig. 5. Once pitting occurs, the decrease in pH inside the pits accelerates the permeation of hydrogen into the steel. Furthermore, stress concentrates at the bottom of pits and consequently SSC will initiate from the bottom of pits. An improvement in pitting resistance of steel can therefore suppress SSC. For this reason Mo, which improves SSC resistance, is added to the 12Cr steel.

However, the reason why pitting occurs in the HAZ but not in the base metal was not clear. The mechanism of pitting occurrence was therefore studied electrochemically and metallographically.

Pitting potentials for HAZ microstructures of the 12Cr steel were measured to evaluate their pitting resistance quantitatively. The synthetic HAZ was made in the Greeble testing machine where electric power was used for heating. The thermal cycles were single cycles with peak temperatures of 700, 1000 and 1400°C and a double cycle with a peak temperature of 1400°C followed by holding at 600°C for 3 min for simulating PWHT. The pitting potentials of the microstructures were measured without applying stress and the results are shown in Fig. 6. The potential of the base metal was found to be the lowest with that of the single cycle HAZ increasing with peak temperature. The single cycle HAZ with a higher peak temperature had a larger amount of Cr effective for pitting resistance since more Cr carbides were dissolved during heating above 700°C with little precipitation during cooling. Reheating to 600°C reduced the pitting potential of the HAZ produced at a peak temperature of 1400°C. This can be explained by the fact that the content of Cr effective for pitting resistance decreased as Cr carbides precipitated during the reheating. These results point to the microstructure of the single cycle HAZ having better pitting resistance than the base metal when stress is not applied.

Sulfide stress cracking occurs when stress is applied. The distribution of stress and strain in the HAZ was measured to clarify the contribution of stress. A round bar specimen was prepared from the simulated HAZ sample subjected to a peak

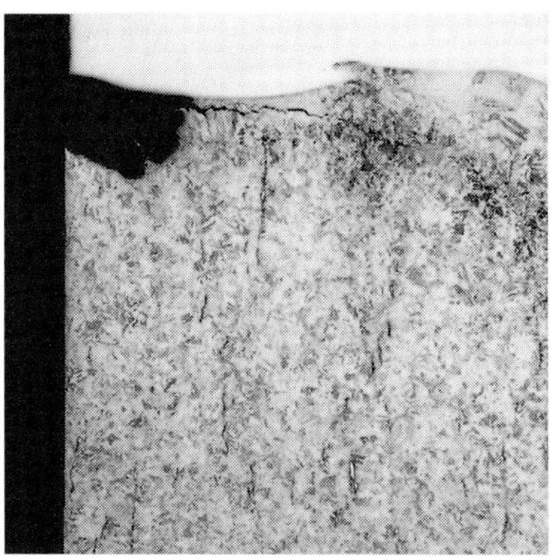

***Fig.* 5** *SSC initiating from pitting in HAZ.*

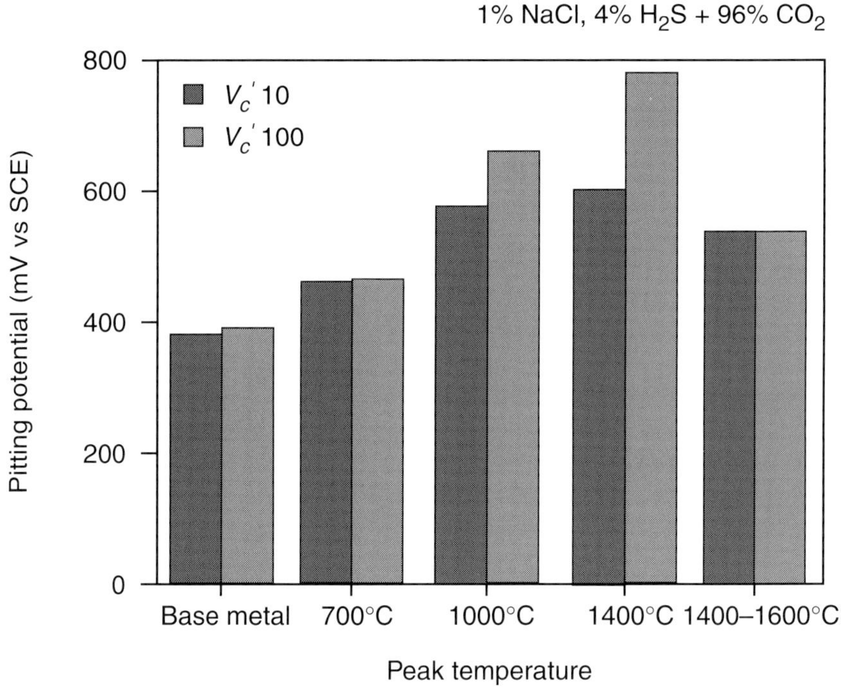

Fig. 6 Pitting potentials of the simulated HAZ. V_c*'10 (100) is the most noble value of potential corresponding to 10 (100)* μAcm^{-2} *(the pitting potential as defined by JIS G 0577).*

temperature of 1400°C. The change in strain was measured by strain gauges mounted on the portions corresponding to base metal and HAZ subjected to peak temperatures of 700 and 1400°C. The results are shown in Fig. 7. Under constant load, the HAZ elongated more than the base metal. This is because the martensitic microstructure, which is formed on quenching, is easy to deform plastically. A passive film on the HAZ, therefore, breaks down more easily than that on the base metal and must be the reason why pitting occurs in the HAZ.

5. Resistance to Hydrogen Embrittlement under Cathodic Protection

Cathodic protection is generally applied to offshore pipelines. Hydrogen can be generated as a result of the cathodic reaction on the outer surface of the pipe and so hydrogen embrittlement due to hydrogen entering the steel is a concern. The possibility of hydrogen embrittlement was studied by simulating cathodic protection of steel in sea water. The test was conducted using the constant load tensile test and the 4-point bend test. The conditions and results of the tests are shown in Table 5. No crack was observed even at potentials of –1200 mV (vs SCE) or lower when more hydrogen was generated. The potential of a zinc alloy used for a sacrificial anode for cathodic protection is generally about –1000 or –1100 mV (vs SCE).

The behaviour of hydrogen in the 12Cr steel with applied cathodic protection was studied electrochemically using the hydrogen permeation test. The results are

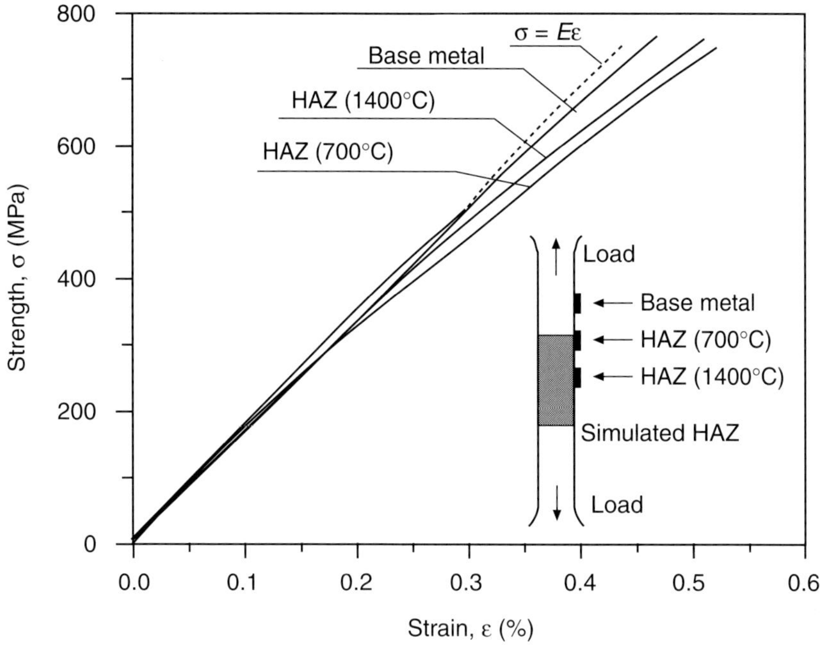

***Fig. 7** Relation between stress and strain for the simulated HAZ.*

shown in Fig. 8. The hydrogen permeation rate during cathodic protection was about two orders of magnitude smaller than that in the low H_2S environment. The result indicates that hydrogen hardly enters the 12Cr steel even at –1600 mV (vs SCE) in sea water.

From these results it was concluded that the cathodic protection of the 11Cr and 12Cr steel pipes in sea water does not cause hydrogen embrittlement.

***Table 5.** Results of hydrogen embrittlement test in cathodic protection environments*

Material		Test method	Applied stress	Solution	Potential (mV vs SCE)	Result
11Cr linepipe	Base metal	CL	551 MPa	Synthetic sea water	–1000	No crack
					–1600	No crack
	Simulated HAZ	CL	551 MPa	Synthetic sea water	–1000	No crack
					–1600	No crack
12Cr line pipe	Welded joint	4PB	551 MPa	3.5% NaCl	–900	No crack
					–1200	No crack
	Welded joint	CL	630 MPa	3.5% NaCl	–1000	No crack
					–1600	No crack

CL: Constant load test. 4PB: 4-point bend test.

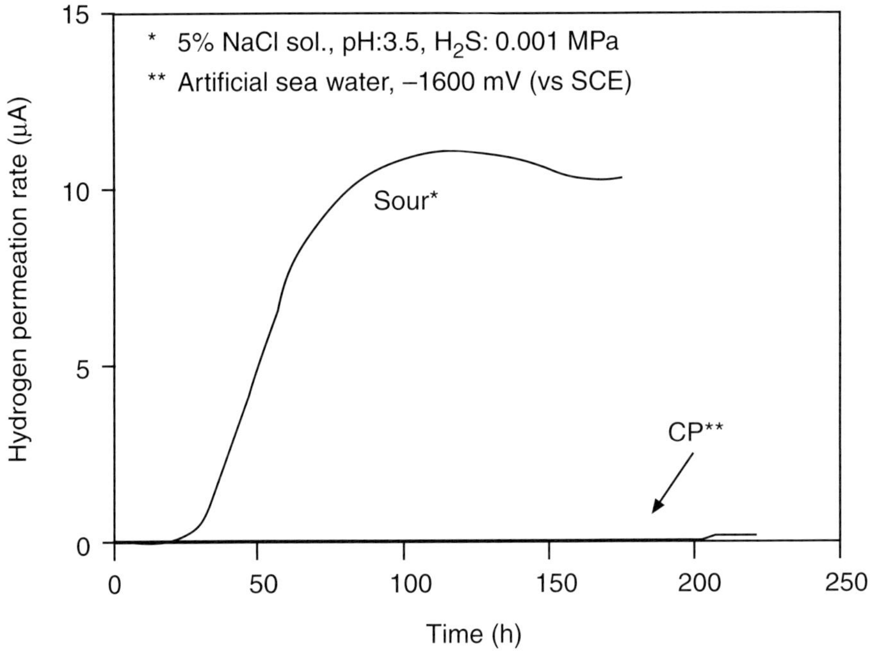

Fig. 8 Hydrogen permeation rate in a sour environment and under cathodic protection conditions.

6. Conclusions

The corrosion performance of two types of weldable martensitic stainless steel seamless pipe, the 11Cr steel pipe (C0.01–Cr11–Ni1.5–Cu0.5–N0.01) for CO_2-containing environments and the 12Cr steel (C0.01–Cr12–Ni5–Mo2–N0.01) for environments containing CO_2 and a small amount of H_2S was studied. The results can be summarised as follows:

1. The 11Cr steel had better CO_2 corrosion resistance than the API 5CT L80 13Cr OCTG steel. This is attributed to the fact that although the two steels had the same amount of the dissolved Cr responsible for CO_2 corrosion resistance, the 11Cr steel contained Ni which is also effective for CO_2 corrosion resistance.

2. Pitting, which can lead to SSC, occurs more easily in the HAZ when stress is applied because the passive film on the HAZ breaks more easily than that on base metal. The quenched martensite of HAZ elongates more when stress is applied.

3. The cathodic protection of the 11Cr and 12Cr steel pipes in sea water does not cause hydrogen embrittlement, since hydrogen hardly enters those steels.

References

1. L. van Bodegon, K. van Gelder and J. A. M. Spanincs, *Corrosion '88*, Paper No. 187, NACE International, Houston, Tx, 1998.
2. T. Ogawa and T. Koseki, *J. Jpn Weld. Soc.*, 1988, **57**, (2), 92.
3. Y. Miyata, M. Kimura, and F. Murase, Kawasaki Steel Technical Report (1998), **38**, 53.
4. M. Watanabe and Y. Mukai, *Sutenresukou-no Yousetsu*, Welding of Stainless Steel, Nikkan Kogyo Shinbunsha, 1971, 62.
5. M. Kimura *et al.*, *Corrosion '97*, Paper No. 22, NACE International, Houston, Tx, 1997.
6. T. Shimizu, H. Oka, and T. Kasuya, *CAMP-ISIJ*, (Current Advances in Materials and Processes, Iron and Steel Institute of Japan) 1992, **5**, (2), 527.
7. K. Tamaki, T. Shimizu, Y. Yamane, and Y. Kitahaba, *Proc. Int. Conf. on Stainless Steels*, ISIJ, Chiba, 1991, 257.

24
Corrosion Properties and Application Limit of Sour Resistant 13% Chromium Steel Tubing with Improved CO_2 Corrosion Resistance

H. ASAHI, T. HARA and S. SAKAMOTO*
Nippon Steel Corporation, Steel Research Laboratories, Chiba, Japan
*Nippon Steel Corporation, Yawata R&D Laboratory, Fukuoka, Japan

ABSTRACT
Corrosion properties of the modified 13%Cr tubing, NT-CRSS-110, were examined to determine conditions for its application in oil and gas production environments. From corrosion test results at elevated temperatures and sulfide stress corrosion (SSC) test results at ambient temperatures, the material was found to be applicable under H_2S partial pressure of 0.1 bar or lower except where the formation water had pH values lower than 3.5. Sulfide stress corrosion test results obtained in a simulated well environment were in agreement with the results in an acetic buffer solution having the same pH and H_2S partial pressure.

1. Introduction

The use of 13%Cr tubing can contribute to cost reductions in oil and gas production [1]. Recently, martensitic 13%Cr tubing with reduced carbon content and additions of alloying elements such as nickel, molybdenum or copper and with corrosion properties greatly improved compared to those of API 5CT 13%Cr tubing, have been developed [2–4] and applied in actual well developments. Nippon Steel has also developed this type of material (NT-CRSS, referred as CRSS in this text) [4] and made it commercially available. It is now very important to clarify the corrosion properties and determine precisely the application limits.

2. Features of the New Material

The steel CRSS contains the major chemical elements shown in Table 1 which also describes API 5CT 13%Cr tubing used as a reference in this work Steel CRSS has the following features [4].

- Very good CO_2 corrosion resistance achieved by a low carbon–13%Cr steel with additions of copper and nickel. Combined additions of copper and nickel transform the corrosion film from the polycrystalline state, which is formed with conventional low carbon–13%Cr steels containing nickel, into an amorphous phase.

Table 1. *Typical chemical compositions of CRSS (mass %)*

	C	Si	Mn	P	S	Cr	Ni	Cu	Mo	N
CRSS	0.017	0.30	0.51	0.017	0.001	12.8	5.90	1.56	1.97	0.017
API	0.190	0.25	0.58	0.011	0.003	12.6	0.13	—	—	

- Improved SSC resistance achieved by an addition of an appropriate amount of molybdenum without allowing the formation of ferrite.

- High strength — up to C110 grade which is attained without adversely affecting the SSC resistance, low temperature toughness, etc.

3. Test Method

3.1. Material

Specimens were mainly machined from seamless pipes (177.8 mm o.d.–12.65 mm thickness) of CRSS. Materials with different yield strengths in C110 grade (YS: 110–125 ksi (758–861 MPa)) were tested.

3.2. Corrosion Tests

3.2.1. SSC test
Four-point bent beam specimens with or without notch as shown in Fig. 1 and tensile type specimens based on NACE TM0177-90A (dia. 6.4 mm) were used. The applied stresses were 100% of the actual yield strength for 4-point bent beam tests and 90% of the actual yield strength for tensile type SSC tests. The test duration was 720 h.

3.2.2. Corrosion test
Block specimens were immersed in a test environment. Corrosion rates were calculated from the weight loss and the test duration.

3.2.3. Test environments
For ambient pressure tests, 1 $molL^{-1}$ acetic buffer solution was used. The test gas bubbled into the test solution was a mixture of H_2S and CO_2. In autoclave (high pressure) tests a simulated formation water and condensed water as indicated in Table 2 were used. A pH value after gas saturation, which was calculated based on Miyasaka's method [5], is indicated for each test condition.

4. Corrosion Properties and Application Limits

4.1. Corrosion Test at High Temperature

Corrosion tests and V-notched four-point bent beam tests were performed under H_2S partial pressure of 0.1 bar and CO_2 partial pressure of 40 bar in the simulated

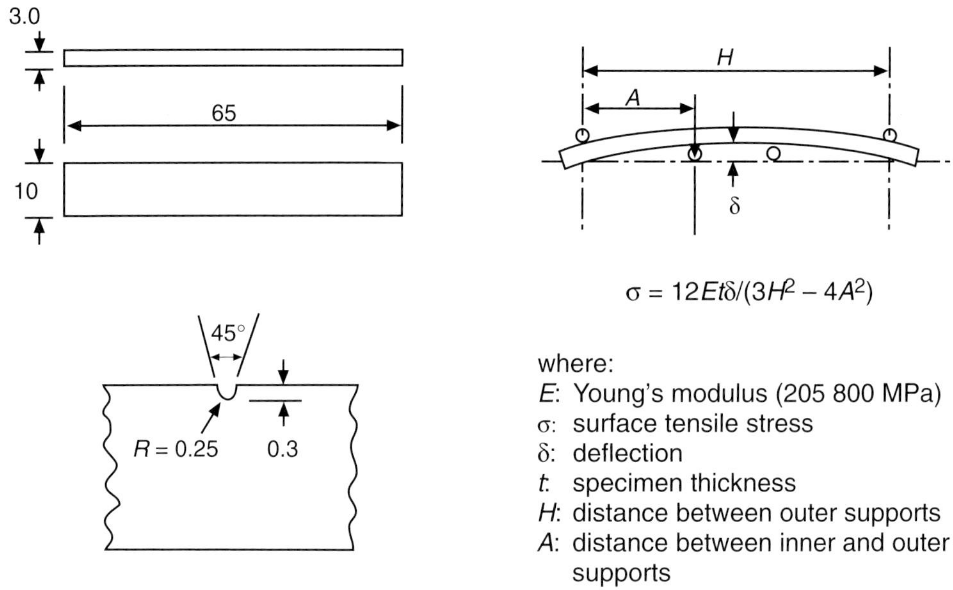

***Fig. 1** Specimen dimensions (mm) and stress calculation formula of 4-point bent beam test.*

***Table 2.** Chemistry of the test solutions (gL^{-1})*

	Cl^- (ppm)	HCO_3^- (ppm)	NaCl	$CaCl_2$	KCl	$MgCl_2$	$NaHCO_3$	CH_3COONa
Formation	68 000	329	102.6 gL^{-1}	5.43 gL^{-1}	2.72 gL^{-1}	1.52 gL^{-1}	0.43 gL^{-1}	0.21 gL^{-1}
Simulated formation	121 300	73	20% NaCl solution + $NaHCO_3$ 100 mgL^{-1}					
Condensed	1000		0.16% NaCl solution					
1.2% NaCl	7200		1.2% NaCl solution					
9.2% NaCl	55 800		9.2% NaCl solution					

formation water at 160°C. The results are given in Table 3. No localised corrosion and no cracking occurred after one month for CRSS. The average corrosion rate was low at 0.027 mm/y, whereas API 5CT 13%Cr was severely corroded.

***Table 3.** Corrosion test results at 160°C*

Test	CRSS	API
Four-point bent beam (V-notched)	No crack	No crack (strain enhanced corrosion)
Corrosion rate (mm/y)	0.027	0.30 (localised corrosion)

4.2. SSC Test Results in Acetic Buffer Solutions

Tensile type SSC tests were carried out under conditions having various pH values, chloride ion concentration and H_2S partial pressures at 24 ± 3°C. The various values of pH were obtained by changing the CH_3COOH: CH_3COONa ratio. Representative results are shown in Figs 2 and 3. As shown in Fig. 2, no SSC occurred in 5% NaCl solution when the pH was 3.5 or greater and H_2S partial pressure was 0.1 bar or lower. Under lower partial pressures such as 0.01 bar, no SSC occurred even at a pH of 3.0. Figure 3 illustrates the effects of chloride ion concentration and pH under an H_2S partial pressure of 0.1 bar. At a very low concentration of chloride ion such as 610 ppm, no SSC occurred at pH values down to 3.0. At 50°C, immunity from SSC appeared under almost the same conditions as at 25°C.

4.3. SSC Test Results in Simulated Well Conditions

Sulfide stress corrosion tests were performed at 24 ± 3°C in an autoclave with solutions saturated with high pressures of CO_2 and H_2S to simulate corrosion conditions in oil and gas production environments. Results are given in Table 4. Sulfide stress corrosion occurred in environments with low pH and high chloride concentrations. The results for 0.1 bar of H_2S are plotted in Fig. 4, and illustrated in Fig. 3. Results from the two evaluations are clearly in agreement. Therefore, SSC test results from acetic buffer solutions could be used for the purpose of determining the application limit of the

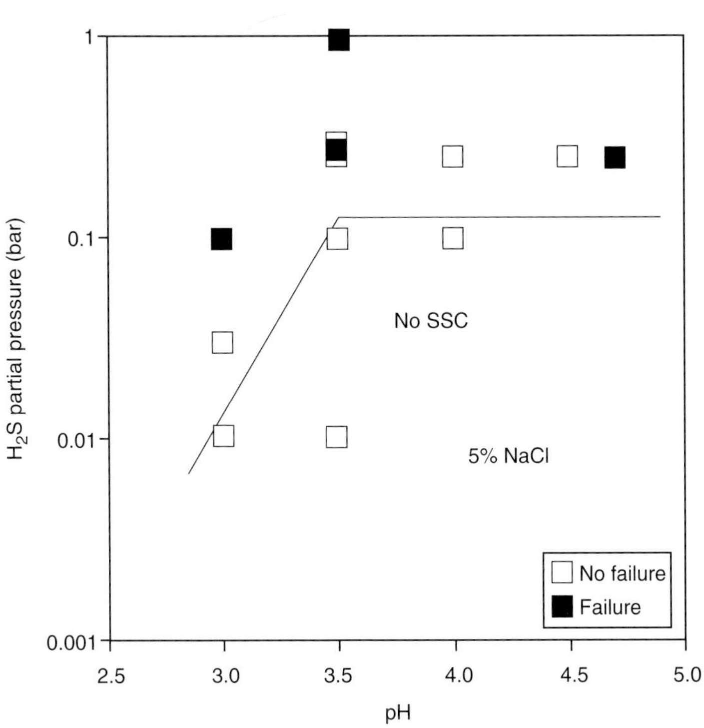

***Fig. 2** Effects of pH and H_2S partial pressure on tensile type SSC results (5% NaCl solution).*

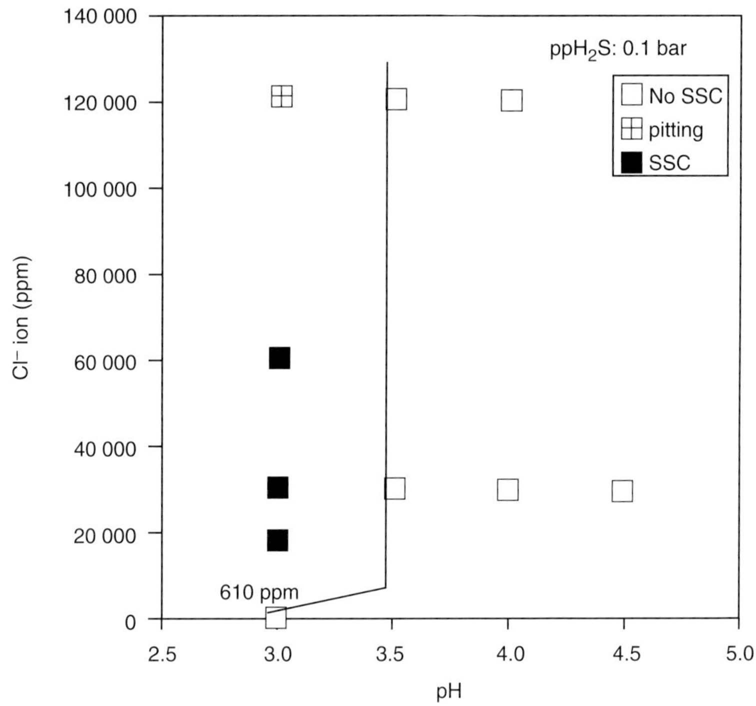

Fig. 3 Effects of pH and chloride ion concentration on tensile type SSC results (H_2S: 0.1 bar, 24 ± 3 °C).

Table 4. SSC test results in high pressure environments

Solution	Environment				Material	
	Cl^- (ppm)	$ppCO_2$ (bar)	ppH_2S (bar)	Cal. pH	CRSS	API
1.2% NaCl	7200	49.98	0.02	2.9	No SSC	SSC
9.2% NaCl	55 800	49.98	0.02	2.7	SSC	SSC
Condensed	1000	44.9	0.10	3.0	No SSC	SSC
Formation	68 000	44.9	0.10	3.3	SSC	SSC
Formation	68 000	20	0.10	3.8	No SSC	SSC

material. It should be noted that, API 5CT 13%Cr suffered from SSC in these environments.

4.4. Application Limit

Solution pH is determined by water chemistry, partial pressure of CO_2 and H_2S gases and temperature. In Fig. 5 examples of calculated pH values of solution are indicated as a function of CO_2 partial pressure and bicarbonate concentration. In gas and gas condensate wells, condensed water which is generated at the dew point contains effectively no or little chemical elements. With increasing CO_2 and H_2S partial

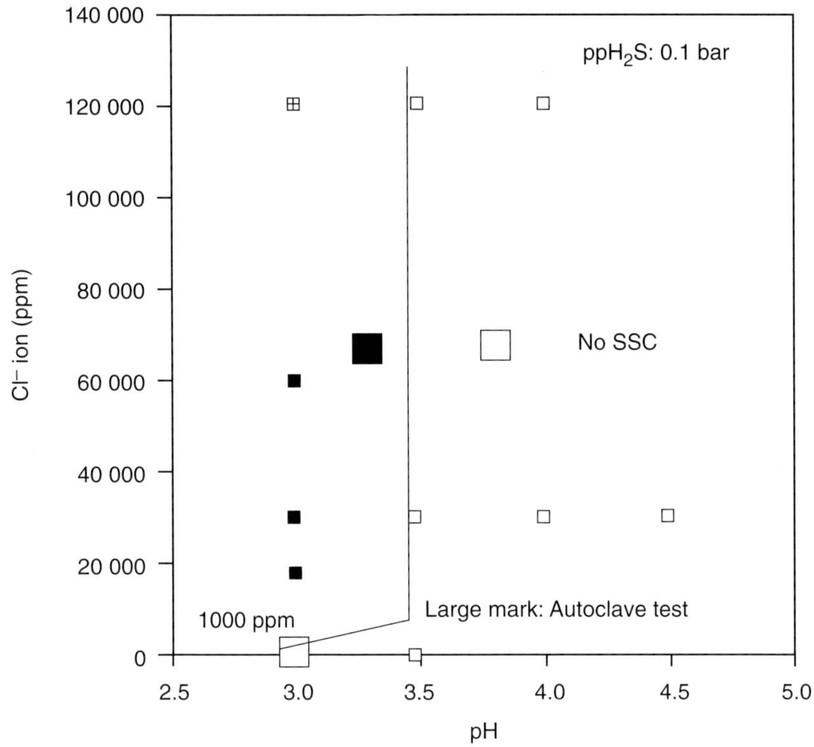

Fig. 4 *Comparison of SSC test results obtained in acetic buffer solutions and in simulated well environments (H_2S: 0.1 bar).*

pressures the pH easily decreases. Therefore, the typical environmental conditions in gas wells are low pH and low chloride ion concentration. In oil wells formation (production) water is present, and this contains much chloride ion, other minerals and, in many cases, buffering elements like bicarbonate. As indicated in Fig. 5, the pH hardly decreases if bicarbonate is contained in water and therefore, the typical environmental conditions in oil wells are high chloride concentrations and rather high pH.

The CRSS material has high SSC resistance at pH of 3.5 or greater in a high chloride ion solution and at pH of 3.0 or greater in a low chloride ion solution under an H_2S partial pressure of 0.1bar or lower. If these conditions are taken to be indicative of well conditions, then the general well conditions where CRSS could be used are as follows: CO_2 partial pressure lower than 20 bar, H_2S partial pressure 0.1 bar or lower and, if the water is formation type, HCO_3^- concentration greater than 60 ppm.

5. Conclusions

1. The 13%Cr CRSS tubing shows good corrosion resistance at temperatures up to 160°C and high SSC resistance at pH of 3.5 or greater in formation type water and at pH of 3.0 or greater in condensed type water under 0.1 bar H_2S.

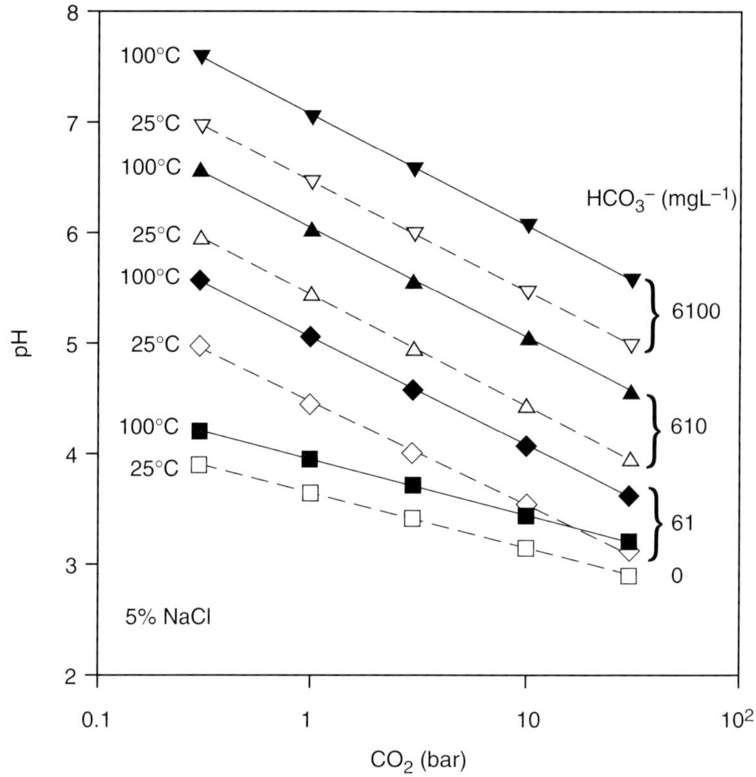

Fig. 5 Effects of CO_2 partial pressure and HCO_3^- concentration on calculated pH.

2. CRSS can be used in many cases when the H_2S partial pressure is below 0.1 bar.

References

1. M. B. Kermani, G. Weighill, T. Penslington and G. Elliot, *Corrosion '95*, Paper No. 96. NACE International, Houston, Tx, 1995.
2. M. Ueda, T. Kushida, K. Kondo and T. Kudo, *Corrosion '92*, Paper No. 55. NACE International, Houston, Tx, 1992.
3. H. Asahi, T. Hara, A. Kawakami and A. Takahashi, *Corrosion '95*, Paper No. 79. NACE International, Houston, Tx, 1995.
4. H.Asahi, T. Hara and M. Sugiyama, *Corrosion '96*, Paper No. 61. NACE International, Houston, Tx, 1996.
5. A. Miyasaka, *Corrosion '92*, Paper No. 5. NACE International, Houston, Tx, 1992.

25
Weldable 13% Chromium Steel: The Development of the Components for a Wet Gas Piping System

J. J. DUFRANE, E. FRANCESCHETTI[*], J. HEATHER[†]
and H. VAN DER WINDEN[§]
Fabrique de Fer, Belgium
[*]Bassi Luigi, Italy
[†]Erndtebrueck Eisenwerk, Gerrnany
[§]Nederlandse Aardolie Maatschappij, The Netherlands

ABSTRACT

For the Nederlandse Aardolie Maatschappij (NAM) Groningen Long Term (GLT) project weldable 13Cr steel was chosen for the wet gas process piping. A technical feasibility study showed that this new material should be able to fulfil all essential technical requirements for the project. A description is given of the more detailed development activities that followed the feasibility study. Aspects covered include plate material properties, longitudinal seam welding, fitting, manufacturing and girth welding with the main attention given to product development, weldability and toughness testing; other aspects such as corrosion testing and cathodic protection have been investigated but will be reported later. After completion of the qualification programme it was concluded that a piping system for wet gas can successfully be realised in weldable 13Cr steel.

1. Introduction

The Nederlandse Aardolie Maatschappij (NAM) has decided to upgrade all their wet gas treatment facilities for the Groningen gas field, situated in the North of Holland. This includes a change in materials from carbon steel with inhibition to weldable 13Cr martensitic stainless steel. This decision was taken on the basis of a life cycle cost evaluation which included, apart from capital expenditure and direct operational costs, aspects such as the minimum manning philosophy and the consequences of unplanned shutdowns and possible leaks. The project will be executed over a 5 to 7 year period, starting in 1998. The design, construction and maintenance of the upgraded installations will be executed by a newly formed consortium headed by the Dutch engineering contractor Stork. Although a considerable length (some 2700 m) of seamless, 8 in. flowlines has already been installed at the first facility, this paper will focus on the larger diameter materials, i.e. class 900 pipe in the range of 12–24 in. dia. for the wet gas piping section of the gas treatment plant.

2. Development Route and Main Uncertainties

Prior to deciding on weldable 13Cr martensitic stainless steel for the wet gas piping, a thorough review was executed to map out all possible risks associated with the choice of this new material. The main concerns included the weldability (seam and girth), the low temperature toughness and the approval to be obtained from the Dutch certifying authority, Stoomwezen. For this application the corrosion properties were less of a risk due to the relatively benign medium and the extensive experience within NAM with the traditional 13Cr tubing that has already been in use for some 10 years in the wells of the Groningen field.

3. Steel and Plate Manufacturing

The plate material was developed and supplied by Fabrique de Fer, Charleroi, Belgium.

Martensitic stainless steels containing 13% Cr, are reputed to be poorly weldable due to ferrite grain coarsening and hardening during welding. In the past, their use has been restricted to non-welded parts of installations. Nevertheless, such materials offered valuable advantages such as:

- a high yield and tensile strengths;
- a corrosion resistance satisfactory for moderately corrosive media;
- a cost level (as a result of their low nickel content) about halfway between carbon steels and austenitic and duplex stainless steels.

Over the last few years a new generation of easily weldable 13% Cr steels has been developed whose main characteristics are described below. Three different versions have been developed depending on field conditions, which are mainly related to the carbon dioxide, hydrogen sulfide and chloride concentrations and to temperature, i.e. a lean alloy design for purely sweet (non-H_2S) conditions and two Mo-containing versions for lightly and mildly sour (H_2S-containing) conditions respectively. The target composition values are given in Table 1.

All three versions are designed to contain no ferrite, i.e. in the as-supplied condition of plate material intended for pipe and fitting welding and fabrication. They all show a strength level meeting the requirements of an X80 line pipe grade (550 MPa specified minimum yield strength). Their impact toughness is designed to meet –20°C (Alloy 1) and –40°C (Alloys 2–3) requirements.

In the frame of the present case study, Alloy 1 (lean alloy) has been investigated particularly because this alloy satisfies the NAM project (corrosion resistance) requirements.

From the plate production viewpoint, the following critical parameters have been emphasised:

- ultra low C, N and S contents;

Table 1. *Target compositions for alloys developed for wet gas service*

	Alloy 1	**Alloy 2**	**Alloy 3**
C	<0.015	<0.015	<0.015
Mn	<2.0	<2.0	<2.0
P	<0.030	<0.030	<0.030
S	<0.002	<0.002	<0.002
Si	0.15	0.15	0.15
Cu	0.40	0.40	0.40
Ni	2.0	4.5	6.5
Cr	11.0	12.0	12.0
Mo	<0.5	1.5	2.5
N	<0.012	<0.012	<0.012

- rolling in controlled temperature ranges in order to prevent grain coarsening; and
- heat treatment by quenching and tempering to provide an ideal fine grained tempered martensite.

The optimisation of the above parameters on plate material from 8 up to 40 mm thick generates the typical characteristics as shown in Table 2.

Most important to note is that the material characteristics are easily recovered after a hot forming fabrication (fittings and heads for instance) by applying the same quenching and tempering sequence as on the original plate material. Furthermore, it became clear that excessive grain coarsening can be prevented by controlling the forging temperature during hot forming operations.

3.1. Main Problems Encountered

3.1.1. Ferrite content

The control of ferrite content, or rather its almost complete suppression, proved to be crucial for obtaining good low temperature properties. To achieve this requires a very good control over all processes that influence the final chemical composition of the product.

3.1.2. Zones with reduced toughness

Particularly in the earlier and thicker materials, small and very narrow zones were sometimes found. These were revealed by a different response to the etching agent

Table 2

$PS_{0.2}$ (MPa)	**UTS (MPa)**	**Elongation (%)**	**Reduction of area (%)**
641	826	18	66

rather than by a clear difference in microstructure. These zones can exhibit somewhat reduced toughness levels as a result of slight segregation arising, most likely, from some interstitial elements. This phenomenon is unlikely to be avoided in steel making, but it is considered that its rare occurrence and very limited size do not present a real threat to the integrity of the material. Nevertheless every practicable effort should be made to suppress such segregation effects.

3.1.3. Optimisation of final heat treatment parameters

Over the period since the first introduction of the steel some fine-tuning of the chemical composition has been executed and this resulted in a lower carbon and a somewhat higher Ni content. This also necessitated an optimisation of the final heat treatment conditions and this resulted in significantly improved toughness properties. The Charpy transition temperature was shifted from a typical value of –20 to –30°C to values of –40°C or even lower.

4. Pipe Making and Seam Welding

The fabrication of longitudinally welded pipe was executed by Erndtebrueck Eisenwerk (EEW) at Erndtebrueck in Germany. The weld preparation is machined onto the longitudinal plate edges and the area next to the weld is preformed to make sure that there are no flat areas adjacent to the weld. The pipes are formed on a 2500 T press. Even with the high actual yield strength of well over 550 MPa, the forming of this material did not present any problem given EEW's forming capabilities.

Welding is executed utilising the double-sided submerged arc welding process. Matching weld consumables with proven acceptable properties are not available at present for the submerged arc welding production welding of pipes. As a consequence EEW has chosen to weld this X80-11 Cr material with 22%Cr9%Ni duplex stainless steel consumables, which, although not identical from the chemical analysis point of view, offer comparable strength values. One of the advantages given by the choice of a duplex stainless steel consumable is the fact that if a fitting, flange or valve in this grade of 13Cr material is not available, a duplex stainless steel part can be used instead. The results of mechanical testing fulfil the requirements set upon this material, although it has been observed that when welding thinner material, where there is a higher degree of base metal dilution, the yield strength of the all weld tensile tests can drop to values as low as approximately 490 MPa. Care has to be taken when welding this material because of a tendency to become magnetised, thus causing arc-blow problems. This phenomenon becomes more apparent as the diameter of the pipes become smaller. For the GLT project all plates have been subjected to an ultrasonic examination for laminations and plate edge discontinuities in accordance with SEL 072-77 table 1, class 3 and table 2, class 1*. The latest revision of the pipe specification calls for 100% radiographic examination and 25% ultrasonic examination for thickness values greater than 11 mm, however this has not been finalised yet.

---

*Stahl und Eisen Lieferbedingungen (European/German standard for NDT inspection of steel products, notably plate).

The feasibility of ultrasonic shear wave examination has been demonstrated on a 16 in. dia., 16 mm thick test piece containing artificial surface notches, side drilled holes and radially drilled holes.

Notches (planar defects) in the heat affected zone can be detected at N5 level*. The signal to noise ratio (> 8 dB at acceptance level) for embedded defects is sufficient in this test piece to perform a reliable examination. It is most likely that the basis for inspection of the longitudinal duplex stainless steel submerged arc weld in X80-11 Cr pipes will be mechanised ultrasonics because this technique will provide the fracture mechanics concept adopted by NAM for this project.

4.1. Main Problems Encountered

4.1.1. Low yield strength values of all-weld metal specimens

During the development of the seam welding procedure, low yield strength values of all-weld metal specimens were obtained while the transverse yield strength was (or appeared to be) over the X-80 level (550 MPa). It appeared that relatively high dilution from the plasma-welded root layer played an important role here. A reduction of heat input of the first pass could prevent this phenomenon. It must be realised however that the 22Cr duplex filler material has already a stainless steel lower yield strength than the plate material which could explain part of the problem.

4.1.2. Arc-blow

The 13Cr material can rather easily be magnetised by the welding operation, which is not too surprising given its inherently magnetic properties. The resulting arc-blow problem seems to be greater at smaller pipe diameters but can in practice be solved by a de-magnetising treatment.

5. Fittings

For this project Bassi Luigi, San Rocco al Porto, Italy will supply the fittings while they also act as the main contractor for the fabrication of the complete pipe spools.

5.1. Forging

The initial forging conditions for fittings were derived from those for high strength carbon steel with some consideration given to the differences in transition temperatures for the 13Cr material. After a limited number of trials it became clear that the forging of fittings does not present particular difficulties in terms of the heating level or forging forces required. Even when varying the forging temperature to lower levels, crack-free forgings were obtained which clearly indicates a good forgeability of the welding 13Cr material.

---

*Notch type and size of reference reflector. N5 indicates the defect is oriented in a plane normal to the plate surface with 5% of wall thickness of the component.

5.2. Heat Treatment

All forgings have to undergo a renewed quench-and-temper (QT) treatment in order to achieve the optimal strength and toughness properties in the final product. The heat treatment conditions are strongly dependent on the chemical composition of the material and have to be established for each grade separately. The raw material manufacturers usually provide the necessary data which may, however, require some fine tuning.

5.3. Seam Welding

For this quench-and-temper material there are two methods to obtain a fitting with a longitudinal seam with adequate properties: (1) use a compositionally matching welding consumable which can undergo the subsequent heat treatment, or (2) apply a small root pass weld of matching consumable, form the fitting, quench-and-temper it and then complete the weld with duplex stainless steel filler metal. Both approaches are technically feasible and the choice depends mainly on economics.

5.4. Problems Encountered

5.4.1. Grain growth

During early trials, when the optimal quenching temperature was not yet established, considerable grain growth occurred with dramatically low toughness test results. An extensive optimisation programme resulted in finding the maximum temperatures during the hot forming and quenching operations to avoid rapid grain growth. So the problem can be prevented when sufficient experience with the material has been gained. Moreover, applying an additional quench-and-temper treatment showed that even a very coarse microstructure (ASTM 2) can be refined again to about ASTM 5 or 6, with excellent toughness (and strength) properties. Thus, the weldable 13Cr material showed an excellent ability to recover from earlier incorrect or non-optimal treatments.

6. Flanges*

The development of 13Cr flanges has progressed to the stage where a 16 in., class 900 test flange has been produced and the basic mechanical testing has been completed. The forging of the flange out of pre-forged slab material could be done without any problems. No excessive loads were required nor did any surface defects occur. The results of the mechanical testing programme show fine-grained microstructures throughout the flange body with fully adequate tensile and hardness results. In general, the Charpy impact values were already very encouraging with very good values at the weld bevel portion. At the middle portion of the ring however, some marginal values were found which prompt further optimisation of the quench-and-temper conditions.

*With due acknowledgement to Brueck GmbH, Ensheim, Germany.

7. Specifications

In the absence of an international material standard for this type of 13Cr steel, a set of specifications was developed based on a combination of duplex stainless steel specifications and mill/manufacturers' specifications. As a result these are still manufacturer oriented, but the process to transform these documents into more universal specifications is ongoing.

8. Girth Welding

The basic philosophy during the feasibility stage has been to use existing, proven welding processes and consumables which should primarily result in a robust, rather than in a fast, welding procedure. The initial approach has been to use the methods and welders qualified for the welding of duplex stainless steel, out of which the manual GTAW (TIG) process was selected because of its high flexibility and excellent results. Initially, the main emphasis for filler metals has been on duplex and superduplex stainless steel consumables but, after this was found to be successful, commercially available, matching (13Cr4Ni) materials were also tried. In general the duplex stainless steel materials gave the best toughness results while the 13Cr4Ni showed the higher strength values. In order to increase productivity, the emphasis shifted mainly to GMAW Gas Metal Arc Welding (MIG (Metal Inert Gas)) and to a limited extent to SMAW Shielded Metal Arc Welding (MMA (Manual Metal Arc)) welding.

At some SMAW welds reduced ductility was noticed which could be caused by insufficient control over oxygen/nitrogen pick-up during welding. This requires further work and possibly the development of improved electrodes for this material rather than the use of the currently available types.

The welding procedure development aimed at avoiding the need for post-weld heat treatment, mainly for cost reasons. It was shown that, when girth welding at usual heat input levels (up to 1.5 kJ mm^{-1}), good mechanical properties are obtained without the need for post-weld heat treatment.

The favoured non-destructive inspection technique is ultrasonic testing, as this can best detect and size all 2-dimensional defects (lack-of-fusion, cracks) which are considered to be the most important threat to the integrity of a weld in 13Cr material.

From the good results obtained by using duplex stainless steel type consumables it can be concluded that the 13Cr materials can be welded to duplex or superduplex stainless steel materials without problems. This offers at this stage of the development an attractive level of flexibility, since a duplex stainless steel equivalent can be used when certain components are not yet available in 13Cr material

In summary, a fully qualified welding procedure has been developed which includes an GTAW (Gas Tungsten Arc Welding) root weld followed by a number of GMAW filler passes while using a matching (13Cr4Ni) consumable. Further improvements could include the use of fully mechanised MIG welding in or not in combination with a root weld deposited according to the Surface Tension Transfer STT* process. The latter could offer clear advantages in depositing a quick and sizeable root pass.

---

*Surface Tension Transfer process, Lincoln Smitweld, Nijmegen, The Netherlands.

9. Toughness Assessment

9.1. Basic philosophy

The material may be subjected to temperatures as low as –20 or even –30°C mainly as a result of cooling by sudden pressure reductions in the gas. Given the martensitic microstructure, the high strength level and the absence of experience it was decided to execute rather extensive fracture toughness testing, at least during the qualification stage. This included apart from Charpy testing, crack tip opening displacement (CTOD) testing and wide plate testing on welding specimens. In this way the best data set could be obtained to enable comparisons with data from other high strength materials.

9.2. Results

On the various 13Cr components and particularly on the welds (seam and girth), extensive testing was executed by the manufacturers, by NAM but mostly by the Belgian Welding Institute in Gent. The main results are summarised in Table 3 which shows typical values.

9.3. Main Problems Encountered

9.3.1. Occasional low Charpy test values

During Charpy testing occasional low values were obtained amongst generally very high values. In several cases this could be attributed to the presence of small defects which acted as initiators in the Charpy specimens. In the other cases however, no such initiators could be identified. This suggests that the material could exhibit a rather wide scatter around the Charpy transition temperature. This behaviour has so far not been clarified, for instance through CTOD tests. The latter tests do not show a similar pattern of scatter and are considered to be a more reliable way of assessing the toughness of these high strength materials.

Table 3. *Typical values from testing of 13Cr components*

Test method	Materials tested	Test temperature	Results
Charpy-V impact	Plate	–20°C	150–200 J
	Weld metal		100–130 J
	Heat affected zone		45–100 J
Wide plate	Weld metal	–20°C	1.0 mm
	Heat affected zone		0.5 mm
Crack tip opening Displacement	Plate	–20°C	0.3–0.4 mm
	Weld metal		0.15–0.25 mm
	Heat affected zone		0.15–0.25 mm

9.3.2. Premature failure of wide plate test specimen
At one of the earlier wide plate tests a sudden, complete rupture occurred at a location far away from the artificial defect in the weld zone and at a load close to the maximum expected value for that specimen. The fracture was completely brittle in nature. Further investigations revealed that the wide plate specimen contained numerous sharp cracks everywhere at its edges with a depth of up to about 1 mm. This had initiated an unstable crack extension which the plate was unable to arrest. The cause of this phenomenon was the oxygen cutting method that had been used for cutting the specimen. After this method had been replaced by a more sophisticated plate preparation technique the problem did not re-occur.

9.3.3. Notch positioning for CTOD tests
Two types of CTOD testing were performed: for the centre of the weld a rectangular (B × 2B) specimen with a through-thickness defect and for the HAZ a surface notch in a square (B × B) specimen. In the latter it proved sometimes difficult to locate the notch at the fusion line as intended; the variance in microstructure at that location, possibly in combination with residual stresses, sometimes caused the fatigue pre-crack to deviate. This could result in an invalid test which then had to be tested again.

10. Certifying Authority Approval

From the very beginning of the project, Stoomwezen have been party to the discussions with the suppliers and manufacturers as, in the end, their acceptance is required for this material which is not yet included in the Stoomwezen Rules. All along the development path, mutually agreed solutions for problems were developed which resulted in an initial Stoomwezen approval given in an early stage of the project. The final, formal approval is expected during the site fabrication stage when all the necessary documentation has become available.

11. Project Status

The status of the project is that the qualification activities have been completed successfully, except for the flange material. During the course of these qualification exercises some problems have emerged as indicated above. The actual production of the pipe spools was to be resumed in August 1998 but since then, some project related, non-13Cr related queries have come up which prompted NAM to postpone the actual production activities for some time. It is anticipated however that the spools will now be fabricated by the end of 1998 or by early 1999.

12. Conclusions and Recommendations

1. Weldable 13Cr pipe and fitting materials are now considered to be qualified for wet gas piping systems (and pipelines).

2. The final heat treatment (quenching-and-tempering) is a dominant activity in the manufacturing route which determines the final properties of the material to a large extent. Effective measures to control and document these activities are essential and need to be developed further.

3. The 13Cr material showed its great potential for recovery from ill-treatment through a repeated quench-and-temper treatment.

4. The toughness properties for weldable 13Cr materials (lean grade) have been improved to well above the original NAM project requirements (min. 70 J at the design temperature of –20°C). Further work on the relevance of Charpy-V impact testing for this very high strength material is however needed.

5. The material has obtained initial Stoomwezen approval, based on an individual production route basis. As the use of weldable 13Cr increases, more data from different sources will enable a more generic acceptance by certifying authorities.

26
Fabricating Pipeline Bundles Using Modified Weldable 13% Chromium Stainless Steel Flowlines

TRICIA BARNETT
Rockwater Ltd, Stoneywood Park, Dyce, Aberdeen, AB21 7DZ, UK

ABSTRACT
Rockwater Ltd fabricate pipeline bundles and tow them to the field using the controlled depth tow method. In recent years pipeline bundle systems have increased in complexity and the flowline materials have become more exotic. Statoil selected modified 13% Cr for high temperature strength and CO_2 resistance for the Gullfaks South Bundle product flowlines after intensive research and a development project. The selected material was a nominal Cr 12, Ni 6 and Mo 2% steel. When fabricating a pipeline bundle, issues such as insulation and heat-up systems, spacer design and manufacture, buoyancy, towhead design, build and attach, and material delivery constraints serve to complicate what would appear to be straightforward pipeline welding activities. Additional issues such as handling of the bare line pipe material and qualification of the welding procedure specifications plus intensive SSCC (sulfide stress corrosion cracking) testing meant that careful planning and execution of the project were necessary.

1. Introduction

Rockwater Ltd has built towed production systems since 1980. The first bundles to be launched from the site at Bridge of Wester near Wick in Caithness, were for the Conoco Murchison project. These bundles consisted of 12 in. carrier pipe containing two 3.5 in. (88.9 mm) o.d. × 6.3 mm w.t. (wall thickness) towed flowline (TFL) pipelines and four hydraulic control lines each being less than 1 in. diameter. At that time the site was 3.5 km long. Since then 31 bundles have been fabricated and launched from Wester Site which now, with a length of just under 8 km and a width of 30 m, is considered to be one of the most unusual fabrication yards in the world.

Fabrication of pipeline bundles is a complicated activity with welding comprising approximately 20% of the fabrication man-hours. This paper will describe towed production systems and how the change in design requirements for flowline systems has increased the complexity of fabrication activities. Emphasis will be on one of the recent projects, namely the Gullfaks South Bundles. These were the most complicated bundles ever built by Rockwater in terms of materials, number and configuration of flowlines and structural and thermal design requirements. Furthermore, it was Rockwater's first project in which modified weldable 13% Cr stainless steel flowlines were fabricated.

2. Controlled Depth Tow Method

The controlled depth tow method (CDTM) involves the onshore construction of multiple flowlines within a single outer carrier pipe terminated at each end by structures designed to accommodate the bundle operationing conditions and transmit the launch and tow forces into the bundle. Once launched from the onshore site, the bundle is transported suspended between two tugs at a controlled depth below the surface to its offshore location. On entry into the field the bundle is lowered to the sea-bed, manoeuvered into location and the carrier pipe flooded in its final position (Fig. 1).

The main purpose of the carrier pipe is to provide buoyancy for the towing operation. It also acts as a protection structure for the inner pipelines and is utilised to maintain the required bundle configuration with the use of spacers, which are attached at 12–15 m intervals. The carrier pipe material has consistently been API 5L X52 grade material. Protection from corrosion is achieved by the use of aluminium alloy anodes and by the application of GFE M912 (a glass flake epoxy coating system). The buoyancy requirements mean that weight control is a crucial activity to successful launching and towing operations. It is essential that every item within the carrier pipe is weighed and recorded on the pipe tracking system so that the action of the bundle during launch and tow can be predicted.

3. History of Bundles at Wester Site

The first bundles to be built at Wester Site consisted of carbon steel flowlines arranged into the configuration noted in Fig. 2. Carbon steels were consistently used for flowline materials at the time. Lately, field developments have involved more sour conditions with higher temperatures and pressures such that alternative materials have been sought. The use of duplex and super duplex stainless steels is well documented as having an expensive capital cost with the savings being made in extended design life and minimal maintenance cost when compared with carbon steels. Rockwater, however offer alternatives in the form of mechanically bonded CRA lined carbon steels. These materials involve a lower capital investment with the same production advantages as the more expensive duplex materials.

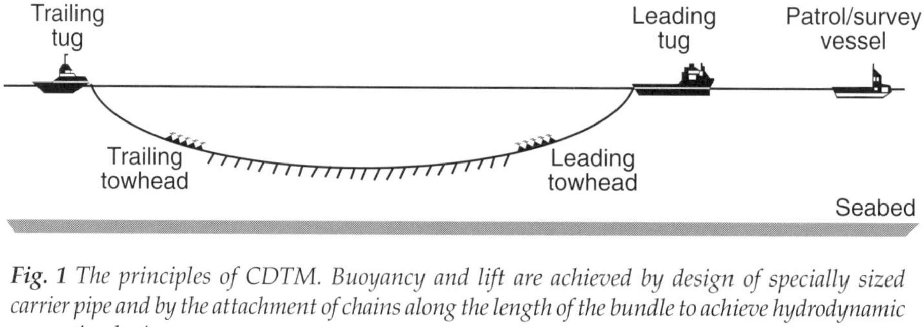

Fig. 1 The principles of CDTM. Buoyancy and lift are achieved by design of specially sized carrier pipe and by the attachment of chains along the length of the bundle to achieve hydrodynamic properties during tow.

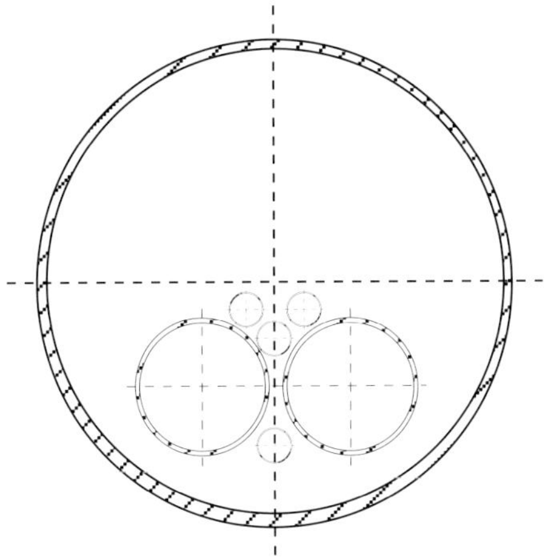

***Fig.* 2** *Conoco Murchison Bundle Cross-Section. All flowlines are carbon steel..*

More recently, developments have been made in weldable martensitic stainless steels for pipeline applications. These were developed for use in non-sour to mildly sour services where high tensile strengths are required at elevated temperatures. Statoil have been the driving force behind the use of these steels and indeed the first application of these steels for Rockwater was for the Gullfaks South Bundles for Statoil. The cross-section of this bundle is shown in Fig 3. For comparison the differences between the first bundle and the latest one are noted in Table 1.

In general, there are no requirements for corrosion coating of the inner flow lines as the carrier pipe is flooded with inhibited water. However insulation systems can be produced using a pipe-in-pipe system as opposed to an inner bundle configuration as shown on the Gullfaks example.

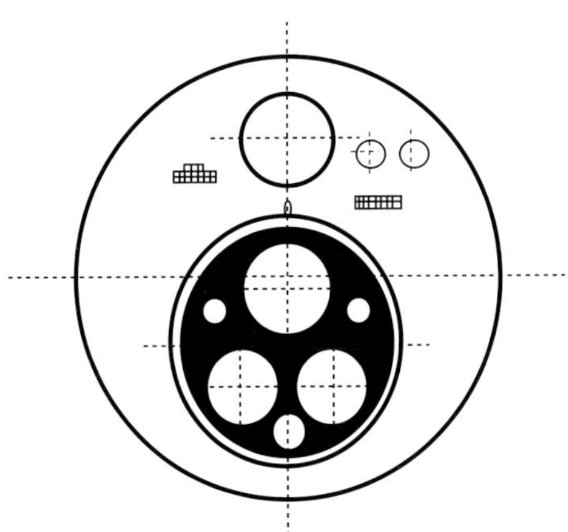

***Fig.* 3** *Statoil Gullfaks Bundle Cross-Section. Various materials are welded side-by-side — complicating the welding operations. Three product lines are held together within an insulated sleeve and with three small diameter hot water and glycol lines to provide a heat-up system to melt hydrates in the event of a shut down (see also Table 1).*

Table 1. *The nature of the most recent (Gullfaks South) pipeline bundles as compared to the first (Murchison) bundles*

	Murchison	Gullfaks
Length of bundle	3 bundles – 2005 m max. individual length	2 bundles – 6435 m max. individual length
Carrier pipe size	12.75 in. o.d. × 6.3 mm w.t.	39 in. o.d. × 11.9 mm w.t.
Number of flowlines	2 TFL pipelines plus 4 hydraulic control lines.	3 production lines; 3 heat-up lines insulated within a 22 in. sleeve plus 1 gas injection line; 2 methanol injection lines; 18 hydraulic control tubing; 12 signal/power lines
Materials	Carbon steel	Super martensitic "13% Cr" stainless steel production lines; all other lines carbon steel except hydraulic control lines which were super duplex
Insulation requirements	None	20 mm thick low-density polyurethane foam held within an inner sleeve to maintain operating temperature of 60°C
Corrosion coating	None except for Carrier Pipe corrosion coating	None except for Carrier Pipe corrosion coating

4. 13% Chromium Stainless Steel Pipelines

The selection of 13% Cr stainless steels for the production flowlines came about when it was realised that the Gullfaks Satellites Development would require such quantities of duplex steels that the market balance in steel pipe production would be heavily influenced and the price would increase accordingly. Furthermore the capacity of the appropriate steel mills would be stretched to the point that delivery schedules would be complicated.

An intensive research and development project led to the development of modified weldable 13% Cr steels from various manufacturers. Rockwater fabricated the Gullfaks South bundle using free-issued steel with the nominal composition as shown in Table 2. Detailed chemical analysis is specified in the material specification [1].

It is documented [2] that these 13% Cr steels have been qualified for welding and have been tested for sulfide stress corrosion cracking in various conditions with and without the use of post weld heat treatment (PWHT). The steel documented in Table 2 was qualified for the Gullfaks field without PWHT and welded principally using a mechanised GMAW system. This process is used in pipeline welding applications for laybarge and reeled spoolbase operations.

Table 2. *Nominal composition (%) of the weldable 13% Cr stainless steel used for the Gulfaks South bundles*

C	Ni	Cr	Mo
0.01	6.0	12.5	2.0

5. Welding Procedure Qualification Requirements

Welding procedures were qualified in accordance with Statoil pipeline welding requirements modified to suit the 13% Cr materials in bundle applications by a Rockwater developed Design Premise. In addition to the usual mechanical testing requirements for pipeline welding procedure qualifications, further tests were required to suit the particular requirements of the Gullfaks field taking into account the pipeline material. Elevated temperature all weld tensile tests were conducted and the 0.2% proof strength was recorded. This value was used as the yield value for the elevated temperature SSCC testing that was performed in accordance with EFC No. 17 [3]. Both constant load and four point bends were tested in formation and condensate water as described in the Statoil specification [1] for the Gullfaks field.

6. Inner Bundle Fabrication Methods

As the emphasis of the paper is on 13% Cr materials in bundles, the description of the fabrication method will focus on the inner bundle section as shown in Fig. 3. It should be noted here, however, that after the inner bundle was completed there were still many other activities to be performed. This included among other items the welding of the gas injection line and two methanol injection lines; the addition of these lines into the bundle formation using spacers and the laying of the coiled control tubing. Afterwards the sheathing of the entire bundle with the seam-welded carrier pipe took place, which arrived on the site in 27 m long spools. Towhead fabrication is subcontracted. The entire Towed Production System is intricately planned such that each item arrives on site ready for installation or attachment at the correct time.

Inner bundle fabrication commenced during the winter months of 1997/98. As the 13% Cr was bare storage and handling control was developed to ensure that exposure to the elements was minimal, thereby minimising the risk of pitting corrosion. This meant that the production lines had to be assembled into the inner bundle configuration, insulated and sleeved inside the fabrication shop. To do this, each production line and heat-up line was fabricated into a 96 m long spool off line and tied in to the bundle. Once all six tie-in welds were completed, the spacers were attached to pull the lines into configuration. When this activity was completed the insulation was installed in the form of 6 m long half shells which encircled the inner bundle. Then the sleeve pipe was pulled over the bundle and welded. When all of these activities were completed the production line was ready for the next set of spools.

7. Welding Procedure Specification

Welding process selection is based on several factors:

- The quality requirements.
- The rate determining activity in the overall fabrication process.
- Availability of skilled labour.
- Pipe dimensions.
- Material requirements.
- Non-destructive examination (NDE) requirements.

The production pipeline dimensions were as noted in Table 3. The majority of the welds were of a size that suited manual welding techniques. The skill of the welders is such that GTAW is a widely used process at the site. Evidence is provided by the large number of small diameter pipelines that have been consistently produced there with high quality and low repair rates.

For the above reasons, the GTAW process was selected as the optimum welding process for the 13% Cr pipelines.

Consumable selection was based on a combination of design and commercial factors. There are no commercially suitable consumables available that incorporate a matching chemistry. Development of a matching filler wire is at present being undertaken by others [4]. Initially a weld development programme was undertaken to assess 2209 type (duplex stainless steel) consumables for elevated temperature yield strength. It was found that 2209 did not have the elevated temperature yield strength required, although it was accepted that it would be a suitable consumable in terms of SSCC resistance. It was therefore decided to opt for a 2509 type (super duplex stainless steel) consumable for the filler passes but maintain the 2209 for the root and second passes. As the super duplex stainless steel consumable was more expensive by a factor of 4, this meant that there would be considerable savings made by using the less expensive 2209 type for the root. Mechanical testing results supported this selection with yield strength values at room and elevated temperatures being within contractual requirements. Mechanical test results are detailed in Table 4.

Sulfide stress corrosion cracking test results all revealed no cracks at magnification of ×50.

Table 3. *Production pipeline dimensions for the Gullfaks satellite bundle*

Description	Dimensions	Materials
Statfjord Production Line	168.3 mm o.d. x 8.8 mm w.t. (11.9 mm w.t. for zone 2 pipe)	Modified super martensitic 13% Cr steel
Brent Production Line	219.1 mm o.d. x 11.5 mm w.t. (14.2 mm w.t. for zone 2 pipe)	Modified super martensitic 13% Cr steel

Table 4. *Typical mechanical testing results for weld procedure qualifications on modified 13% Cr pipeline materials*

Yield Strength (MPa)	Tensile Strength (MPa)	WCL Charpy –40°C (J)	FL Charpy Test –40°C (J)	Maximum Hardness (HV10)
630 at 25°C 590 at 100°C	864 at 25°C 810 at 100°C	191 average (188; 202; 183)	266 average (262; 263; 273)	317

WCL = weld centre line. FL = fusion line.

The bevel angle was selected to reduce welding times and decrease dilution effects for the strength and microstructure of the weld metal. The weld configuration is shown in Fig. 4.

Using this process with the method as described, an optimum fabrication process was developed. The overall welding quality was very high with a nominal repair rate of less than 0.5% by length of weld.

9. Future Work

Rockwater are contracted to build more bundles for Statoil over the next two years. Each bundle contains modified 13% Cr material and in each case elevated temperature tensile tests and SSCC testing will be required for weld procedure qualifications. Pipe dimensions will increase and service conditions may preclude the need for PWHT. This will be determined by the SSCC test in each project. Automated welding systems are being investigated in a search for the optimum welding process to suit the most efficient fabrication method within the parameter limits of the materials and bundle design. The larger pipe dimensions will suit a submerged arc welding process that would be utilised for double joints. Tie-in welds may be performed by an orbital GTAW system.

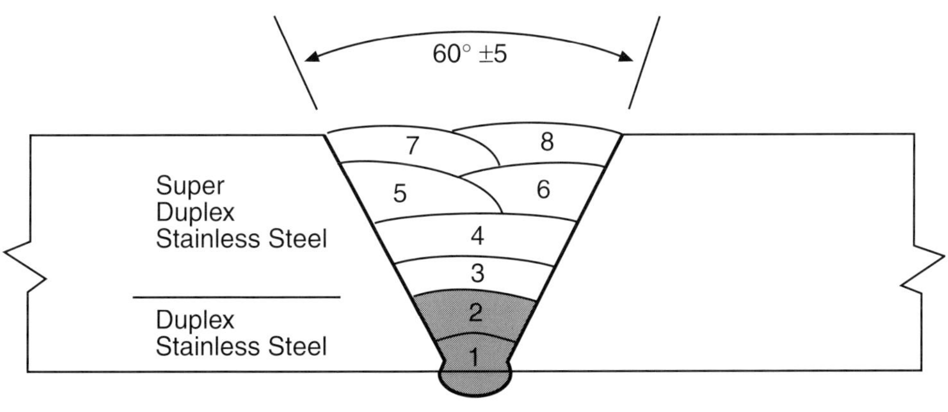

Fig. 4 *Typical weld configuration and pass sequence.*

References

1. Statoil Specification for Seamless Line pipe - 13% Cr steel.
2. J. Enerhaug and P. E. Kvaale, Qualification of Welded Super 13% Cr Martensitic Stainless Steels for Sour service Applications – A Statoil paper.
3. *Corrosion Resistant Alloys for Oil and Gas Production: Guidance on General Requirements and Test Methods for H_2S Service*. Publication No. 17 in the European Federation of Corrosion publication series. Published by The Institute of Materials, London, 1996, 84 pp.
4. H. Heuser, E. Perteneder and J. Tosch, Welding of 13% Cr Supermartensitic Stainless Steels (the present state of development of matching filler metals) – from *Pipelines '98*.

27
Corrosion Resistance of 13% Chromium Stainless Steel Welded Joints in Flow Line Applications

M. UEDA*, H. AMAYA, H. HIRATA, K. KONDO, Y. MURATA† and Y. KOMIZO
Corporate R&D Laboratories, Sumitomo Metal Industries Ltd., Japan
*Wakayama Steel Works, Sumitomo Metal Industries Ltd., Japan
†Amagasaki Factories, Sumikin Welding Industries Ltd., Japan

ABSTRACT
The mechanical properties and sulfide stress corrosion (SSC) resistance of weldable super 13Cr stainless steel welded joints were investigated in flow line application. The joints welded by Gas Tungsten Arc (GTA) and Gas Metal Arc (GMA) using matching or over-matching weld metal, consisting of super duplex stainless steel, without any preheating and post weld heat treatment (PWHT), showed strength and toughness and good SSC resistance. However, in the welded joint with the under-matching weld metal that would be adequate for conventional 22Cr duplex stainless steel, SSC occurred near the fusion line. These SSC results could be explained by the stress concentration, obtained by FEM calculation, near the fusion line with the under-matching weld metal. Such stress concentration did not exist in the welded joint with the matching or over-matching weld metal. From these results, the super duplex stainless steel was recommended as a suitable material in the welding of super 13Cr stainless steel pipes.

1. Introduction

Martensitic stainless steels, API 5CT 13Cr have been widely used for Oil Country Tubular Goods (OCTG) as having good corrosion resistance to CO_2, environments with and without a small amount of H_2S [1]. However, the application of this type of material for flow lines was difficult because preheating prior to welding and post weld heat treatment (PWHT) had to be applied to the welded joints to prevent toughness reduction and sulfide stress cracking (SSC) at the heat affected zone (HAZ) with high hardness. Therefore, conventional 22Cr duplex stainless steel (UNS S3 1803) has been used for flow lines with no preheating and no PWHT, even though the material cost was high and the corrosion resistance of the welded joint was over specification for flow lines in environments where API 5CT 13Cr steel has been used as OCTG.

Recently, a number of new 13Cr stainless steels with improved resistance to SSC and CO_2 corrosion have been developed and used as OCTG [2]. Weldability of these steels and corrosion resistance of welded joints have been investigated and a weldable Super 13Cr stainless steel of 550 MPa (80 ksi) grade, with ultra low C–12Cr–6Ni–2.5Mo–Ti, has been developed by Sumitomo Metal Industries Ltd [3]. Welded joints of this new steel with no preheating or PWHT required, because of its low C content (below 0.01 mass%) and increased Mo content, have superior resistance to SSC and CO_2 corrosion compared with that of API 5CT 13Cr OCTG.

In this paper, the weldability without any preheating and PWHT of super 13Cr stainless steel joints welded by using Gas Tungsten Arc (GTA) and Gas Metal Arc Methods (GMA) with duplex stainless steel welding materials is investigated for line pipe application. Then, in order to obtain a good combination for mechanical properties and corrosion resistance between base metal and weld metal, the effect of the strength of weld metal on the stress distribution in welded joints was analysed by the Finite Element Method (FEM) calculation. The relationship between the stress distribution and SSC resistance of the welded joint is discussed.

2. Experimental

Materials used: Weldable super 13Cr stainless steel pipes which had the chemical compositions and the tensile properties shown in Table 1 were used in this study. These materials were melted and manufactured in mill as seamless pipe of 168.3 mm (o.d.) and 12 mm wall thickness (w.t.), and 273.1 mm (o.d.) and 12 mm (w.t.), respectively. They were water quenched and tempered to provide a 0.2% yield strength of 550 MPa (80 ksi) grade.

Welding materials and procedures: The chemical compositions of the welding materials used are shown in Table 2. Those welding materials which consisted of super duplex stainless steels had tensile strengths of about 900 MPa, compared with about 800 MPa for conventional 22Cr duplex stainless steel. The welding material was solid type wire of composition Cr25–Ni9–Mo3–W2–N0.25 for GTA and metal cored Cr23–Ni9–Mo4–W2–N0.17 for GMA. Resistance to blowholes was obtained by reducing the nitrogen content, depression of intermetallic compound precipitation by reducing the Cr content and increasing pitting resistance by adjusting the PREW (Pitting Resistance Equivalent (mass%); = Cr + 3.3(Mo + 0.5W) + 16N) [8]. The GTA and GMA girth welds were provided on pipes of 168.3 mm and 273.1 mm o.d., respectively. The welding conditions used are shown in Table 3. The interpass temperature was controlled at under 150°C, and preheating and PWHT were not applied.

Experimental procedures: The side bend test, the tensile test at room temperature, measurement of hardness, the Charpy impact test at –30°C and observation of microstructure were carried out for joints welded by GTA and GMA using super duplex stainless steel weld materials. Radiographic inspection was also conducted for the GMA

Table 1. *Chemical compositions and tensile properties of materials used*

Mark*	C	Si	Mn	P	S	Ni	Cr	Mo	Ti	N	YS (MPa)	TS (MPa)
No. 1	0.007	0.31	0.45	0.018	0.001	6.20	12.05	2.53	0.070	0.004	567	883
No. 2	0.010	0.24	0.46	0.020	0.001	6.42	11.96	2.42	0.077	0.010	579	906

* No. 1, 168.3 dia. × 12w.t. (mm); No. 2, 273.1 dia. × 12w.t. (mm)

Table 2. Chemical compositions of welding materials used

Mark*	Type	C	Si	Mn	P	S	Ni	Cr	Mo	W	N	TS (MPa)
GTAW-1	Solid	0.017	0.42	0.53	0.010	0.008	9.3	25.7	3.1	2.3	0.26	900
GTAW-2	Solid	0.007	0.40	1.61	0.008	0.001	8.4	22.1	3.3	—	0.10	780
GMAW*	MC**	0.025	0.45	0.95	0.022	0.004	9.4	23.3	4.3	2.1	0.17	898

*Chemical composition of weld metal; **Metal cored.

Table 3. Welding conditions applied

Method	Number of passes	Welding current (A)	Welding voltage (V)	Travelling speed (mm s^{-1})
GTAW–1, 2	10	130	15	1.67
GMAW	5	140–150	23–25	6.7–10

welded joints. The parallel part of each tensile test specimen was 6 mm dia. and 40 mm in length with the weld metal in the centre. In the Charpy impact test specimen, the notch was machined at the fusion line. The SSC resistance of welded joints was evaluated by the constant load test using the specimen as shown in Fig. 1. The fusion line in the specimen was located in the centre of the parallel part. The stress applied to the specimen was the yield strength of the base metal taking into account the residual stress on the welded joints. The corrosion tests were carried out in a simulated formation water containing 68 000 ppm Cl^- controlled at pH 4.5, and with 0.04 bar H_2S + 0.96 bar CO_2; the testing period was 720 h.

Finite element calculation method (FEM): The stress distribution of the welded joint in two cases was analysed by the FEM for the test specimen including weld metal, fusion line and HAZ as shown in Fig. 1. The mesh for the FEM analysis had 623 nodes and 528 elements. The minimum mesh size was 0.1 mm for the longitudinal direction and 0.5 mm for the radius direction. In Case I it was assumed that the specimen had the weld metal of conventional 22Cr duplex stainless steel. The tensile strength results shown in Tables 1 and 2 showed that the weld metal in this case under matched the parent metal. In Case II it was also assumed that the specimen had weld metal of the super duplex stainless steel. The weld metal in this case was almost even matching. The calculation was carried out based on the stress and strain curve for each material and the increase of hardness in HAZ was taken into the calculation.

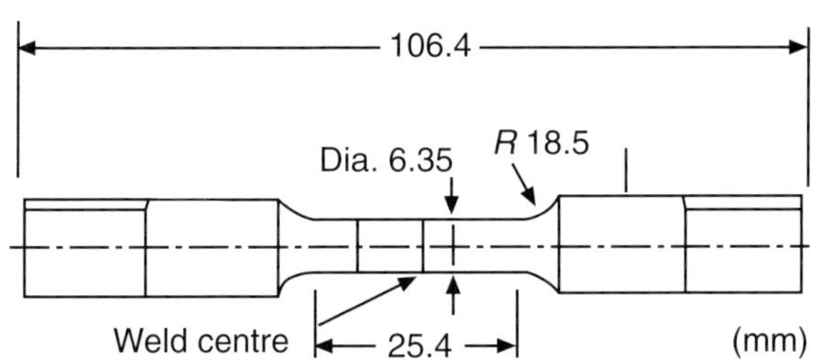

Fig. 1 Schematic illustration of constant load test specimen.

3. Results and Discussion

Mechanical properties of welded joints: The result of radiographic inspection of GMA welded joints is shown in Table 4. In the weld metal, there was no noticeable weld defect and it was evaluated as 1st class according to the Japanese Industrial standard (JIS) Z3106. The result of side bend tests is shown in Table 5. In the side bend tests, no cracking was observed in weldments of GTA and GMA welded joints. The tensile strengths of the welded joints at room temperature are shown in Table 6. The GTA and GMA welded joints had high strengths of more than 620 MPa which is the specification value of API 5L X80. The fractured portion of one specimen of the joints welded by GMA with super duplex stainless steel welding material, was weld metal because the weld metal with tensile strength of 888 MPa was almost evenly matched to the base metal with tensile strength of 906 MPa. In GTA welding with conventional 22Cr duplex stainless steel welding material, the fractured portions were again weld metal because the weld metal with tensile strengths of 770 and 765 MPa was under matched to the

Table 4. *Radiographic inspection result of GMA welded joint (Japanese Industrial Standard Z3106)*

Type I (Blowhole)	Type II (Lack of fusion)	Type III (Crack)	Type IV (Tungsten inclusion)
Dia. 1.00 mm × 2 (1st class)	None (1st class)	None (1st class)	None (1st class)

Table 5. *Side bend test result of welded joints*

Welding method	Side bend test	
	Specimen No. 1	Specimen No. 2
GTAW-1	Satisfactory	Satisfactory
GMAW	Satisfactory	Satisfactory

Table 6. *Tensile test result of welded joints at room temperature*

Welding method	Specimen No.	Tensile strength (MPa)	Fractured portion
GTAW-1	1	889	Base metal
	2	885	Base metal
GTAW-2	1	770	Weld metal
	2	765	Weld metal
GMAW	1	888	Weld metal
	2	894	Base metal

base metal. The maximum hardness in the HAZ of both GTA and GMA welded joints was under 330 (Vickers hardness number) and the increase of hardness in the HAZ was less than 30 as shown in Table 7. The absorbed energies of GTA and GMA welded joints in Table 8 were more than 200 J and 100 J respectively which were sufficient for the actual welding fabrication. The weld metals in both GTA and GMA welded joints consisted of α- and γ-phase and there was no noticeable coarsening in the HAZ.

Stress distribution in welded joints by FEM calculation: The stress distribution of the welded joint with the weld metal of conventional 22Cr duplex stainless steel was calculated by the FEM. The result for Case I in Fig. 2 was considered as under-matching. The result corresponding to Case II of the stress distribution in the welded joint with the weld metal of super duplex stainless steel was almost even matching. In Case I, a stress concentration existed in the base metal near the fusion line. On the other hand, no stress concentration was found in the base metal in Case II.

SSC resistance of the welded joints: The welded joint with the weld metal of conventional 22Cr duplex stainless steel failed by SSC before 720 h (Table 9) and the fractured portions were near the fusion line. No SSC after 720 h was observed in GTA and GMA welded joints with the weld metal of super duplex stainless steel. The FEM calculation result

Table 7. *Maximum hardness in heat affected zone of welded joint*

Welding method	Maximum hardness in heat affected zone	Average hardness of base metal
GTAW-1	313 HV	290 HV
GMAW	328 HV	315 HV

Table 8. *Charpy impact test result of welded joints at –30 °C*

Welding method	Absorbed energy $_vE_{-30}$ (Joules)		
GTAW-1	241	218	233
GMAW	105	121	156

Table 9. *Corrosion test result of joints welded by GTA welding with weld metals of conventional 22 Cr duplex stainless steel and super duplex stainless steel*

Weld Metal	SSC test result
Super duplex (GTAW-1)	No Failure, No Failure
22Cr duplex (GTAW-2)	Failure (230 h), Failure (609 h)

Testing environment: 68 000ppm Cr, pH 4.5, 0.04 bar H_2S + 0.96 bar CO_2. Applied stress: 100% YS of base metal.

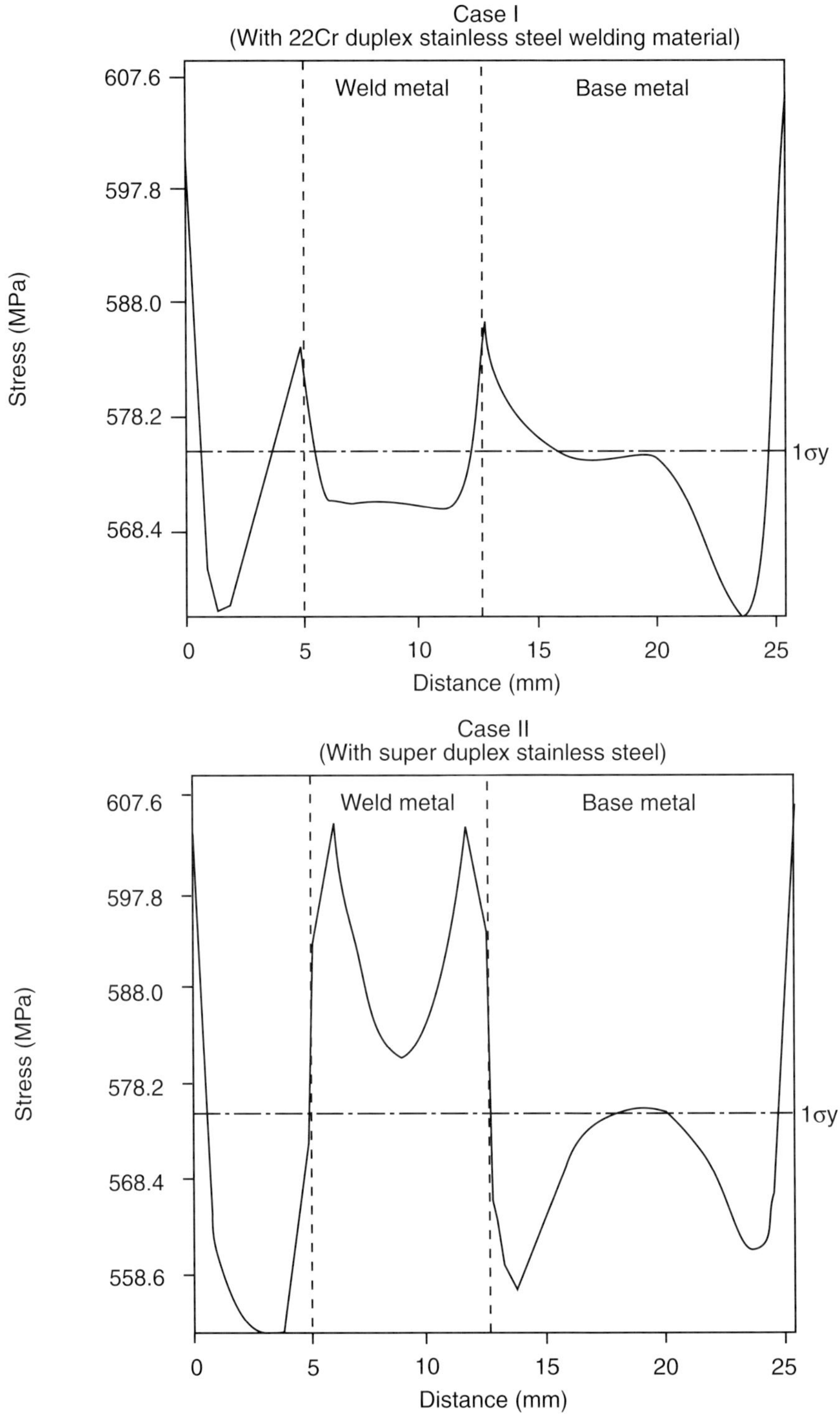

Fig. 2 *The stress distribution of the welded joints.*

suggested that the welded joint with the under-matching weld metal had inferior SSC resistance because the stress near the fusion line was higher than the yield strength of the base metal. Therefore, even or over-matching weld material is recommended for welding super 13Cr stainless steel without preheating and PWHT.

4. Conclusions

The mechanical properties and corrosion resistance of super 13Cr stainless steel welded joints was investigated for flow line applications. The results obtained can be summarised as follows:

1. GTA and GMA welded joints with adequate strength and toughness. and good SSC resistance were completed by using welding material of super duplex stainless steel without any preheating or PWHT.

2. SSC occurred near the fusion line in welded joints made with under-matching weld metal of conventional 22Cr duplex stainless steel, but did not occur in the welded joint with matching or over-matching weld metal of super duplex stainless steel. These SSC results can be explained by the stress concentration near the fusion line as revealed by FEM calculation for the welded joint with under-matching weld metal.

3. Based on this investigation, the super duplex stainless steel welding materials are recommended for welding super 13Cr stainless steel pipes.

5. Acknowledgements

The authors wish to thank Sumitomo Metal Industries, Ltd for allowing publication of this research. The assistance and discussion of co-workers in laboratories are gratefully acknowledged.

References

1. A. Baudoin, K. Barbin and J. Skogsberg, 'Experiences with 13Cr for mitigating CO_2 corrosion in the oilfield case histories: the Gulf of Mexico and inland gas wells', *Corrosion '95*, Paper No. 639, NACE International, Houston, Tx, 1995.
2. M. Ueda, T. Kushida, K. Kondo and T. Kudo, Corrosion resistance of 13Cr–5Ni–2Mo martensitic stainless steel in CO_2 environment containing a small amont of H_2S', *Corrosion '92*, Paper No.55, NACE International, Houston, Tx, 1992.
3. M. Ueda, H. Amaya, K. Kondo, K. Ogawa and T. Mori, Corrosion resistance of weldable super 13Cr stainless steel in H_2S-containing CO_2 environments, *Corrosion '96*, Paper No. 58, NACE International, Houston, Tx, 1996.

28
Slow Strain Rate Testing of Low Carbon Martensitic Stainless Steels

Th. BOELLINGHAUS, H. HOFFMEISTER and S. DIETRICH*
Institute for Failure Analysis and Failure Prevention, ISSV e.V., Hamburg, Germany
*Laboratory for Welding Technology, Federal Armed Forces University, Hamburg, Germany

ABSTRACT
As a result of increased H_2S levels in recently developed oil and gas fields the sensivity to sour service conditions of materials needs to be verified. Slow strain rate tests (SSRT) of low carbon martensitic chromium steels, containing 12 and 15 wt% chromium as well as 3.7 and 4.4 wt% nickel and 0.8 and 1.5 wt% molybdenum have been performed in modified NACE TM0177-90 standard solutions at strain rates from $1{\cdot}10^{-4}$ to $5{\cdot}10^{-7}s^{-1}$. Based on low strain rates, the tolerance fields for environmental cracking with respect to pH and H_2S show a superior cracking resistance of the higher alloyed as compared to the lower alloyed steels at low H_2S levels, but a greater susceptibility at higher H_2S levels as a result of pitting corrosion.

1. Introduction

The application of low carbon martensitic stainless steel as a pipeline material, in particular for flowlines in recently developed North Sea oil and gas fields, will subject such materials to sour environments containing H_2S up to 5% at bulk pH of 4.5 [1]. The lack of service experience so far requires the evaluation of tolerance fields against environmetal cracking at these and even worse conditions in critical regions as, for example, notches and crevices. Additionally, such materials might be strained during fabrication welding and pipe laying as well as during service of pipelines. An internal pressure change of 0.6 kPa s^{-1} for instance, causes a strain rate of $1.3 \cdot 10^{-7}s^{-1}$ [2] which may provide local plastic deformation in notches. Thus, sour service frameworks for low carbon martensitic stainless steels should not only be based on constant loading [3,4]. Slow Strain Rate Testing (SSRT) represents a suitable method for determining the cracking resistance of a material subjected to H_2S-containing environments at time-dependent loading [5,6]. Slow strain rate testing of two commercially available alloys has been conducted in NACE standard electrolytes to provide a rapid screening [7] of their mechanical behaviour and tolerance to hydrogen assisted cracking in sour environments prior to full scale component testing (see this volume page 286).

2. Experimental

Standard samples [8] of steel A and B with the chemical composition and mechanical properties listed in Table 1 were machined from seamless pipes. The specimens were encapsulated in an environmental cell and mounted in a four column mechanically driven tensile test machine, providing actual constant strain rates down to $8 \cdot 10^{-8}$ s^{-1} for standard specimens by analogous velocity control. As a first series, the steels were tested at low pH in the NACE standard solution at various strain rates. For security reasons, the different H_2S saturation levels of the electrolytes were prepared by partial mixture with totally saturated solutions. Contents of H_2S were controlled by an H_2S sensitive electrode. For long test durations the electrolytes were renewed if a variation of the H_2S content above 10% occurred. As a second series, the pH was adjusted up to 5.5 by NaOH. The pH of the buffered solutions did not vary significantly during the test periods. Both materials were tested in the as-delivered state at a constant strain rate of $1 \cdot 10^{-1}$ s^{-1} until fracture. After rupture, one part of the specimens was taken for microstructural and SEM analysis, the other part to determine the average hydrogen concentration in the gauge length by carrier gas hot extraction.

3. Results and Discussion

The effect of strain rates and relative saturation of H_2S on 0.2% proof stress and tensile strength is shown in Fig. 1(a), for steel A. Compared to the mechanical properties obtained by testing in air at strain rates of $1 \cdot 10^{-5}$ s^{-1} and $1 \cdot 10^{-4}$ s^{-1}, a severe reduction of the tensile strength was found from 850 down to 630 MPa with increasing H_2S up to 50%, at the lowest strain rate of $5 \cdot 10^{-7}$ s^{-1}. At increasing strain rates, the reduction in tensile strength caused by H_2S was less severe and at $1 \cdot 10^{-4}$ s^{-1} the strength levels appeared to be independent of H_2S.

The results of the 0.2% proof stress levels reveal a particular effect of H_2S at $5 \cdot 10^{-7}$ s^{-1}. For example, with 50% H_2S, fracture already appeared at 0.2% proof stress after extensive pre-cracking. At higher strain rates, the effect of H_2S and even of strain rates was less pronounced thus providing proof stress levels close to those measured in air.

As a more significant indication of environmental effects on ductility, Fig. 1(b) reflects the strain rate and H_2S effects on reduction of area at fracture relative to results in air [9,10]. As expected from Fig. 1(a), rather low ductility was found at the $5 \cdot 10^{-7}$ s^{-1} strain rate already at a low H_2S relative saturation level of 1%. Increasing

Table 1. *Chemical composition (wt%) and mechanical properties of the test materials as delivered*

Type	C	Cr	Ni	Mo	Mn	Cu	Si	Ti	R_m (MPa)	$R_{p0.2}$ (MPa)	A(%)	Z(%)
A	<0.03	11.6	3.7	0.8	0.3	≤0.1	0.2	≤0.15	880	775	24	68
B	<0.03	15.0	4.4	1.5	0.6	<0.1	0.2	<0.15	900	640	27	72

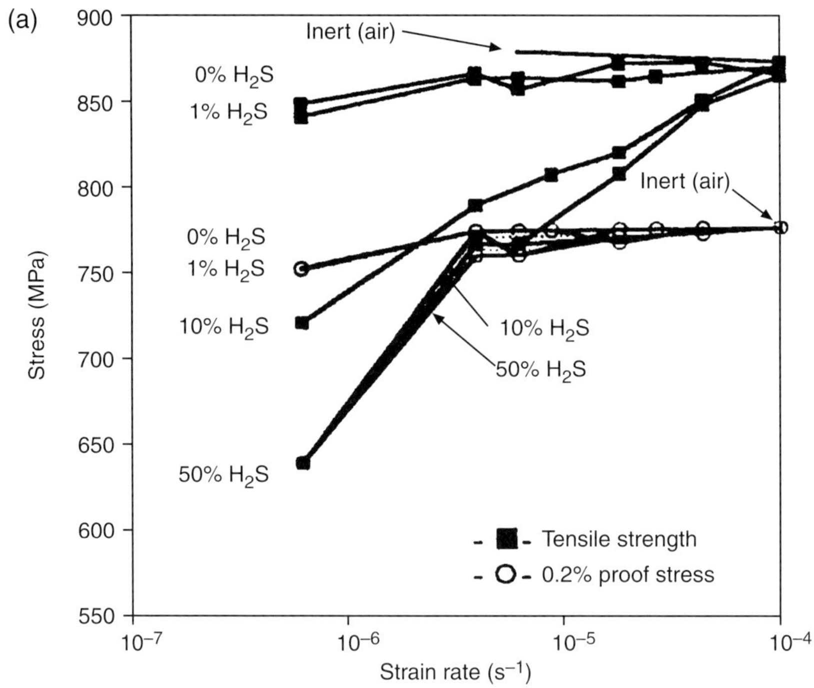

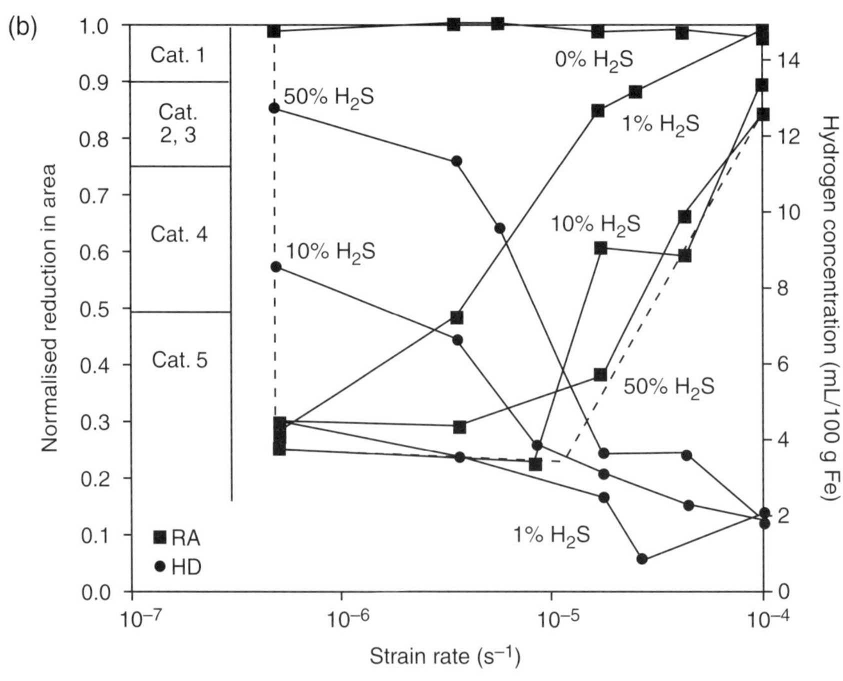

Fig. 1 *Performance of steel A in standard NACE solution of a pH of 2.7 at various strain rates. (a) Tensile properties at various H_2S levels. (b) Reduction of ductility at various H_2S levels.*

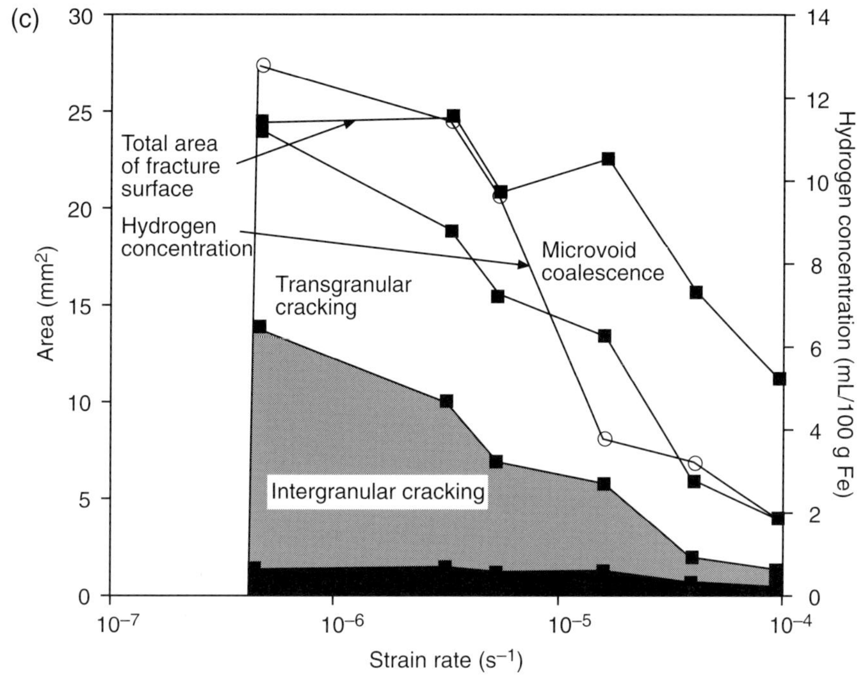

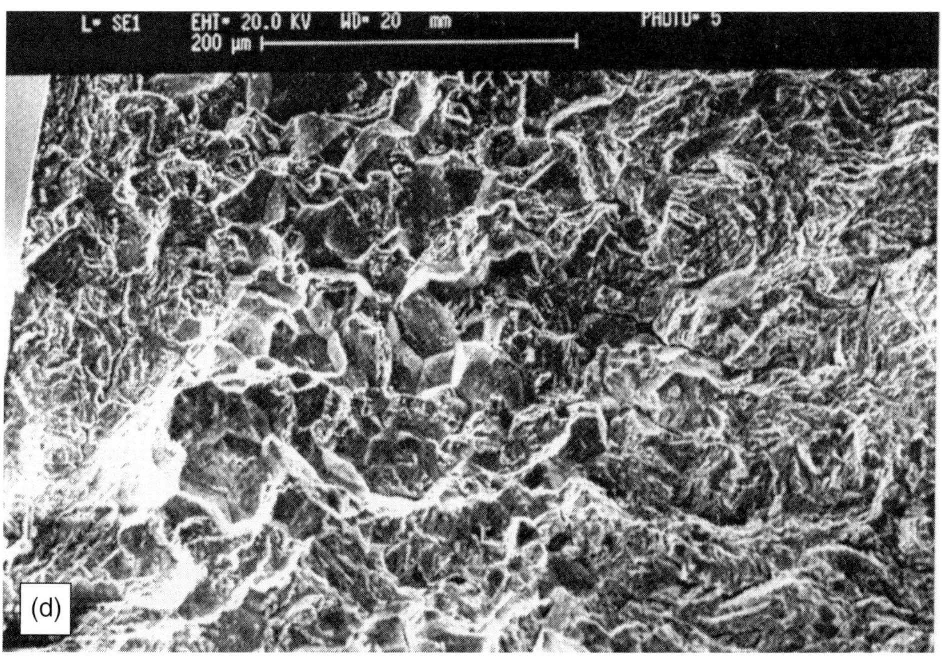

***Fig. 1** (continued)Performance of steel A in standard NACE solution of a pH of 2.7 at various strain rates. (c) Areas of different fracture modes at 50% H_2S. (d) Fracture topography at $\varepsilon' = 3 \cdot 10^{-6}$ s^{-1}, 50% H_2S.*

the strain rate provided the expected gains in relative ductility of up to almost 1 at 1% H_2S, but only to 0.8 to 0.9 at 10 and 50% H_2S at the high strain rate of $1 \cdot 10^{-4}$ s^{-1}.

For additional failure characterisation, Fig. 1(c) shows the strain rate effect on fracture topographies at the 50% relative H_2S level, based on typical fracture appearances as shown in Fig. 1(d). With increasing strain rate and reduced total fracture area, the area of microvoid coalescence (dimple) cracking increased at the expense of transgranular as well as intergranular cracking as an indication of environmentally (hydrogen assisted) cracking. The start of the circumferential crack revealed transgranular fissuring (Fig. 1d), together with evolution of, presumably, hydrogen bubbles, as seen in Fig. 2(a). The type of environmental cracking was finally confirmed from the total hydrogen contents as a function of H_2S as well as strain rates (Fig. 1b, c). At low strain rates and longer test periods, the hydrogen pickup from the standard NACE test solution under free corrosion conditions ranged from more than 12 mL/100 g Fe at 50% down to 4 mL/100 g Fe at 1% relative H_2S saturation. Strain rates of $1 \cdot 10^{-4}$ s^{-1} and short electrolyte exposure times provided hydrogen levels as low as 2 mL/100 g Fe, irrespective of the H_2S saturation.

The observed effect of strain rates and H_2S saturation of the NACE solution at a pH of 2.7 thus has to be attributed to hydrogen pick up and its detrimental effect on reduction of ductility. This is confirmed in Fig. 3 which compares normal tensile tests on the same steel after charging with hydrogen at various pH and H_2S levels [11].

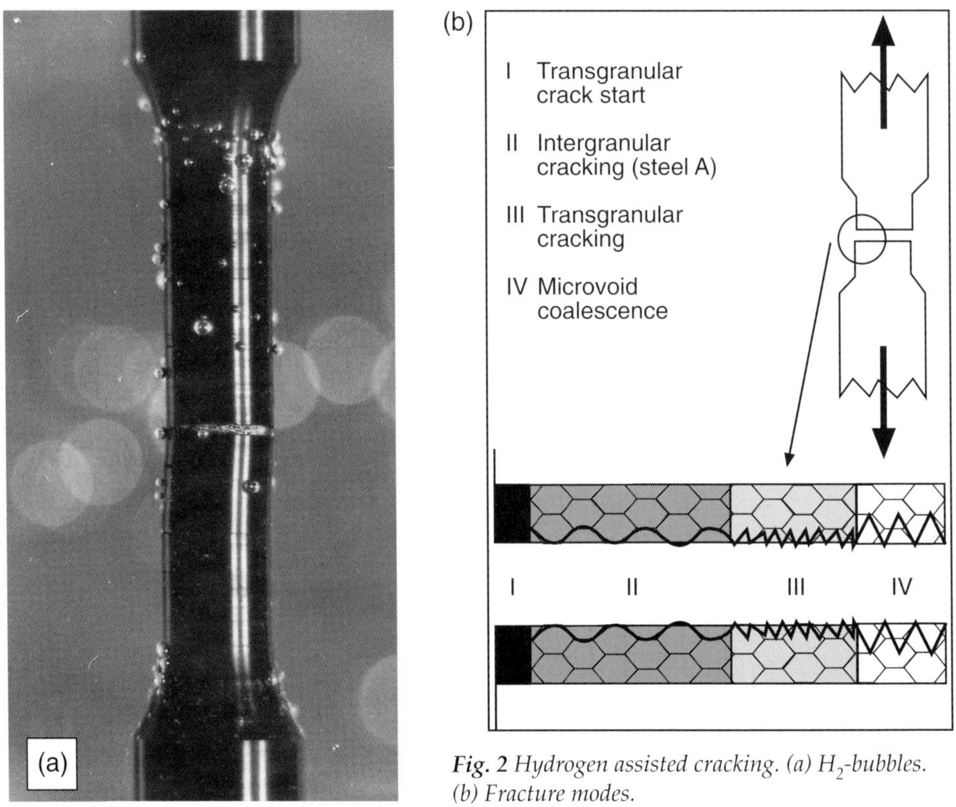

***Fig. 2** Hydrogen assisted cracking. (a) H_2-bubbles. (b) Fracture modes.*

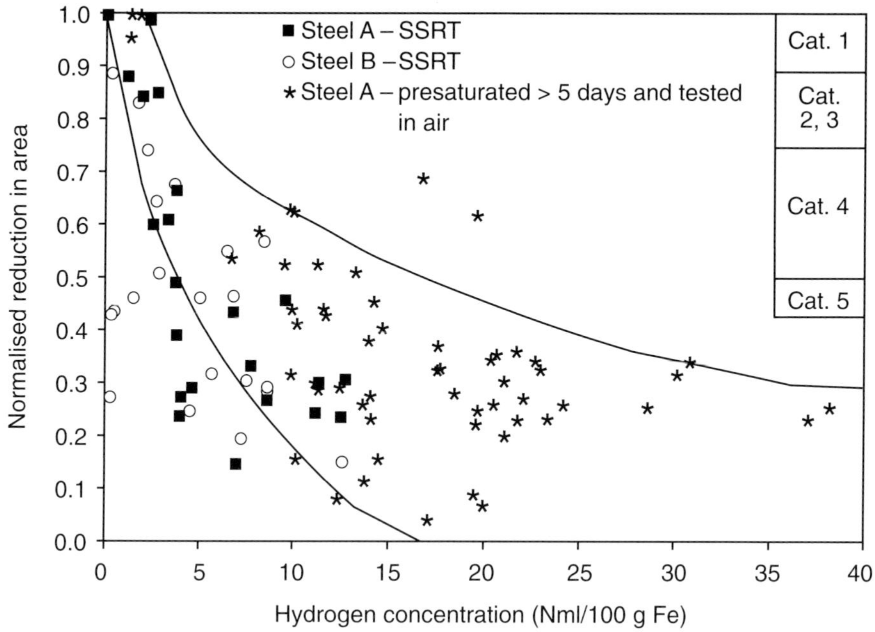

Fig. 3 Area reduction dependence on hydrogen contents.

For the same ductility, the precharged tensile test pieces revealed higher homogeneous hydrogen contents as compared to the SSRT pieces. Presumably, this could be due to the inhomogeneous hydrogen distribution in the SSRT pieces, leading to lower mean levels, with hydrogen accumulations at propagating cracks.

The SSRT results for steel B are given in Figs 4 (a–d). With a higher tensile strength and lower 0.2% proof stress, the effect of increasing H_2S saturation compared to steel A in air is again most pronounced at the low strain rate of $5 \cdot 10^{-7}$ s^{-1} and a long exposure time which led to tensile strength reduction from 900 to 625 MPa at 50% H_2S saturation. As with steel A, increasing strain rates provided increases in fracture strength with less influence of H_2S levels. At the strain rate of $1 \cdot 10^{-4}$ s^{-1}, no significant effect of H_2S was revealed. The 0.2% proof stress values of steel B were less influenced by H_2S and strain rates than for steel A within the scatterband which included the in-air test results. However, at 50% H_2S saturation and with the low strain rate of $5 \cdot 10^{-7}$ s^{-1}, fracture occurred again at a stress close to the 0.2% proof stress, indicating a similar severe reduction in ductility as shown for steel A.

Consequently, a relative ductility reduction by 50% H_2S saturation as well as by low strain rates of $5 \cdot 10^{-7}$ s^{-1} down to 0.15 was found for steel B (Fig. 4b), accompanied by a fracture topography revealing mostly transgranular crack propagation (Fig. 4c and 4d). However, contrary to steel A, steel B demonstrated better ductilities of 0.3 and 1.0 at $5 \cdot 10^{-7}$ s^{-1} strain rate in the case of exposure to 10 and 1% H_2S relative saturation, respectively. In particular, increasing the strain rate and reducing the exposure time is providing the improvement of ductility obtained in the case of steel A. At the same time, for 50% H_2S saturation, Fig. 4(c) shows the increase in ductile dimple fracture at the expense of transgranular hydrogen assisted environmental cracking.

It should be noted that the high ductilities at 1% H_2S saturation shown in Fig. 4(b)

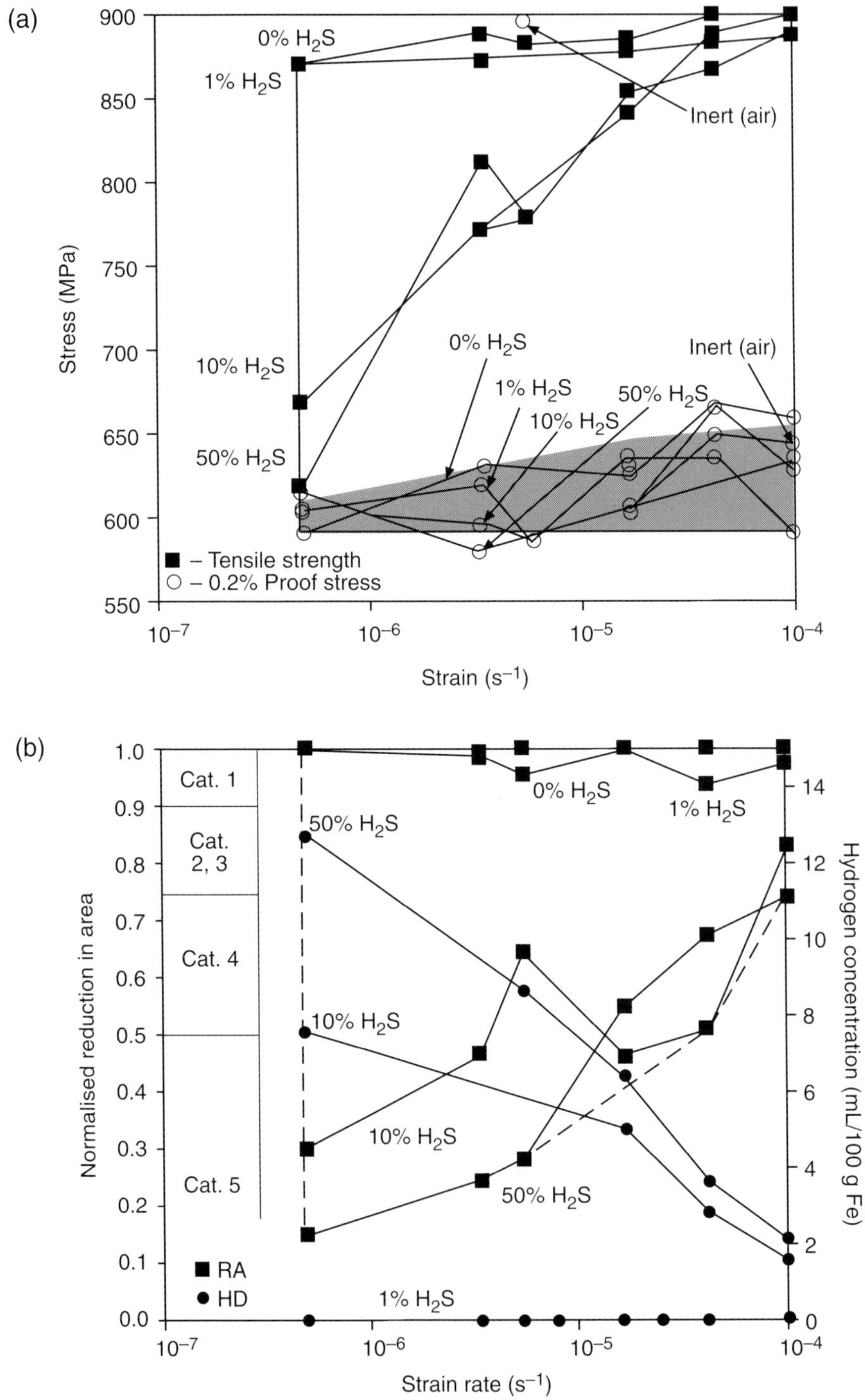

Fig. 4 *Performance of steel B in standard NACE solution at a pH of 2.7 at various strain rates. (a) Tensile properties at various H_2S levels. (b) Reduction of ductility at various H_2S levels.*

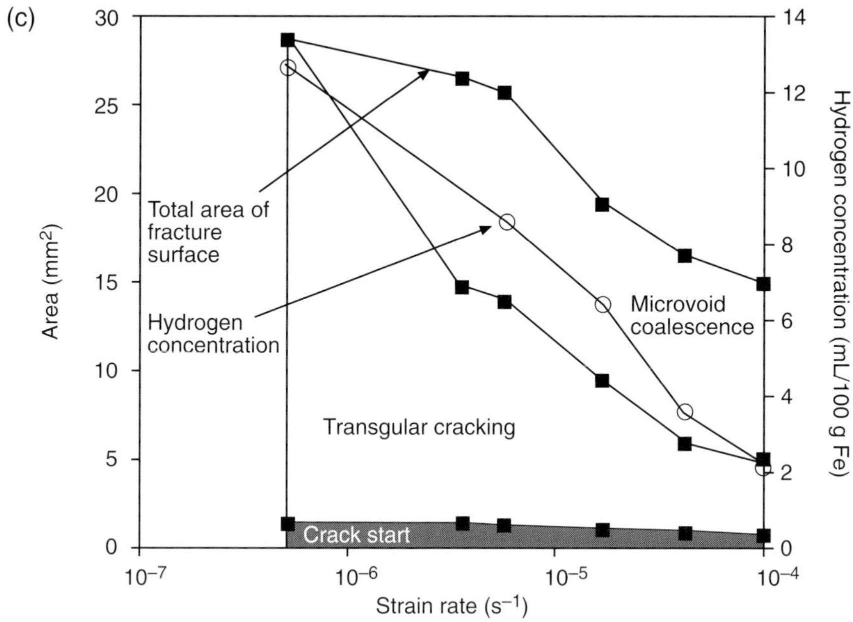

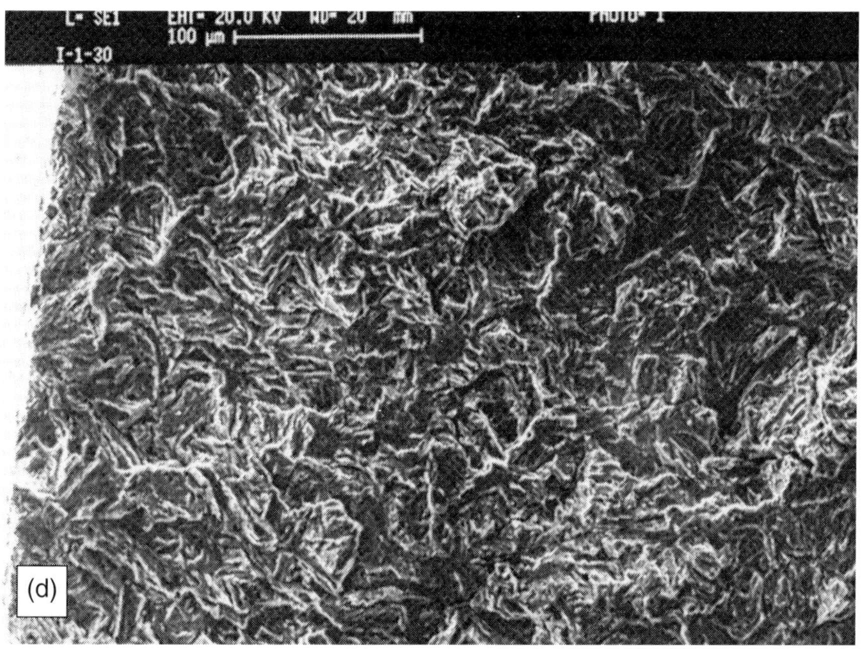

Fig. 4 (Continued) Performance of steel B in standard NACE solution at a pH of 2.7 at various strain rates. (c) Areas of different fracture modes at 50% H_2S. (d) Fracture topography at $\varepsilon' = 3 \cdot 10^{-6}\ s^{-1}$, 50% H_2S.

can be explained by the apparent reluctance of the higher alloyed steel B to pick up hydrogen from this solution. Total average hydrogen levels were close to 0, in qualitative agreement with literature results [12].

For steel B, Fig. 3 shows basically the same effect of total hydrogen on normalised reduction of area as compared to steel A and tests for precharged steel A.

As a result of the investigation of the strain rate effect on H_2S induced hydrogen cracking of both materials at pH 2.7, steel A shows less resistance to stress corrosion cracking than steel B, which exhibits retarded hydrogen pick up. This is demonstrated, in particular, at low strain rates, for instance $5 \cdot 10^{-7}\ s^{-1}$, which is even lower than the value recommended in the EFC guidelines [9,10].

Another result is that, for low strain rates, hydrogen pick up from a 50% saturated H_2S NACE solution reduced the 0.2% proof stresses to the minimum boundary levels as derived from design considerations for pipelines [13].

On the basis of the 'worst case' conditions that have been identified, i.e. a strain rate of $5 \cdot 10^{-7}\ s^{-1}$ and 50% H_2S saturation, the effect of increased pH of the base standard NACE solution on relative reductions of area of both steels can be demonstrated (Fig. 5). Starting with the low ductilities at pH 2.7, steel A showed an increase up to 'category 1' ductility levels [7] at a pH of about 5, while, steel B, after an initial increase at a pH of 4, indicated an intermediate stage at relative ductilities of around 0.4 up to a pH of 5, afterwards increasing to full ductility levels at a pH of 5.5. From visual observation of the test piece surface, this intermediate pH/ductility stage of steel B is to be attributed to the effect of pitting, appearing at a pH of 4 and operating as crack initiation sites, as demonstrated in Fig. 6. This assumed alternative crack initiation mechanism is supported by the low pH inside the pits at higher bulk pH and is still associated with rather low ductilities, outside the scatterband of the established hydrogen effect on ductility given in Fig. 3. Figure 5 is in accordance with literature results of recent SSRT tests [7] indicating a similar pH-ductility slope for another low carbon 13% Cr martensitic steel at low H_2S levels.

Using the 'category 1' criteria [7] for constant levels of saturated H_2S and increasing pH, Fig. 7(a) and (b) represent respective performance diagrams for the two steels, in modified NACE solutions at the critical slow strain rate of $1 \cdot 10^{-6}\ s^{-1}$ and respective hydrogen levels. For steel A, the lower H_2S levels of around 1% saturation seem to allow less of a pH-reduction (down to a pH of 3.5) without failure compared with steel B (down to a pH of 2.5), presumably due to increased hydrogen pick up capacity of steel A. However, at higher H_2S levels of around 10 to 50% saturation, steel B seems to exhibit more sensitivity to environmental cracking, which was supported by local pitting corrosion serving as crack initiation sites for hydrogen assisted stress corrosion cracking. Consequently, at 50% H_2S saturation, steel A may tolerate a reduction of pH down to 4.5, but steel B to only 5.5, at room temperature. The reasons for this behavior are still under investigation and will be discussed in another publication.

4. Conclusions

From the present work on slow strain rate testing of martensitic low carbon chromium steels the following conclusions are drawn:

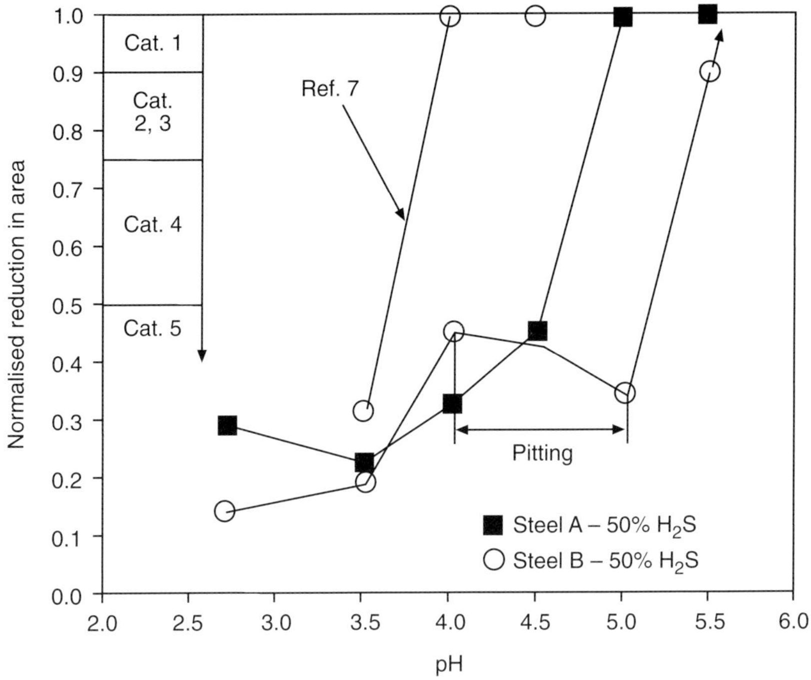

Fig. 5 *Reduction of ductility dependence on pH.*

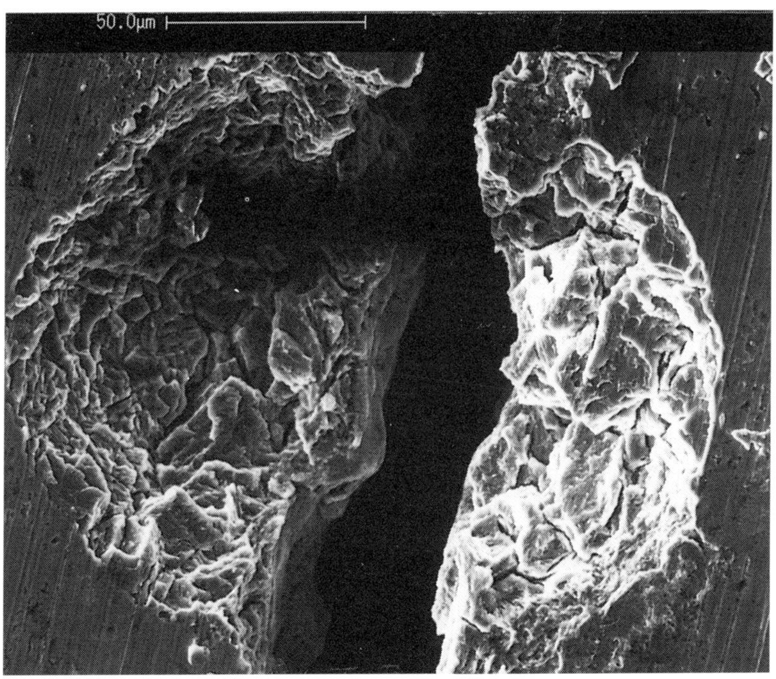

Fig. 6 *Pit as crack initiation site in Steel B.*

(a)

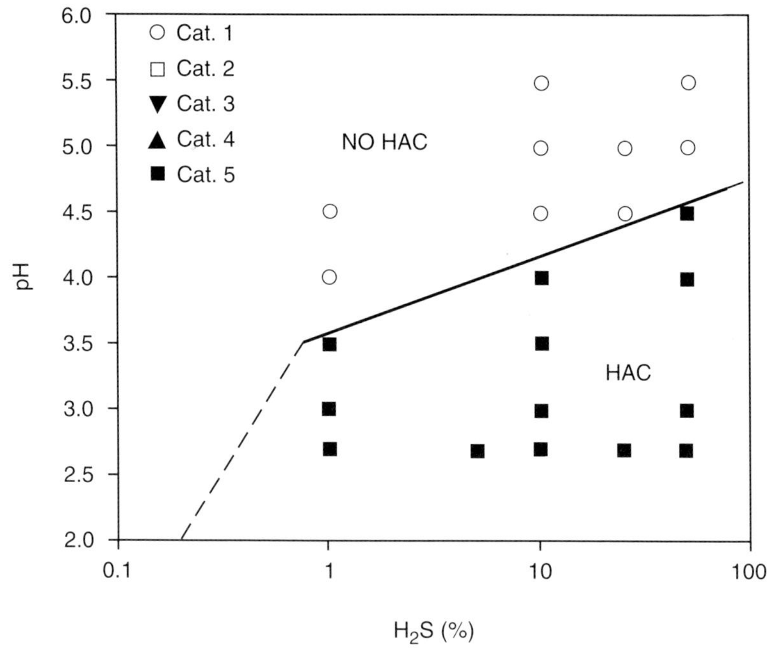

(b)

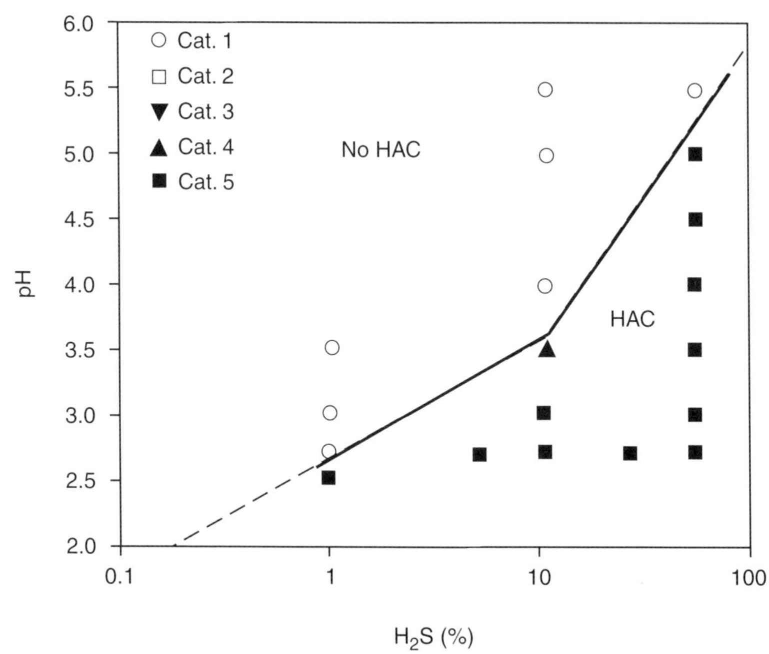

Fig. 7 *H_2S-pH tolerance of low carbon martensitic stainless steels (HAC = hydrogen assisted cracking) at SSRT in NACE solutions. (a) Steel A. (b) Steel B.*

- stress corrosion testing at slow strain rates down to $5 \cdot 10^{-7}\ s^{-1}$ is providing a suitable method for evaluation of the tested materials on exposure to sour service environments;

- the effect of strain rates and H_2S activity as well as pH levels on mechanical properties and cracking behavior is best reflected by degradation of ductilities due to pick up of hydrogen, leading to higher crack sensitivity for steel A at low H_2S as compared to steel B;

- at higher H_2S levels, steel B is more susceptible to local pitting at increasing pH, and has a reduced crack resistance

- the reasons for the different behaviour of the two steels are still under investigation.

5. Acknowledgements

The authors are grateful to SINTEF as well as to Norsk Hydro A. S. for providing the materials.

References

1. J. Enerhaug, S. L. Eliassen and P. E. Kvaale, *Corrosion '97*, Paper No. 60. NACE International, Houston, Tx, 1997.
2. R. W. Revie *et al.*, *Pipeline Technol., Vol. 1*, Conf. Proc., Oostend, 1995, pp. 571–576.
3. M. B. Kermani *et al.*, *Corrosion '91*, Paper No. 21. NACE International, Houston, Tx, 1991.
4. H. Asahi, T. Hara and M. Sugiyama, *Corrosion '96*, Paper No. 61. NACE International, Houston, Tx, 1996.
5. R. N. Parkins, *SSRT for the Evaluation of EIC*, R. D. Kane, Ed. pp. 7–21, 1992. Publ. ASTM, West Conshohocken, Pa. USA.
6. R. P. Gangloff, B. P. Somerday and D. L. Cooke, Understanding crack tip strain rate effects in hydrogen embrittlement for improved Fitness-for-Service modelling, *Corrosion '96 (Research Topical Symposia)*, NACE International, Houston, Tx, 1996, pp. 161–175.
7. C. J. B. M. Joia, R. D. Kane *et al.*, *Corrosion '97*, Paper No. 48. NACE International, Houston, Tx, 1997.
8. NACE Standard TM 0177-96.
9. *Guidelines on Materials Requirements for Carbon and Low Alloy Steels for H_2S-Containing Environments in Oil and Gas Production*. Publication No. 16 in the European Federation of Corrosion Series, published by The Institute of Materials, London, 1995, 56pp.
10. *Corrosion Resistant Alloys for Oil and Gas Production: Guidance on General Requirements and test methods for H_2S Service*. Publication No. 17 in European Federation of Corrosion Series published by the Institute of Materials, London, 1996, 84 pp.
11. Th. Boellinghaus, H. Hoffmeister and L. Reuter, Material properties of as delivered and quenched modified martensitic stainless steels dependant on hydrogen concentration. *Proc. Int. Conf. on Supermartensitic Stainless Steels*, Brussels, 27–28 May, 1999 (in print).
12. M. Iino *et al.*, *Hydrogen problems in Steels*, Conf. Proc., ASM, Ohio, 1982, pp. 159–167.
13. P. S. Jackman and H. Everson, *Corrosion '95*, Paper No. 89. NACE International, Houston, Tx, 1995.

29
On-line Sulfide Stress Cracking Monitoring of 13% Cr Pipe Welds at Realistic Weld Restraint Conditions in the Instrumented Restraint Cracking (IRC) Test

Th. BOELLINGHAUS, H. HOFFMEISTER and M. LITTICH
Institute for Failure Analysis and Failure Prevention e. V., Holstenhofweg 85, D-22043 Hamburg, Germany

ABSTRACT
With the extended application of 13% Cr LC (low carbon) corrosion resistant pipe materials in the oil and gas industry the realistic assessment of, in particular, sour service performance of welded joints is becoming increasingly important. Apart from the restricted information provided by small scale standard test piece procedures the failures of fabrication welds are also characterised by the time to cracking after welding or exposure to the corrosive environment and by the respective multiaxial strain/stress conditions at the location of the crack start. In the present paper these fabrication-relevant parameters are identified by on-line monitoring of restraint forces and restraint moments of pipe segments welded under realistic shrinkage restraint conditions in the Instrumented Restraint Cracking (IRC) Test. This procedure has already been sucessfully applied to the assessment of hydrogen assisted cracking resulting from welding. As a first step towards full pipe welding, pipe segments were partial penetration GTAW (TIG) welded and tested in free corrosion conditions at room temperature. The results from tests at various pH and H_2S levels indicated higher cracking tendencies in three different North Sea formation waters than in the standard NACE solution. Finally, stress distributions of complete orbital pipe welds were demonstrated and related to failure stresses in realistic conditions.

1. Introduction

Two types of stresses result from cooling after welding. During free shrinkage, residual stresses are caused by the stress/strain interaction with the surrounding material. However, the actual construction also provides a specific shrinkage restraint which results in much higher reaction stresses, as illustrated schematically in Fig. 1 [1]. Thus, during cooling after welding a complex stress and strain state in the weld joint region will be established, particularly at fabrication induced notches. Such notches may provide crack initiation sites according to their geometry and the stress–strain distribution from restrained thermal shrinkage after welding. As a typical fabrication-relevant test procedure, the Instrumented Restraint Cracking (IRC) test has been frequently applied for assessment of weld induced hydrogen cold cracking [2] and development of crack free welding procedures. The test facility is providing realistic restraints typical of those found in actual structures [3,4].

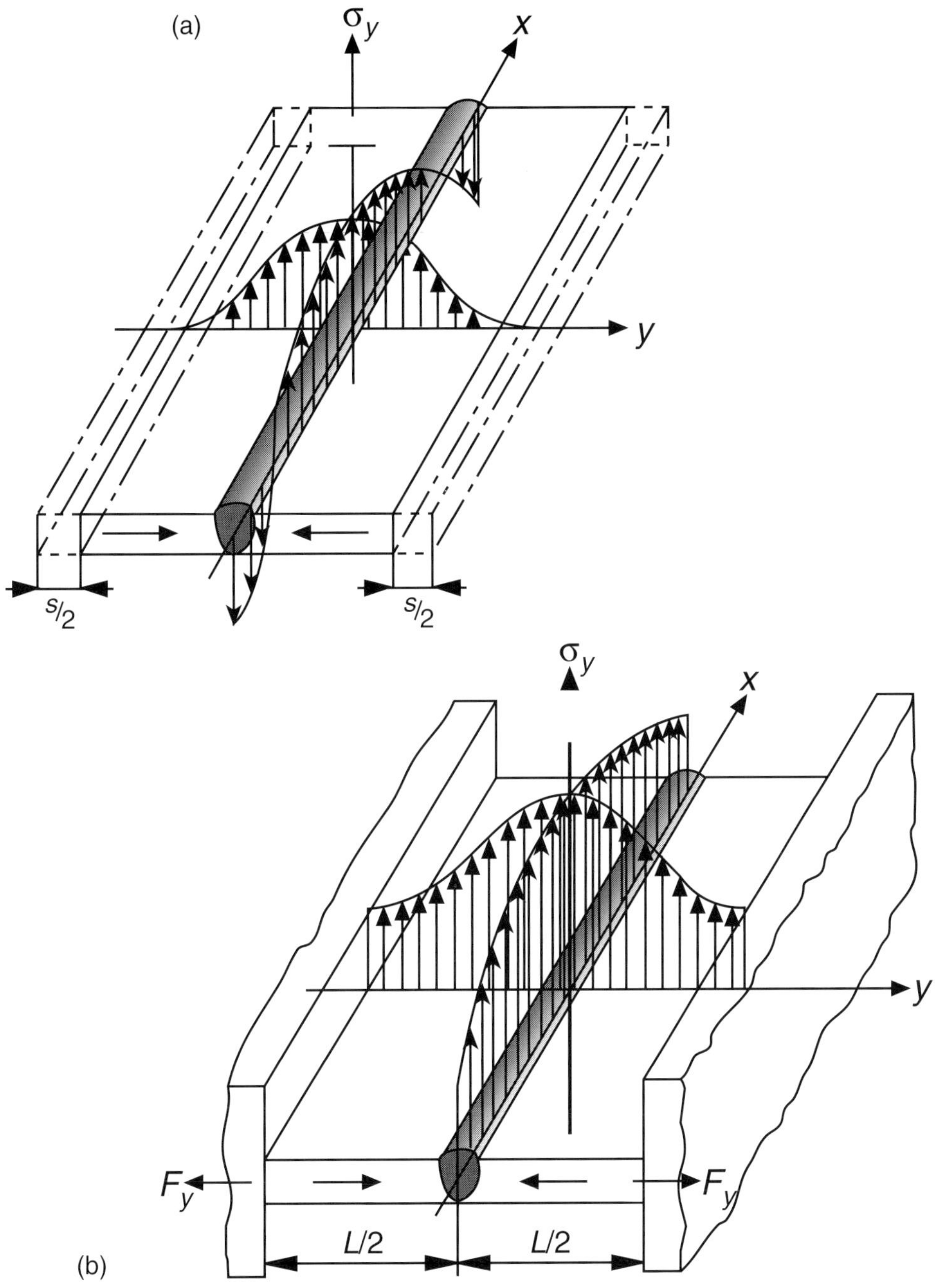

Fig. 1 *Schematic illustration of stresses transverse to the welding direction (a) at free shrinkage, and (b) under shrinkage restraint.*

The test also provides realistic welds as applied in specific fabrication cases. For on-line monitoring of mechanical loading conditions after welding the thermally induced reaction force F_y and the reaction moments M_x and M_z (Fig. 2) are continuously registered and evaluated with respect to stress and time to crack initiation, as indicated by release of signals. The procedure is regarded as a full scale test serving for identification of critical combinations of design, fabrication parameters and operating conditions of a specific component. In the present work it is applied to evaluation of resistance to sulfide stress cracking of LC (low carbon) 13Cr pipe welds in various environments.

2. Experimental

For each test weld two pipe segments were symmetrically welded to anchor plates fastened by screw bolts to the clamps of bolt flanges which were connected to a rigid frame, as shown in principle in Fig. 2. During the test period, on-line monitoring of the reaction force as well as the reaction moments was carried out by continuous registration of the signal from strain gauges located at the bolt flanges. The reaction force and moments, F_y, M_x and M_z were calibrated by a sequence of increasing loads in the y-axis direction and assignment of the specific loads to the signals produced by respective straining and bending of the bolt flanges.

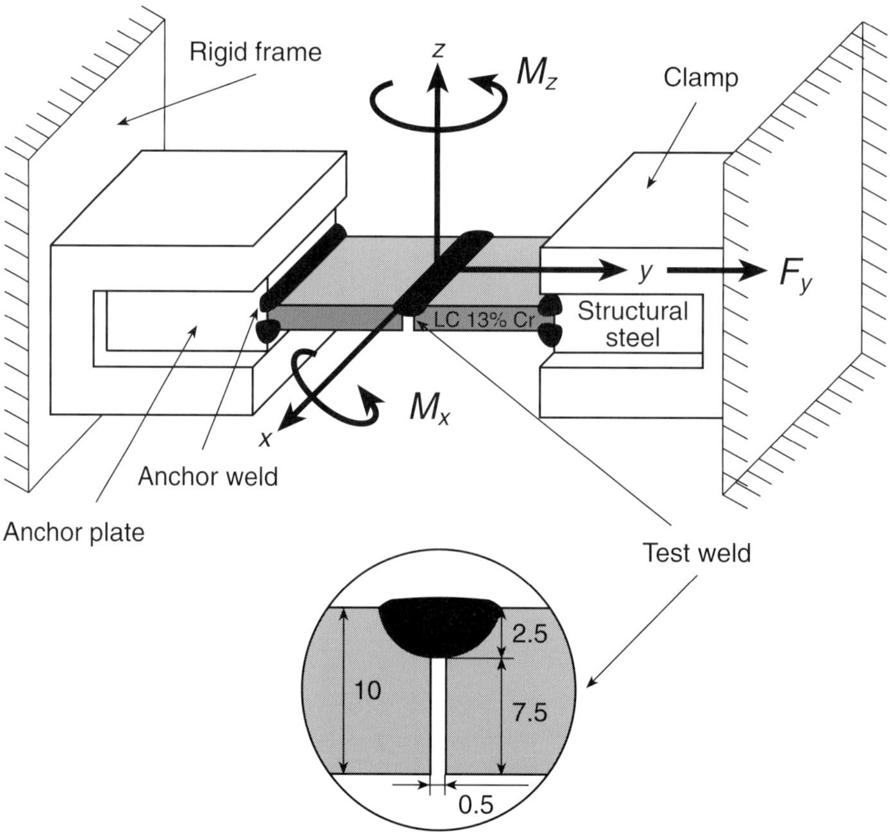

Fig. 2 Schematic illustration of the IRC test set up for TIG welding of pipeline segments.

Pipe segments of a modified 13% Cr steel (Fig. 2) with C 0.025, Cr 13.3, Ni 3.85, Mo 0.96 (wt%) were partial penetration TIG welded with a gap of 0.5 mm to achieve high local stresses and crack starting at the root. The shielding gas was pure argon. After welding, the specimens were enclosed by an environmental cell. The electrolyte was supplied to the cell 18 h after welding. Different pH and H_2S levels in the NACE solution (NaCl 50, CH_3COOH, 5 gL^{-1}) and three different North Sea formation waters (Table 1) were used at room temperature.

The cracking tests were conducted using the above procedure at a restraint intensity of 6 and 7 kN mm^{-2} [1, 2]. During welding of the pipe segments the reaction moments M_x and M_z were scattered only marginally around zero, and so they did not contribute significantly to the reaction stress build-up in the pipeline segments. For those experiments, the nominal reaction stresses σ_y, could be calculated only from the transverse reaction force F_y and the weld metal area. Depending on the restraint intensity the final stresses ranged between 200 and 300 MPa, at the starting point of corrosion testing.

3. Results

A typical IRC test record of the transverse reaction force F_y and the calculated nominal reaction stress σ_y, for welding such pipeline segments is shown in Fig. 3. As a result of the cooling after welding, the reaction stress increases at first to approximately 70% of the final level.

The martensite transformation at low temperatures around 200°C was marked by a significant drop of the reaction stress before it increased to its final level of 275 MPa (Fig. 3).

After 22 h this weld was exposed to the NACE test solution at 10% H_2S saturation. Cracking initiated only after a further 5000 s. It was followed by slow crack growth as indicated by a significant drop of the reaction force down to zero when the weld finally cracked through.

Figure 4 shows that the time to crack initiation, identified by the first drop of the reaction force during exposure to the test medium, is reduced by increasing the H_2S level of the NACE solution. By microscopic inspection and scanning electron miscroscopy (SEM) investigation it was found that the cracks started at the root notches directly beneath the fusion line. Crack propagation took place through the weld metal centre to the topside and then in the longitudinal direction of the welds towards both ends of the weld.

As shown in Fig. 5 for the 50% H_2S saturated NACE solution, the time to crack initiation significantly increases with the pH of the modified NACE solution – as

Table 1. *Chemical compositions of formation water types (pH adjusted by NaOH and HCl)*

Type	NaCl	$SrCl_2$	$BaCl_2$	$CaCl_2$	$MgCl_2$	KCl	$NaHCO_3$	$MgSO_4$	CO_2
FW I	30.15	0.13	0.14	0.94	–	1.05	1.45	–	Sat.
FW II	92.13	2.40	3.75	16.48	1.50	2.73	0.45	0.07	Sat.
FW III	32.89	0.42	0.11	8.70	1.08	0.51	0.21	0.05	Sat.

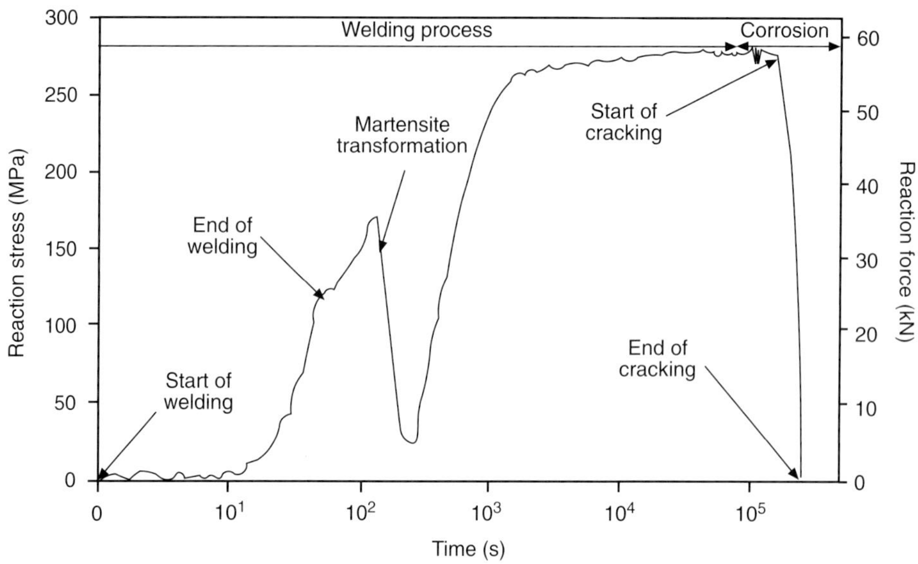

Fig. 3 *IRC test tecord for LC 13Cr pipeline segment weld exposed to 10% H_2S sat. NACE solution 22 h after welding.*

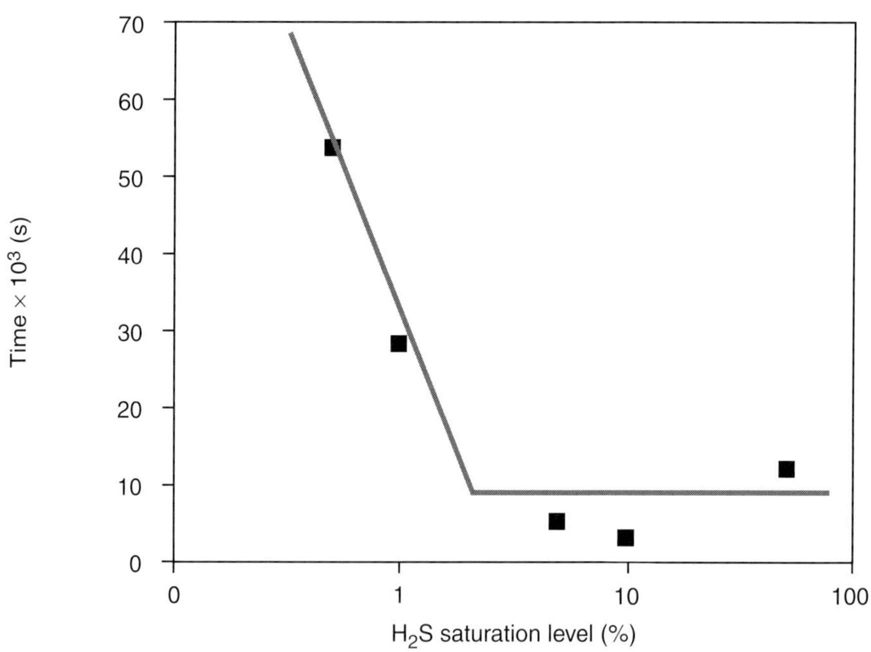

Fig. 4 *Time to crack initiation dependent on H_2S saturation of the NACE solution at a pH of 2.7.*

adjusted by sodium hydroxide. At pH above 4.5 no cracking has been observed in this type of electrolyte which is consistent with slow strain rate test (SSRT) results of this steel in the same environment [5], and also at other H_2S levels. However, it has to be remembered that the weld metal microstructure differs from that of the base material in the slow strain rate test.

The decrease of the reaction force with time indicates the crack propagation. It was therefore plotted vs time, as shown in Fig. 6. The diagram makes it obvious that crack incubation is followed by a period of slow crack growth, indicating typical hydrogen crack propagation controlled by hydrogen uptake and diffusion and local stress increasing with crack length. This is also indicated by the slope of the curves, which is representative of the cracking velocity. The slope increases with the H_2S saturation level in the solution. However, total rupture has been observed only in the 10% H_2S saturated solution.

The fracture surfaces (Fig. 7a) exhibited intergranular cracking along the prior austenite grain boundaries (Fig. 7b) in the regions where the crack started, followed by transgranular fracture during crack propagation (Fig. 7c) and ductile rupture of the rest of the weld metal area (Fig. 7d).

Such fracture modes have also been revealed by slow strain rate testing of this steel in the same environment [5]. Inter- and transgranular fracture have also been found with pre-charged tensile specimens of the same steel [6]. Usually, such experiments also revealed intergranular cracking at higher, and transgranular cracking at lower hydrogen concentrations.

Figure 8(a) is a sketch of the weld microstructure and the predominant crack paths. A cross section through the weld metal, the fusion line, the HAZ and the base material

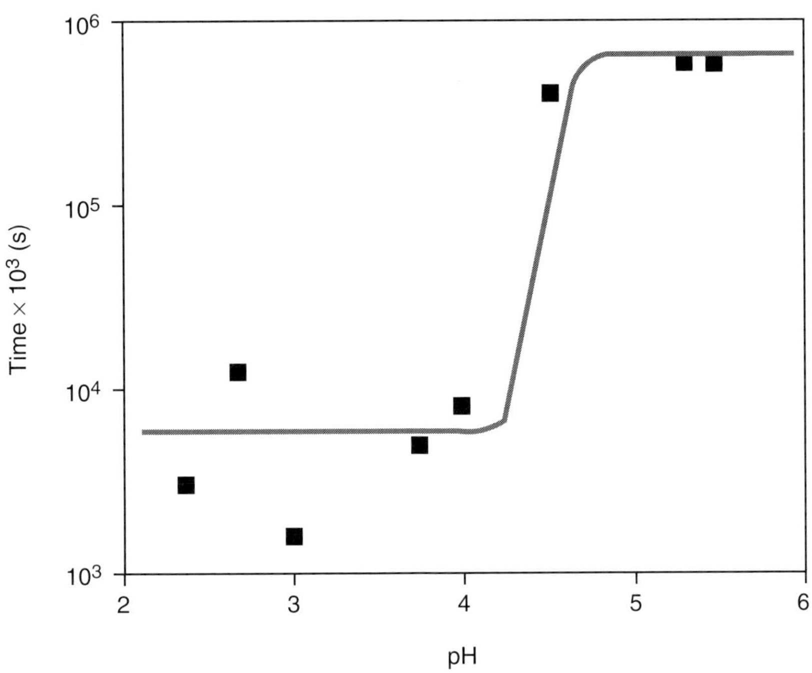

Fig. 5 Time to crack initiation dependence on pH of the 50% H_2S saturated NACE solution.

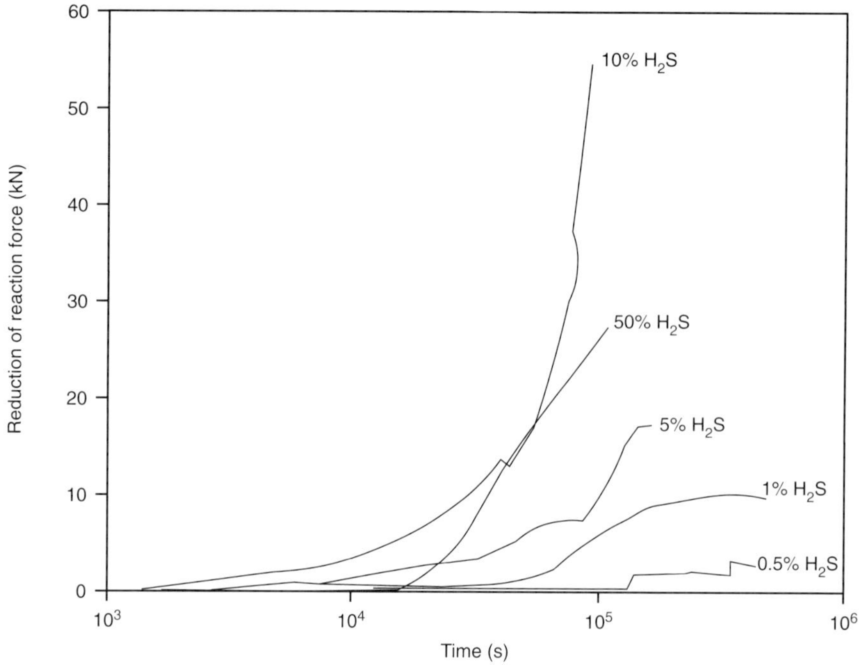

Fig. 6 *Reduction of reaction force vs exposure time to the NACE solution.*

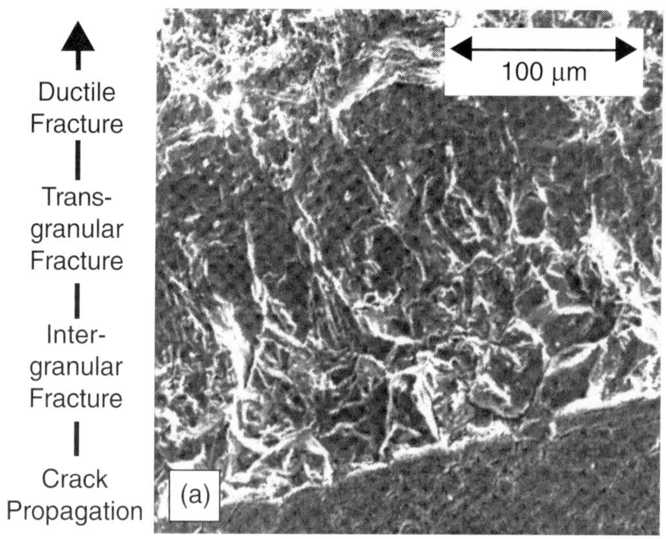

Fig. 7 *Typical fracture surfaces of the welds exposed to the NACE solution at various H_2S saturation levels. (a) Overview.*

is shown in Fig. 8(b). Small amounts of δ-ferrite have been observed in the weld microstructure forming a network alongside the fusion line (Fig. 8c). With some

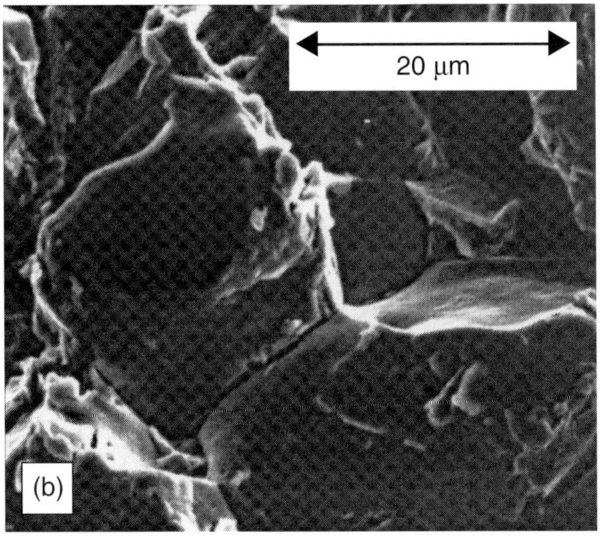

Fig. 7 *(continued). Typical fracture surfaces of the welds exposed to the NACE solution at various H_2S saturation levels. (b) Intergranular Zone.*

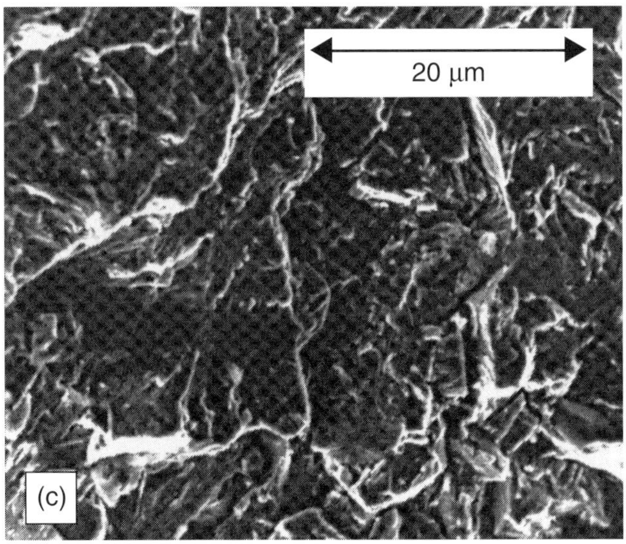

Fig. 7 *(c) Transgranular Zone.*

specimens, side cracks at the root notches followed the fusion line, and a typical example is shown in Fig. 8(d). However, the main cracks always propagated through the centre of the weld metal.

Figure 9 summarises the results with respect to the effect of pH and H_2S on weld metal cracking. As compared to slow strain rate testing of the base material at a

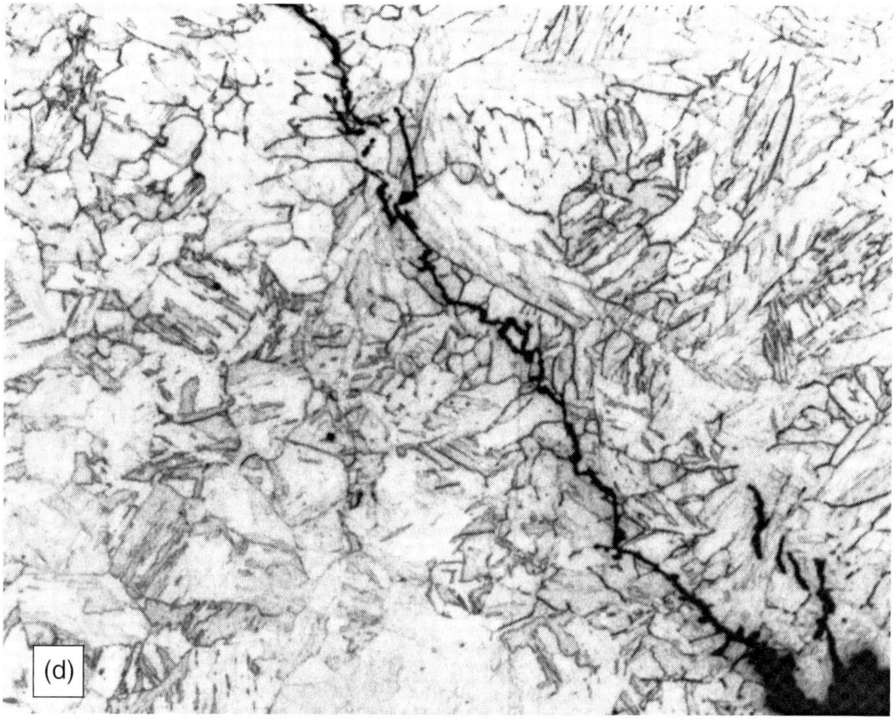

Fig. 8 (continued) (d) Microstructure of TIG welds of pipeline segments exposed to NACE. Side crack at root notch (200×).

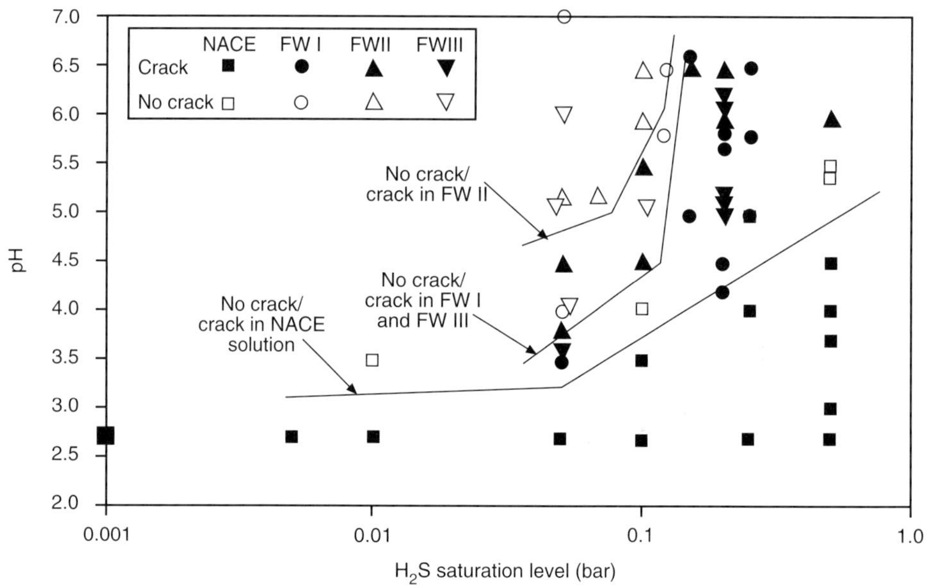

***Fig. 9** pH–H_2S tolerance of LC 13% Cr pipeline segment welds in NACE solution as compared to three different North Sea formation water compositions (1 bar = 100 % saturation).*

The tolerance to pH and H_2S of weld metal cracking is considerably lower for the formation waters than for the NACE solutions. In such electrolytes sporadic cracking also appears at nominal non-sour service conditions as indicated by the EFC guideline for low alloyed steels [7]. This is particularly the case for the formation water type II, representing the chemical composition at the Aasgard field [8].

As a first step towards full scale testing, full pipes have been orbitally welded in the IRC test, as described earlier [1]. In orbital welding of pipes the stress distribution is also significantly affected by the reaction moments. The reaction stresses therefore have to be calculated from the monitored signals of F_y, M_x and M_z.

For a 87 mm dia., 4 mm wall thickness pipe piece, Fig. 10(a) shows the on-line registration and Fig. 10(b) the respective orbital stress distribution transverse to the weld at a restraint intensity of 3.2 kN mm^{-2}. It is typically characterised by various levels of stress depending on local weld metal thickness and mechanical loading distributions (Fig. 10b). Twenty hours after welding the pipe was filled with 50% H_2S (0.5 bar) saturated NACE solution at a final transverse average stress level of approximately 250 MPa. A crack started approximately 12 h after exposure to the solution at the 7 o'clock position where also the highest local stresses of about 310 MPa occurred (Fig. 10b). As a specific feature at orbital welds, circumferential crack propagtion is identified by respective changes of the reaction moments M_x and M_z.

This means that the location of crack initiation and the direction of propagation can be determined from the changes of the reaction moments with time. As visualised

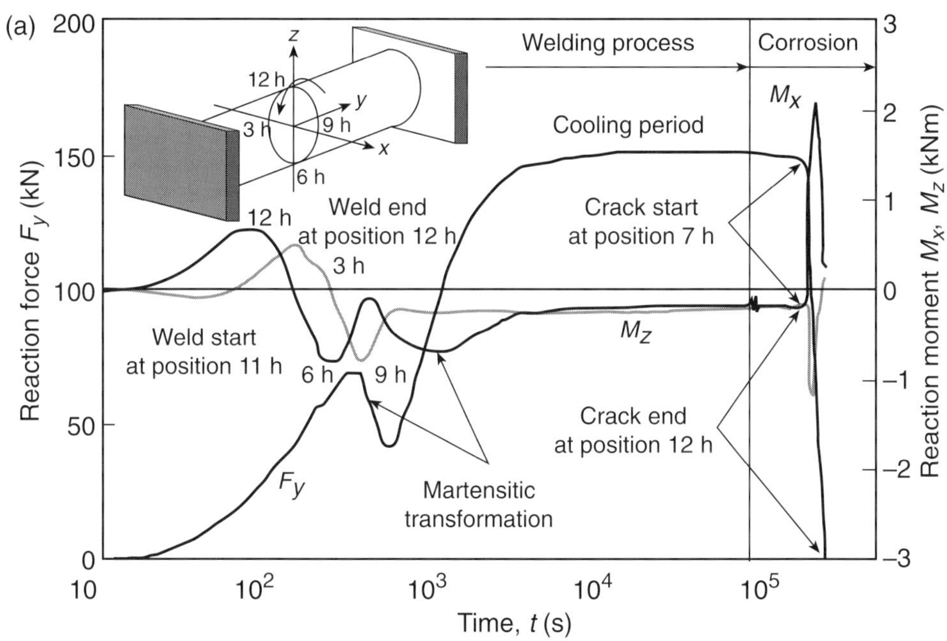

Fig. 10 IRC test data for a 4 in. LC 13% Cr steel pipe weld in 50 % H_2S satd NACE solution. (a) IRC test record of the welding and corrosion process.

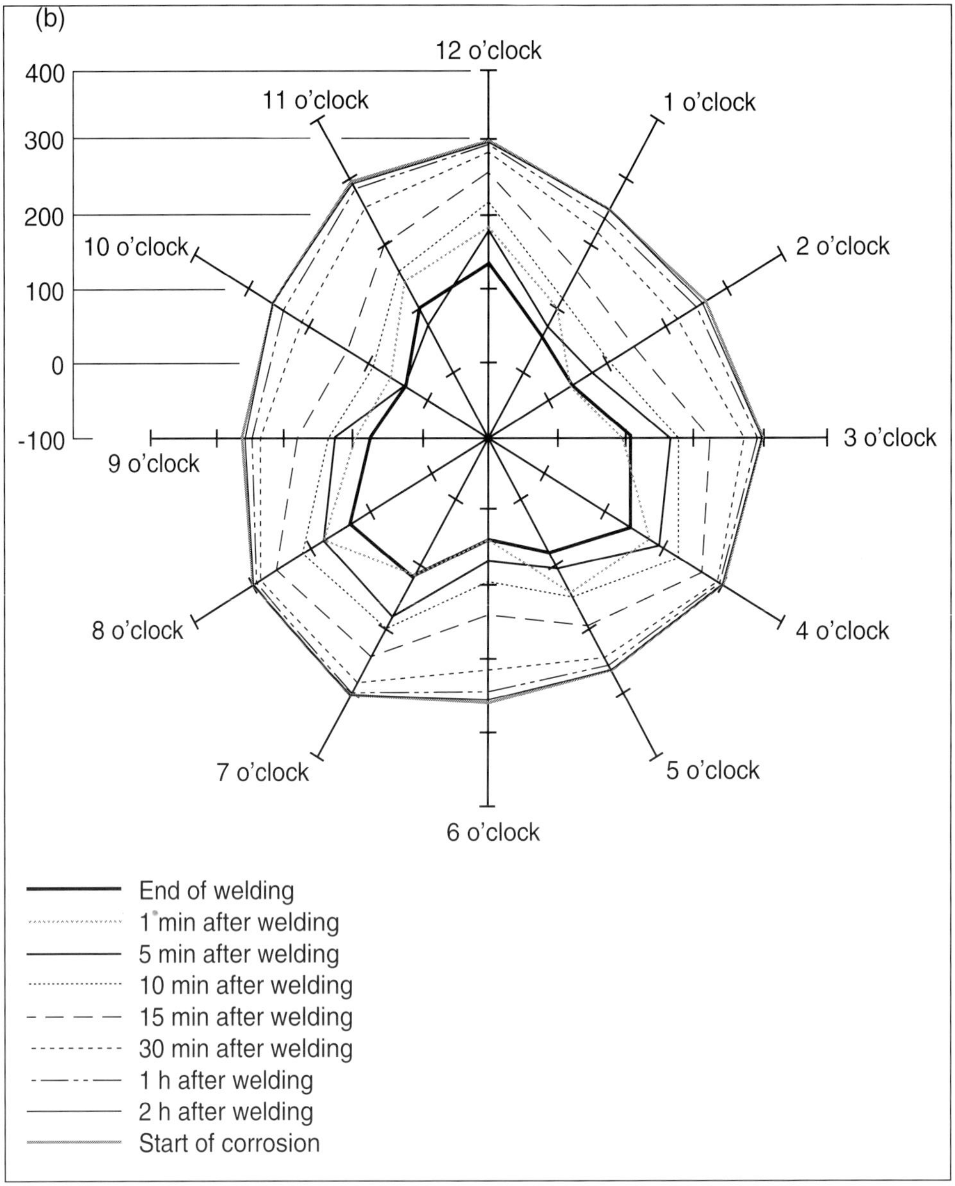

Fig. 10 (continued) IRC test data for a 4 in. LC 13% Cr steel pipe weld in 50 % H_2S sat. NACE solution. (b) Circumferential stress distribution (MPa).

in Fig. 10(c), the crack in this pipe propagated in both directions until the weld finally cracked through at the 12 o'clock position.

As a further example, a pipe piece with a chemical composition of C 0.01, Cr 12.4, Ni 5.2, Mo 2.0 (wt%) and with a diameter of 165 mm and 10 mm wall thickness has been TIG welded by the same procedure.

As can be seen from the corrosponding IRC test record in Fig. 11(a), the amplitudes

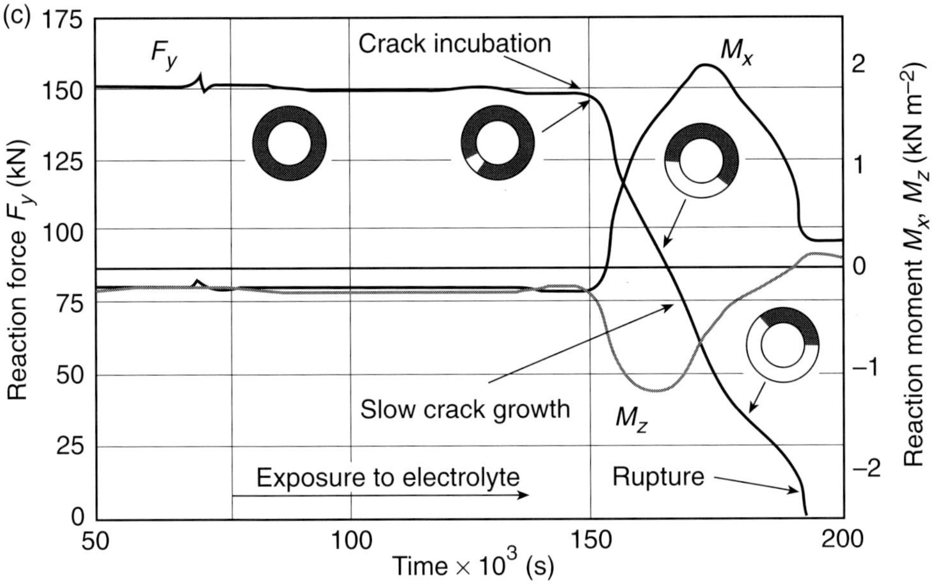

Fig. 10 (continued) (c) IRC test record during crack propagation.

of the moments are increased with the larger diameter. (Note that the welding direction changed as compared with the weld above.) Because downhill welding is always assigned to the positions between 1 and 5 the reaction moment M_z is now

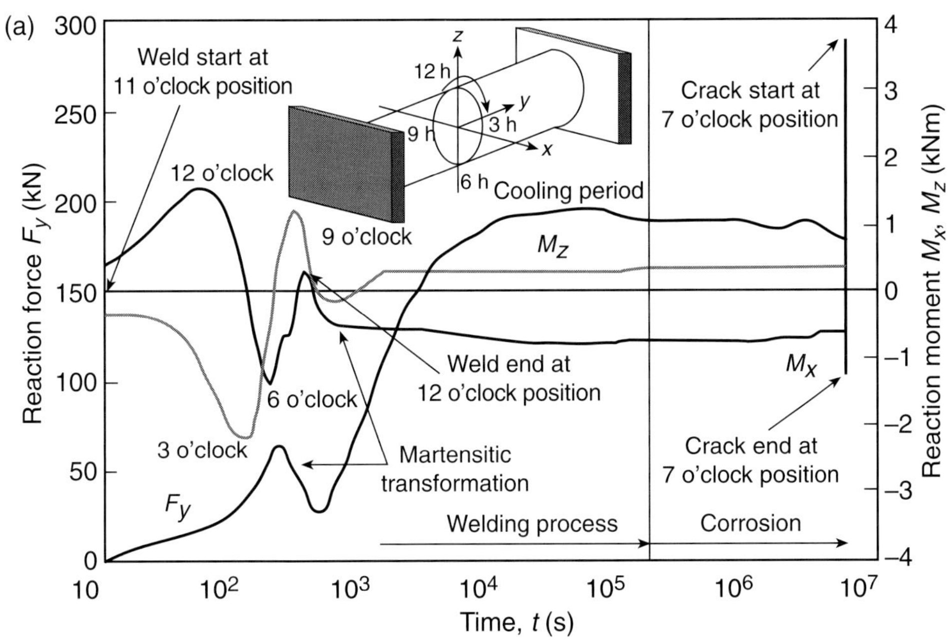

***Fig. 11** IRC test data for a 6 in. LC 13% Cr steel pipe weld in 50% H_2S sat. NACE solution. (a) IRC test record of the welding and corrosion process.*

decreasing during downhill welding and reaches its minimum at the 3 o'clock position.

The larger heat input over the circumference causes only a slightly higher reaction force. Due to the larger load-carrying weld metal area, the reaction stresses were only around 150 MPa at the 5 o'clock position (Fig. 11b) and have to be considered as the main reason why this specimen did not crack over a long period during exposure to the same electrolyte (50% H_2S saturated NACE solution). After 500 h (3 weeks) the frame of the IRC test set up was heated in order to increase the stresses in the weld. The specimen itself, however, was kept at room temperature.

As seen in the respective IRC test record shown in Fig. 11(c), the transverse reaction force F_y was increased by 100 kN during this procedure. The moments were only marginally affected by the heating of the frame. After two hours the weld started cracking at the 1 o'clock position, where high reaction stresses had also been detected after welding (Fig. 11b). Because of the imposition of an external load by heating the frame no crack incubation period was observed, i.e. as compared to the previously tested pipe. Again, the crack propagated in both directions, preferentially towards the 5 o'clock position, causing corresponding deviations of the reaction moments M_x and M_z. The decrease of the reaction force F_y indicates slow crack growth, as already shown for the pipe segments in Fig. 6. As a result of stopping the heating of the frame, the crack stopped at the 5 o'clock positionat a final reaction force of 100 kN.

Heating of the frame was conducted in order to show the capabilities of the IRC test to simulate hazardous operational procedures that would lead to stress increments in the girth welds during pipeline service. Such pipeline welds can also be subjected to the very slow strain rates that occur during operation using time-dependent heating of the IRC test frame [9]. Further development of the IRC test for testing pipelines at higher internal pressure and temperature is under way.

4. Conclusions

The Instrumented Restraint Cracking (IRC) test has been applied as a full scale test for assessment of environmentally assisted weld joint cracking of LC 13% Cr martensitic steels. The following conclusions are drawn.

1. The IRC test is a suitable method for testing welds with the mechanical conditions produced by the thermal cycle at realistic restraint intensities that are related to the surrounding construction.

2. From on-line monitoring of the reaction force and the reaction moments nominal reaction stresses transverse to the welding direction can be evaluated and the time of crack initiation can be determined. Additonally, crack propagation can be followed by observation of the decreasing reaction force and the changes in the respective moments.

3. Instrumental Restraint Cracking (IRC) testing of welded pipeline segments revealed a significant higher tendency to sulfide stress cracking in North Sea formation water compositions than in the standard NACE solution. The reasons for such behaviour are to be investigated further.

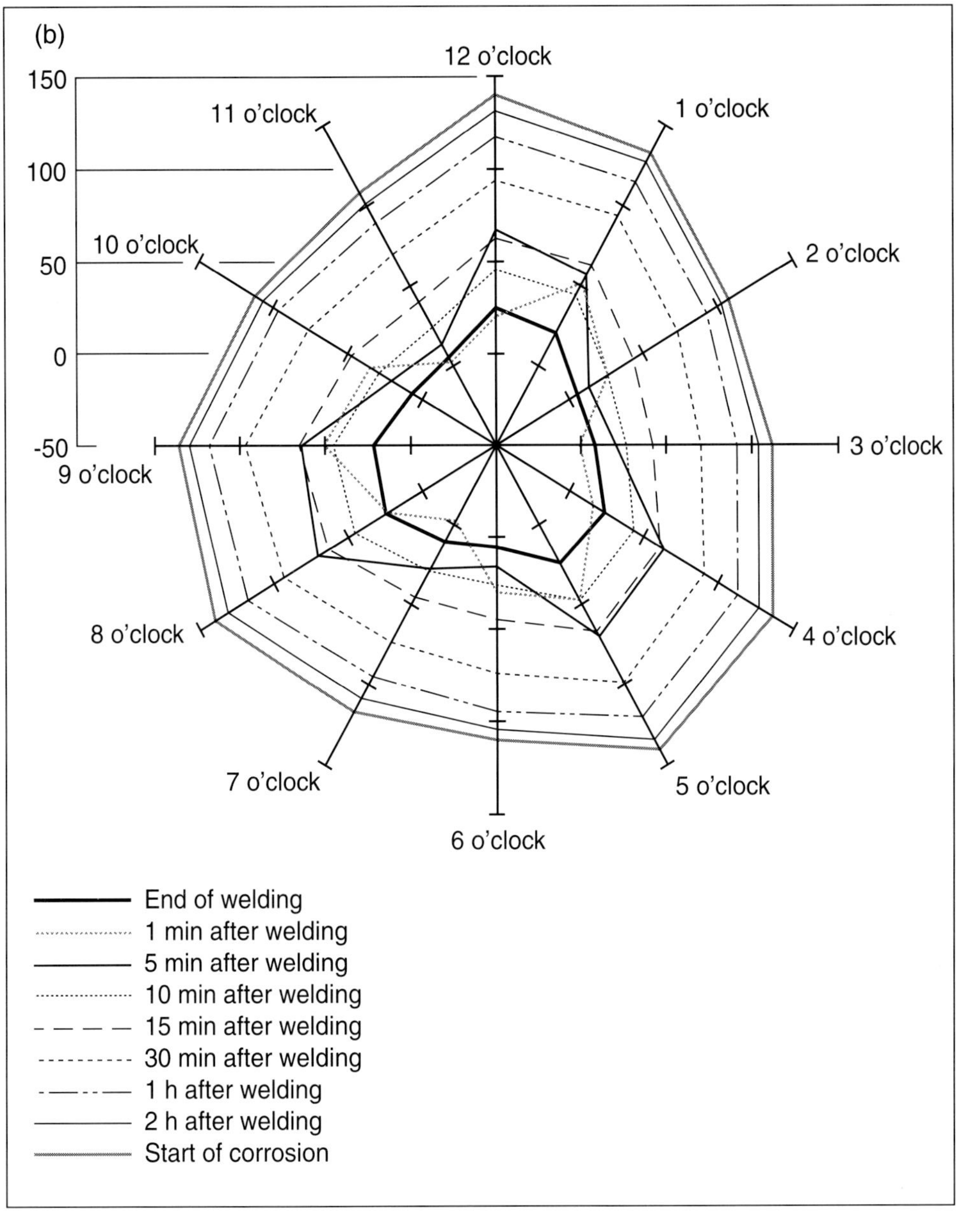

Fig. 11 *(continued) (b) Circumferential stress distribution (MPa).*

4. First IRC tests of completely orbital welded pipelines have shown the capabilities of this method to determine time and location of crack initiation and propagation.

5. By heating of the test frame operational procedures causing additional stresses in the girth welds can also be simulated in the IRC test.

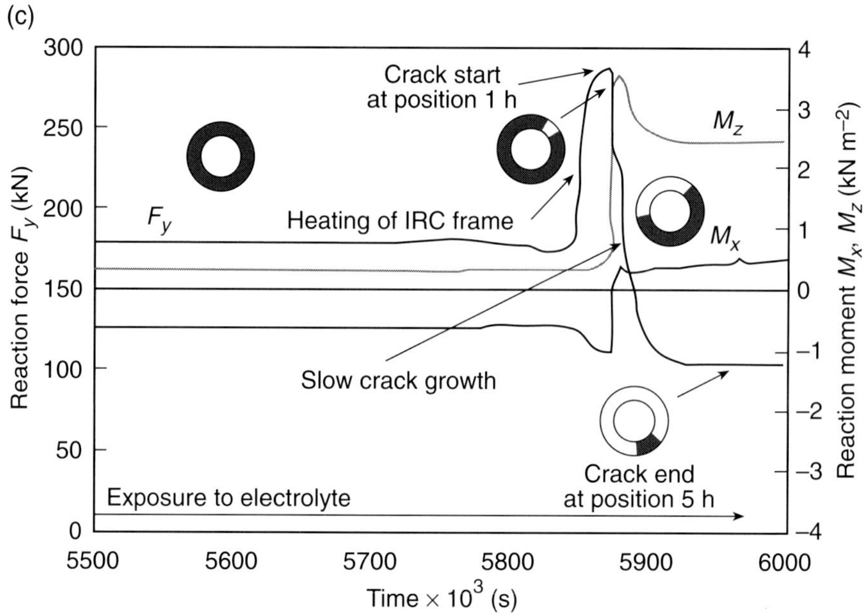

***Fig. 11** (continued) IRC test data for a 6 in. LC 13% Cr steel pipe weld in 50% H_2S sat. NACE solution. (c) IRC test record during crack propagation.*

References

1. Th. Boellinghaus, H. Hoffmeister and M. Littich, Application of the IRC-test for assessment of reaction stresses in tubular joints with respect to hydrogen assisted weld cracking, International Institute of Welding, IIW Doc. No. 1900-98, to be published in *Welding in the World.*
2. H. Hoffmeister, Concept and procedure of the IRC-test for assessing hydrogen assisted weld cracking, *Steel Res.*, 1986, **57**, (7), 344–347, and International Institute of Welding, IIW-Doc No. IX-1369-85.
3. J. Lereim, G. Eide, C. M. Braathen, H. Hoffmeister and E. Ruyter, Cracking risk assessment of highly restrained offshore steel joints of the SNORRE TLP structure by on site restraint measurements and IRC testing, *Proc. Int. Conf. on Offshore Mechanics and Arctic Engineering*, OMAE, Stavanger, 1991, Volume III, Part B, pp.495–500.
4. Th. Boellinghaus, H. Hoffmeister and A. Schwager, Calculation of restraint intensities at large offshore steel structures by Finite Element Analysis, in *Mathematical Modelling of Weld Phenomena 3*, Eds H. Cerjak and H. K. D. H. Bhadeshia. Published by The Institute of Materials, London, 1997, pp.624–651.
5. Th. Boellinghaus, H. Hoffmeister and S. Dietrich, *Slow Strain Rate Testing of Low Carbon Martensitic Stainless Steels*, this volume, p.274.
6. Th. Boellinghaus, H. Hoffmeister and L. Reuter. Material properties of as-delivered and quenched modified martensitic stainless steels dependent on hydrogen concentration. *Proc. Int. Conf. on Supermartensitic Stainless Steels,* Brussels, 1999 (to be published).
7. *Corrosion Resistant Alloys for Oil and Gas Production: Guidance on General Requirements and Test Methods for H_2S Service.* Publication No. 17 in European Federation of Corrosion Series. Published by The Institute of Materials, London, 1996, 84pp.

8. J. Enerhaug, S. Eliassen and P. E. Kvaale, Qualification of welded super 13% Cr martensitic stainless steels for sour service applications, *Corrosion '97*, Paper No. 60, NACE International, Houston, Tx, 1997.
9. R. W. Revie, W. Zheng, F. A. MacLeod and D. Kiff, Methodology for realistic full-scale testing of line pipe for stress corrosion cracking, *Proc. Int. Conf. On Pipeline Technology*, Oostend, 1995, Vol. I, pp. 571–576.

30
Effect of Hydrogen Sulfide Partial Pressure, pH and Chloride Content on the SSC Resistance of Martensitic Stainless Steels and Martensitic Precipitation Hardening Stainless Steels

D. D. VITALE
Dresser-Rand Turbo Products Division, Olean, NY, USA

ABSTRACT
Centrifugal compressor applications require the use of martensitic stainless and martensitic precipitation hardening stainless steels at high hydrogen sulfide partial pressures. These materials do not perform well when tested with standard NACE TM0177 test solutions. This paper describes the effect of hydrogen sulfide partial pressure, pH, and chloride content on their SSC resistance and explains their successful field operational experience. Environmental limits are determined for several materials and heat treat conditions.

1. Introduction

Martensitic stainless steels and martensitic precipitation hardening steels have been used successfully in centrifugal compressor sour gas applications for many years. However, with conventional SSC testing environments, the threshold stress value may be well below yield strength. These materials are accepted in NACE Standard MR0175 but the standard states 'These materials (martensitic stainless steels) may, however exhibit threshold stress levels in NACE Standard TM0177 that are lower than those for other materials included in this standard'.

The purpose of the project was twofold:

- To evaluate the effect of pH, hydrogen sulfide partial pressure, and chloride content on the SSC resistance, and

- To evaluate the SSC resistance at higher strength levels to determine the operating envelope for gas streams containing hydrogen sulfide, but outside the MR0175 definition of sour service.

There have been many papers presented recently on the SSC resistance of super 13% chromium stainless steels, but the testing conditions have included high chloride content with low hydrogen sulfide partial pressure. In centrifugal compressor applications, the chloride content is low or reported as zero and the hydrogen sulfide partial pressure may often be well in excess of one bar.

Three materials were tested: 17-4PH is a precipitation hardening martensitic stainless steel containing chromium 17 and nickel 4%; 15-5PH is consumable vacuum arc remelted precipitation hardening martensitic stainless steel having an overlapping chemistry with 17-4PH and identical heat treatment and tensile strength, yield strength, and hardness; and F6NM is a low carbon martensitic stainless steel containing chromium 13 and nickel 4%. Chemistries of the alloy classes are listed in Table 1.

2. Test Procedure

The test requirements are modelled on the principles set forth in *European Federation of Corrosion Publication Number 17* [1]. Materials are evaluated under expected service environmental and loading conditions. The environmental variables to be evaluated are pH, chloride concentration (Cl^-), temperature and partial pressure of H_2S. Figure 1 illustrates a method of presenting material service limits. The suggested test load for constant load tests is 90% of the actual test material yield strength.

All specimens were prestressed to 100% of the 0.2% offset yield strength, load held for 2 min, and the load released. Testing was performed with the pre-stress to duplicate material conditions and behaviour in impellers after overspeed. At overspeed, portions of the impeller yield in the plastic zone [2].

The 17-4PH stainless steel elastic limit is 35–40% of the 0.2% offset yield strength. After pre-stress to the 0.2% yield strength, release of load, and re-load, the elastic limit is approximately 90% of the 0.2% offset yield strength.

All testing was performed at *InterCorr* International, Inc., Houston, Tx.

2.1. Atmospheric Pressure SSC Procedure

Standard SSC specimens were machined according to the specifications of NACE Standard TM0177-96, Method A with a 0.250-in. gauge diameter. The specimens were machined parallel to the rolling direction or grain direction of the supplied materials. Prior to exposure, the specimens were degreased with toluene and rinsed with acetone. Following cleaning, the specimens were handled with gloves only.

The solution differed from the standard TM0177-96, Method A, Solution A in that both chloride content and pH were primary variables in the testing conducted. The chloride content ranged from 100 to 10 000 ppm and was achieved with the addition

Table 1. *Alloy chemistries*

Alloy	%C	%Mn	%P	%S	%Si	%Cr	%Ni	%Cu	Nb + Ta%	%Mo
17–4PH	0.07 max.	1.00 max.	0.04 max.	0.03 max.	1.00 max.	15.5–17.5	3.00–5.00	3.00–5.00	0.15 0.45	
15–5PH	0.07 max.	1.00 max.	0.04 max.	0.03 max.	1.00 max.	14.0–15.50	3.50–5.50	2.5–4.5	0.15–0.45	
F6NM	0.05 max.	0.50–1.00	0.03 max.	0.03 max.	0.60 max.	11.50–14.0	3.5–5.5			0.4–0.8

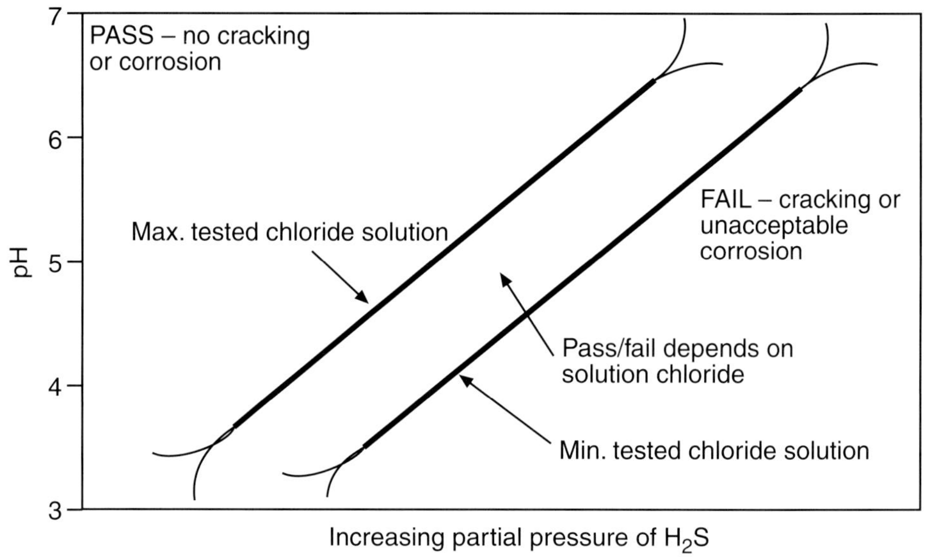

Fig. 1 *Influence of pH,* p_{H_2S} *and* Cl^- *on SSC resistance.*

of NaCl. The solutions were buffered with additions of acetic acid and sodium acetate trihydrate to maintain a constant pH during testing.

The specimens were loaded into independent glass test vessels dry and fixed into standard NACE TM0177 proof ring loading devices. Loading was accomplished with the use of the proof ring. The procedure consisted of loading the specimen dry to 100% of the actual yield strength of the material and holding at the stress for a 2 min period, after which the specimen was fully unloaded. The stress was re-applied to 100% of the actual yield strength. The vessel was deaerated with dry nitrogen and deaerated test solution was introduced to the vessel. The vessel was again purged with nitrogen according to the requirements of NACE Standard TM0177-96, Method A. Following deaeration, the solution was purged at ambient temperature and pressure with the specified test gas incorporating the desired H_2S partial pressure. The gas composition ranged from 1 percent H_2S, balance CO_2 to 100% H_2S. Following saturation, the gas purge rate was reduced to a few bubbles per minute throughout the 720 h test duration or until failure, whichever came first. Following exposure, the gauge surface was cleaned and examined at 10× magnification for surface cracking.

2.2. High Pressure SSC Procedure

Long SSC specimens were machined for conducting the high pressure SSC evaluations. The specimens measured 7.25-in. (184.2 mm) in length and contained a 1.0 in. (25.4 mm) by 0.188 in. (4.78 mm) diameter gauge section. The specimens were machined parallel to the rolling direction or grain direction of the supplied materials. Prior to exposure, the specimens were de-greased with toluene and rinsed with acetone. Following cleaning, the specimens were handled with gloves only.

As in the atmospheric pressure SSC procedure the solution differed from the standard TM0177-96, Method A, Solution A in that both chloride content and pH were primary variables in the testing conducted. The chloride content in these high pressure tests ranged from 0 to 10 000 ppm and was achieved with the addition of NaCl. The solutions were buffered with additions of acetic acid and sodium acetate trihydrate to maintain a constant pH during testing. A software model was used to calculate the starting pH of the test solution, such that on loading with the acid gases it would yield the desired pH. This difference was minor, but was incorporated into the test protocol. (For example, for a desired pH of 3.50, the solution was buffered to pH 3.54 and subsequently loaded with 220 psia H_2S and 220 psia CO_2 which acted to decrease the pH to the desired level of 3.50.)

The specimens were loaded into a 0.5L Alloy C-276 autoclave such that the specimen ends protruded out from each end of the autoclave for attachment to the load train. Sealing was made directly on the test specimen with the use of PTFE seals. Due to the low level of chlorides being evaluated, the autoclave was glass bead blasted prior to each test and the associated valves, fittings and tubing were either replaced or cleaned.

Following specimen insertion, the autoclave was deaerated with multiple alternate vacuum and N_2 fill cycles. The autoclave was subsequently evacuated once more by vacuum and deaerated test solution was drawn into the vessel. The vessel was again purged with N_2 to remove any air, that may have been introduced during the solution transfer. Another vacuum was pulled and the system was backfilled with the respective H_2S partial pressure following by the addition of the CO_2 partial pressure as required.

Loading of the specimen was accomplished with the use of the proof ring. The procedure consisted of loading the specimen to 100% of the actual yield strength of the material and holding at the stress for a 2 min period, after which the specimen was fully unloaded. The stress was reapplied to 100% of the actual yield strength. The testing was conducted at ambient temperature, 75 ± 5°F (23.9 ± 2.4°C). Conditions were maintained for 720 h or until failure. Following exposure, the gauge surface was cleaned and examined at 10× magnification for surface cracking.

Mechanical properties and heat treat condition of the various materials are listed in Table 2.

3. Results

1. Testing was initiated with the 17-4PH, H1150 material. Various levels of chloride, hydrogen sulfide partial pressure, and pH were evaluated in an effort to establish operating parameters for the material. Gaugh [3] reported time to failure of less than 100 h with this material stressed to 25 ksi in the standard TM0177 solution. Longer times to failure were reported for material in the H1150+1150 and H1150M conditions when stressed at 25 ksi. When stressed at 75 ksi, all material conditions had failure in less than 50 h.

The results of the SSC tests are shown in Table 3. Initial testing was with 1000 and 10 000 ppm chloride and pH 3.5 and pH 4.0. Samples A1–A11 established a no-failure limit of 1.5 psi (0.1 bar) H_2S with 1000 ppm chloride and pH 3.5. Sample A3

Table 2. *Mechanical properties and heat treat condition*

Material	Form	Heat treat condition	Tensile strength, ksi (MPa)	Yield strength, ksi (MPa)
17–4PH, H1150 + 1150	Plate	Solution anneal plus double age at 1150°F (621°C)	143 (987)	109 (752)
17–4PH, H1150	Plate	Solution anneal plus age at 1150°F (621°C)	144 (987)	116 (800)
17–4, H1150M	Plate	Solution anneal plus age at 1400°F (760°C) plus 1150°F (621°C)	134 (925)	102 (704)
15–5PH, H1150 + 1150	Bar	Solution anneal plus double age at 1150°F (621°C)	135 (932)	109 (752)
F6NM, NACE*	Forging	Austenitise plus temper at 1250°F (676°C) plus 1125°F (607°C)	114 (787)	99 (683)
F6NM,	Forging	Austenitise plus temper at 1050°F (566°C)	129 (890)	123 (849)

* heat treated to NACE MR0175 recommendation.

had no failure, but samples A5, A8, and A10 with 10 000 ppm chloride failed at equal or lower H_2S partial pressures.

After this series of testing, it was decided to use a more realistic atmosphere for compressor applications and 100 ppm chloride was chosen. Chloride contact is often reported as 'none', but it was felt that to be conservative, there should be some chloride in the test atmosphere to represent instances where some chloride could be present in the water.

Atmospheric testing concluded with determining a no-fail limit of 15 psi (1 bar) H_2S, 3.5 pH, and 100 ppm chloride (sample A9). Note the failure of sample A12 with 300 ppm chloride.

High pressure testing determined a no-fail limit of 105 psi (7 bar) H_2S, 3.5 pH and 100 ppm chloride (sample A16). Failure occurred on increasing H_2S to 220 psi (15 bar) at 3.5 pH and 100 ppm chloride (sample A17) or 50 ppm chloride (sample A19). No failure occurred at 220 psi (15 bar) H_2S, 3.5 pH, and 0 chloride (sample A18).

The data summarising the effects of chloride level are graphically represented in Figs 2 and 3.

2. Conditions H1150+1150 (samples B1-B9) and H1150M (samples D1-D3) are the two approved material heat treat conditions for 17-4 PH in NACE Standard MR0175.

Samples B1-B3 repeated the atmospheric testing results of condition H1150. Sample B2 had no failure, but B1 with 300 ppm chloride failed and B3 with pH 3.0 failed.

High pressure testing also established the no-fail limit of 105 psi (7 bar) H_2S, 3.5 pH and 100 ppm chloride (sample B7) and 220 psi (15 bar) H_2S, 3.5 pH, and 0 chloride (sample B4).

Failure occurred with 105 psi (7 bar) H_2S, 3.25 pH and 100 ppm chloride when testing at 100% of yield strength (sample B8) and at 80% of yield strength (sample B9).

Sample pair B6 and B7 shows no effect of the presence of carbon dioxide with 105 psi (7 bar) H_2S, 3.5 pH and 100 ppm chloride. Sample B6 with 105 psi (7 bar) CO_2 and Sample B7 with 0 psi CO_2 have similar results. The results are summarised in Fig. 4.

Table 3. Test results

Sample number	Material	pH	pH_2S psia (bar)	pCO_2 psia (bar)	Chloride ppm	Stress, ksi (MPa)	Time to failure (h)
A1	17–4PH, H1150	4.0	0.15 (0.01)	14.85 (0.99)	1000	116 (800)	No-fail, 2 samples
A3	17–4PH, H1150	3.5	1.5 (0.1)	13.5 (0.9)	1000	116 (800)	No-fail, 2 samples
A5	17–4PH, H1150	3.5	1.5 (0.1)	13.5	10000	116 (800)	33.8
A6	17–4PH, H1150	3.5	15 (1)		10000	116 (800)	9.3
A7	17–4PH, H1150	3.5	15 (1)		1000	116 (800)	51.3
A8	17–4PH, H1150	3.5	0.15 (0.01)	14.85 (0.99)	10000	116 (800)	112
A9	17–4PH, H1150	3.5	15 (1)		100	116 (800)	No-fail
A10	17–4PH, H1150	4.0	1.5 (0.1)	13.5 (0.9)	10000	116 (800)	92.6
A11	17–4PH, H1150	4.0	15 (1)		1000	116 (800)	24
A12	17–4PH, H1150	3.5	15 (1)		300	116 (800)	163
A16	17–4PH, H1150	3.5	105 (7)	105 (7)	100	116 (800)	No-fail
A17	17–4PH, H1150	3.5	220 (15)		100	116 (800)	11.2
A18	17–4PH, H1150	3.5	220 (15)		0	116 (800)	No-fail
A19	17–4PH, H1150	3.5	220 (15)		50	116 (800)	4.9
B1	17–4PH, 1150 + 1150	3.5	15 (1)		300	109 (752)	68
B2	17–4PH, 1150 + 1150	3.5	15 (1)		100	109 (752)	No-fail
B3	17–4PH, 1150 + 1150	3.0	15 (1)		100	109 (752)	18.7
B4	17–4PH, 1150 + 1150	3.5	220 (15)	220 (15)	0	109 (752)	No-fail
B5	17–4PH, 1150 + 1150	3.5	220 (15)		100	109 (752)	18
B6	17–4PH, 1150 + 1150	3.5	105 (7)	105 (7)	100	109 (752)	No--fail
B7	17–4PH, 1150 + 1150	3.5	105 (7)		100	109 (752)	No-fail
B8	17–4PH, 1150 + 1150	3.25	105 (7)		100	109 (752)	15
B9	17–4PH, 1150 + 1150	3.25	105 (7)	105 (7)	100	87* (600)	4

Table 3. Test results (continued)

Sample number	Material	pH	pH_2S psia (bar)	pCO_2 psia	Chloride, ppm	Stress, ksi (MPa)	Time to failure (h)
C1	15–5PH, 1150 + 1150	3.5	105 (7)		100	109 (752)	No fail
C2	15–5PH, 1150 + 1150	3.25	105 (7)	105 (7)	100	109 (752)	1.7
D1	17–4PH, H1150M	3.25	105 (7)	105 (7)	100	102 (704)	15
D2	17–4PH, H1150M	3.5	220 (15)	220 (15)	100	102 (704)	7
D3	17–4PH, H1150M	3.5	105 (7)	105 (7)	100	102 (704)	No fail
E1	F6NM**	3.5	220 (15)	220 (15)	100	99 (683)	1
E2	F6NM**	3.25	105 (7)	105 (7)	100	99 (683)	2
E3	F6NM**	3.5	105 (7)	105 (7)	100	99 (683)	No fail
F1	F6NM	3.5	1.5 (1)		100	123 (849)	1.5
F2	F6NM	3.5	1.5 (0.1)	13.5 (0.9)	100	123 (849)	2
F3	F6NM	5.5	1.5 (0.1)	13.5 (0.9)	100	123 (849)	No fail
F5	F6NM	4.5	0.15 (0.01)	14.85 (0.99)	100	123 (849)	No fail

* B9 tested at 80% of yield strength. ** heat treated to the requirements of NACE Standard MR0175.

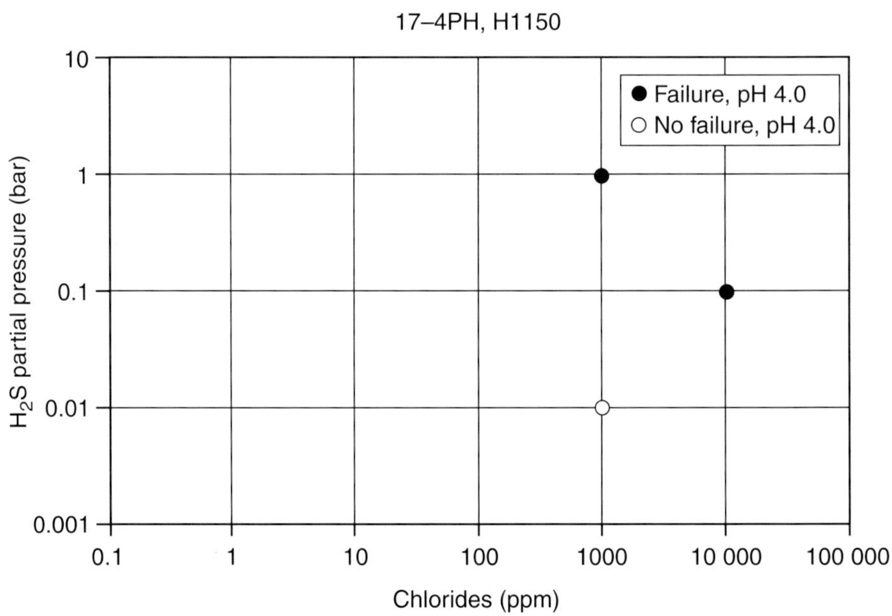

***Fig. 2** Influence of chlorides and* p_{H_2S} *on SSC resistance.*

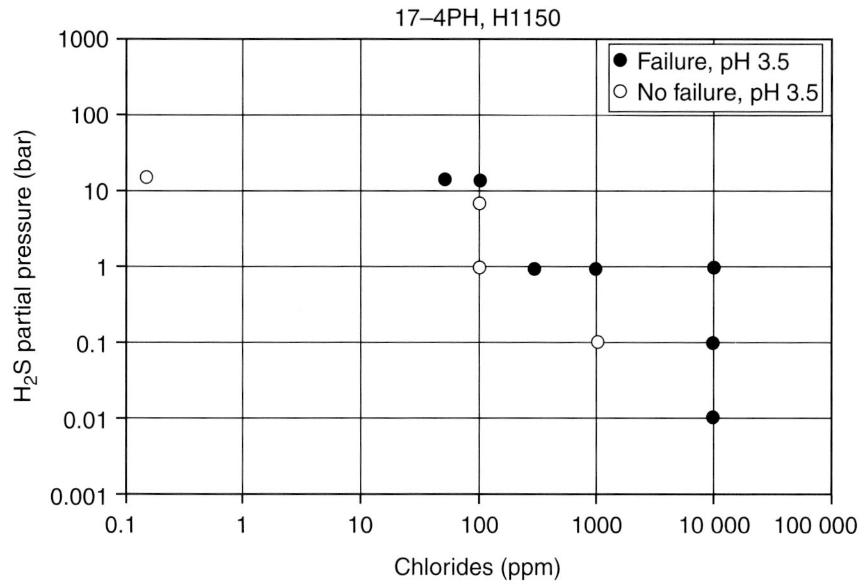

***Fig. 3** Influence of chloride, and p_{H_2S} on SSC resistance.*

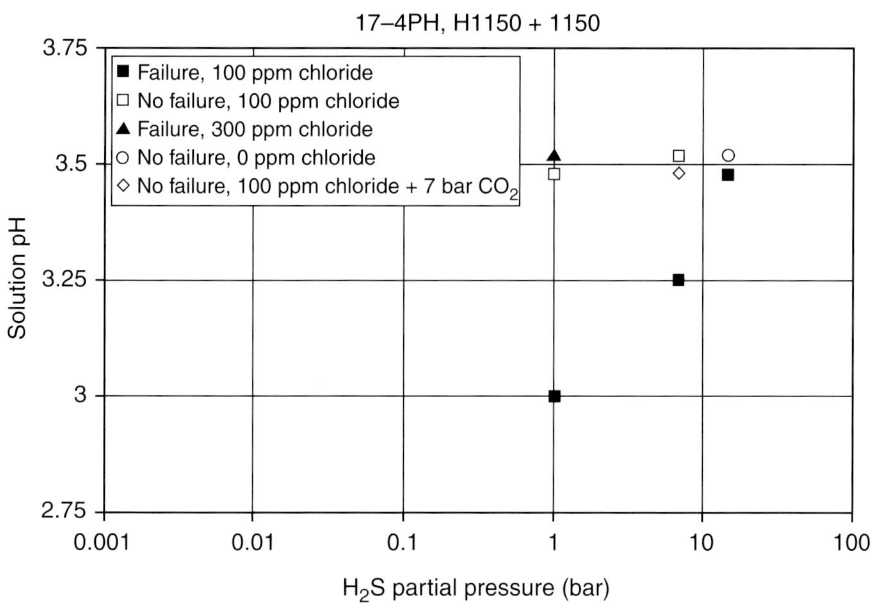

***Fig. 4** Influence of p_{H_2S} and pH on SSC resistance.*

Samples D1-D3 established a no-fail limit of 105 psi (7 bar) H_2S, 3.5 pH and 100 ppm chloride for 17-4PH in the H1150M condition (sample D3, no fail, samples D1 and D2, failure). The results are shown in Fig. 5.

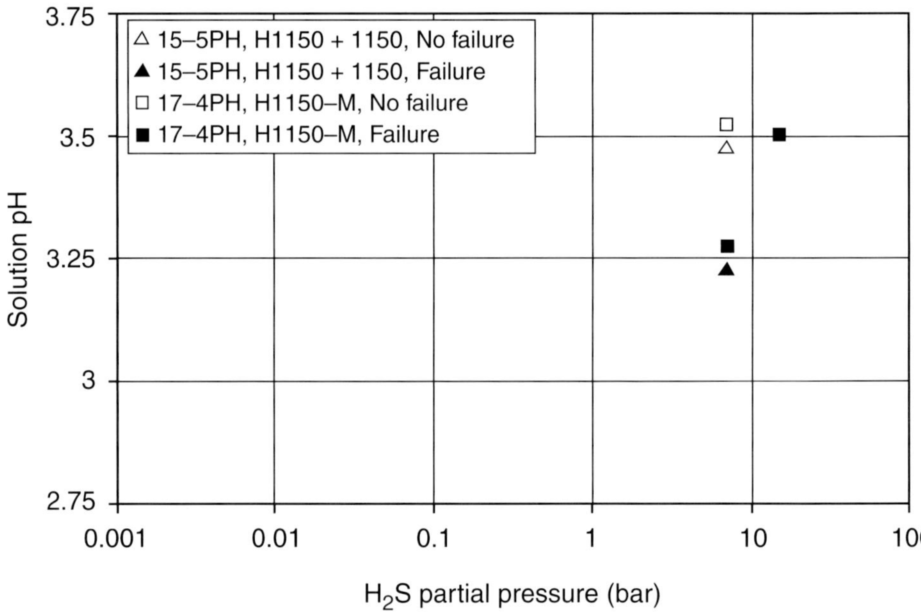

Fig. 5 *Influence of* p_{H_2S} *and pH on SSC resistance.*

3. The no-fail limit for 15-5PH, H1150 + 1150, is also 105 psi (7 bar) H_2S, 3.5 pH and 100 ppm chloride (sample C1, no fail, sample C2, fail). The results are shown in Fig. 5.

4. Samples E1-E3 represent F6NM heat treated to the requirements of NACE Standard MR0175. Crawford [4] reports no failure at 60% of yield strength for this material when stressed in the NACE TM0177 solution. Based on this history, it was anticipated that the F6NM might give greater resistance to SSC than 17-4PH.

Once again, the no failure limit was 105 psi (7 bar) H_2S, pH 3.5 and 100 ppm chloride. Sample E3 had no failure and samples E1 and E2 failed. The results are shown in Fig. 6.

5. Samples F1-F5 represent a higher strength heat treatment for F6NM with a 123 ksi yield strength. The no failure limit was 1.5 psi (0.1 bar) H_2S, pH 5.5, 100 ppm chloride (sample F3), and 0.15 psi (0.01 bar) H_2S, pH 4.5, 100 ppm chloride (sample F5). The results are shown in Fig. 6.

4. Conclusions

1. Chloride content has a significant effect on the SSC resistance of the martensitic stainless and martensitic precipitation stainless steels tested.

2. The pH range 3.0–3.5 is very critical to the SSC resistance of all materials tested.

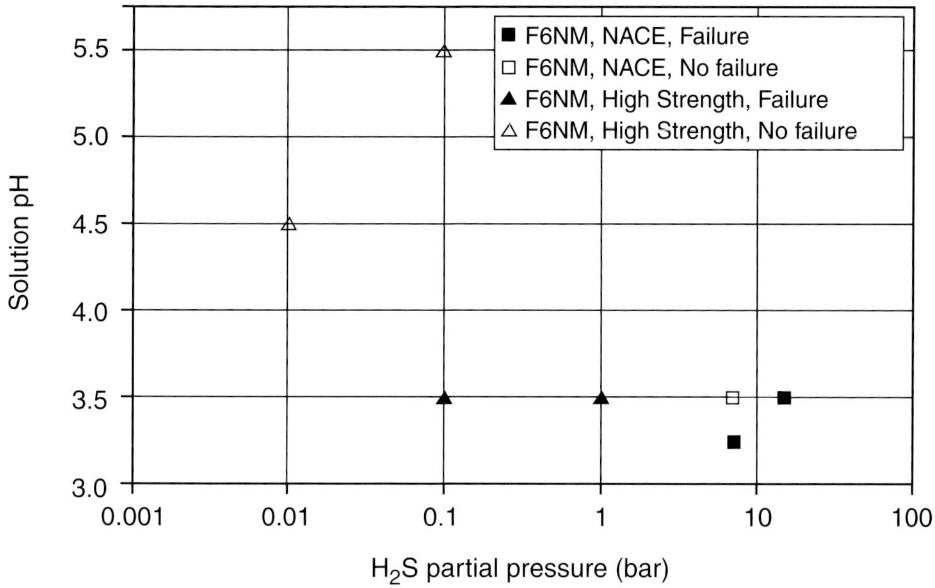

***Fig. 6** Influence of* p_{H_2S} *and pH on SSC resistance.*

3. No failure limits of 105 psi (7 bar) H_2S, pH 3.5 and 100 ppm chloride were determined for 17-4PH, condition H1150, 17-4PH condition 1150 + 1150, 17-4PH, condition H1150M, 15-5PH, condition 1150 + 1150.

4. A no failure limit was observed to be 105 psi (7 bar) H_2S, pH 3.5 and 100 ppm chloride for F6NM when heat treated according to the requirements of NACE Standard MR0175.

5. Higher strength F6NM has significantly less SSC resistance, with no-fail limits of 1.5 (0.1 bar) psi H_2S, pH 5.5, 100 ppm chloride, and 0.15 psi (0.01 bar) H_2S, pH 4.5, 100 ppm chloride.

References

1. European Federation of Corrosion Publications Number 17, *A Working Party Report on Corrosion Resistant Alloys for Oil and Gas Production: Guidance on General Requirements and Test Methods for H_2S Service*, The Institute of Materials, London, UK, 1996, p. 19–31.
2. D. W. Cameron, P. R. Geise and J. S. Abbott, 'Establishing overspeed limits for centrifugal compressor impellers', *Dresser-Rand Techn. J.*, 1996, **2**, 89–109.
3. R. R. Gaugh, 'Sulfide stress cracking of precipitation-hardening stainless steels', *Corrosion '77,* Paper No. 109, NACE International, Houston, Tx, 1977.
4. J. D. Crawford, 'CA-6NM — An Update', *Steel Foundry Facts*, 1974, Issue 313.

31
Evaluation of 13% Chromium Martensitic Stainless Steel in H_2S-Containing Environments by using the Contact Electric Resistance and Impedance Techniques

K. SAARINEN and J. HILDÉN*
VTT Manufacturing Technology, P.O. Box 1704, FIN-02044 VTT, Finland

ABSTRACT

Martensitic 13Cr steels are widely used in oil and gas industry applications. Reported failure cases are typically consequences of localised corrosion or sulfide stress cracking (SSC) of constructions and components. The initiation of localised corrosion and SSC is considered to require the breakdown of the passive film and so the role of stability of surface films in the initiation of corrosion has become studied more as new *in situ* techniques have been evolved.

In the Contact Electric Resistance (CER) technique two specimens immersed in the test solution are brought into contact and then disconnected at a selected frequency. As the surfaces are disconnected they interact with the environment. When the surfaces are brought into contact the contact electric resistance of the film, R_f, is calculated after every cycle so that changes in electronic conductivity of the surface layer in the thickness, i.e. z-direction, can be monitored on-line.

In this work the resistance of the surface film was studied in a sodium chloride (50 gL^{-1}) solution. The experimental variables were the amount of hydrogen sulfide and the temperature of the test solution. It was concluded that in a chloride solution at room temperature a partial passivation of polished specimen surfaces takes place. H_2S rapidly forms a surface layer which, however, changes or dissolves within a few hours. Passivation of the specimen surface does not take place at 90°C and active dissolution occurs.

1. Introduction

Corrosion of steels in sour gas environments is a complex phenomenon involving a number of reactions between the environment and the construction materials resulting in different kinds of corrosion products [1]. For passivating alloys the reduction of sulfur is the most feasible cathodic reaction that will accelerate localised corrosion [2]. The anodic dissolution rate is observed to increase and the passivation of corrosion resistant alloys to be delayed significantly as the hydrogen sulfide content

*P.O. Box 1703, FIN-02044 VTT, Finland.

of the environment increases [3]. Beside hydrogen sulfide other chemical species present in the solution also have to be considered [4].

In sour wells oil and natural gas are often contaminated by hydrogen sulfide, the content of which may vary from a few parts per million up to 30% [5]. In the presence of hydrogen sulfide the atomic hydrogen that is evolved on the surface of the steel by corrosion reactions tends to enter the steel and cause embrittlement leading to cracking. This phenomenon is called sulfide stress corrosion cracking (SSC). The SSC susceptibility is enhanced by chlorides and the results obtained by Fliethmann *et al.* [6] indicate that the cracking mechanism is a mixed mechanism of hydrogen- and chloride-induced SSC.

The initiation of localised corrosion and SSC is considered to result from the breakdown of the passive film and so the role of stability of surface films in the initiation of corrosion has become increasingly recognised as new *in situ* techniques have been evolved. Passivation behaviour of martensitic stainless steel improves considerably as the nickel content is increased to a value of about 4% [7]. The aim of this work is to study the passivation behaviour of 13Cr martensitic stainless steel by using different electrochemical techniques.

2. Experimental

The material studied was a cast 13Cr martensitic stainless steel with 4% nickel. The nominal composition of the steel studied is given in Table 1.

Studies were conducted in a modified NACE TM0177 solution. The NaCl content of the test solution was 5% and the pH was about 5. The test solution was deaerated with a gas mixture of Ar + 3% H_2. Testing temperatures were 25°C and 90°C and the partial pressure of H_2S was 100 mbar at room temperature and 20 mbar at the temperature of 90°C. The studies were carried out according to the test matrix described in Table 2.

The specimen surfaces were polished using 4000 grit SiC emery paper. A hydrogen saturated Pd electrode was used as a reference electrode. The Pd electrode is assumed to behave as a reversible hydrogen electrode (RHE). All the potentials are reported versus the RHE scale.

Table 1. *Nominal chemical composition of the 13Cr martensitic stainless steel*

Material	C	Mn	S	P	Cr	Ni	Mo	N
CA6NM	0.01	1	0.005	0.02	13	4	0.5	0.01

Table 2. *The test matrix of the polarisation curve, CER and impedance measurements*

Test environment	T (°C)	NaCl (%)	$p\,H_2S$ (mbar)
Room temperature, no H_2S	25	5	—
Room temperature, H_2S added	25	5	100
High temperature, no H_2S	90	5	—
High temperature, H_2S added	90	5	20

2.1. Contact Electric Resistance (CER) Test Method

In the CER method two identical specimen surfaces are brought into contact and then disconnected at a chosen frequency. When the surfaces are disconnected, they interact with the test environment. When the surfaces are in contact, the resulting change in the contact resistance *R* is recorded. This is done by passing direct current through the surfaces and measuring the resulting voltage. The contact resistance *R* is calculated using Ohm's law. The resulting resistance is that of the surface layer in the through thickness direction, i.e. *z*-direction.

The design of the CER-instrument developed for measurements in high temperature high pressure process environments is shown in Fig. 1. The same design was also used in this study.

2.2. Impedance Measurements

A separate sample, 5 mm in diameter, was used for impedance measurements, which were performed using a Solartron ECI 1287/FRA 1260 system in a frequency range 0.01 Hz–30 kHz at an a.c. amplitude of 10 mV (rms).

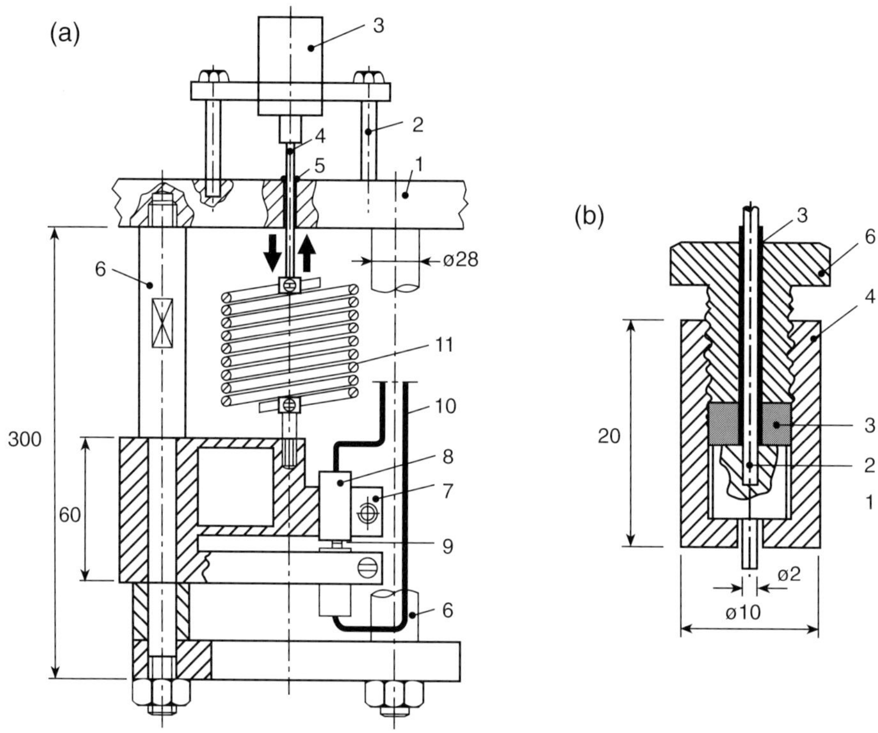

***Fig. 1** Design of the CER instrument for measurements at high temperature high pressure processes. (a) **1** flange, **2** supporting frame, **3** step motor, **4** pull rod, **5** seal, **6** supporting frame, **7** stiff spring, **8** specimen holder, **9** specimen, **10** connecting wires, **11** soft spring. Specimen holder with fixed specimen. (b) **1** specimen, **2** connecting wire, **3** insulation, **4** Zr holder, **5** Zr-ring, **6** Zr screw.*

3. Results and Discussion

3.1. Room Temperature, no H_2S

The CER measurement was started at room temperature with a polished specimen surface. Contact resistance increased within one hour by more than two orders of magnitude to the value of about 0.005 Ωcm^2 (Fig. 2) and the corrosion potential increased by about 100 mV in half an hour. This indicates that an oxide layer is formed on the polished surfaces and passivation takes place to some extent. The resistance value, however, remains so low that the surface layer can be considered as electronically conductive and electrons are thereby easily transferred through it.

A two step process is dominating the impedance spectra, illustrated in Fig. 3. The spectra are interpreted as two consecutive electrochemical reaction steps of metal dissolution. This means that ions are also transferred through the surface layer. Corresponding capacitance and resistance values are calculated in Table 3. Both steps are associated with high capacitance values C_1 and C_2, which indicates that they are interfacial reactions and probably take place at metal–film and film–electrolyte interfaces. The charge transfer resistance R_2 of the slower step at a lower frequency is much higher and is probably controlling the dissolution of the test material.

It is concluded from the CER and impedance measurement results that the surface layer formed on a fresh 13Cr martensitic stainless steel surface is not a protective one and therefore cannot prevent hydrogen from entering the metal.

3.2. Room Temperature, H_2S Added

As H_2S was added the CER test results behaved similarly to that described above. However, the contact resistance increased within half an hour by about four orders

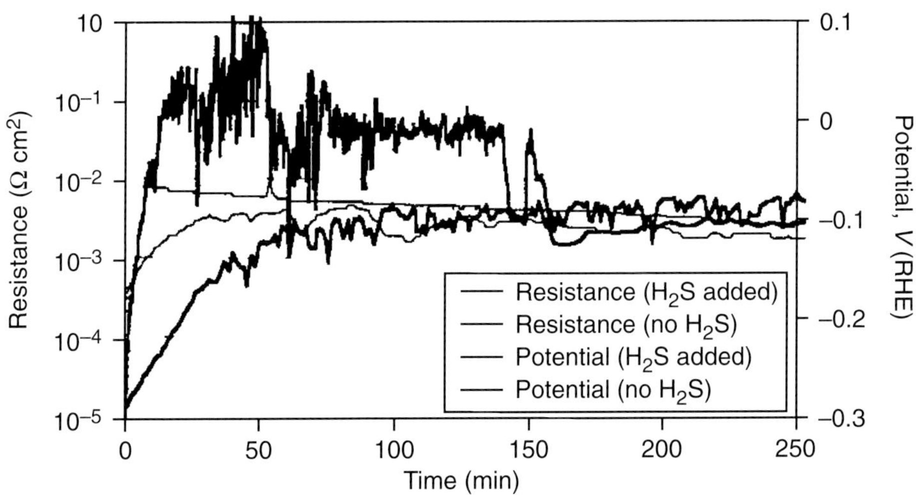

Fig. 2 *Contact resistance and corrosion potential of the martensitic 13Cr steel at room temperature as a function of time.*

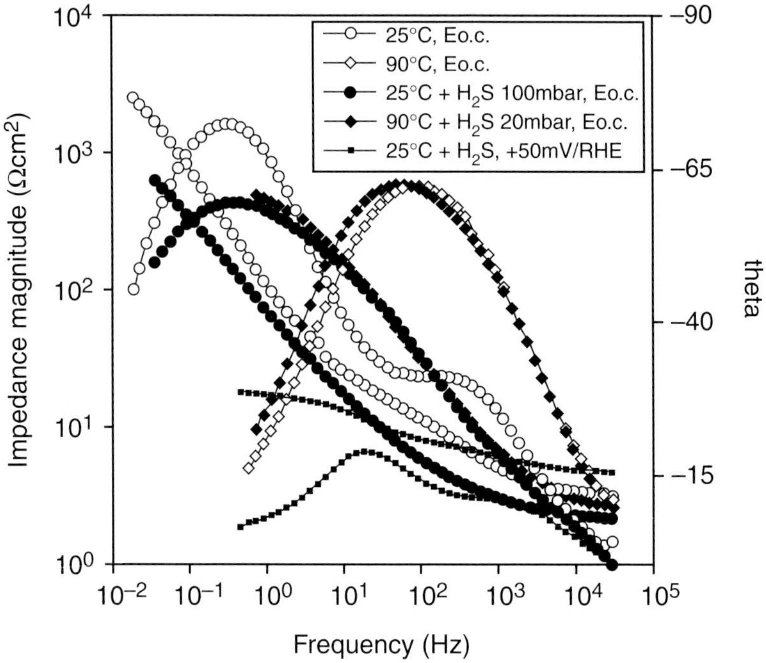

Fig. 3 Impedance spectra results for the martensitic 13Cr steel.

of magnitude (Fig. 2) indicating that a thicker sulfide-containing surface layer is formed. However, since rather large fluctuations were observed in the resistance values, the surface layer was not stable and after about 2.5 h immersion the resistance value decreased, indicating that the sulfide-containing surface layer was dissolving or that its electronic resistance had changed.

The impedance measurements also showed high capacitance values when H_2S is present (Table 3). The faster process step was not so clearly discerned in the spectrum as in the test without H_2S, but the calculated results indicate its existence. The presence of H_2S lowered the charge transfer resistance values indicating an increase in surface

***Table 3.** Fitting parameters of the impedance spectra of the 13Cr martensitic steel*

Test parameters Temperature/H_2S	Capacitance C_1 (μF cm^{-2})	Charge transfer R_1 (Ωcm^2)	Capacitance C_2 (μF cm^{-2})	Charge transfer R_2 (Ωcm^2)
Room temp., no H_2S	165	3.4	3930	856
Room temp., H_2S added	721	2.9	2913	389
Room temp., H_2S added E = + 50 mV (RHE)	300	0.3	1786	2.6
High temp., no H_2S			75	102
High temp., H_2S added			81	128

area due to formation of an heterogeneous surface layer or to increased dissolution rate of the test material.

As the specimen was polarised to +50 mV (RHE) (about 150 mV positive to the corrosion potential) the charge transfer resistance dropped two orders of magnitude indicating rapid dissolution of the test material. The CER showed anodic film formation and contact resistance of about 0.1 Ωcm^2 at +50 mV (RHE) (Fig. 4). The anodic film formed is not protective and at anodic potentials the charge transfer resistance is small and dissolution accelerates. This interpretation is supported by the marked potential dependence of the charge transfer process. Accelerated dissolution was also reflected in an increasing polarisation current.

3.3. High Temperature, no H_2S

The specimen surface was reduced by polarising it cathodically. The CER measurement was then performed under open circuit conditions. No changes in the resistance value were noted and it remained at a low value (0.0002 Ωcm^2) for several hours. This indicates that if the oxide layer is destroyed, no repassivation will take place.

At 90°C only one reaction step is shown in the impedance spectra and the faster step is probably moved above the measured range. The capacitance measured at 90°C is close to the value expected for the double layer capacitance, and the charge transfer resistance is much lower than at 25°C. This indicates that most probably active dissolution is taking place. The combined CER and polarisation measurement showed a film formation at anodic potentials. The layer is, however, not protective since it dissolves at a high rate.

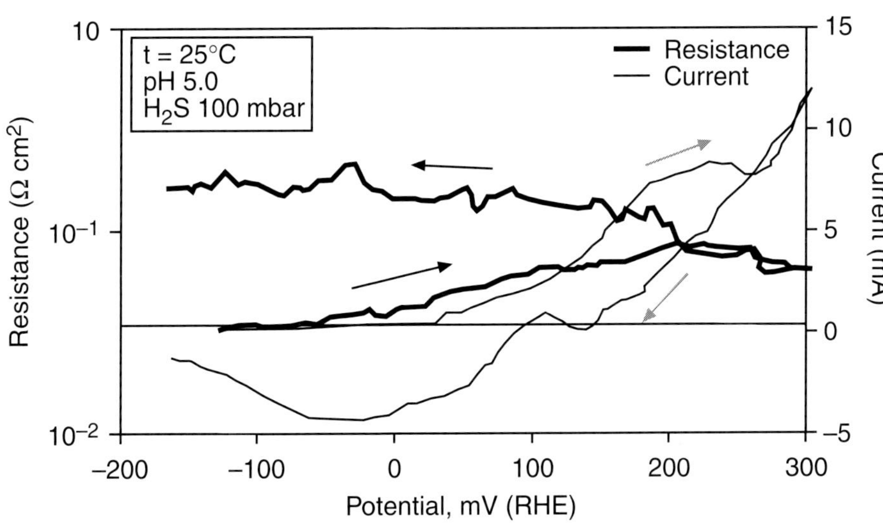

Fig. 4 *Current and contact resistance as a function of potential of the martensitic 13Cr steel after 20 h immersion at room temperature.*

3.4. High Temperature, H_2S Added

After cathodic reduction the contact resistance was measured at open circuit. Results showed that no repassivation will take place and the contact resistance was about 0.0004 Ωcm^2 for several hours.

The impedance spectrum was almost identical compared to that measured without H_2S. At high temperature the specimen was dissolving rapidly and hydrogen sulfide had no great influence on this phenomenon. This is in line with the CER results which showed that repassivation did not take place, with or without H_2S.

4. Conclusions

In this work the passivation behaviour of 13Cr martensitic stainless steel was studied by using CER- and impedance techniques. According to the results:

- Partial passivation of a polished specimen surface takes place at room temperature in chloride solution.
- Hydrogen is able to enter the metal through the partially passivated surface.
- H_2S formed a surface layer rather rapidly on the polished surface at room temperature, which however changed or dissolved within a few hours.
- The presence of H_2S lowered the charge transfer resistance values indicating increased surface area or dissolution rate of the test material.
- Passivation of the specimen surface does not take place at 90°C and active dissolution occurs.

5. Acknowledgements

This study was financed by the European Coal and Steel Community (ECSC), Nordberg-Lokomo Oy Lokomo Steels and by VTT Manufacturing Technology. Authors acknowledge these organisations for the financial support and for providing the test material.

References

1. G. I. Ogundele and W. E. White, 'Some observations on the corrosion of carbon steel in sour gas environments', *Corrosion*, 1986, **42**, 7, 398–408.
2. A. Miyasaka *et al.*, 'Environmental aspects of SCC of high alloys in sour environments', *Corrosion*, 1989, **45**, 9, 771–780.
3. A. Miyasaka *et al.*, 'Prediction of critical environments for active-passive transition of corrosion resistant alloys in sour environments', *Iron Steel Inst. Jap. Int.*, 1991, **31**, 2, 194–200.

4. Z. A. Foroulis, 'Role of solution pH on wet H_2S cracking in hydrocarbon production', *Corros. Prev. & Control,* 1993 (August), 84–89.
5. R. Garber *et al.*, 'Sulfide stress cracking resistant steels for heavy section wellhead components,' *J. Mater. Eng. Syst.*, 1985, 91–103.
6. J. Fliethmann *et al.*, 'Autoklaven-Untersuchungen der Spannungsribkorrosion von Fe–Cr–Ni–Legierungen in $NaCl/CO_2/H_2S$-Medien', *Werkst. Korros.*, 1992, **43**, 467–474.
7. N. De Cristofaro, 'Passivity and passivity breakdown of 13%Cr, 15%Cr and 13Cr5Ni2MoN stainless steels in chloride containing solutions', *EUROCORR '97*, 1997, Trondheim, Norway, (this volume p.322).

32
Passivity and Passivity Breakdown of 13%Cr, 15%Cr and 13Cr5Ni2MoN Stainless Steels in Chloride-Containing Solutions

N. de CRISTOFARO
Centro Sviluppo Materiali S.p.A., P. O. Box 10747. 100-00128 Roma-EUR, Italy

ABSTRACT
The corrosion and passivation behaviour of three martensitic stainless steels was examined in chloride-containing solutions using Electrochemical Impedance Spectroscopy (EIS), Electrochemical Noise (EN) and potentiodynamic measurements. The characteristics of polarisation curves show that corrosion resistance of the examined steels falls in the following order 13Cr5Ni2MoN, 15%Cr and 13%Cr. The 13Cr5Ni2MoN steel showed a pitting corrosion resistance higher than the conventional 15%Cr and 13%Cr stainless steels. EIS results showed that 13Cr5Ni2MoN steel remained passive and free of pitting corrosion after 504 h immersion in the NaCl solution. The 13%Cr and 15%Cr conventional martensitic steels showed a decay in the film protection after about 864 h of immersion in the chloride-containing solution. In the same electrochemical conditions the 13%Cr steel was more noisy than the 15%Cr and 13Cr5Ni2MoN steels. The noise resistance was found to increase with increasing molybdenum content in the steels.

1. Introduction

Recently, several types of modified martensitic steels have been developed [1,2]. The field application of this kind of steels has been increasing due to good corrosion behaviour in relatively low pH environments and their lower cost compared to duplex 22%Cr steels. In our previous work [1] the pitting corrosion resistance of the conventional 13%Cr and 15%Cr steels in neutral chloride- and sulfide-containing environments was studied and compared with the behaviour of the supermartensitic 13Cr5Ni2MoN steel. The supermartensitic 13Cr5Ni2MoN steel showed higher pitting corrosion resistance than the conventional martensitic steels.

The aim of this work is to study the corrosion and passivation behaviour of three martensitic stainless steels in acidic chloride-containing solutions using potentiodynamic techniques, impedance and electrochemical noise measurements.

2. Experimental

The chemical compositions of the examined martensitic stainless steels are shown in Table 1. Specimens of martensitic stainless steels of circular shape were axially

***Table 1.** Chemical composition (wt%) of the martensitic stainless steels*

Steels	Cr	Ni	C	Mo	Mn	Si	P	S	Cu	N	Al
13% Cr	13.07	0.19	0.20	0.02	0.81	0.27	0.012	0.002	0.10	n.d.	0.035
15% Cr	15.10	1.42	0.11	0.56	0.79	0.30	0.014	0.001	0.02	0.08	0.020
13Cr5Ni2MoN	13.29	4.75	0.02	1.62	0.73	0.27	0.014	<0.002	0.08	0.08	0.030

n.d. = not determined.

embedded in an epoxy resin holder to expose an area of 1 cm^2. The steel surface was mechanically polished down to 1000-grit emery paper, rinsed with distilled water and then dried in air. All the measurements were made with a conventional glass cell containing 0.2 L of NaCl solution, pH 1.7, at 23°C. The potential of the working electrode was measured against a saturated calomel electrode (SCE). The counter electrode was a large area platinum foil. Polarisation curves were obtained at a potential sweep rate of 0.001 V s^{-1}. Before starting each polarisation curves, the electrodes were electro-reduced at –1.0 V for 90 s. Pitting potential values were derived from the polarisation curves. Each alloy was subjected to a minimum of three complete scans. Electrochemical Impedance Spectroscopy (EIS) measurements were carried out with a three electrode set-up using a Solartron model 1250 Frequency Response Analyser and a Solartron 1286 Electrochemical Interface with a software Omega Pro. The measurements were performed at the free corrosion potential between 2.10^4 Hz and 10^{-3} Hz. The alternating amplitude of potentials was equal to $\pm$ 0.01 V.

Electrochemical Noise (EN) studies were carried out using a Solartron model 1286 potentiostat as zero resistance ammeter (ZRA) with a software Omega Pro. A two-electrode set-up including two steel specimens and a reference electrode was employed in order to measure simultaneously both the potential and current fluctuations.

3. Results

Polarisation Curves. Figure 1 shows a superimposed set of the anodic polarisation curves for martensitic stainless steels obtained from tests in the chloride-containing solution. The response of 15%Cr and 13Cr5Ni2oN steels displays the classical features of a material undergoing an active-passive transition. Above the potential of about 0.350 V the 13Cr5Ni2MoN reaches the passivation state. The passivation of 13Cr5Ni2MoN takes place at a smaller critical current density (6.7×10^{-2} A m^{-2}) than for the corresponding 15%Cr steel. (0.7 A m^{-2}). The presence of molybdenum decreases the critical current density and promotes the passivation process. At potentials more positive than 0 V (E_p) a significant current increase related to the nucleation and growth of pits on the steel surface is observed. The polarisation curve obtained for 13%Cr steel shows a different behaviour. The dissolution rate for 13%Cr is higher than for the other steels and the active–passive transition is not present. Under the same experimental conditions, a higher pit density was observed on the 13%Cr steel surfaces compared with the other examined steels. The results show clearly that corrosion resistance of the supermartensitic 13Cr5Ni2MoN steel and 15%Cr steels is higher than that of the conventional 13%Cr.

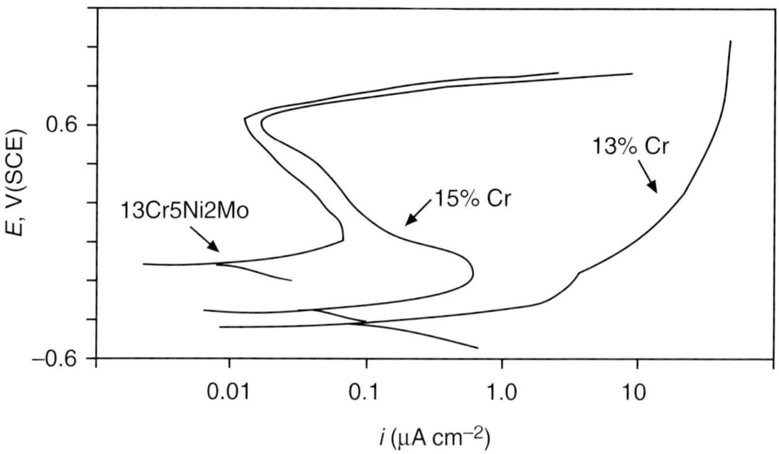

Fig. 1 Anodic polarisation curves of stainless steels in 0.7M NaCl solution, scan rate 0.001 V s^{-1}, 230°C.

Electrode Impedance Measurements. Figure 2 shows characteristic Nyquist plots of 13%Cr stainless steel obtained at the open circuit corrosion potential (OCP) as a function of immersion time. After 15 min of immersion in the NaCl solution, the OCP was about –0.550 V (Table 2). The impedance diagram showed one resistive-capacitive loop and the initiation of an inductive contribution after 6 h of immersion (a in Fig. 2). After 6 h of immersion, (b, Fig. 2) the size of the loop increased, the inductive contribution disappeared and the OCP shifted slightly to more noble values (–0.500 V). A similar pattern to that observed after 6 h of immersion was seen after 96 h immersion in the chloride-containing solution (c, Fig. 2) and the impedance diagrams were always growing. The shape of the impedance plot was modified after 504 h of exposure in the chloride-containing solution (d, Fig. 2); the presence of an important inductive loop can be seen. The OCP shifts slightly to more negative values (–0.540 V). The EIS diagrams obtained for 15%Cr stainless steel in the NaCl solutions as a function of time are shown in Figs 3 and 4. After 15 min of immersion the EIS plot (a, Fig. 3) shows one resistive-capacitive loop and the initiation of an inductive contribution. The diameter of the semi-circle grows with immersion time until 96 h of exposure in the NaCl solution (b, Fig. 3; Fig. 4). An inductive loop, distortions of the semi-circle and lower impedance values were observed after 504 h of immersion in the electrolyte (c, Fig. 3). The OCP did not change significantly with exposure time. After testing, pitting was observed on 13%Cr and 15%Cr steel surfaces. Figures 5, 6 and 7 display a set of typical EIS spectra of 13Cr5Ni2MoN steel in the chloride-containing solution as a function of time. During the first hours of immersion in the NaCl solution, the impedance plots showed an additional low frequency response (a in Fig. 5). Both semi-circles grew with the immersion time (b, Fig. 5). The shape of the impedance spectra changed after 504 h of exposure to the chloride-containing solution. The separation of the two loops is more difficult to distinguish and a flattening of the impedance diagram (Fig. 6) appeared. An important OCP ennoblement and only one capacitive loop were observed after 864 h of exposure time. After the experiment, the steel surfaces did not show pitting. For the analysis

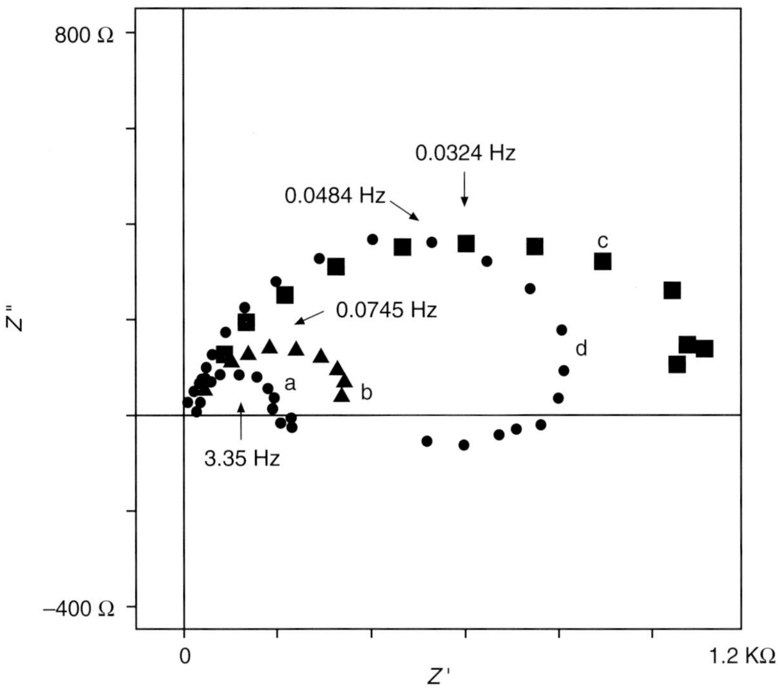

Fig. 2 *EIS plots of 13%Cr martensitic stainless steel, 0.7M NaCl, pH 1.7, 23°C; (a) after 15 min, (b) 6 h, (c) 96 h and (d) 504 h.*

of experimental data only the first loop was taken into account. The charge transfer resistance R_{ct} was determined at the intersection of the semi-circle with the real axes. Capacitance values C_d were obtained from the impedance measured at the frequency f of maximum phase angle according to:

Table 2. *Parameters from EIS measurements, 0.7 M NaCl, pH 1.7, 23°C*

Stainless steel	*t* (h)	OCP	R_{ct} (Ω)	C_d (F cm^{-2})
13%Cr	0.25	−0.550	196	2.4×10^{-4}
	6	−0.500	875	2.6×10^{-3}
	96	−0.470	1300	3.0×10^{-3}
	504	−0.540	480	6.9×10^{-3}
15%Cr	0.25	−0.500	220	1.6×10^{-4}
	6	−0.480	851	8.9×10^{-4}
	96	−0.485	5400	1.9×10^{-5}
	504	−0.485	856	4.4×10^{-4}
13Cr5Ni2MoN	0.25	−0.415	120	3.0×10^{-3}
	6	−0.425	130	1.1×10^{-3}
	504	−0.185	1170	5.9×10^{-4}
	864	−0.145	48990	2.6×10^{-4}

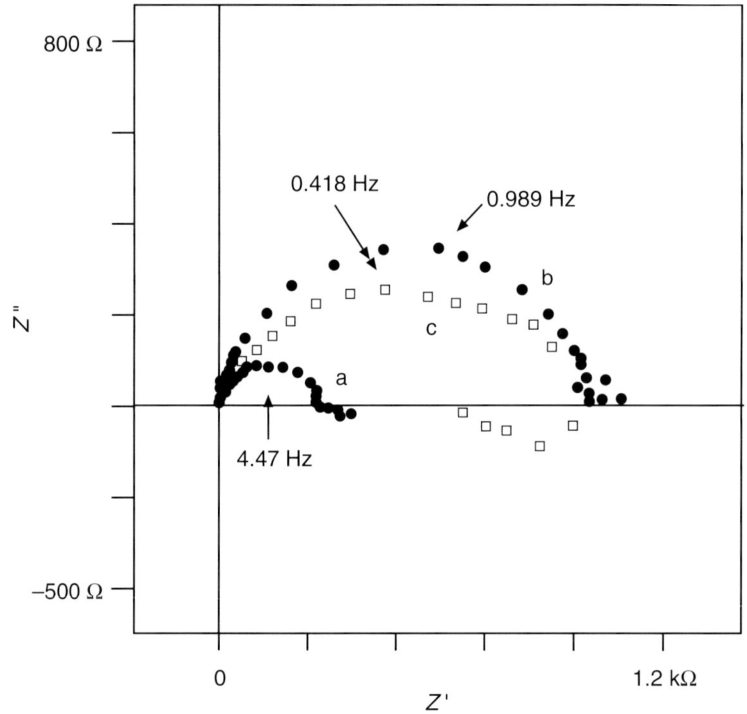

***Fig. 3** EIS plots of 15%Cr martensitic steel, 0.7M NaCl, pH 1.7, 23°C; (a) after 15 min, (b) 6 h and (c) 504 h.*

$$C_d = 1/2\ \pi f Z \qquad (1)$$

where Z is the impedence modulus.

For 13%Cr and 15%Cr steels (Table 2) it can be seen that R_{ct} initially increases with time up to 96 h and then decreases. This decrease in R_{ct} was not observed for the 13Cr5Ni2MoN steel. While for 13%Cr the C_d value increases with immersion time, for 15%Cr steel a noticeable increase in C_d is observed after 504 h of immersion time in the NaCl solution. The C_d of 13Cr5Ni2MoN steel decreases continuously with immersion time, only after 864 h of exposure is a slight increase in the C_d observed.

Electrochemical Noise Measurements. Figure 8 shows typical current–time records after 15 min of immersion in the NaCl solution. Bursts of noise were observed on 13%Cr stainless steel approximately 8 min after recording the time record. Incipient localised attack was detected by microscope examination at the manganese sulfide inclusion interfaces. The corrosion activity on 15%Cr was very low and in the case of 13Cr5Ni2MoN steel was buried in the background noise. The noise resistance R_n was calculated as the ratio of the standard deviation of voltage fluctuations divided by the same quantity relative to the current fluctuations. A higher R_n was found for 13Cr5Ni2MoN (11.5 kΩ) steel compared with 15% Cr (4.32 kΩ) and 13% steels Cr (1.68 kΩ).

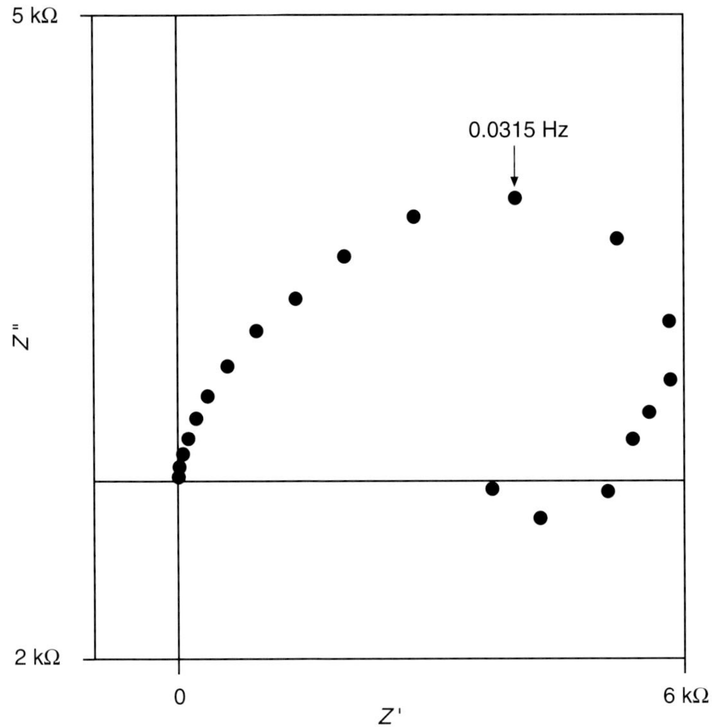

Fig. 4 *EIS plot of 15%Cr martensitic steel, 0.7M NaCl, 23°C, pH 1.7, after 96 h.*

4. Discussion

After exposure of the steels to chloride-containing solution, potential and impedance evolution show a gradual change in the characteristics of the passive film. The impedance diagrams of 13%Cr and 15%Cr stainless steels exhibited during the first hours of immersion only one capacitive loop which can be interpreted as a parallel circuit of a resistor corresponding to ionic conduction and a capacitance corresponding to dielectric properties of the oxide film. Similar findings have been observed for 18%Cr stainless steels in acidic or chloride solutions [3]. The increase in the impedance values and R_{ct} with the immersion time indicates a thickening of the passive film. Evidently, the steels became more passive as the immersion time increases. After an immersion time of 504 h, a noticeable inductive loop appears; R_{ct} decreases and C_d increases. This fact suggests that the corrosion resistance of the steels decreases because more sites on the surfaces of the steels can be activated — probably as a result of a change in the composition and structure of passive film before passivity breakdown. After the experiment the presence of very small pits was observed on the specimen surfaces by microscopy. With the occurrence of pits, inductance occurs in the low frequency region; also, the observed decrease in impedance values, after 504 h of immersion, indicates an increase in the corrosion rate. The greater R_{ct} values of 15%Cr steel compared with 13%Cr steel indicates that charged species would

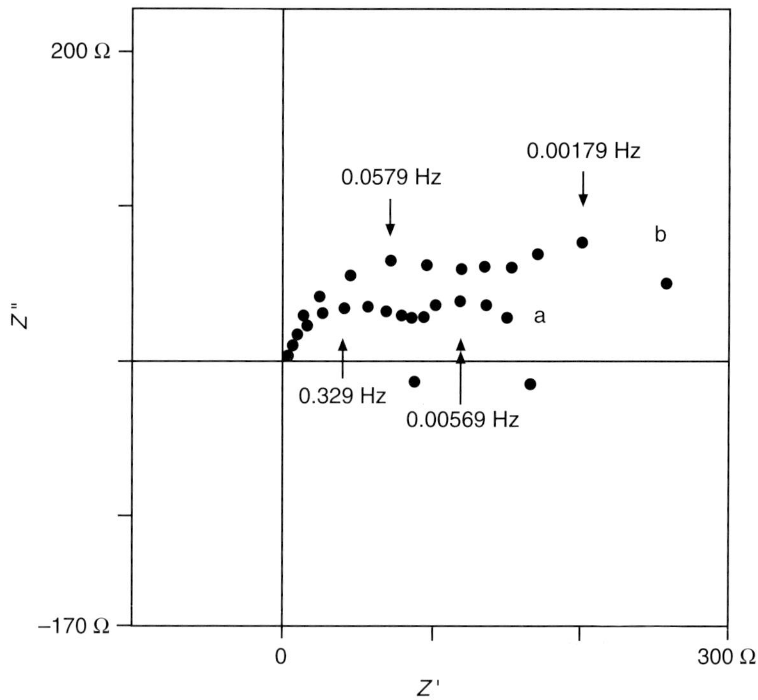

Fig. 5 *EIS plots of 13Cr5Ni2MoN steel, 0.7M NaC1, pH 1.7, 23°C; (a) after 15 min and (b) 6 h.*

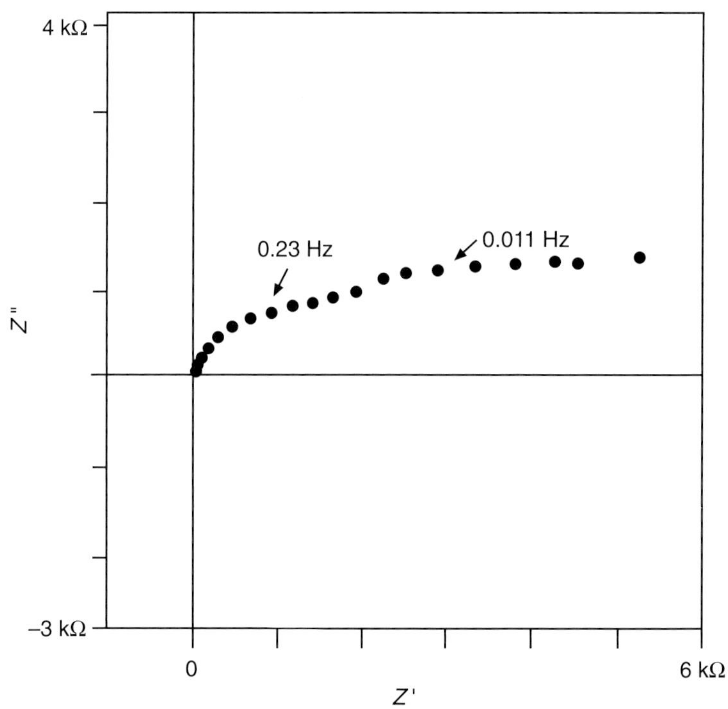

Fig. 6 *EIS plot of 13Cr5Ni2MoN steel, 0.7M NaC1, pH 1.7 23°C after 504 h.*

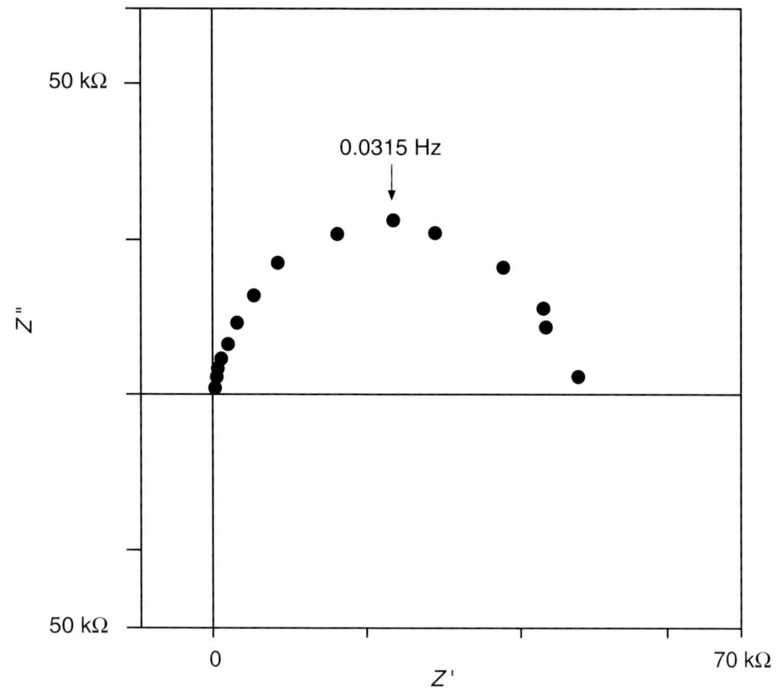

Fig. 7 *EIS plot of 13Cr5Ni2MoN steel, 0.7M NaCl, pH1.7 23°C after 864 h.*

penetrate through the film with more difficulty. When 13%Cr and 15%Cr steels were subjected to anodic polarisation, the electrochemical behaviour was very different. The dissolution current density of 13%Cr steel was higher than that of the 15%Cr steel. This behaviour suggests that the film formed on the 13%Cr steel surface is less protective than that grown on 15%Cr steel. While 13%Cr and 15%Cr steels suffer pitting corrosion after 504 h of immersion in the NaCl solution, this phenomenon has not been observed on 13Cr5Ni2MoN steel. Furthermore, a fairly different behaviour, with respect to 13%Cr and 15%Cr steels, was shown by the impedance plots. Thus, the 13Cr5Ni2MoN martensitic steel exhibits, during the first hours of exposure, two separated loops but the inductive loop that was observed for 13%Cr and 15%Cr for 13Cr5Ni2MoN steel was not present. This is possibly due to the higher molybdenum content in the steel. The continuous increase of impedance values and film resistance with the immersion time indicates an increase of film thickness. The second loop became more important and as a consequence the flattening of the impedance plot occurred. After 864 h of immersion in the chloride-containing solution 13Cr5Ni2MoN martensitic steel was still in the passive state as indicated by the high value of R_{ct} and by the ennoblement of OCP as a function of immersion time.

5. Conclusions

The passivation behaviour of 13Cr5Ni2MoN steel in 0.7M NaCl solution, pH 1.7 was superior to that of conventional 13%Cr and 15%Cr steels. Polarisation curves showed

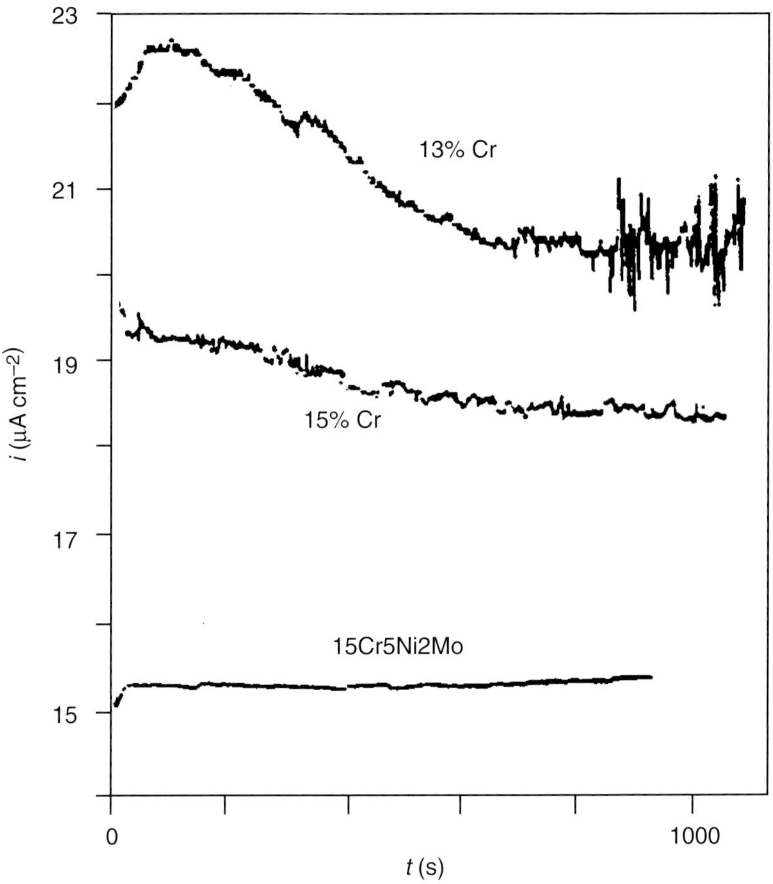

Fig. 8 EN current of steels in 0.7M NaCl, pH 1.7, 230°C.

lower dissolution current density, a smaller critical current density and a higher pitting corrosion resistance than those for the other stainless steels examined for Electrochemical Impedance Spectroscopy results showed that 13Cr5Ni2MoN steel remained passive and free from pitting corrosion after 864 h of immersion in the NaCl solution, whereas conventional 13%Cr and 15%Cr martensitic steels showed a decay of film protection after about 96 h of immersion in this solution.

Electrochemical noise measurements have shown a higher corrosion activity of 13%Cr steel compared to that of the 15%Cr and 13Cr5Ni2MoN steels.

6. Acknowledgements

The author wishes to thank Dalmine SpA for the materials supplied and for the grant to carry out the research.

References

1. M. Barteri, N. De Cristofaro, L. Scoppio, G. Cumino and G. Della Pina, *Corrosion '95*, Paper No. 77. NACE International, Houston, Tx, USA, 1995
2. M. Ueda, T. Kushida and T. Mori, *Corrosion '95*, Paper No. 80. NACE International, Houston, Tx, USA, 1995.
3. C. Gabrielli, M. Keddam, H. Takenouti, Vu Quang Kinh and F. Bourelier, *Electrochim. Acta*, 1980, **24**, 61.
4. I. Annergren, M. Keddam, H. Takenouti and D. Thierry, *Electrochim. Acta*, 1993, **38**, 763.

33
Serviceability of 13% Chromium Tubulars in Oil and Gas Production Environments*

M. S. CAYARD and R. D. KANE
InterCorr International, Inc., Houston, Tx, USA

ABSTRACT
The current literature concerning the serviceability of conventional and modified 13Cr tubulars in oil and gas production environments is reviewed. The serviceability in sour environments is considered in relation to the variables which affect corrosion performance, namely system pH, chlorides, temperature and H_2S/CO_2. At low temperatures, SSC is of primary concern. At elevated temperatures (> 100°C), SSC is generally not a problem. In this regime, mass loss corrosion and localised attack dictate the serviceability. Mass loss corrosion increases with a decrease in pH and in general, increases with increases in temperature, H_2S and CO_2 partial pressures. Susceptibility to localised corrosion was found to be highest at intermediate temperatures in the vicinity of 150°C.

1. Introduction

Due to the increased development of deep hot CO_2 wells and the impact of corrosion on conventional low alloy steel tubulars, the more corrosion resistant 13Cr martensitic type stainless steels have been used because of their good corrosion resistance and economic benefits, being about half the price of duplex stainless steels. The usage of 13Cr tubing has not been limited to sweet applications, but has also been selected for use in numerous sour (containing H_2S) applications.

In general, two 13Cr martensitic stainless steel chemistries are available from a variety of steel manufacturers. First, a conventional 13Cr chemistry which roughly mimics an API 5CT tubular chemistry with the addition of 12–14 wt% Cr. The second steel chemistry is that of the modified or enhanced API 13Cr. These modified 13Cr chemistries vary among the steel manufacturers, but generally have a low carbon content, Cr 11.5–13.5, Mo 1–2, and Ni 4–6 wt% and in some steels Cu 1–2 wt%. The comparative chemistries of the conventional and modified 13Cr materials are provided in Table 1. The API 5CT L80 and AISI 420 stainless steel chemistries are shown for comparison.

This paper is intended to provide guidance on the serviceability limits for 13Cr tubulars in oil and gas production. The parameters affecting the performance of 13Cr tubulars in both sweet and sour service are discussed and their respective implication to the overall integrity addressed.

Table 1. *General composition of L80 and various Cr-containing martensitic stainless steels*

Grade	C	Mn	P	S	Si	Cr	Ni	Mo	Cu
API 5CT L80	0.43	1.90	0.030	0.030	0.45	—	0.25	—	0.35
AISI 420	0.15	1.00	0.040	0.030	1.00	12.0–14.0	—	—	—
Conventional 13Cr	0.22	1.00	0.020	0.010	1.00	12.0–14.0	0.50	—	0.25
Modified 13Cr	0.03	0.50	0.020	0.005	0.50	11.5–13.5	4.0–6.0	1.0–2.0	1.0–2.0

2. Parameters Affecting 13Cr Performance

The major parameters affecting the performance of 13Cr tubulars in oil and gas production include pH, Cl^-, temperature and the partial pressures of H_2S and CO_2. Additional considerations reported in the literature include the role of acetates, flow velocity effects and acidising resistance [1]. However, the effects of these three variables are beyond the scope of this paper.

2.1. System pH

The system pH simply provides a measure of the hydrogen ion availability in the solution. The lower the pH the greater the amount of atomic hydrogen available potentially to enter the steel. With respect to 13Cr alloys, the ingress of atomic hydrogen into the steel can lead to hydrogen embrittlement which in the presence of H_2S is referred to as sulfide stress cracking (SSC). An SSC domain chart, developed by Kermani *et al.* [2], now serves as a guideline for specifying materials requirements for carbon and low alloy steels in H_2S environments and has been adopted by the European Federation of Corrosion (EFC) [3]. This chart, shown in Fig. 1, illustrates SSC domains as a function of solution pH and H_2S partial pressure. Domain 1 indicates a safe region where SSC is not expected to occur. Domain 2 is referred to as the transition region where materials should be qualified under some test programme. Domain 3, indicates a region where SSC would be expected to occur and hence requires sour service materials.

Sulfide stress cracking data on both conventional and modified 13Cr published in the literature [4–10] has been compiled and plotted as pH vs H_2S partial pressure. The data collected were all taken at 25 °C for steels stressed from 80 to 100% of the specified minimum yield strength (SMYS) in brine solutions ranging from 30 000 to 60 000 ppm Cl^-. These data are presented in Fig. 2. The EFC SSC domain guidelines are plotted on the figure for reference. Cracking occurred more readily at low pH levels and at higher H_2S partial pressures as expected. Some failures were noted in Domain 2, however no failures were observed in Domain 1. Another interesting point is the number of no-fail results observed in Domain 3. This relates to the quality and control of the 13Cr material which was tested. The large majority of the no-fail data in this domain was associated with the modified 13Cr chemistry. Hence, the modified 13Cr steels do appear to offer substantial benefit in the sour regime which makes up Domain 3.

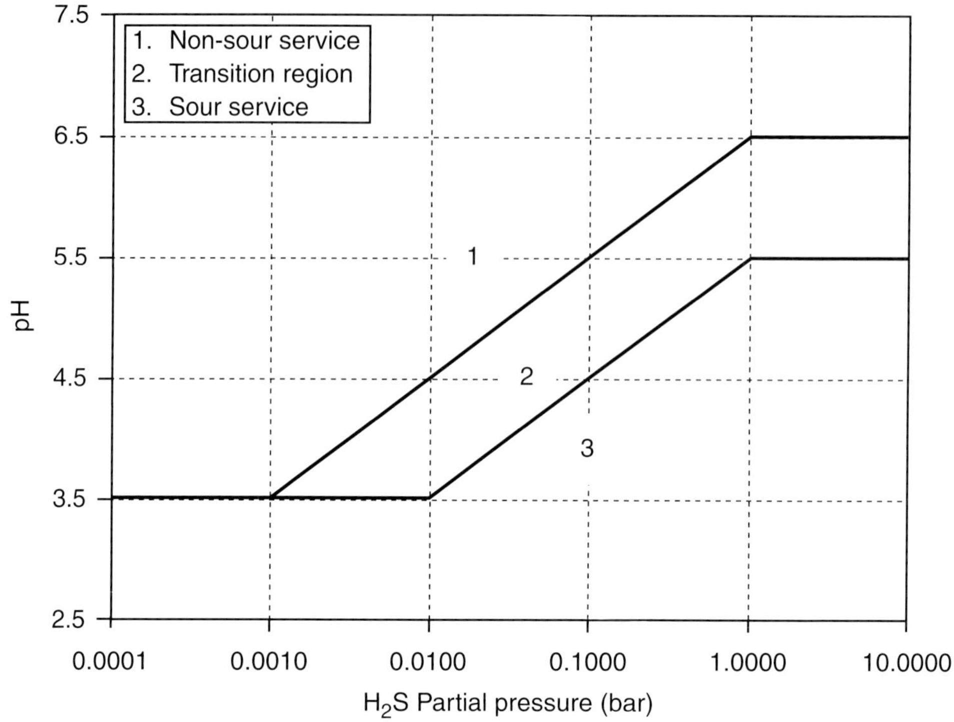

Fig. 1 *SSC domains as a function of pH and H_2S partial pressure.*

System pH also plays a direct role in the corrosion rate of 13Cr steels in oil and gas production. While the system pH is a function of the partial pressures of H_2S and CO_2, bicarbonate concentration, weak acids and temperature, it does, in itself, provide a trend with respect to corrosion rate. Data from the literature [11–13] combined with data collected at *InterCorr* International, Inc. up to November 1995 [14] were plotted. The data set represents a range of conditions with respect to H_2S, CO_2, temperature and Cl^- concentrations. The data for conventional and modified 13Cr are shown in Figs. 3 and 4, respectively. The magnitude in the corrosion rate at low pH values varies between the conventional and modified chemistries as expected. However, the two plots do indicate that above a pH of approximately 4.2 the corrosion rates remain quite low.

2.2. Chlorides

The role of chlorides on the corrosion/cracking performance of martensitic stainless steels simply relates to the breakdown of the corrosion films which are normally protective to the material. Increased chlorides result in an increased tendency for localised corrosion resulting from the local breakdown of the protective film. Modified 13Cr chemistries enhance the resistance to localised attack as a result of the addition of 1 to 2% Mo which effectively increases the pitting resistance of the material.

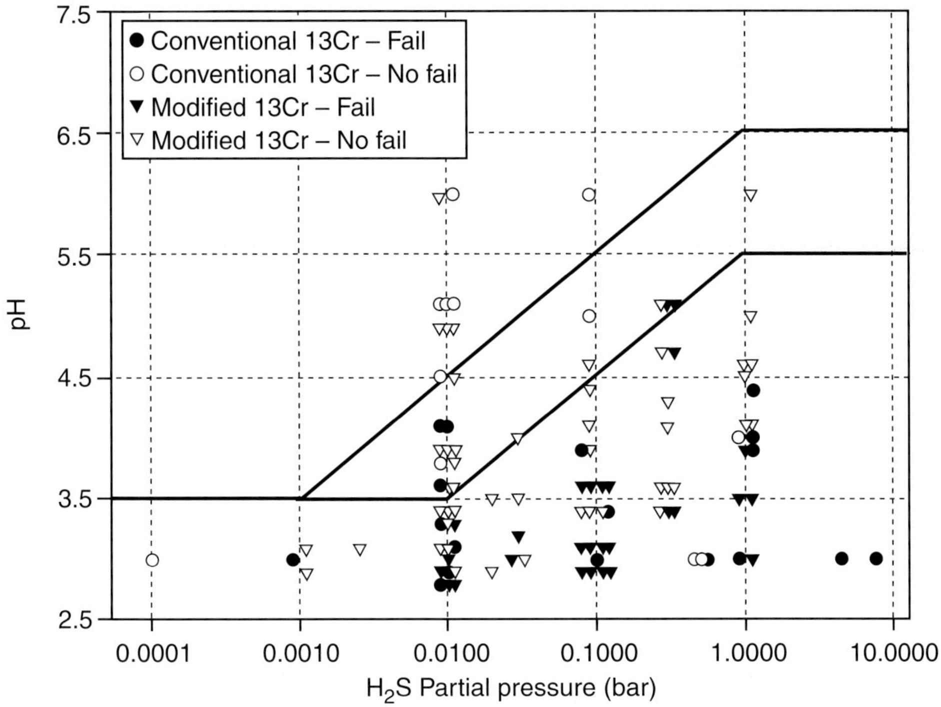

Fig. 2 SSC performance of conventional and modified 13 Cr.

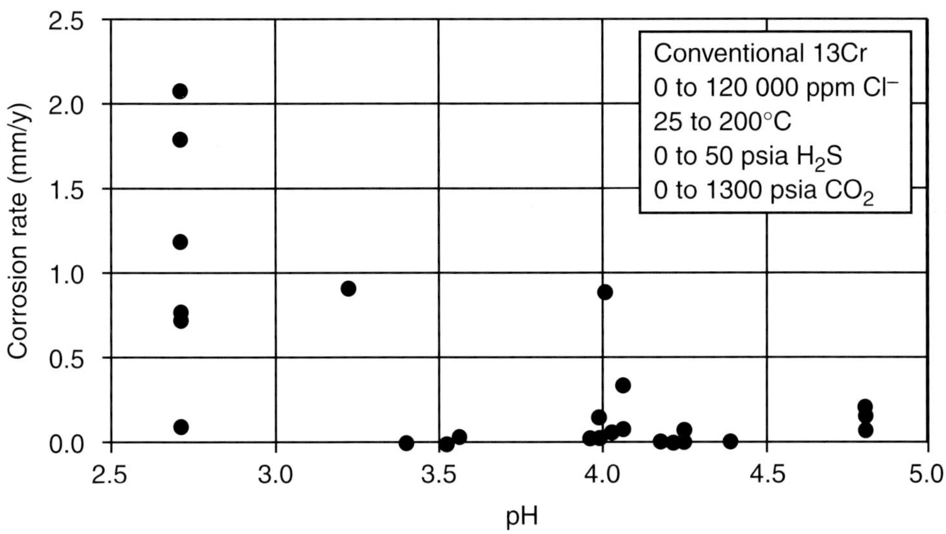

Fig. 3 Corrosion rate vs system pH for conventional 13 Cr.

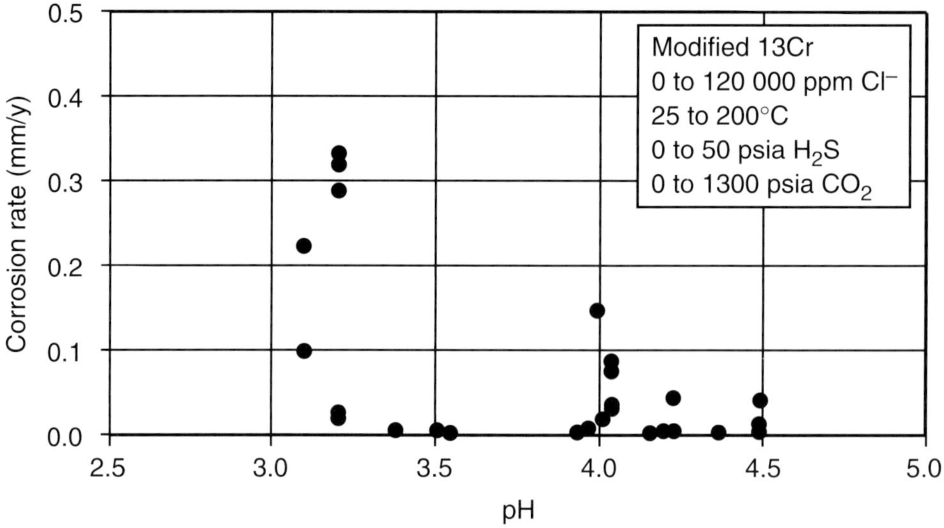

Fig. 4 *Corrosion rate vs system pH for modified 13 Cr.*

There exists a synergy between localised corrosion, produced by the presence of chlorides, and susceptibility to SSC. Atomic hydrogen, produced by localised corrosion, diffuses into the steel and can lead to SSC in the presence of applied or residual tensile stress and a susceptible temperature (generally < 100°C). Sulfide stress cracking is almost always preceded by some form of localised attack on martensitic stainless steels. The cracking is due to a combination of the increased hydrogen charging and the stress concentration associated with the localised corrosion (pitting or crevice attack). If the localised attack is eliminated, susceptibility to SSC will also generally diminish.

2.3. Temperature

At low temperature, limits on the performance of 13Cr are based on susceptibility to SSC. Generally, conventional 13Cr data indicate excellent serviceability at chloride levels lower than 6000 ppm. Above 6000 ppm Cl^- the susceptibility to SSC is based on pH and H_2S partial pressure [4]. The modified 13Cr steels do offer an advantage with respect to SSC. They can tolerate higher chloride levels and exhibit higher threshold stresses than conventional 13Cr exposed to the same environment [9].

At elevated temperatures, the concern with respect to serviceability is focused on mass loss corrosion and localised attack. Mass loss corrosion rates of both conventional and modified 13Cr materials increase with an increase in temperature. This behaviour is driven largely by the mechanism of CO_2 corrosion. For conventional tubulars, the rule of thumb is that below 65°C corrosion caused by CO_2 is seldom a problem. Above 65°C, the corrosion rate is a function of the CO_2 partial pressure, pH and temperature. The corrosion rates of martensitic stainless steels are drastically reduced from their conventional steel counterparts in the same environments. Based on a review of the literature [7,9,15–17], the corrosion rates for conventional 13Cr double for every 25°C rise in temperature above 80°C. For the modified 13Cr materials, corrosion rate doubles for every 25°C rise in temperature above 150°C. Hence the modified 13Cr chemistries offer more protection between 80°C and 150°C.

The other aspect at elevated temperatures involves the susceptibility to localised attack. The dependence of temperature on the localised corrosion performance of 13Cr in H_2S/CO_2 environments is not simply linear as in the case of mass loss corrosion. The 13Cr materials, both conventional and modified, exhibit increased susceptibility to localised attack at intermediate temperatures in the presence of H_2S. Data collected at 95, 150 and 200°C in a simulated sour oil well environment indicated higher pit penetration rates at the intermediate temperature (see Fig. 5). This behaviour may relate to the transition from oxide passivity to a pseudo-passive sulfide film. However, no work has been done to verify this speculation. This increased localised attack at intermediate temperatures has also been reported by Kane *et al.* [18].

2.4. H_2S and CO_2 Partial Pressures

The role of H_2S and CO_2 on the serviceability of 13Cr tubulars is simple. At low temperatures, H_2S, in combination with Cl^-, acts to promote cracking due to local poisoning and increased hydrogen charging. The limits of H_2S for the use of 13Cr in production environments have been imposed simply due to the associated cracking at low temperatures. At elevated temperatures (> 100°C), cracking via SSC is generally not observed. Limits are imposed strictly based on the general or localised corrosion behaviour. The presence of H_2S and CO_2 at elevated temperatures acts to reduce the system pH and increase the corrosion rate as discussed above.

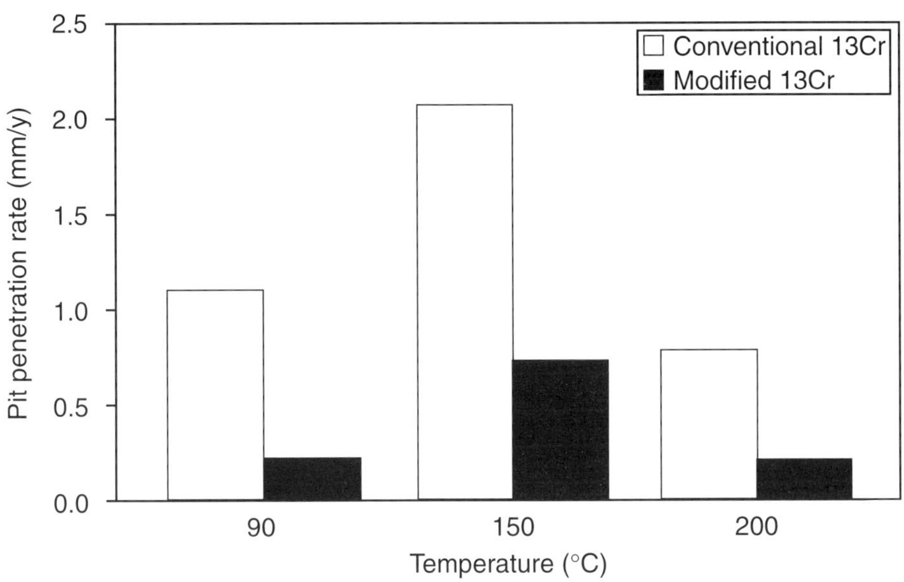

Fig. 5 *Localised corrosion performance vs temperature for the conventional and modified 13Cr.*

3. Summary

The serviceability of 13Cr in sour environments has been considered with respect to the variables which affect corrosion performance, namely, system pH, chlorides, temperature and H_2S/CO_2. Based on data collected from a variety of environments, reduced severity of corrosion was observed at pH values exceeding approximately 4.2. The role of chlorides on the performance of 13Cr steels was simply described as their ability to break down protective films and lead to localised attack. Serviceability with respect to temperature was broken down into two categories, low temperature (< 100°C) and high temperature. At low temperatures, SSC is of primary concern. At elevated temperatures (> 100°C), SSC is generally not a problem. In the high temperature domain, mass loss corrosion and localised attack dictate the serviceability. Mass loss corrosion increases with an increase in temperature. Susceptibility to localised corrosion was found to be highest at intermediate temperatures in the vicinity of 150°C. The role of H_2S and CO_2 on the serviceability of 13Cr steels was related to their ability to reduce the system pH. Furthermore, H_2S, in combination with chlorides increases the susceptibility to localised attack, which at low temperatures can lead to SSC.

References

1. S. F. Biagiotti, Jr. and J. S. Reichman, "Justifying the use of 13Cr steels for corrosive CO_2 operations", *Corrosion '95*, Paper No. 81. NACE International, Houston, Tx, USA, 1995.
2. M. B. Kermani, D. Harrop, M. L. R. Truchon and J. L. Crolet, "Experimental limits of sour service for tubular steels," *Corrosion '91*, Paper No. 21. NACE International, Houston, Tx, USA, 1991.
3. *Guidelines on Materials Requirements for Carbon and Low Alloy Steels for H_2S Containing Environments in Oil and Gas Production*. European Federation of Corrosion Publication Number 16, The Institute of Materials, London, 1995.
4. L. J. Klein, "H_2S cracking resistance of type 420 stainless steel tubulars," *Corrosion '84*, Paper No. 211. NACE International, Houston, Tx, USA, 1984.
5. M. F. Galis and J. J. Damman, "13 percent chromium steels for slightly sour service," *Corrosion '91*, Paper No. 20. NACE International, Houston, Tx, USA, 1991.
6. T. Mori, T. Okazawa, M. Ueda and T. Kushida, "Corrosion performance of super 13Cr steel in CO_2 environments containing a small amount of H_2S," *Proc. 2nd NACE Asian Conf. – Corrosion Asia 1994*. NACE International, Houston, Tx, USA, 1994.
7. H. Asahi, T. Hara, A. Kawakami and A. Takahashi, "Development of sour resistant modified 13Cr OCTG," *Corrosion '95*, Paper No. 79. NACE International, Houston, Tx, USA, 1995.
8. M. Ueda, T. Kushida and T. Mori, "Evaluation of SSC resistance on super 13Cr stainless steel in sour applications," *Corrosion '95*, Paper No. 80. NACE International, Houston, Tx, USA, 1995.
9. Y. Miyata, Y. Yamane, O. Furukimi, H. Niwa and K. Tamaki, "Corrosion of new 13Cr martensitic stainless steel OCTG in severe CO_2 environment," *Corrosion '95*, Paper No. 83. NACE International, Houston, Tx, USA, 1995.
10. H. Asahi, T. Hara and M. Sugiyama, "Corrosion performance of modified 13Cr OCTG," *Corrosion '96*, Paper No. 61. NACE International, Houston, Tx, USA, 1996.

11. G. Fierro, G. M. Ingo and F. Mancia, "XPS investigation on the corrosion behaviour of 13Cr martensitic stainless steel in CO_2–H_2S–Cl^- environments," *Corrosion '88*, Paper No. 215. NACE International, Houston, Tx, USA, 1988.
12. K. Tamaki, "A new 13Cr OCTG for high temperature and high chloride CO_2 environments," Corrosion '89, Paper No. 469. NACE International, Houston, Tx, USA, 1989.
13. T. F. Illson and R. Greenwood, "Autoclave testing of 13Cr and modified 13Cr stainless steels in simulated field conditions," *Corrosion '95*, Paper No. 87. NACE International, Houston, Tx, USA, 1995.
14. M. S. Cayard and R. D. Kane, Compilation of 13Cr Data Generated at *InterCorr* International, Inc. through November 1995.
15. M. Ueda, T. Kushida, K. Kondo and T. Kudo, "Corrosion resistance of 13Cr–5Ni–2Mo martensitic stainless steel," *Corrosion '92*, Paper No. 5. NACE International, Houston, Tx, USA, 1992.
16. S. Hashizume, K. Masamura and Y. Ishizawa, "Corrosion data on laboratory performance of Ttype 420 for OCTG," *Corrosion '95*, Paper No. 77. NACE International, Houston, Tx, USA, 1995.
17. M. Ueda and A. Ikeda, "Effect of microstructure and Cr content in steel on CO_2 corrosion," *Corrosion '96*, Paper No. 13. NACE International, Houston, Tx, USA, 1996.
18. R. D. Kane, D. Abayarathna and M. J. Schofield, "Final Report: Safe Use Limits for Stainless Steels in Production of Oil and Gas Containing H_2S and CO_2," Joint Industry Project, *InterCorr* International, Houston, Tx, USA, 1995.

Part 4

Corrosion Resistant Alloys

34
Localised Corrosion of some Selected Corrosion Resistant Alloys in the Presence of Very High Salinity

T. CHELDI and L. SCOPPIO
Eni Agip Exploration and Production, Italy
Centro Sviluppo Materiali S.p.A., Italy

ABSTRACT
Oil well reservoir water can have very high salinity values of more than 200 gL^{-1} as NaCl, and have weakly 'sweet' environments (pCO_2 = 0.2 MPa) and medium temperatures (100–110°C). Experimental work was carried out to investigate the corrosion resistance of some CRAs (Corrosion Resistant Alloys) in these conditions and to verify their suitability for tubing strings in wells. Autoclave exposures and potentiodynamic tests were conducted in simulated environments to verify the resistance to uniform and localised (pitting, crevice) corrosion. Results showed that duplex and superduplex stainless steels present good resistance to pitting and crevice corrosion. Supermartensitic steels are affected by rare pitting; potentiodynamic test results revealed significant differences between the vaious tested supermartensitic steels showing differing pitting resistances.

1. Introduction

In Africa, several oil fields have similar environmental aggressiveness which is mainly characterised by the very high salinity of the reservoir water. Well conditions of some selected oil fields are summarised in Table 1.

Based on literature data [1–3], the available material selection expert systems [4,5]

Table 1. *Some selected field conditions characterised by very high salinity reservoir water*

Geographical Area	B.H.T. (°C)	B.P.P. (MPa)	B.H.P. (MPa)	CO_2 in gas phase at BPP (%mol.)	Formation water salinity (gL^{-1})	*in situ* pH
Algeria (on-shore)	90	14	32	1	250–330	4.5
Congo (off-shore)	100–120	26–29	26–32	0.8–1	190	4.6
Congo (off-shore)	80–90	13–20	22	3–3.5	250–300	3.7–3.9(1)
Libya (on-shore)	120	21	21	3.8	250	/ /(2)

B.H.T. = bottom hole temperature; B.P.P. = bubble point pressure; B.H.P. = bottom hole pressure.
(1) *In situ* pH values have been calculated on the basis of dissolved CO_2, ignoring any buffering effects from bicarbonate ions as no data were available on water chemical composition.
(2) No data were available on water chemical composition.

and manufacturers engineering field diagrams and guidelines, it was concluded that materials as noble as duplex stainless steel 22 or 25%Cr or higher, had to be selected to avoid corrosion in such environments. Carbon steel use is considered a high risk because of the potentially high corrosivity of the produced water; the consequent high corrosion rates mainly depend on the water cut, water chemical composition and on the oil filming efficiency. Other solutions to corrosion mitigation, such as internally coated carbon steels have been considered to be unsafe for production environments where damage to the coatings during wire line operations cannot be excluded in advance [6].

Martensitic stainless steels are normally considered safe up to a medium value of 80–150 gL^{-1} NaCl depending on the temperature [4]. The following empirical formulae are also used to provide limits for the applicability of martensitic and supermartensitic stainless steels, in the absence of H_2S, depending on salinity and temperature:

↓ **if** | **No** | **No**

NaCl (gL^{-1}) < 130–2/3 T(°C) → NaCl (gL^{-1}) < 200–8/15 T(°C) → higher resistant steels

↓ | ↓

yes | **yes**

martensitic SS13Cr | supermartensitic SS S13Cr or 15Cr

Nickel Development Institute (NIDI) guidelines [7] limits the use of martensitic 13%Cr to a maximum temperature of about 80°C with 250 gL^{-1} NaCl depending on CO_2 content; the temperature limit decreases if the NaCl content is increased.

The behaviour of the modified martensitic and duplex stainless steels at high chloride contents (> 20 % NaCl) has not yet been studied in detail [8]. In 200 gL^{-1} NaCl, low H_2S (0.005 MPa) conditions, pitting occurs in the range 70–130°C on martensitic steel and slight general corrosion and severe general corrosion are observed in the range 130–180°C and above 180°C respectively [6].

Experimental work has been carried out to investigate the corrosion behaviour of the most used CRAs (Corrosion Resistant Alloys) for well completions in slightly sweet environments, i.e. as characterised by low CO_2 contents, medium temperatures of up to 120°C, very high salinity values > 200 gL^{-1}, with the objective of selecting the most cost-effective CRAs.

2. Experimental

2.1. Materials

All samples used in this investigation were taken from production tubing of oil country tubular goods. For this study martensitic and some supermartensitic stainless steels provided by different steel makers (indicated as 'A', 'B' and 'C' in Table 2), were chosen. Supermartensitic stainless steels (SM) were characterised, from the corrosion resistance point of view, by a slight difference in Mo and Ni contents.

Duplex 22%Cr (D22), 25%Cr (D25) and a superduplex stainless steel (SD) alloyed with 0.7% Cu and 0.7% W were also tested. Carbon steel (CS) was selected as a reference material. The chemical compositions of the tested materials are shown in Table 2.

Table 2. *Chemical composition of the tested steels (wt%) (— nil or not tested)*

Steel	UNS	C	Mn	Cu	Ni	Cr	Mo	W	N
CS		0.290	0.67	0.17	0.10	1.07	0.53	—	0.01
13Cr	S42000	0.200	0.81	<0.05	0.19	13.70	0.02	—	<0.05
15Cr 'B'	S42500	0.110	0.76	<0.05	1.55	14.87	1.55	—	0.08
15Cr 'C'	S42500	0.110	0.76	<0.05	1.55	14.59	0.50	—	0.08
SM 'A'	S41424	0.020	0.45	0.08	5.50	13.50	2.06	—	0.05
SM 'B'	S41425	0.019	0.73	0.09	4.75	13.29	1.62	—	0.08
D22	S31803	0.023	1.41	0.24	5.40	21.80	2.90	—	0.13
D25	S31260	0.023	1.25	0.24	6.60	24.40	2.70	—	0.18
SD	S32760	0.020	0.80	0.70	7.50	25.00	3.80	0.70	0.27

2.2. Corrosion Performance

The resistance to general and localised corrosion such as pitting and crevice was investigated in two conditions simulating oilfield environments of increasing chloride contents. Two methods were utilised — namely autoclave long term exposure and electrochemical potentiodynamic tests.

2.3. Autoclave Testing

Mass loss measurements were obtained from flat coupon specimens taken from the mid-thickness of the tubes. Specimens were ground with No. 800 emery paper, rinsed with deionised water and then cleaned with acetone. The average corrosion rate in mm/year was calculated following the ASTM G-1 test procedure.

Localised corrosion was investigated using flat coupons and Anderson multicrevice specimens assembly according to ASTM G-78. After exposure in autoclaves the number of corroded sectors for each surface of the sample was determined.

The chemical composition of the solutions and autoclave testing conditions used to simulate the oilfield environments are shown in Table 3. Tests were carried out at a total pressure of 10 MPa obtained by adding nitrogen.

2.4. Electrochemical Potentiodynamic Tests

2.4.1. Pitting

Pitting corrosion behaviour was studied by means of the potentiodynamic technique complemented with optical microscopy.

Table 3. *Environmental conditions chosen for the autoclave exposure tests*

Environment	NaCl solution (gL^{-1})	Temperature (°C)	Exposure time (h)	p_{CO_2} (MPa)	P_{tot} (MPa)
a	330	115	720	0.2	10
b	315	115	720	0.2	10

Critical current density (i_{cp}), passivity potential (E_{pp}), passive current density (i_p), pitting potential (E_p), transpassivity (E_t) and corrosion potential (E_c) were determined for each test.

Specimens were polarised from a cathodic potential E_c = –1.0 V in the anodic direction until pitting occurred. Pitting potential values were taken as the last value at which the current was as low as that of a completely passive specimen.

The environment for the pitting corrosion experiments was a deaerated NaCl solution (330 gL^{-1}). Measurements were made at 90°C with a conventional glass cell.

A scanning rate of 1 mVs^{-1} was chosen for all the tests. Tests were considered completed when either the transpassivity E_t or the pitting potential E_p value was reached.

2.4.2. Crevice

The potentiodynamic method developed by Crolet *et al.* [9] was used in order to assess the crevice corrosion resistance of the selected materials. This is based on determining a depassivation pH, that is, the pH at which the passive film no longer offers protection against corrosion. Polarisation curves were a deaerated 330 gL^{-1} NaCl solution at 90°C at various pH values. The height of the active anodic peak was plotted as a function of pH. The pH values at which the active peak reached a nominal value of 0.1 Am^{-2} was taken as the depassivation pH [10].

3. Results and Discussion

3.1 Autoclave Exposures

3.1.1. Mass loss

The average corrosion rates of the flat coupons exposed to the environmental (see Table 3) conditions in the autoclave are reported in Table 4. Higher mass loss values were obtained in the simulated oilfield solution ***a*** where the chloride content was 330 gL^{-1}.

Carbon steel coupons were affected by general corrosion and high values of the mass loss were obtained: 450–459 mm/year and 172–210 mm/year for solution ***a*** and ***b*** respectively.

As expected [11,12], the corrosion rate of supermartensitic and duplex stainless steels was practically nil, i.e. lower than 4 mm/year for all the coupons exposed in both environments (Table 4).

3.1.2. Crevice corrosion

Carbon steel was not considered since it was affected by general corrosion.

Specimens of duplex and superduplex stainless steel exposed in the simulated environment were only slightly affected by crevice corrosion. (Fig. 1).

Specimens of 13%Cr, 15%Cr and supermartensitic steels were prone to crevice corrosion (4–6 out of 80 corroded sectors).

However, it is worthy of note, that in all the examined cases, the affected sectors were only partially corroded.

3.1.3. Pitting corrosion

Table 5 summarises the visual inspection of the pitting corrosion specimens after the test completion.

Table 4. *Results of the autoclave exposure tests: mass loss determination*

Steel	Environment a		Environment b	
	Multicrevice corrosion rate (μm/y)	Flat coupons corrosion rate (μm/y)	Multicrevice corrosion rate (μm/y)	Flat coupons corrosion rate (μm/y)
CS	450	459	210	172
13Cr	Not tested	Not tested	59	94
15Cr 'B'	Not tested	Not tested	24.7	17.7
15Cr 'C'	Not tested	Not tested	17.7	16.5
SM 'A'	1.3	1.4	0	0
SM 'B'	0.5	0.7	0	0.1
D22	Not tested	Not tested	4.2	3.4
D25	0.2	0	0	0
SD	0.1	0	0	0

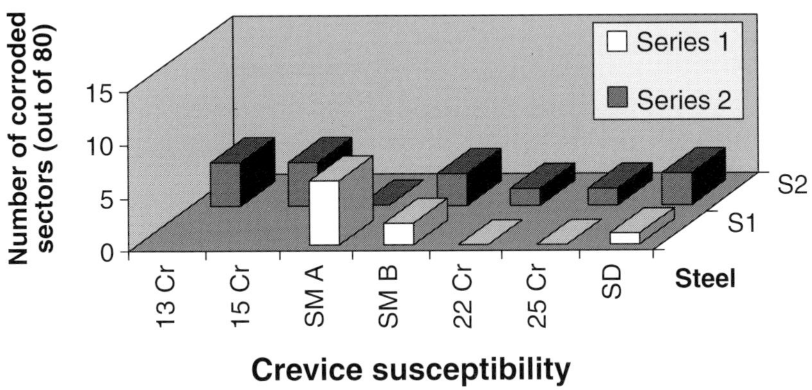

Fig. 1 *Crevice corrosion susceptibility as a function of the environment salinity. Number of corroded sectors out of 80 (series 1 = 330 gL^{-1} NaCl (**a**), series 2 = 315 gL^{-1} NaCl (**b**)).*

All the specimen surfaces were very shiny except for carbon steel and martensitic stainless steel specimens.

Diffuse pitting was seen on 13%Cr steel.

Supermartensitic stainless steel specimens exposed to environment ***a*** were slightly prone to pitting: both SM 'A' and SM 'B' showed very few pits on at least one of the specimen surfaces.

Pit depth was evaluated by means of the stereomicroscope in order to determine the seriousness of the localised corrosion attack. Pits were mainly 5–10 μm deep and as a consequence the pitting corrosion rate was estimated to be 60–120 μm/y.

Samples of duplex (D25) and superduplex stainless steels (SD) showed good resistance to pitting corrosion.

Table 5. Pitting corrosion susceptibility of the tested steels exposed in the autoclave

	Environment a		Environment b	
Steel	pitting (side 1)	pitting (side 2)	pitting (side 1)	pitting (side 2)
CS	General corrosion	General corrosion	General corrosion	General corrosion
13Cr	Not tested	Not tested	Diffuse	Diffuse
15Cr 'B'	Not tested	Not tested	Rare	Rare
15Cr 'C'	Not tested	Not tested	Rare	Very rare
SM 'A'	No	Very rare	No	Very rare
SM 'B'	Very rare	Rare	No	Very rare
D22	Not tested	Not tested	No	No
D25	No	No	No	No
SD	No	No	No	No

3.2. Potentiodynamic Tests

Potentiodynamic tests were performed on some duplicate material (see Table 6).

Duplex stainless steel (D25) had the highest observed E_p value, namely 991 mV, confirming an excellent resistance of the passive film to localised breakdown by pitting.

Superduplex stainless steel (SD), although characterised by a higher PREN, had an E_p value (804 mV) lower than D25, possibly due to the presence of some inclusions in the microstructure.

Both supermartensitic stainless steel SM 'A' and SM 'B' pitting potential values lower than those of duplex stainless steels, the mean values being 30 and –47 mV, respectively. A significant difference of around 100 mV was observed between the two steels; SM 'A' showing the better pitting resistance. The anodic polarisation curves of the tested steels are shown in Fig. 2.

Figure 3 shows the ranking of the tested steels based on the depassivation pH value. The higher the pH at which the passive film breaks down, the less resistant is the material to crevice corrosion.

Decidedly higher crevice corrosion resistance was shown by the stainless duplex 25%Cr steel (D25). Some difference was observed between the two supermartensitic stainless steels in that SM 'A' was less susceptible to crevice than SM 'B' (Fig. 3).

The depassivation pH of 15%Cr steels was approximately one pH unit greater than the values for supermartensitic steels.

Table 6. Potentiodynamic test potential values (E_{corr}, E_p) and current density (I_{pass})

Steel	E_{corr} (mV)		E_p (mV)		I_{pass} (mAm^{-2})	
SM 'A'	–650	–680	10	50	19	17
SM 'B'	–563	–720	–50	–44	17	19
D25	–650	–700	991	991	15	15
SD	–617	–630	804	804	13	13

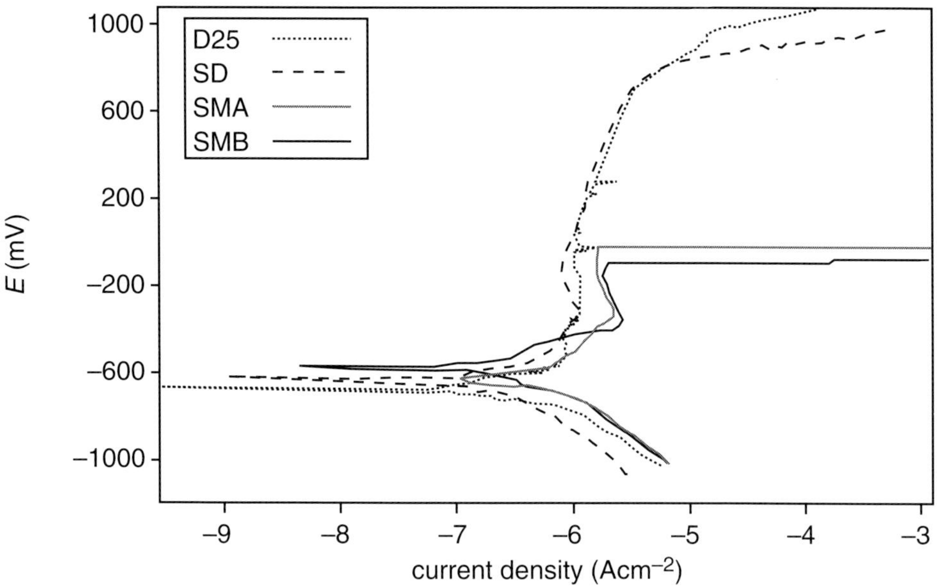

Fig. 2 *Anodic polarisation curves of supermartensitic steels (SM 'A' and SM 'B'), duplex (D25) and superduplex steels (SD).*

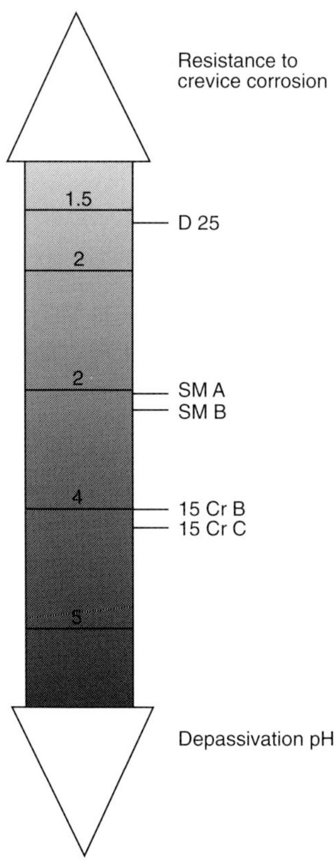

Fig. 3 *Critical depassivation pH of the tested steels in 330 gL^{-1} NaCl, 90°C.*

3.3. Discussion

All the stainless steel specimens examined after autoclave exposure in simulated oil field environments showed no general corrosion.

Martensitic stainless steels showed a high susceptibility to localised corrosion in these environments. The 15Cr steels showed some tendency to localised corrosion with pitting and crevice corrosion being detected. Critical depassivation pH values are in fact higher than those of supermartensitic stainless steels.

Supermartensitic stainless steels present a slight susceptibility to localised attack and proneness to pitting and crevice corrosion should be taken into account. Resistance to pitting attack of steel 'A' was slightly superior to that of supermartensitic stainless steel 'B'.

In particular, the positive effect of a higher molybdenum content of the supermartensitic stainless steel 'A' is shown by potentiodynamic tests, highlighting the different film self-repairing capacity of the steels.

Duplex 25%Cr and superduplex stainless steels tested at 115°C and in high chloride solutions (concentration up to 330 gL^{-1} NaCl) gave good localised corrosion resistance. No susceptibility to pitting corrosion was observed and passive film resistance to pitting attack was satisfactory. As regards crevice corrosion resistance although a few examples of attack were observed in autoclave tests an excellent critical depassivation pH was found.

4. Conclusions

Experimental work has been carried out to investigate the corrosion behaviour of the most used CRAs for well completions in environments characterised by low CO_2 contents, medium temperatures up to 120°C, very high salinity values of > 200 gL^{-1}, with the purpose of selecting the most cost-effective material. Based on the results the following conclusions have been drawn:

- 13%Cr and 15%Cr steels are not to be considered safe and applicable in the tested environments because of their proneness to localised corrosion.

- supermartensitic stainless steels can be used up to a salinity value of about 300 gL^{-1} in a temperature range of 100–120°C with a low content of CO_2. However, in these environmental conditions supermartensitic stainless steels are not completely immune to localised corrosion.

- duplex and superduplex stainless steels have higher corrosion resistance.

It is worthy of note that these environmental conditions are the 'border line' and consequently the chemical composition which characterises the various martensitic steels will affect the corrosion resistance mainly towards localised corrosion.

5. Acknowledgements

The authors would like to thank ENI-Agip Division for the kind permission to publish the results presented in this paper and steel makers for the supply of tubular materials for testing.

References

1. M. Barteri *et al.*,"Conservative Criteria for Application of Stainless Alloys in Gas & Oil Sour Wells" *Innovation Stainless Steels Conference*, Florence, Italy, 1993.
2. M. Schofield *et al.*, "Comparison of Duplex and Modified 13Cr Martensitic Stainless Steels in High Temperature, Slightly Sour Conditions" *Proc. Int. Conf. on Duplex Stainless Steels'97*, The Hague, The Netherlands, 97.
3. M. Verneau *et al.*, "Evaluation of the Corrosion Resistance of Duplex Stainless Steel in H_2S and Chloride Containing Media" *Proc. Int. Conf. EUROCORR '96*, Paper 6, Nice, France. Cefracor/Soc. Chim. Ind., Paris 1996.
4. A. Kopliku *et al.*, "An Expert System to Assist Corrosion Engineers in Material Selection for Well Completion", *Corrosion '97*, Paper No. 327, NACE International, Houston, Tx, 1997.
5. S. Srinivasan *et al.*, "Selection of Martensitic and Duplex Stainless Steels for Oil and Gas Production Service", *Corrosion '92*, Paper No. 56, NACE International, Houston, Tx, 1992.
6. D. Condanni *et al.*, "Carbon Steel Tubing Internally Coated with Resins: Mechanical and Corrosion Testing for Application in Corrosive Environments, *Corrosion '97*, Paper No. 65, NACE International, Houston, Tx, 1997.
7. B. D. Craig, "Selection Guidelines for Corrosion Resistant Alloys in the Oil and Gas Industry, NiDi Technical Series No. 10 073.
8. C. D. S. Tuck *et al.*, "The Influence of Copper on Pitting Corrosion of Duplex Stainless Steels in Saline and Acid Media" *Proc. Int. Conf. on Duplex Stainless Steels '97*, The Netherlands, 1997.
9. J .L. Crolet, J. M. Defranoux, L. Seraphin and R. Tricot, *Metaux-Corros. Ind.*, 1975, **599**, 262.
10. J. Oldfield, "Test Techniques for Pitting and Crevice Corrosion Resistance of Stainless Steel and Nickel base Alloys in Chloride-containing Environments", NiDi Technical series No. 10016.
11. S. Huizinga *et al.*, "Limitation for the Application of 13Cr Steel in Oil and Gas Production Environments", *Corrosion '97*, Paper No. 39, NACE International, Houston, Tx, 1997.
12. P. J. Cooling *et al.*, "The Application Limits of the Alloyed 13%Cr Tubular Steels for Downhole Duties", *Corrosion '98*, Paper No. 94, NACE International, Houston, Tx, 1998.

35
High-Strength Corrosion Resistant Nickel-Base Alloys for Oilfield Applications

E. L. HIBNER, C. S.TASSEN and P. W. RICE*
Inco Alloys International, Inc., USA
*Halliburton Energy Services, USA

ABSTRACT
In selecting materials for corrosive sour oilfield environments, the materials of choice must be reliable and cost-effective. Materials have to meet criteria for corrosion resistance and mechanical properties in service environments for the required service life. Both precipitation hardened nickel-base alloys and cold-worked solid solution nickel-base alloys offer many advantages such as high-strength, toughness, galling resistance and excellent corrosion resistance. The mechanical properties and corrosion resistance of high-strength nickel-base Corrosion Resistant Alloys (CRAs) are discussed.

1. Introduction

The selection of Corrosion Resistant Alloys (CRAs) for oilfield applications can be a complex procedure. If done improperly, an alloy selection can lead to poor performance of a CRA in a specific sour gas environment. Different alloy selection methods are used by individuals and companies when choosing a CRA for specific service environments [1]. A recognised selection procedure is to review the literature for corrosion data that, in general, apply to anticipated field conditions. Then, a group of alloys is selected that represents a range of alternatives. A test programme, simulating the subject field environment, is sometimes initiated. A final CRA selection is made for a specific application based on test results and an economic analysis of cost effective alternatives.

While more detailed testing and analysis are often required, guideline tables and diagrams are often used before extensive efforts are made to make a final alloy selection for a specific oilfield application. This paper presents corrosion data for cold-worked austenitic alloys including INCOLOY® alloy 825 (UNS N08825), INCO® alloy 25-6MO (UNS N08926), INCONEL® alloy 625 (UNS N06625), INCO alloy C-276 (UNS N10276), alloy 28 (UNS N08028) and age-hardened alloys such as INCOLOY alloy 925 (UNS N09925), INCONEL alloy 718 (UNS N07718) and INCONEL alloy 725 (UNS N07725)†.

†INCOLOY®, INCONEL® and INCO® are registered trademarks of the INCO group of companies.

2. Data Review

Tables 1–9 list sour oilfield environments from a literature review in which cold-worked Oil Country Tubular Goods (OCTGs) and age-hardened CRAs have either been recommended or where corrosion testing has validated their use. Results are generally based on stress corrosion cracking and sulfide stress cracking hydrogen embrittlement data.

Table 1. *Environments in which Alloy 825 has been reported as acceptable*

Reference→	[2]	[3]	[3]	[3]	[5]	[7]
Cl^- (ppm)	151750	NRE*	NRE	NRE	100 000	150 000
pH	–	–	–	–	3.5	3.5
Temp. (°C)	200	175	220	230	205	205
H_2S (MPa)	6.0	1.4	0.7	0.2	0.69	0.69
CO_2 (MPa)	–	NRE	NRE	NRE	2.76	2.76
S°	–	0	0	0	0	0

– no data available; NRE no reported effect; 0 determined as zero; S° elemental sulfur.

Table 2. *Environments in which Alloy 825 has been reported as acceptable*

Reference→	[2]	[3]	[3]	[3]
Cl^- (ppm)	151 750	NRE	NRE	NRE
pH	–	–	–	–
Temp. (°C)	200	230	190	150
H_2S (MPa)	6.0	1.0	3.5	NRE
CO_2 (MPa)	–	NRE	NRE	NRE
S°	–	0	0	0

Table 3. *Environments in which Alloy C-276 has been reported as acceptable*

Reference→	[2]	[3]	[3]	[6]	[7]	[7]
Cl^- (ppm)	NRE	NRE	NRE	151 750	121 400	121 400
pH	–	–	–	3.1	3.0	3.1
Temp. (°C)	260	205	230	230	230	230
H_2S (MPa)	66.0	NRE	1.0	0.83	6.9	0.5
CO_2 (MPa)	–	NRE	NRE	NRE	4.8	4.8
S°	–	0	0	Yes	Yes	0

Yes = known to be present.

Table 4. Environments in which Alloy 25-6MO has been reported as acceptable

Reference→	[2]	[2]	[2]	[3]	[3]	[8]
Cl^- (ppm)	121 400	12 140	12 140	NRE	NRE	60 700
pH	–	–	–	–	–	3.3
Temp. (°C)	250	200	150	150	170	120
H_2S (MPa)	0.0	0.14	0.27	0.3	0.1	0.7
CO_2 (MPa)	–	–	–	NRE	NRE	1.4
S°	–	–	–	0	0	0

*NRE = No reported effect.

Table 5. Environments in which Alloy 28 has been reported as acceptable

Reference→	[2]	[2]	[3]	[3]	[3]
Cl^- (ppm)	37 634	15 175	NRE	NRE	NRE
pH	–	–	–	–	–
Temp. (°C)	100	204	175	220	230
H_2S (MPa)	0.5	1.31	1.4	0.7	0.2
CO_2 (MPa)	–	–	NRE	NRE	NRE
S°	–	–	0	0	0

*NRE = No reported effect.

Table 6. Environments in which Alloy G-3 has been reported as acceptable

Reference→	[3]	[3]	[3]	[9]
Cl^- (ppm)	NRE	NRE	NRE	151 750
pH	–	–	–	3.3
Temp. (°C)	230	190	150	220
H_2S (MPa)	1.0	3.5	NRE	2.1
CO_2 (MPa)	NRE	NRE	NRE	2.1
S°	0	0	0	0

*NRE = No reported effect.

For references [2] and [3], H_2S limits are based on the presence of a significant concentration of chloride salts in the aqueous phase. It is recognised that alloys exposed to environments with little or no chloride may be able to tolerate higher H_2S partial pressures. Appropriate testing and/or available test data are required to identify the effects of these environments.

Most alloys have demonstrated corrosion resistance at 230°C, depending on the chloride concentration, H_2S partial pressure, and the presence of elemental sulfur [4]. The cold-worked alloys ranked by corrosion resistance are alloy C-276/ > alloys

***Table 7(a)**. Environments in which Alloy 718 has been reported as acceptable*

Reference→	[4]	[4]	[4]	[4]	[4]	[4]
Cl^- (ppm)	151 750	151 750	60 700	151 750	151 750	151 750
pH	3.13	3.13	3.19	3.13	3.13	3.14
Temp. (°C)	148.9	148.9	148.9	121.1	148.9	148.9
H_2S (MPa)	0.34	2.76	2.76	0.69	0.69	0.34
CO_2 (MPa)	4.83	2.76	1.38	4.83	4.83	4.83
S°	–	–	–	–	–	–

*NRE = No reported effect.

***Table 7(b)**. Environments in which Alloy 718 has been reported as acceptable*

Reference→	[4]	[2]	[3]	[3]	[3]	[3]
Cl^- (ppm)	151 750	91 050	NRE	NRE	NRE	NRE
pH	3.15	–	–	–	–	–
Temp. (°C)	148.9	150	175	205	220	230
H_2S (MPa)	0.17	1.4	1.4	1.0	0.7	0.2
CO_2 (MPa)	4.83	–	NRE	NRE	NRE	NRE
S°		–	0	0	0	0

*NRE = No reported effect.

***Table 8(a)**. Environments in which Alloy 925 has been reported as acceptable*

Reference→	[4]	[4]	[4]	[4]	[4]	[4]	[4]
Cl^- (ppm)	151 750	151 750	60 700	151 750	91 050	151 750	151 750
pH	3.13	3.13	3.19	3.13	3.13	3.13	3.14
Temp. (°C)	148.9	148.9	148.9	121.1	148.9	148.9	148.9
H_2S (MPa)	0.34	2.76	2.76	0.69	0.69	0.69	0.34
CO_2 (MPa)	4.83	2.76	1.38	4.83	4.83	4.83	4.83
S°	0	0	0	0	0	0	0

*NRE = No reported effect.

***Table 8(b)**. Environments in which Alloy 925 has been reported as acceptable*

Reference→	[4]	[2]	[3]	[3]	[3]	[3]	[9]
Cl^- (ppm)	151 750	91 050	NRE	NRE	NRE	NRE	300 000
pH	3.15	–	–	–	–	–	3.1
Temp. (°C)	148.9	150	175	205	220	230	182
H_2S (MPa)	0.17	1.4	1.4	1.0	0.7	0.2	6.8
CO_2 (MPa)	4.83	–	NRE	NRE	NRE	NRE	2.9
S°	0	–	0	0	0	0	0

*NRE = No reported effect.

Table 8(c). *Environments in which Alloy 925 has been reported as acceptable*

Reference→	[9]	[9]	[9]	[9]
Cl^- (ppm)	99 000	Saturated	Condensed	63 000
pH	5.0	3.3	3.4	3.2
Temp. (°C)	177	199	105	190
H_2S (MPa)	6.2	2.3	0.3	2.5
CO_2 (MPa)	3.1	1.5	0.9	3.3
S°	0	0	0	0

*NRE = No reported effect.

Table 9. *Environments in which Alloy 725 has been reported as acceptable*

Reference→	[3]	[3]	[3]	[10]	[10]	[10]	[10]
Cl^- (ppm)	NRE	NRE	NRE	100 000	250 000	250 000	151 750
pH				3.3	3.0	3.0	3.1
Temp. (°C)	230	190	150	220	205	175	175
H_2S (MPa)	1.0	3.5	NRE	1.4	4.1	8.3	2.1
CO_2 (MPa)	NRE	NRE	NRE	1.4	4.8	4.8	4.8
S°	0	0	Yes	Yes	0	0	Yes

*NRE = No reported effect; Yes = known to be present.

625 and G-3 > alloy 825 > alloy 28 > alloy 25-6MO. The precipitation hardened alloys ranked by corrosion resistance are alloy 725 > alloy 925 > alloy 718.

Tables 1–9 show service environments where the literature has indicated acceptable corrosion resistance for the austenitic alloys listed.

The precipitation hardened alloys are used at different strength levels depending on the application, but generally alloy 925 is used at 758 MPa (110 ksi) specified minimum yield strength and alloys 725 and 718 are used at 827 MPa (120 ksi) specified minimum yield strength.

The cold-worked solid solution alloys are also used at various strengths. Of the standard OCTGs, alloys 825 and 28 are used at 758 MPa (110 ksi) specified minimum yield strength and alloys G-3 and C-276 are used at 862 MPa (125 ksi) specified minimum yield strength.

3. Summary

Ultimately, it is the user's responsibility to establish the acceptability of an alloy for a specific environment. This paper presents data from a literature review intended for use in selecting materials for corrosive sour oilfield environments. A group of alloys that represents a range of alternatives can be selected for testing in an environment simulating the field environment under study. A final CRA selection is made for a specific application based on test results and an economic analysis of cost effective alternatives.

The manufacturers of equipment and components will also have a data bank of previous service recommendations which can be an excellent aid in determining the candidate alloys for particular service conditions.

References

1. B. D. Craig, Selection Guidelines for Corrosion Resistant Alloys in the Oil and Gas Industry, NiDi Technical Series No. 10 073, Toronto, Ontario, Canada, July, 1995.
2. L. M. Smith *et al.*, Material Selection for Gas Processing Plant, Stainless Steel Europe, 1995, Jan./Feb., p.21–31.
3. Draft No. 1, Sulfide Stress Cracking and Stress Corrosion Cracking Resistant Metallic Materials for Oilfield Equipment, NACE International, Houston, Tx, USA, 1997.
4. R. B. Bhavsar and E. L. Hibner, Evaluation of testing techniques for selection of corrosion resistant alloys for sour gas service, *Corrosion '96*, Paper No. 59, NACE International, Houston, Tx, USA, 1996.
5. E. L. Hibner *et al.*, Effect of alloy content vs PREN on the selection of austenitic oil country tubular goods for sour gas service, *Corrosion '98*, Paper No. 106, NACE International, Houston, Tx, USA, 1998.
6. Inco Alloys International Inc., Technical Bulletin on Corrosion Resistant Alloys for Oil and Gas Production 1996.
7. R. H. Moeller *et al.*, Large diameter cold-worked C-276 for downhole equipment, *Corrosion '91*, Paper No. 30, NACE, Houston, Tx, USA, 1991.
8. Standard Materials Requirement MR0175, Sulfide Stress Cracking Resistant Metallic Materials for Oilfield Equipment, NACE International, Houston, Tx, USA, 1998.
9. Field Data from Halliburton Energy Services.
10. Laboratory Test Data from Inco Alloys International Inc.

36
Effect of Alloy Nickel Content vs Pitting Resistance Equivalent Number (PREN) on the Selection of Austenitic Oil Country Tubular Goods for Sour Gas Service

E. L. HIBNER, C. S. TASSEN and J. W. SKOGSBERG*
Inco Alloys International, Inc., USA
*Chevron Research and Technology Co., USA

ABSTRACT
Traditionally, in the selection of Oil Country Tubular Goods (OCTG) for sour gas service, Corrosion Resistant Alloys (CRAs) are screened first by their pitting resistance equivalent number (PREN) and then by environmental cracking data generated in sour brine environments. The theory is that a pit occurs first, which provides a stress-raiser for initiation of anodic chloride stress corrosion cracking (SCC). Among the primary CRAs currently used in the cold worked condition for OCTG in sour gas wells are alloy 825 (UNS N08825) and alloy 28 (UNS N08028). While alloy 28 has a somewhat higher PREN than alloy 825, alloy 825 has a significantly higher nickel content. Slow strain rate (SSR) tests conducted in severe sour brine environments showed that the higher nickel content of alloy 825 results in better stress corrosion cracking resistance than that exhibited by alloy 28. The effect of nickel content on chloride SCC resistance of austenitic alloys was originally reported by Copson in 1959 [4]. This suggests that in some cases for austenitic alloys, the nickel content of the CRA may be more important than the PREN in OCTG selection.

1. Introduction

Corrosion resistant alloys (CRAs) have historically been screened in the selection of OCTG applications by a required minimum Pitting Resistance Equivalent Number (PREN) based on environmental severity and maximum operating temperature [1,2]. Typically, the Pitting Resistance Equivalent Number [1] is calculated using the following equation (eqn 1):

$$\text{PREN} = \%\text{Cr} + (3.3 \times \%\text{Mo}) + (11 \times \%\text{N}) + 1.5\,(\%\text{W} + \%\text{Nb}) \qquad (1)$$

The theory is that pitting occurs first in the sour brine environments, which provides stress raisers for anodic chloride stress corrosion cracking (SCC) to occur. This logic suggests that the elimination of pitting, by use of an alloy with sufficient PREN for the environmental severity, will prevent the initiation of SCC. However, the effect of alloying elements such as nickel on anodic chloride SCC resistance is not addressed.

For quite some time, alloy 825 and alloy 28 have been successfully used as cold worked OCTG in sour gas wells [3]. While alloy 825 has a PREN of 31 compared to a PREN of 38 for alloy 28, alloy 825 has a significantly higher nickel content of 42% compared to only 31% for alloy 28. The beneficial effect of nickel content on chloride SCC resistance of austenitic alloys has been well established [4]. In the case of alloys 825 and 28, the nickel content of the CRA may be more important than the PREN in OCTG selection.

2. Experimental

Table 1 displays the chemical composition of alloy 825 and alloy 28. Material for testing came from commercially produced cold worked Oil Country Tubular Goods. Slow strain rate specimens were machined from the mid-wall area of the tube.

Slow strain rate tests (SSR) were conducted in accordance with the latest draft of the NACE International, Technical Activities Committee T-1F-9 standard on "Slow Strain Rate Test Method for Screening Corrosion Resistant Alloys (CRAs) for Stress Corrosion Cracking in Sour Oil Field Service." The strain rate was 4×10^{-6} s^{-1}.

Pitting tests were conducted on 25 mm × 50 mm specimens suspended in 3.2 L of corrodent contained in a 4 L autoclave.

C-ring specimens were prepared for testing as per the American Society for Testing and Materials (ASTM) Standard Practice G38. Duplicate C-rings were stressed to 95% of the longitudinal yield strength of the tubes and suspended in autoclaves. After 90 days, the autoclaves were opened and the samples were removed and carefully examined in an attempt not to disturb the corrosion scale. They were then returned to the autoclave for another 90 days of exposure.

All autoclaves were evacuated and deaerated before adding the deaerated solutions. All specimens were suspended in the environment so that electrical isolation was maintained. Reagent grade chemicals and high purity gases were used. Temperature was measured in °F and converted to °C, and pressures were measured in psi and converted to MPa. After testing, specimens were examined under 20× magnification for localised corrosion.

3. Results and Discussion

In review, the most common pass/fail criteria for SSR testing is a ratio of Time To Failure (TTF), % Reduction of Area (%RA) and/or % Elongation (%El) measured in a simulated oilfield environment relative to the same parameter in an inert environment (air or nitrogen). These are referred to as "critical ratios". If the ratios

Table 1. *Nominal chemical composition (wt %)*

	Ni	Cr	Mo	Cu	Al	Ti	Fe
Alloy 825 (UNS N08825)	42.0	21.5	3.0	2.2	0.1	0.9	Bal.
Alloy 28 (UNS N08028)	31.0	27.0	3.5	1.0	–	–	Bal.

are below 0.90, the specimen is examined using the Scanning Electron Microscope (SEM) for evidence of ductile or brittle fracture of the primary fracture surface. Ductile behaviour passes and brittle behaviour fails. All specimens are examined for secondary cracking in the gauge length, away from the primary fracture. The absence of secondary cracking is indicative of good stress corrosion cracking resistance and passes. The presence of secondary cracks fails. An inert (air) SSR test is conducted along with two or more environmental SSR tests for each test lot [5]. The commonly accepted criteria of TTF, %RA and % El critical ratios of ≥ 80 mm typically represent pass behaviour in SSR tests and are based upon results obtained earlier for cold worked solid solution nickel-based alloys [6,7].

Tables 2 to 5 display SSR test data in environments containing from 100 000 ppm to 150 000 ppm Cl^- (as NaCl), with H_2S content varied from 0.207 to 0.690 MPa and with CO_2 content varied from 2.76 to 4.83 MPa at 121.1°, 148.9°, and 204.4°C. Alloy 825 displayed excellent SCC resistance, exhibiting critical ratios for TTF, %RA and %El of ≥ 0.85 in all environments, with no secondary cracking. Alloy 28 exhibited similar behaviour to alloy 825 when tested in the environments at 121.1° and 148.9°C. However, when tested in the two environments at 204.4°C, alloy 28 displayed unacceptable SCC resistance. That is, alloy 28 exhibited critical ratios significantly below the minimum 0.80 critical ratio for TTF, %RA and %El. With the higher nickel

Table 2. *SSR Tests data, environment: 100 000 ppm Cl^- (as NaCl) + 0.207 MPa H_2S + 4.83 MPa CO_2 at 121.1°C*

	TTF ratio	%RA ratio	%El ratio	Secondary cracking
Alloy 825	0.97	0.85	0.96	No
	0.98	0.93	0.93	No
	avg. 0.98	**avg. 0.89**	**avg. 0.98**	
Alloy 28	0.83	0.97	0.83	No
	0.86	0.97	0.83	No
	avg. 0.85	**avg. 0.97**	**avg. 0.83**	

Table 3. *SSR Tests data, environment: 150 000 ppm Cl^- (as NaCl) + 0.517 MPa H_2S + 2.76 MPa CO_2 at 148.9°C*

	TTF ratio	%RA ratio	%El ratio	Secondary cracking
Alloy 825	1.08	0.99	1.02	No
	1.10	1.04	1.04	No
	1.07	1.05	1.00	
	avg. 1.08	**avg. 1.03**	**avg. 1.02**	
Alloy 28	1.02	0.90	0.99	No
	1.01	0.92	0.97	No
	1.09	1.04	1.09	
	avg. 1.04	**avg. 0.95**	**avg. 1.02**	

Table 4. *SSR Tests data, environment: 100 000 ppm Cl^- (as NaCl) + 0.690 MPa H_2S + 2.76 MPa CO_2 at 204.4°C*

	TTF Ratio	%RA ratio	%El ratio	Secondary cracking
Alloy 825	1.06	0.98	1.08	No
	1.00	0.95	1.00	No
	avg. 1.03	**avg. 0.97**	**avg. 1.04**	
Alloy 28	0.88	0.49	0.84	No
	0.82	0.49	0.71	No
	avg. 0.86	**avg. 0.49**	**avg. 0.78**	

Table 5. *SSR Tests data, environment: 150 000 ppm Cl^- (as NaCl) + 0.690 MPa H_2S + 2.76 MPa CO_2 at 204.4°C*

	TTF ratio	%RA ratio	%El ratio	Secondary cracking
Alloy 825	1.04	0.97	0.98	No
	1.05	0.94	1.02	No
	1.04	0.98	1.00	
	avg. 1.04	**avg. 0.96**	**avg. 1.00**	
Alloy 28	0.94	0.55	0.78	No
	0.79	0.33	0.62	No
	avg. 0.87	**avg. 0.44**	**avg. 0.70**	

content of alloy 825 relative to alloy 28, the nickel content of the CRA may be more important than the PREN in OCTG selection.

Table 6 displays the pitting test data for alloy 825 and alloy 28 evaluated in a severe sour brine environment containing 100 000 ppm Cl^- (as NaCl) + 0.690 MPa H_2S + 2.76 MPa CO_2 at 400°F (204°C) for 30 days. Both alloys exhibited similar pit densities and general corrosion rates, but pit penetration was three times higher for alloy 28 than was observed for alloy 825. No beneficial effect of the higher PREN of alloy 28 was indicated.

Table 7 lists the composition of the three severe sour brine environments used in conducting long term C-ring tests of commercially produced alloy 825 OCTG from

Table 6. *Pitting test data for alloy 825 and alloy 28 evaluation in 100 000 ppm Cl^- (as NaCl) + 0.690 MPa H_2S + 2.76 MPa CO_2 at 204°C for 30 days*

	Pit density (pits/cm^2)	**Max. depth (mm)**	**Corrosion rate (mm/y)**
Alloy 825	1.0	0.013	<0.03
	0.2	0.013	<0.03
Alloy 28	1.1	0.038	<0.03
	0.8	0.038	<0.03

Table 7. Composition of environments used in conducting long term C-ring tests

Environment	NaCl	H_2S (MPa)	CO_2 (MPa)	Total Pressure*	Temp. (°C)
A	25%	0.0069	2.07	5000	218
B	25%	0.069	2.07	5000	218
C	25%	0.69	2.07	5000	218

* Bal. nitrogen.

three different manufacturers. Basically, the environments contained 25% NaCl with 2.07 MPa CO_2 and with H_2S contents of 0.069. 0.069 and 0.69 MPa at 218°C. No SCC was observed.

4. Conclusions

- Based on SSR data Alloy 825 (UNS N08825) was consistently more stress corrosion cracking resistant in severe sour brine environments than alloy 28 (UNS N08028).
- In a severe sour brine environment, no beneficial effect of the higher PREN of alloy 28 was indicated. That is, alloy 825 (PREN 31) was as resistant to pitting as alloy 28 (PREN 38).
- In the case of alloy 825 and alloy 28, the nickel content of the alloy may be more important than the PREN in OCTG selection.

References

1. *SOCRATES: Comprehensive Material Selection Tool for Oil and Gas Service*, CLI International, Houston, Tx, 1997.
2. G. Schmitt *et al.*, "Performance of CRA in concentrated brines at 130° to 180°C — Effect of H_2S, elemental sulfur and brine composition", *Corrosion '97*, Paper No. 33. NACE International, Houston, Tx, 1997.
3. "Survey of CRA Tubular Usage", prepared by NACE Work Group T-1F-9d, Houston, Tx, 1995.
4. H. R. Copson, "Effect of Composition on Stress Corrosion Cracking of Some Alloys Containing Nickel", in *Physical Metallurgy of Stress Corrosion Fatigue*, T. Rhodin, Ed. Interscience Publishers, New York, NY, USA, pp.247–269, 1959.
5. E. L.Hibner, "Improved SSR Test for Lot Acceptance Criterion,", in *Slow Strain Rate Testing for the Evaluation of Environmentally Induced Cracking: Research and Engineering Applications*, ASTM STP1210, R. D. Kane, Ed. American Society for Testing and Materials, West Conshohocken, Pa, 1993, p.290.
6. H. E. Chaung, M. Walkins and G. A. Vaughn, "Stress corrosion cracking resistance of stainless alloys in sour environments," *Corrosion '85*, Paper No. 277. NACE International, Houston, Tx, 1985.
7. M. Watkins, H. E. Chaung and G. A. Vaughn, "Laboratory testing of SCC resistance of stainless alloys," *Corrosion '87*, Paper No. 283. NACE International, Houston, Tx, 1987.

37
Effect of Grain Size on Stress Corrosion Cracking Resistance of Alloy G-3 (UNS N06985) OCTG in Sour Gas Environments

E. L. HIBNER and C. S. TASSEN
Inco Alloys International, Inc., USA

ABSTRACT

Limited corrosion data exist on the effect of grain size on the stress corrosion cracking (SCC) resistance of Oil Country Tubular Goods (OCTG) in sour gas environments. Among the primary Corrosion Resistant Alloys (CRAs) currently used in the cold worked condition for OCTG in sour gas wells is Alloy G-3 (UNS N06985). Slow strain rate (SSR) tests conducted in a severe sour brine environment representative of Mobile Bay Alloy conditions showed that the grain size of Alloy G-3 OCTG has no discernible effect on stress corrosion cracking resistance. Also, no effect of grain size on intergranular attack (IGA) susceptibility was observed in the Streicher Test, ASTM Standard Test Method G28A.

1. Introduction

With the increased use of CRAs due to sour gas, there is a need to discern possible effects of grain size and/or microstructure on the corrosion resistance of OCTGs. Published data on the effect of grain size or microstructure on environmentally assisted cracking resistance of OCTGs have been limited to low alloy steels [1–3].

Alloy G-3 is among the CRAs currently used in severe sour gas well service such as Mobile Bay in the Gulf of Mexico. A study was initiated to determine if grain size has an effect on the SCC resistance and IGA susceptibility of extruded, annealed and cold worked Alloy G-3 OCTGs.

2. Experimental

Table 1 displays the nominal chemical composition of Alloy G-3. Material for testing came from commercially produced Oil Country Tubular Goods in the extruded,

Table 1. *Nominal chemical composition (wt%)*

	Ni	Cr	Mo	Cu	Co	W	Fe	Min. YS
Alloy G-3 (UNS N06985)	Bal.	22.2	7.0	1.9	2.0	0.8	19.5	120 ksi (827 MPa)

annealed and cold worked (tube reduced) condition. Slow strain rate (SSR) specimens were machined from the tube wall so as to maximise grain size variance.

Slow strain rate tests were conducted in accordance with the latest draft of the NACE International, Technical Activities Committee T-IF-9 standard on "Slow Strain Rate Test Method for Screening Corrosion Resistant Alloys (CRAs) for Stress Corrosion Cracking in Sour Oil Field Service". The strain rate was 4×10^{-6} s^{-1}.

The Streicher Test for intergranular attack susceptibility was conducted in accordance with the American Society for Testing and Materials (ASTM) Standard Test Method G28A.

3. Results and Discussion

In review, the most common pass/fail criteria for SSR testing are a ratio of Time to Failure (TTF), % Reduction of Area (%RA) and/or % Elongation (%El) measured in a simulated oil field environment relative to the same parameter in an inert environment (air or nitrogen). Depending on the alloy and the environment, a ratio of 0.90 or greater typically passes. If the ratio is between 0.80 and 0.90, the specimen is examined under the Scanning Electron Microscope (SEM) for evidence of ductile or brittle fracture of the primary fracture surface. A ratio below 0.80 typically fails. All specimens are examined for secondary cracking in the gauge length, away from the primary fracture. The absence of secondary cracking is indicative of good Stress Corrosion Cracking (SCC) resistance and passes. The presence of secondary cracks fails. In most cases, one inert SSR test is conducted along with two or three environmental SSR tests for each acceptance lot. If all the environmental tests pass the TTF, %RA and/or %El ratios (as required by specification) and exhibit no secondary cracking, the test lot of material passes.

ASTM Grain Size readings, taken from cross-sectional metallographic mounts of tested SSR specimens in the centre area representing the 3.94 mm gauge diameter, are listed in Table 2. Heats 1, 2 and 3 exhibited a fine, normal grain size distribution with an average ASTM grain size of 6 to $7\frac{1}{2}$ and with maximum grain sizes as large as 2 to 6. Heats 4 and 5 displayed coarse, wide range and bimodal grain size distribution with an average grain size of 4 to $7\frac{1}{2}$ and with maximum grain sizes as large as 0 to $4\frac{1}{2}$. This significant variation of heat to heat grain size was due to differences in processing history of the affected heat.

Table 3 lists the critical ratios for Alloy G-3 OCTG evaluated in deaerated 25% NaCl + 1.03 MPa H_2S + 1.03 MPa CO_2, tested at 218°C. This sour brine environment represents severe conditions in Mobile Bay.

Comparing Tables 2 and 3, heats listed as 1 to 5 are the same. Alloy G-3 displayed excellent SCC resistance, exhibiting critical ratios for TTF, %RA and %El of ≥ 0.92 in the severe environment, with no secondary cracking. No effect of grain size on slow strain rate test results was clearly discernible.

When evaluated in the Streicher Test for intergranular attack susceptibility, Alloy G-3 exhibited very low corrosion rates of 0.27 to 0.36 mm/y. The Chemical Process Industries allow a maximum corrosion rate of 0.61 mm/y for Alloy G-3 pipe and tubing. Again, no effect of grain size was clearly discernible.

Table 2. *ASTM grain size readings taken from cross-sectional metallographic mounts of tested SSR specimens, in the centre area representing the 3.94 mm gauge diameter*

Heat	SSR test type	Average grain size	Grain size distribution	As large as
1	Air	$6\frac{1}{2}$	Normal	5
	Environment	7	Normal	$5\frac{1}{2}$
		6	Normal	5
2	Air	$7\frac{1}{2}$	Normal	6
	Environment	$7\frac{1}{2}$	Normal	$5\frac{1}{2}$
		$7\frac{1}{2}$	Normal	2
3	Air	6	Normal	4
	Environment	$7\frac{1}{2}$	Normal	$3\frac{1}{2}$
		6	Normal	4
4	Air	$7\frac{1}{2}$	Bimodal, 70% $7\frac{1}{2}$, 30% 2	0
	Environment	6	Normal	$4\frac{1}{2}$
		$4\frac{1}{2}$	Bimodal, 80% $2\frac{1}{2}$, 20% 8	0
5	Air	4	Bimodal, 90% 3, 10% 8	1
	Environment	4	Wide range, 2 to 8	0
		$4\frac{1}{2}$	Normal	1
		$5\frac{1}{2}$	Wide range, $2\frac{1}{2}$ to 8	1

Table 3. *Critical ratios for alloy G-3 OCTG evaluated in deaerated 25% NaCl + 1.03 MPa H_2S (at 218°C) + 1.03 MPa CO_2 (at 218°C), tested at 218°C*

Heat	TTF Ratio	%El Ratio	%RA Ratio	Secondary Cracking
1	0.96; 0.92	0.96; 0.92	0.94; 0.97	No; No
2	1.07; 1.07	1.06; 1.08	1.00; 0.99	No; No
3	1.00; 0.96	1.00; 0.96	1.05; 1.04	No; No
4	0.92; 1.10	0.92; 1.10	0.97; 1.00	No; No
5	0.97; 1.10; 0.97	0.96;1.11; 0.93	0.95; 0.94; 0.97	No; No; No

4. Conclusions

No statistically significant effect of grain size on corrosion resistance of Alloy G-3 OCTG was discernible. The material exhibited excellent SCC resistance in slow strain rate tests in a severe sour brine environment similar to Mobile Bay type conditions and excellent resistance to intergranular attack in the Streicher Test.

References

1. T. Kaneko *et al.*, "Influence of microstructure on SCC susceptibility of low-alloy, high strength Oil Country Tubular Goods", *Corrosion*, 1989, **45** (1), 2–6.
2. A. Ikeda *et al.*, "Corrosion behavior of high-alloy Oil Country Tubular Goods for deep sour gas well", *Corrosion '84*, Paper No. 20. NACE International, Houston, Tx, 1984.
3. H. Morikawa *et al.*, "Metallurgical Solution to Hydrogen Embrittlement of High-Strength OCTG for Sour Service", *Proc. Symp. on Current Solutions to Hydrogen Problems in Steels*, pp. 219-224, Metals Park, OH, USA, ASM, 1982.

Part 5

Galvanic Corrosion

38
Galvanic Corrosion — Principles and Practice for Use of Titanium

D. K. PEACOCK
Head of European Technical Services (Industrial) TIMET
Chairman, Titanium Information Group

ABSTRACT
Titanium is normally the cathode in mixed metal structures. Guidelines for design and application based on the principles of avoidance or minimising galvanic corrosion do not necessarily address all of the conditions likely to be met in service. The offshore oil and gas industry in particular has provided a wide range of differing environments in which metal couples have to perform, frequently with different media on either side of the coupled component. Metals galvanically close to titanium may perform satisfactorily with titanium in one environment, but suffer corrosion in another. Titanium may also be coupled to materials more noble than itself. The increasing use of composites adds a new dimension to the concept and practice of galvanic coupling. The data base of field experience continues to grow and qualifications continue to be added to the basic guidelines in the light of practical findings. Recommendations for installation, inspection and maintenance of mixed metal systems often prove inadequate to prevent trouble over longer periods of service.

1. Introduction

An extremely wide variety of materials is in use today with the aim of providing optimum performance of components and systems in all fields of engineering. In certain working environments the incompatibility of dissimilar materials may result in problems, among the most common of which is galvanic corrosion. The basic simplicity of the concept of galvanic corrosion suggests that potential problems in design and application should be easily avoided. The abundant evidence of galvanic corrosion in everyday experience suggests however that this is not the case, and instances of failure continue to be reported. Many of these arise from an unexplored or unexpected influence of the working environment and not, as is more frequently supposed, simply due to the inappropriate or accidental coupling of dissimilar metals. When changes are made to the specification of metals and alloys or other materials including composites in existing systems, attention to galvanic compatibility within the working environment is of paramount importance. Experience shows that galvanic couples which are compatible under one set of operating conditions cannot be assumed to be safe even in what appear to be similar environments.

The increasing use of titanium to remedy basic material corrosion problems in a variety of hostile environments has in some instances led to further problems. Titanium is most frequently the noble metal in the structure and, as in offshore oil and gas extraction equipment and in maritime vessels, sea water and chlorides are omnipresent to provide an electrolytic path of relatively low resistance. Galvanic corrosion of less noble metals to which the titanium has been coupled whether due to oversight, accident, or unknowingly inappropriate installation or operational practice, continue to serve well to provide guidelines for correct practice. New designs should seek to minimise or eliminate the risks of galvanic corrosion from the start. This paper reviews the factors influencing galvanic corrosion, points to where problems may arise, and reviews strategies to avoid damage.

1.1. Requirements for Galvanic Corrosion

Galvanic corrosion may occur when two metals of sufficiently different electrode potential are coupled together directly or via an electrically conductive path in a working environment which provides an electrolytic path between the exposed areas of the one metal and the other. If one or other of these three principal requirements; potential difference, electrical connection, electrolytic path, is absent galvanic corrosion will not occur [1].

Galvanic corrosion in the presence of the three key requirements, may be indicated, but may not necessarily occur either at all, or to the extent predicted from the difference in potentials. Equally well, what are judged to be safe, 'passive' couples may under certain circumstances become active with substantial damage to one or other of the metals concerned.

Secondary factors which may modify the rate of galvanic corrosion include the following.

1.1.1. Relative cathode/anode areas

The relative surface area of the noble (cathodic) metal to that of the less noble (anodic) metal is frequently the dominant control of galvanic corrosion rate. As shown in Table 1 [2], if the cathode area is small and the anode is large, the damage to the less noble metal may be minimal. If, on the other hand, the cathode area is large, and the anode area is small, the corrosion of the anodic metal may be most severe, including deep pitting, due to concentration of the corrosive attack. It is for this reason that it is essential to take care when coupling titanium to a less noble metal if only that metal is coated. Any coating defects, damage or breakdown in localised areas will immediately cause rapid attack of the less resistant metal unless cathodic or chemical protection is available or unless the adjoining titanium structure is also coated, thus effectively reducing the area of the cathode [3].

In situations where the total surface area of titanium is large in relation to the adjoining base metal, it is important to establish what percentage of the exposed titanium will be effective in the galvanic cell. Tube in shell condensers and heat exchangers are a particular case in point, and the implications of relative areas are now well understood in terms of the effect of titanium tubes fitted in non-titanium tube sheets. The rapidity and severity of damage to unprotected tubeplates led to an objective evaluation of the effective area, which for titanium and other corrosion resistant materials proved in most cases to be the full length of the condenser tube, and certainly in excess of 12 m (40 ft) for standard 25.4 mm (1 in.) diameter tubes.

Table 1. *Acceleration factors for corrosion due to dissimilar metal coupling*

Coupled metal or alloy M	Uncoupled corrosion rate (mm/y)	Acceleration factors according to ratio of coupled area to titanium Ti:M		
		10:1	1:1	1:10
Zinc	0.05	35	4	2
Carbon steel	0.15	5	2	1
Nickel aluminium bronze	0.02	2	1	1
70/30 Cu Ni + 1% Fe	0.02	5	1	1
Inconel 625	<0.005	3	2	1
Titanium	<0.001	1	1	1

1.1.2. Nature and conductivity of the electrolytic path

Sea water and chloride brines are efficient electrolytes and galvanic currents will 'throw' over considerable distances for a metal couple exposed in sea water with substantially higher weight loss on the less noble metal as compared to exposure buried in soil or to the atmosphere.

1.1.3. Nature and kinetics of the anode and cathode reactions

The reactions taking place at the surface of each metal in the electrolyte concerned may stimulate or stifle a particular reaction, make more base or ennoble either metal and either increase, reduce or even eliminate galvanic corrosion. Important influences in this respect in sea water, are biofilming, the level of dissolved oxygen, chlorination, temperature, pollutants such as sulfides, illumination and water velocity over the anode and cathode surfaces [4]. The US Navy has reported that calcareous deposits reduce the effects of galvanic coupling. Titanium is easily polarised. At the most recent Sea Horse Conference in August 1998 at Wrightsville Beach, N.C., USA, a US Navy spokesman stated that this factor frequently means that titanium is less harmful in a galvanic couple than its potential would suggest.

1.1.4. The corrosion potentials of each metal in the environment

It follows from the preceding paragraph that the electrode potential of a metal, or of a galvanic couple has neither an absolute nor unvarying value. Potentials can change with time and conditions by several hundreds of millivolts. For many years data such as those shown in Table 2 which detail the potentials of various metals and alloys in flowing sea water, have been used to guide the designer towards the avoidance of galvanic corrosion. Revisions to this series have been developed over the past few years by several organisations. As shown in Table 3, different potentials and different relations of potentials, metal to metal, usually apply in other media [5]. Of even more importance, potentials in a single medium may not be constant, and may be variable under different conditions within an operating cycle. Most recently, work, which compares the spread of potential values for a range of metals and alloys in natural sea water at 10°C (50°F) and 40°C (104°F) shows that titanium and all of the corrosion resistant alloys (CRA) studied become more electronegative with temperature increase, and a much reduced potential difference exists between the CRAs and some of the less corrosion resistant metals at higher temperature [6].

Table 2. *Galvanic series based on potential measurements in flowing sea water at 35°C*

Material	Steady state electrode potential (v)
Graphite	+0.25
Zeron 100 super duplex steel	–0.01
316 stainless steel passive	–0.05
22% Cr duplex SAF2205	–0.07
Monel 400	–0.08
Hastelloy C	–0.08
Titanium	–0.1
13% Cr stainless steel passive	–0.15
316 stainless steel active	–0.18
Nickel aluminium bronze	–0.2
70–30 cupro-nickel	–0.25
13% Cr stainless steel active	–0.52
Carbon steel	–0.61
Aluminium	–0.79
Zinc	–1.03
Magnesium	–1.7

Table 3. *Potential measurements in various oil and gas industry environments*

Material	UNS	Sweet well brine[1]	Sour well brine[2]	Heavy brine packer fluid[3]	Acidising fluid[4]
Titanium*	R50400	–0.11	–0.28	–0.09	+0.37
Duplex 55	S31803	–0.17	–0.30	–0.17	–
Inconel 718	N07718	–0.16	–0.29	–0.19	+0.1
Low alloy steel	G41300	–0.17	–0.37	–0.29	–0.21
9 Cr steel	S50400	–0.26	–0.41	–0.29	–0.20

*Titanium Alloy ASTM Grade 19;

[1] 5% NaCl with 80 bar CO_2 at 150°C;

[2] 25% NaCl with 0.07 bar H_2S and 80 bar CO_2 at 200°C;

[3] Uninhibited, deaerated $CaCl_2$ brine 1200 gL^{-1}, 26 bar CO_2;

[4] Inhibited 15% HCl at 120°C.

2. Metallurgical Structure and Surface Stability

Titanium and its alloys are protected by a highly stable, substantially inert, tenacious and permanent titanium oxide film. This provides outstanding resistance to corrosion in a wide range of conditions in mildly reducing, neutral and oxidising environments. Titanium is immune to corrosion under many of the conditions normally encountered in marine service. The oxide film forms as equally on welds, heat affected zones and cast structures as on parent metal, and on the industrial alloys as on commercially pure titanium. In situations where the oxide film is maintained or its formation supported, titanium will survive erosion, cavitation, pitting and crevice corrosion over a wide range of pH and temperature. Commercially pure titanium is substantially a single phase alloy of metallurgical simplicity. Industrial alloys, such as Ti–6Al–4V, have a nominally two phase structure, but there is no electrochemical heterogeneity between the phases to give concern about selective corrosive attack [7].

3. Reducing Acid Attack of Titanium

Circumstances do exist where the titanium oxide film can be attacked notably by reducing acids. If the film cannot repair itself the titanium will also be attacked. If in these conditions the titanium is coupled to a nominally less corrosion resistant metal, such as iron or copper based alloys, the titanium will become the anode in the bi-metal system and the rate of attack will be substantially increased. One important requirement is that titanium equipment should not be cleaned with uninhibited reducing acid solutions, especially in a mixed metal system.

4. Micro-Galvanic Protection of Titanium against Crevice Corrosion

Localised corrosion caused by reducing acidic conditions created in threaded joints, tight crevices and under scale or other deposits is a controlling factor in the application of titanium as it is for many corrosion resistant alloys. Titanium is however one of the most resistant metals to crevice corrosion and attack will not occur on commercially pure titanium or industrial alloys below 70°C (158°F) regardless of solution pH. In chloride environments the crevice corrosion of titanium is more dependent upon temperature than chloride concentration and will normally not occur below 80°C (176°F). As solution pH rises the critical temperature also rises. Only at temperatures close to the boiling point do sea water and neutral brines develop reducing acidic conditions, where localised attack of titanium may occur.

Historically the most common resort, where crevice corrosion is a potential threat, has been to specify one or other of the titanium alloy grades containing palladium, or ruthenium, or ruthenium and nickel, or nickel and molybdenum. Alloys with 0.05% Pd or 0.1% Ru serve wholly to defeat all threats of crevice corrosion and under deposit attack at temperatures up to at least 250°C (480°F) [8] . Alloyed with titanium, palladium forms a compound identified as Ti_2Pd. In conditions where corrosion of titanium is initiated, this compound is also dissolved, but the palladium is almost immediately re-precipitated onto the titanium surface. These areas of palladium act as cathodic depolarisers which effectively raise the electrode potential of the metal into the passive region, restoring and maintaining the titanium oxide film and halting corrosion.

5. Crevice Corrosion and Activation

Care must be taken in mixed metal systems which include alloys of substantially lower resistance to crevice corrosion than titanium. These may survive without problems in an uncoupled situation but may be susceptible to activation by pitting or crevice corrosion when coupled to a more noble metal. Care must therefore be taken to check compatibility in the intended working environment. For example 22% chromium duplex stainless steel may be coupled to titanium in mildly sour oil and gas brines without attack, but not in sea water, unless cathodic protection is provided to overcome the susceptibility of the steel to crevice corrosion.

6. Biofilming

Titanium is immune to microbiologically influenced corrosion [9]. Biofilming, a natural consequence of immersion in untreated natural sea water at temperatures up to 40°C (104°F), can have a significant effect on galvanic corrosion by depolarising the cathodic reaction, thus increasing the potential at the cathode by up to +300 mV SCE. Titanium the high grade stainless steels and nickel alloys quickly accept biofilms. Copper based alloys typically take much longer for films to become established if at all. The biofilm takes time to develop, its stability and permanence depending on velocity, temperature, chlorination and mechanical or chemical cleaning routines. The potential of the filming metal changes accordingly. Water velocities between 3 and 5 ms^{-1} (10–16 ft s^{-1}) in continuous flow systems, which are completely safe for titanium with respect to erosion/corrosion, serve to minimise biofouling. Heavier biofilms which may develop from cathodic protection (or galvanic corrosion) can act as a barrier to diffusion. Films of carbonates and biomaterial formed on a titanium alloy deliberately charged at –1.1 V in natural sea water served effectively to limit the diffusion of hydrogen to the metal surface as compared to similar charging in synthetic sea water [10].

7. Effect of Chlorination

Regular or shock chlorination, used to control biofouling, will also increase to varying extents the electrode potential of titanium and most 'passive' corrosion resistant metals and alloys. The reduction of dissolved oxygen, the normal reaction at the cathode, is replaced by the reduction of hypochlorite to chloride. Caution is required that potentials of metals in a couple do not switch over, or that the difference of potential is not widened to the point where galvanic corrosion between supposedly compatible metals becomes possible especially at higher temperatures, or because of variations of chloride concentration within the system or chloride access to the cathodic areas. The relative potential of efficient cathodes such as graphite under chlorination appears to be a problem for some stainless steels, which may suffer crevice corrosion against graphite seals and gaskets [11]. The ennoblement of titanium effectively eliminates the possibility of corrosion, including crevice corrosion in, what are for titanium, 'passive' environments such as sea water.

8. Effect of Oxygen and Temperature

Oxygen dissolved in the electrolyte frequently plays a significant role in determining the rate and severity of galvanic corrosion. The cathodic reaction of (dissolved) oxygen reduction will normally control the overall corrosion reaction rate. Metals such as titanium which have high oxygen reduction overvoltages cause less galvanic damage when coupled to carbon steel than those like copper with lower oxygen reduction overvoltages. Although copper is significantly less noble than titanium with reference to steel, coupling of steel to copper causes weight loss on the steel some seven times greater than when coupling to titanium [4].

The potential of titanium does not alter significantly with a reduction of dissolved oxygen in sea water. This behaviour of titanium is in sharp contrast to the performance of most stainless steels, which typically move to around –400 mV SCE in fully deaerated sea water (<5 ppb). At ambient temperature and up to 150°C (300°F), the median potential for titanium is some 50 mV higher in deaerated sea water than in aerated sea water. At around 150°C (300°F) the potential in deaerated water starts to fall more rapidly to a position more than 200 mV below that for aerated water at 190°C (375°F). Although this potential shift may cause problems with certain galvanic couples, similar negative shifts are observed for other materials such as copper–nickel alloys and stainless steels .

9. Effect of Velocity

Increasing the velocity of the electrolyte adjacent to the cathode or anode surfaces may create a combination of effects including rapid renewal of dissolved oxygen, uniform distribution of chlorination, removal or reduction of micro- and macrobiofouling, breaking down of passive barrier films, erosion corrosion if the electrolyte contains abrasive particulates (e.g. sand in sea water). The overall observed effect is that increases in velocity normally result in an increase in current density, metal loss at the anode in galvanic corrosion [12]. The increase is progressive up to 30 ms^{-1} (98 ft s^{-1}), but even at this velocity the oxide film on titanium continues to exhibit excellent erosion resistance and stability.

10. Effect of Sulfides

Hydrogen sulfide present in sea water or other aqueous electrolytes can dramatically affect the galvanic corrosion rate for example by stimulation of the cathodic production of hydrogen. For this reason, titanium and its alloys must not be coupled to carbon steel, aluminium, zinc or active stainless steels at temperatures above 75°C (167°F) in sour sulfide containing aqueous environments. The titanium will absorb hydrogen and this may lead ultimately to failure by embrittlement.

10.1. Coupling Titanium to Galvanically Compatible Metals

There are a number of corrosion resistant alloys which possess potentials close to those of titanium. In neutral, slightly reducing and oxidising environments titanium may be directly coupled to alloys such as super duplex stainless steels, 6Mo austenitic steel, 22% chromium duplex steel (but not in sea water), Alloy 625, C-276 and the like without fear of damage to either metal in the couple. Titanium may also be connected directly with metals and alloys (such as Type 410) stainless steels which are galvanically compatible when in their passive condition in a specific environment. These alloys may become activated, for example by local corrosion or pitting but the added effect of coupling to titanium is small. The primary consideration must be to ensure that the alloys selected are appropriate and compatible for the service environment.

10.2. Coupling Titanium to Less Corrosion Resistant Metals

Titanium should not be coupled directly to magnesium, and care is required when coupling to zinc, and to aluminium in sea water. This requirement applies in particular to systems where these metals are used as sacrificial anodes. Magnesium anodes must not be used in conjunction with titanium, their potentials are too negative. The anodes are likely to experience accelerated corrosion and, in the process, titanium may pick up hydrogen which is generated as the cathodic product of the corrosion reaction. Careful design of the system and positioning of zinc and aluminium anodes is necessary for their successful use in conjunction with titanium. When the adjacent titanium components are thin walled (such as heat exchanger tubing) or are heavily stressed critical parts, anodes must be selected or regulated to produce negative potentials not more negative than –0.85 V SCE. Aluminium and zinc sacrificial anodes delivering more negative potentials may however be used when the adjoining titanium parts are under low levels of stress and are of heavier section. e.g. 6 mm (0.24 in.) wall thickness or more. A review of the cathodic protection system is essential when a significant area of titanium replaces steel in a system operating in a corrosive environment. In less aggressive conditions, such as sea water, titanium may be coupled to copper based alloys and carbon steel, but effective protection such as chemical inhibition or by sacrificial anodes or impressed potential must be provided for these less noble metals. Galvanic corrosion of less resistant metals may be harmful to titanium, as the cathode, if conditions lead to the uptake of hydrogen. However, it is only in certain environments, and under specific conditions, that titanium alloys will absorb hydrogen. In addition, absorption of hydrogen does not necessarily cause a problem or limit service life. The surface condition of titanium components is an important control, the normal oxide film is a barrier to hydrogen diffusion. Carbonate films formed by cathodic charging in sea water protect against hydrogen absorption. Concern that iron embedded or smeared on the titanium surface may accelerate hydrogen uptake is limited to certain chemical plant at high temperatures, and in subsea environments, iron, rust and rust stains on the surface of titanium pose no problem. Hydrogen uptake can normally be avoided by the proper design of equipment and control of operating conditions.

10.3. Coupling Titanium to More Corrosion Resistant Metals

Metals and materials such as graphite and carbon fibre composites which are even more corrosion resistant than titanium and which raise the corrosion potential into the passive region for titanium, may be coupled to it, and will maintain the resistance of titanium in reducing as well as neutral and oxidising environments. This protection may not be provided in the few cases where in very strong reducing acid conditions (e.g. in concentrated sulfuric and hydrochloric acids and hydrofluoric acid) the oxide film on titanium is attacked and cannot be maintained or restored.

11. Avoiding Galvanic Corrosion

Galvanic corrosion should be avoided by selection of materials in design, and by protection of adjoining less noble metals in the system. Techniques include coating the titanium in the neighbourhood of the joint to reduce the effective cathode/anode ratio; electrical isolation of the titanium components through the use of non conducting gaskets and sleeved bolts; installation of short, easily replaced, heavy wall sections of the less noble metal; and chemical corrosion inhibition of the active metal.

11.1. Guidelines for the Control of Impressed Current Cathodic Protection Schemes

1. Do not exceed –0.85 V(SCE).

2. The system must have automatic potential control.

3. The system must have good reference electrodes placed in positions determined by survey to monitor accurately and reliably the impressed potential.

4. System anodes must not be placed closer than 750 mm (30 in.) to the titanium.

12. Summary

The simple concept of galvanic corrosion is shown in practice to be a complex and continuously evolving science. New working environments and new materials and material combinations serve further to justify the continuing research and reporting required to minimise future problems and avoid failures.

References

1. Titanium Information Goup, Data Sheet No. 6, 'Connecting Titanium to Other Metals' Revision 1 October 1997.
2. I. R. Scholes *et al.*, 'Bimetallic Corrosion in Seawater', 7th European Congress on Metal Corrosion, London, *Soc. Chem. Ind.*, 1977, 161–169.
3. L. Lunde, 'Properties of Titanium Alloys — Corrosion and Galvanic Compatibility' Conference on Titanium in Practical Applications, Trondheim, Norway, 1990.
4. J. A. Beavers *et al.*, 'Corrosion of Metals in Marine Environments', Metals and Ceramics Information Centre, Report 86-50, 1986, Columbus, Ohio, USA.
5. S. M. Wilhelm, 'Galvanic Corrosion in Oil and Gas Production Part 1 Laboratory Studies' *Corrosion '92*, Paper No. 480, NACE International, Houston, Tx, 1992.
6. A. J. Griffith and A. Turnbull, 'Galvanic Coupling of Corrosion Resistant Alloys in Seawater at Elevated Temperatures', *Stainless Steel World*, 1998, Jan/Feb 3 8–39.
7. R. W. Schutz and D. E. Thomas, 'Corrosion of Titanium and Titanium Alloys', in *Metals Handbook*, 9th Edition, 1987, Vol. 13, p.669–706.
8. D. K. Peacock, and J. S. Grauman, 'Crevice and Under Deposit Corrosion Resistance of Titanium Alloys in Highly Aggressive Environments'; Dechema, Frankfurt, Germany, Achema Conference '97, DECHEMA, Frankfurt, Germany.
9. D. K. Peacock, 'Titanium — A Natural Solution to Microbiologically Influenced Corrosion', *UK Corrosion '91*, Manchester, UK. Institute of Corrosion, Leighton Buzzard, UK, 1991.
10. J. S. Grauman and D. K. Peacock, 'Performance of a High Strength Beta Titanium Alloy Under Tensile Load with Cathodic Charging of Hydrogen', *Titanium '95 Science and Technology*, p.1844 –1851. The Institute of Materials, London, UK, 1995.
11. F. Egan, 'Service Experience of Super Duplex Stainless Steel in Seawater', *Stainless Steel World*, 1997, Dec., 61–65.
12. J. A. Davis and G. A. Gehring, 'Galvanic Corrosion as a Function of Applied Potential in High Velocity Seawater', *Corrosion '76*, Paper No. 75, NACE International, Houston, Tx, 1976.

39

Performance of Cold-worked Duplex Stainless Steels in Oilfield Environments under Cathodic Charging Currents Appropriate to Galvanic Coupling Conditions

A. J. GRIFFITHS and A. TURNBULL

Centre for Materials Measurement and Technology, National Physical Laboratory, UK

ABSTRACT

Slow strain rate testing of solution-annealed and cold-worked 22 Cr duplex stainless steels (DSS) has been conducted. The charging currents and environments were designed to simulate coupling of duplex stainless steel tubulars to carbon steel casing assuming a leak of production fluid into the annulus. To provide a framework for evaluating the results, hydrogen permeation measurements were also conducted.

In H_2S-environments, the cold-worked material was less resistant to cracking than the solution-annealed material. The resistance to cracking did not correlate simply with hydrogen uptake measurements suggesting that film rupture may be important.

1. Introduction

Duplex stainless steel (DSS) tubing is often used in oil and gas wells in conjunction with a carbon steel casing and consequently these materials may be galvanically coupled. Ideally, the annulus between the tubing and the casing should be filled with inhibited sea water or man-made brine. However, in practice, the joints on the DSS tubing may leak allowing the production fluid, containing sour acid brine, to enter the annulus. Hydrogen produced by the coupling reaction will be absorbed into the DSS and may cause embrittlement. Failures of heavily cold-worked DSS tubulars due to coupling to carbon steel have been reported [1].

The aim of our work is to provide guidance on the environmental conditions for which DSS can be coupled to carbon steels without the risk of hydrogen embrittlement. Coupling tests were conducted previously to establish the currents and potentials of coupled specimens [2,3]. In the present study, hydrogen permeation tests were conducted to assess the relative aggressivity of combinations of pH, H_2S and coupling current in terms of hydrogen uptake, and slow strain rate tests were conducted to evaluate the impact on cracking resistance.

2. Experimental Details

2.1. Materials

Two type UNS 31803 solution-annealed materials, SAF 2205 (nominal Cr22 Ni 5%) and 22 Cr duplex solution-annealed materials, SAF 2205 (nominal Cr22, Ni5%) and 22 Cr duplex stainless steel (DSS), were used for the permeation testing.

Slow strain rate tests were conducted for the 22 Cr solution-annealed DSS ($\sigma_{0.2}$ = 552 MPa) and a cold-worked DSS ($\sigma_{0.2}$ = 1068 MPa).

2.2. Test Environment

The test environment for the cold-worked material was deaerated 250 gL^{-1} NaCl with 4 gL^{-1} CH_3COONa, based on an EFC guidelines document [4]. A lower chloride level of 16.5 gL^{-1} NaCl was used previously for tests with solution-annealed material [2,3] but the chloride content did not affect the hydrogen uptake and resistance to cracking under cathodic charging conditions. The pH was adjusted to 3.5 using HCl. All tests were carried out at 80°C at a range of applied cathodic current densities (2 μAcm^{-2} to 1 $mAcm^{-2}$) designed to simulate coupling to carbon steel.

2.3. Hydrogen Permeation Testing

The permeation testing was conducted using a two-compartment cell in accordance with the British standard [5]. Cathodic charging was applied potentiostatically with the potential increased stepwise after attaining a steady-state permeation current at each applied potential.

2.4. Slow Strain Rate Testing

The slow strain rate tests were conducted according to ISO 7539 Part 7. A displacement rate of 1.27×10^{-5} mm s^{-1} was used. Tests were conducted also in silicone oil at 80 ± 1°C.

3. Results and Discussion

3.1. Hydrogen Uptake Measurements

Hydrogen uptake is plotted as a function of charging current density, i, in Fig. 1. Hydrogen uptake was defined in terms of C_t, the total hydrogen content on the charging side of the membrane, and C_0, the sub-surface concentration of hydrogen atoms in the lattice per unit volume of material on the charging side of the membrane. Hydrogen uptake was similar for the two solution-annealed materials in H_2S-free environments. Hydrogen uptake of cold-worked DSS in these environments is of the same order [6].

At low charging currents, hydrogen uptake was independent of pH and H_2S and proportional to $i^{1/2}$. As the charging current density increased, a critical value was reached in H_2S-free solution above which hydrogen uptake was independent of the

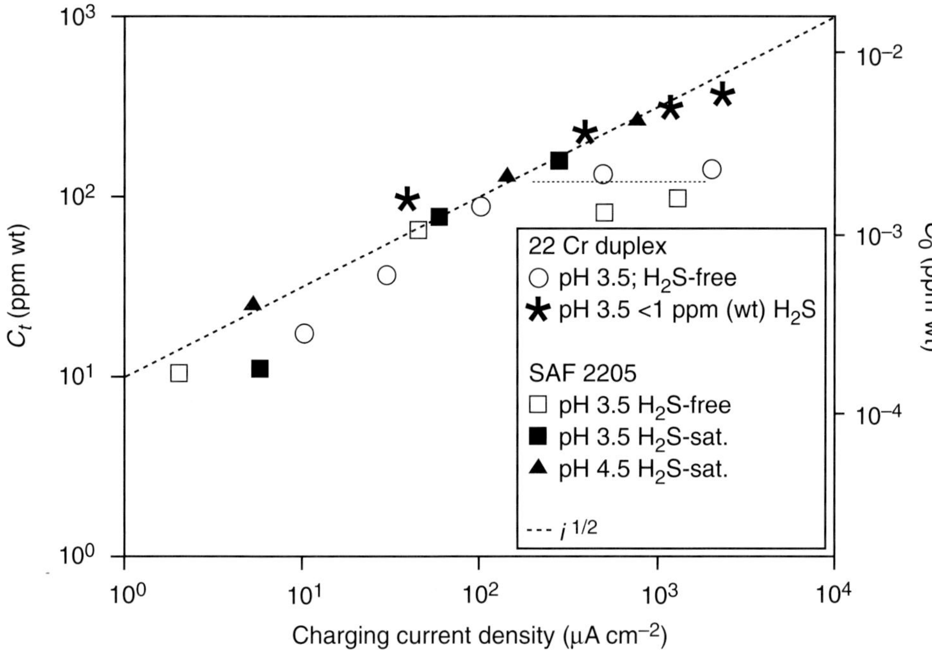

Fig. 1 *Hydrogen content vs charging current density for 22Cr duplex stainless steel in acid brine at 80°C.*

charging current. This behaviour is readily explained in terms of a limiting surface coverage associated with increasing efficiency of hydrogen recombination (as more hydrogen atoms are produced on the surface the probability of being in close proximity increases). In H_2S-saturated solution, hydrogen uptake remained proportional to $i^{1/2}$. In this region, H_2S increased hydrogen uptake by up to a factor of 4. The continued $i^{1/2}$ dependence in H_2S environments reflects the ability of H_2S to impede the hydrogen atom recombination reaction.

The distinctive feature of the permeation data is the similarity in hydrogen contents generated by very low H_2S solution (< 1 ppm) with that for the saturated solution. Under cathodic polarisation or galvanic coupling conditions, only a monolayer of H_2S on the steel surface may be required to retard the recombination reaction and, correspondingly, to enhance hydrogen uptake. Clearly, this has important implications for the definition of sour service.

The data in Fig. 1 can be used as a basis for estimating hydrogen uptake in duplex stainless steels due to coupling to carbon steels. Galvanic coupling tests were undertaken previously [2] to measure the coupling current for a duplex stainless steel/carbon steel couple in buffered NaCl solution of pH 3.5 at 80°C. In H_2S-free environments the coupling current was 100 μA cm^{-2} corresponding to a hydrogen content of 100 ppm (wt). In H_2S-saturated environments the coupling current was 550 μA cm^{-2} corresponding to a hydrogen content of 230 ppm (wt).

A maximum hydrogen content of 30 ppm (wt) has been reported [1] in a cold-worked duplex tubular which had been coupled to carbon steel casing in service at

120 °C. The hydrogen contents were measured 6 months after the tubular came out of service and it is estimated that the hydrogen content at the charging surface during service would have been about 50 ppm (wt). These values cannot be compared directly with the values measured in the present work but they indicate that large concentrations of absorbed hydrogen are attainable.

3.2. Slow Strain Rate Testing

3.2.1. Solution-annealed material

The normalised plastic strain-to-failure in the aqueous environment relative to that in silicone oil is plotted vs hydrogen uptake at the input surface in Fig. 2. In H_2S-free environments there was no reduction in the strain-to-failure relative to oil but microcracks were observed on the surface of the specimens.

The results in H_2S-saturated environments generally show a reduction in the relative strain-to-failure with increase in charging current. The most significant result is that obtained for a charging current density of 20 μA cm^{-2}, corresponding to a total hydrogen content at the charging surface of 45 ppm (wt). The relative strain-to-failure was 0.7.

An effect of H_2S on cracking resistance would generally not appear surprising. However, Fig 1 indicates the same extent of hydrogen uptake in solutions with and without H_2S at the charging current used. This might be inferred to indicate that the impact on strain-to-failure would be similar. However, the permeation tests were conducted in the absence of strain and consequently do not take into account enhanced hydrogen uptake due to mechanical film rupture during slow strain rate

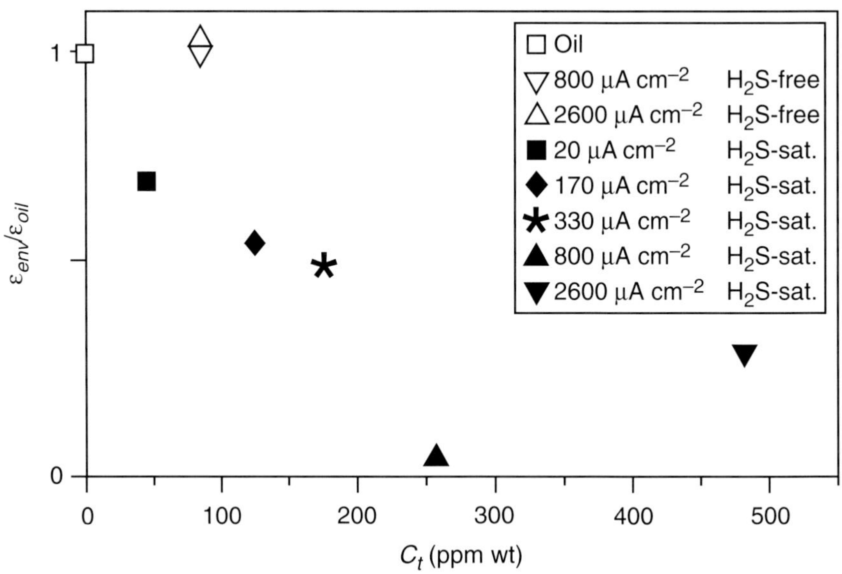

Fig. 2 Normalised plastic strain-to-failure vs hydrogen uptake for 22Cr duplex stainless steel in acid brine at 80°C.

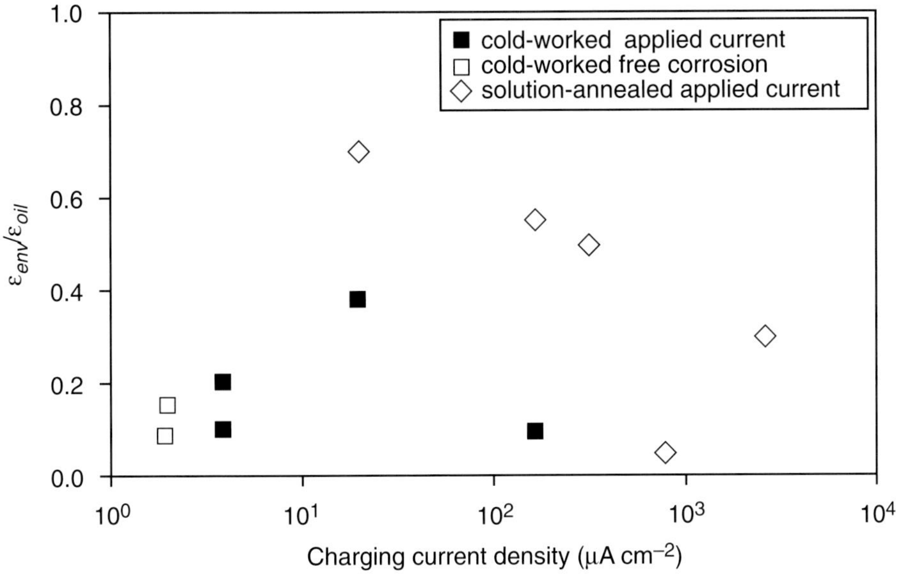

Fig. 3 *Normalised plastic strain-to-failure vs charging current density for 22Cr duplex stainless steel in acid brine at 80°C.*

testing. Hydrogen uptake will be enhanced only locally but this may be sufficient to cause cracking in this area. The extent of hydrogen uptake under plastic straining conditions would depend critically on the refilming characteristics of the material. The effect of H_2S on this process and particularly the characteristics of the films formed may be the explanation for the enhanced aggressivity relative to the H_2S-free solution.

3.2.2. Cold-worked material

The normalised plastic strain-to-failure in the aqueous environment relative to that in silicone oil is plotted vs charging current density in Fig. 3. In this figure, the charging current at the free corrosion potential was estimated to be 2 μA cm^{-2} (from extrapolation of a plot of potential vs current to the free-corrosion potential) but this reflects the passive current and not localised charging via pits. Pitting of the cold-worked material occurred also at a net applied charging current of 2 μA cm^{-2} (i.e. 4 μA cm^{-2} total charging current density). There appears to be some increase in the resistance to cracking of this cold-worked DSS as the charging current is raised to 20 μA cm^{-2} but this needs to be verified.

The observation of pitting is compatible with a mechanism involving failure by hydrogen embrittlement since the effect of pitting will be to enhance the local uptake of hydrogen because of breakdown of the passive film and the increased acidity associated with a pit solution. Indeed, the apparent increase in the resistance to cracking as the charging current is raised to 20 μA cm^{-2} may be associated with suppression of pitting and reduced local uptake of hydrogen. As the current is increased still further hydrogen uptake via the passive film becomes progressively more significant. Of course, a change in mechanism to chloride stress corrosion cracking can be invoked also to explain the failure at the free corrosion potential.

Further work involving testing of the solution-annealed DSS under free corrosion conditions is required to resolve this issue.

4. Conclusions

1. Under cathodic charging conditions, simulating coupling to carbon steel, an effect of H_2S on hydrogen uptake by duplex stainless steel was observed only above a critical charging current density, consistent with H_2S acting as a hydrogen recombination poison. Hydrogen uptake in solutions containing about 1 ppm H_2S was equivalent to that for saturated H_2S implying that sour service conditions may prevail even at very low levels of H_2S for galvanically coupled DSS.

2. In H_2S-free environments there was no reduction in the strain-to-failure of solution-annealed material. In H_2S-saturated environments, cracking occurred at charging currents as low as 20 $\mu A\ cm^{-2}$ which is readily attainable by coupling to carbon steel. The resistance to cracking did not correlate simply with bulk hydrogen uptake suggesting that enhanced local hydrogen uptake due to film rupture during slow strain rate testing may be significant. The extent of hydrogen uptake under plastic straining conditions appears to depend on the presence of H_2S.

3. The results emphasise the importance of dynamic plastic straining in test methods for sour service evaluation of duplex stainless steels.

4. In H_2S-saturated environments, the cold-worked material was less resistant to cracking than the solution-annealed material. Failure of the cold-worked material occurred under freely corroding conditions and was considered to be triggered by pitting corrosion.

5. Further tests with low H_2S are desirable both under free corrosion conditions and with cathodic charging, to simulate galvanic coupling, in order to define limits of sour service.

5. Acknowledgements

This work was carried out as part of the 'Materials Degradation in Aggressive Environments Programme', a programme of underpinning research financed by the United Kingdom Department of Trade of Industry.

The authors are grateful to Sandvik Steel for providing the cold-worked duplex stainless steel.

References

1. P. Sentance, 'Hydrogen embrittlement of cold worked duplex stainless steel oilfield tubulars', *Duplex Stainless Steels '91*, page 895, Les editions de physique, France, 1991.

2. A. J. Griffiths and A. Turnbull, 'Preliminary investigation of hydrogen uptake and cracking in 22 Cr duplex stainless steel under galvanic coupling conditions', *Corrosion '96*, Paper No. 72. NACE International, Houston, Tx, 1996.

3. A. J. Griffiths and A. Turnbull, 'Impact of coupling to carbon steel on hydrogen uptake and cracking of duplex stainless steel in oilfield environments', *Proc. 11th European Conf. on Fracture*, EMAS, West Midlands, UK, 1996.

4. *Corrosion Resistant Alloys for Oil and Gas Production: Guidelines on General Requirements and Test Methods for H_2S service*, European Federation of Corrosion Publications Number 17, ISBN 1-86125-001-0, The Institute of Materials, London, UK, 1996.

5. Method of measurement of hydrogen permeation and the determination of hydrogen uptake and transport in metals by an electrochemical technique, BS7886: 1997, ISBN 0 580 27127 7, BSI, London, UK, 1997.

6. A. Turnbull, A. Griffiths and T. Reid, 'Hydrogen embrittlement of duplex stainless steels – simulating service experience,' *Corrosion '99*, Paper No. 148. NACE International, Houston, Tx, 1999.

40
Galvanic Corrosion in Oil and Gas Environments

T. HARA, H. ASAHI and H. KANETA*
Nippon Steel Corp., Steel Research Lab., Japan
*Japan National Oil Corp., Petroleum Engineering Lab., Japan

ABSTRACT

Galvanic corrosion behaviour in sour and sweet environments for combinations of materials from carbon steel to corrosion resistant alloys (CRAs) was investigated. The galvanic effect index (GEI) is defined as the ratio of the corrosion rate of a coupled to an uncoupled material. GEI increased with an increase in the corrosion rate of the 'uncoupled specimen' and was at most 2.0 in any sweet and sour environments for equal areas of anode and cathode. GEI increased with increasing surface ratio but was almost independent of surface ratio at temperatures higher than 120°C. Galvanic corrosion is not serious in oil and gas environments.

1. Introduction

In oil and gas wells, the conditions of corrosive environments vary with depth. By using a tubing string composed of different tubing materials (combination string), the most suitable material required for each depth can be chosen. Although a combination string would realise cheaper material costs, galvanic corrosion would be a concern. Extensive research on galvanic corrosion has been carried out but there has been little research on galvanic corrosion in oil and gas well environments [1–6]. In the present study, in order to clarify combinations of materials that can be applied without pronounced galvanic corrosion, galvanic corrosion behaviour was investigated for combinations of typical tubing materials in various simulated oil and gas environments, such as sweet and sour environments, and taking into account effects of flow velocity, surface ratio and distance.

2. Experimental Procedure

2.1. Materials

Table 1 lists the chemical compositions and the mechanical properties of materials used in this study. All the materials were properly heat-treated.

2.2. Test Environment

Deaerated 5% sodium chloride solution (5% NaCl) was used in this test with CO_2 partial pressures from 0.4 MPa to 4 MPa and H_2S partial pressures from 0.001 to

Table 1. *Chemical compositions and mechanical properties of tested steels (mass%)*

Steel	C	Si	Mn	Ni	Cu	Cr	Mo	Others	YS (MPa)	TS (MPa)
P	0.28	0.22	1.28	—	—	—	—	Ti,Al	885	985
SP	0.23	0.13	0.38	—	—	0.56	0.74	Ti,Al,Nb,B	778	856
T0	0.19	0.25	0.58	0.13	—	12.6	0.01	Al	537	735
T3	0.02	0.26	0.46	4.54	1.51	12.7	1.43	Al,N	805	905
D	0.02	0.43	1.78	5.30	—	21.4	2.78	N	503	733
625	0.02	0.32	0.17	61.6	—	21.9	9.04	Al,Ti,	415	859

0.01 MPa and test temperatures from 25 to 180°C. Test durations were 1 or 2 weeks for the immersion test and 96 h for the loop test.

2.3. Immersion Test

Corrosion rates were determined from the weight loss, after descaling, divided by the immersion time and the specimen surface area. Specimens were mechanically polished with No. 320 emery paper. Figure 1 shows a 'coupled specimen' and an 'uncoupled specimen'. A 'coupled specimen' was joined by a bolt and a nut made of polytetrafluoroethylene (PTFE). To investigate the effect of surface ratio, a length of a material suspected to be a cathode was varied up to 335 mm, to provide anode to cathode ratios of 1:10. 'Coupled specimens' and 'uncoupled specimens' were put in an autoclave and the test solution added. The solution was deaerated thoroughly by nitrogen gas bubbling after the autoclave was sealed. Test gases were introduced into the autoclave after the solution reached the test temperature.

2.4. Flow Loop Test

A flow loop (shown in Fig. 2) was used to study the effect of flow velocity on corrosion rate. Liquid at a high temperature was circulated through the tubular specimens at a

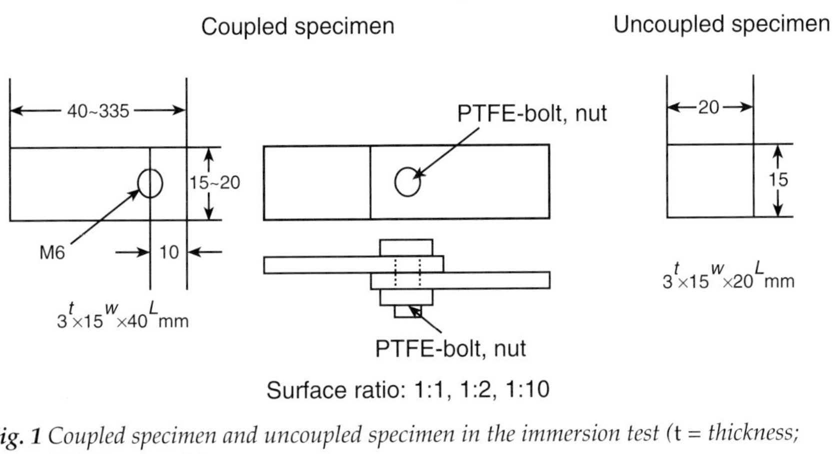

Fig. 1 *Coupled specimen and uncoupled specimen in the immersion test* (t = *thickness;* w = *width;* L = *length).*

maximum velocity of 20 ms^{-1}. The concentration of dissolved oxygen in the test solution was controlled to below 10 ppb before the temperature of the solution was raised. Figure 3 shows the shapes of a 'coupled specimen' and an 'uncoupled specimen'. A 'coupled specimen' was connected by the thread. The ratio of anode to cathode area was 1:1.

2.5. Measurements of Galvanic Current and Coupled Potential

Galvanic current and coupled potential were measured using a high pressure and high temperature autoclave. A pressure-balanced outer Ag/AgCl (0.1M KCl) reference electrode was used. The specimen was coated with silicone resin, after mechanical polishing, leaving an area of 1 cm^2 for measurement. Just before putting into an autoclave, the specimen was pickled in 50% sulfuric acid at 60°C to remove the air-formed film. The two specimens were connected after the potential of each specimen reached the steady state. Measurements of galvanic current and coupled potential were then started. The test apparatus is illustrated in Fig. 4.

3. Results and Discussions

3.1. Immersion Test Results

Table 2 shows corrosion rates of the 'uncoupled specimen' and those of each of the two materials composing the 'coupled specimen' when the surface area ratio was 1:1. The corrosion rate of the 'uncoupled specimen' for the low alloy steel (SP) was 154 $mg/dm^2/day$ (mdd), while that coupled to the 13Cr steel (T0) was 307 mdd in a sour environment at 70°C.

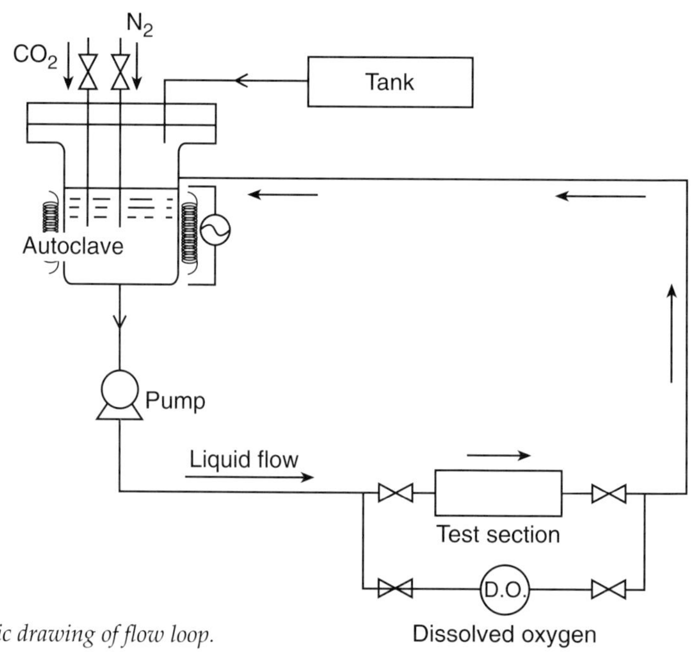

Fig. 2 Schematic drawing of flow loop.

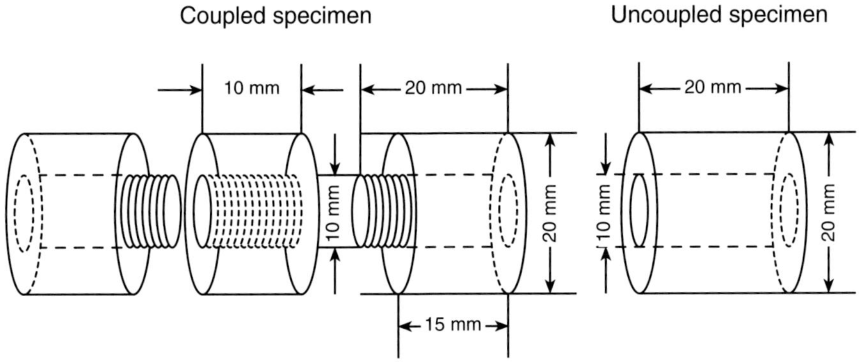

Fig. 3 *Coupled specimen and uncoupled specimen in the loop test.*

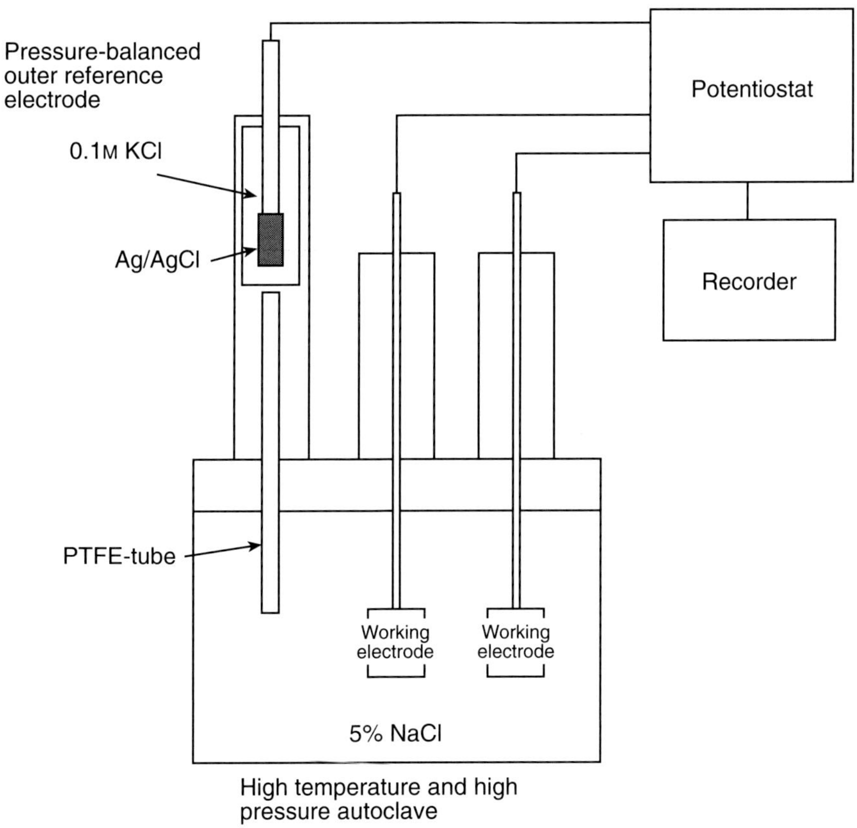

Fig. 4 *Test apparatus of measurements of galvanic coupled current density and coupled potential.*

The corrosion rate of a less noble metal is increased by an amount corresponding to that shown as (3) in Fig. 5 when in contact with a more noble metal. Since it is important to evaluate how much weight loss is increased due to galvanic corrosion,

Table 2. *Galvanic effect in CO_2 and H_2S environments*

	Test temperature (°C)	H_2S partial pressure (MPa)	CO_2 partial pressure (MPa)	Test duration (h)	Corrosion rate of less noble material in coupling A (mdd)	Corrosion rate of noble material in coupling (mdd)	Corrosion rate of less noble material B (mdd)	Corrosion rate of noble material (mdd)	Galvanic effect A/B
T3-T0	120	0.001	4	168	105	4	72	1	1.5
T0-SP	25	0.001	0.4	168	46	4	37	4	1.2
T0-SP	70	0.001	0.4	168	307	5	154	6	2.0
D-T3	180	0.001	4	168	100	2	77	1	1.3
D-T3	180	0.001	4	336	79	1	70	1	1.1
T3-T0	120	—	4	168	102	6	86	6	1.2
D-T3	120	—	4	336	4	0.2	4	1	1.0
D-T0	120	—	4	336	90	0.5	78	1	1.2
T3-T0	120	—	4	336	96	6	78	5	1.2
T0-SP	120	—	4	336	345	76	170	78	2.0

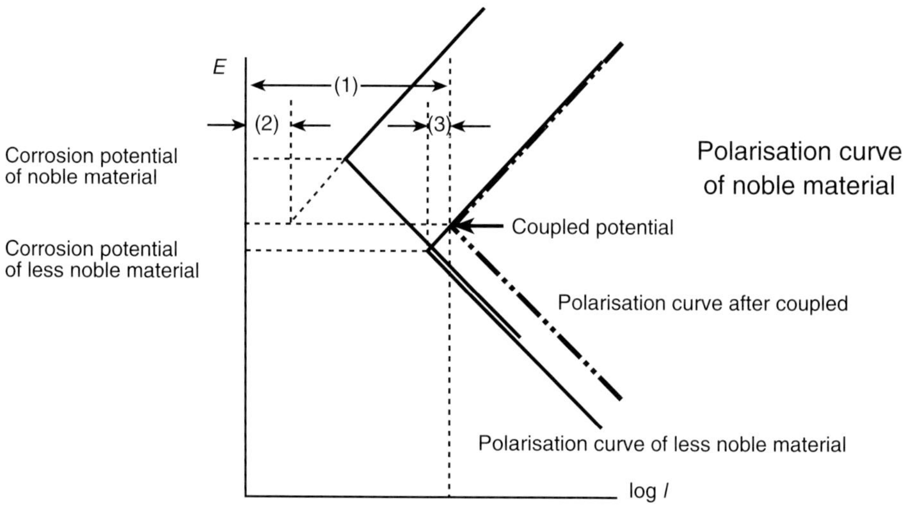

(1)–(2): Galvanic couple current density; (3): Increase in current density of less noble material due to galvanic effect

$$\text{Galvanic effect index (GEI)} = \frac{\text{Corrosion rate less noble in coupling: (1)}}{\text{Corrosion rate less noble: (1)–(3)}}$$

Fig. 5 *Schematic diagram explaining galvanic corrosion.*

a galvanic effect index (GEI) was introduced. The GEI is defined as a ratio of weight loss of a sample composed of a 'coupled specimen' to that of an 'uncoupled specimen' made of the same material, that is, how many more times a material is corroded by connecting to another kind of material. The ratios of A to B (the GEI), are listed in Table 2. As indicated in Table 2, for example, the GEI for the combination of low alloy steel (SP) and the 13Cr steel (T0) was 1.2 at 25°C and 2.0 at 70°C. Figure 6 illustrates the relationship between the corrosion rate of a less noble material uncoupled and coupled. The GEI was at most 2.0 for equal anode and cathode area and the GEI increased with increase in the weight loss of the less noble uncoupled material. In addition, the GEI is theoretically negligibly small when coupled materials are passive. Galvanic test specimens corroded uniformly and did not selectively corrode near the contact site.

3.2. Loop Test Results

The effect of flow velocity on galvanic corrosion was investigated. For example, the corrosion rate of the low alloy steel (SP) of an 'uncoupled specimen' was 3879 mdd while that of a 'coupled specimen' with the 13Cr steel (T0) was 4116 mdd in a sweet environment at 17 ms^{-1} at 70°C, that is, the galvanic effect (A/B) was 1.1. Figure 7 illustrates a relationship between a corrosion rate of less noble uncoupled material and that of less noble material on coupling. Although flow velocity increased the

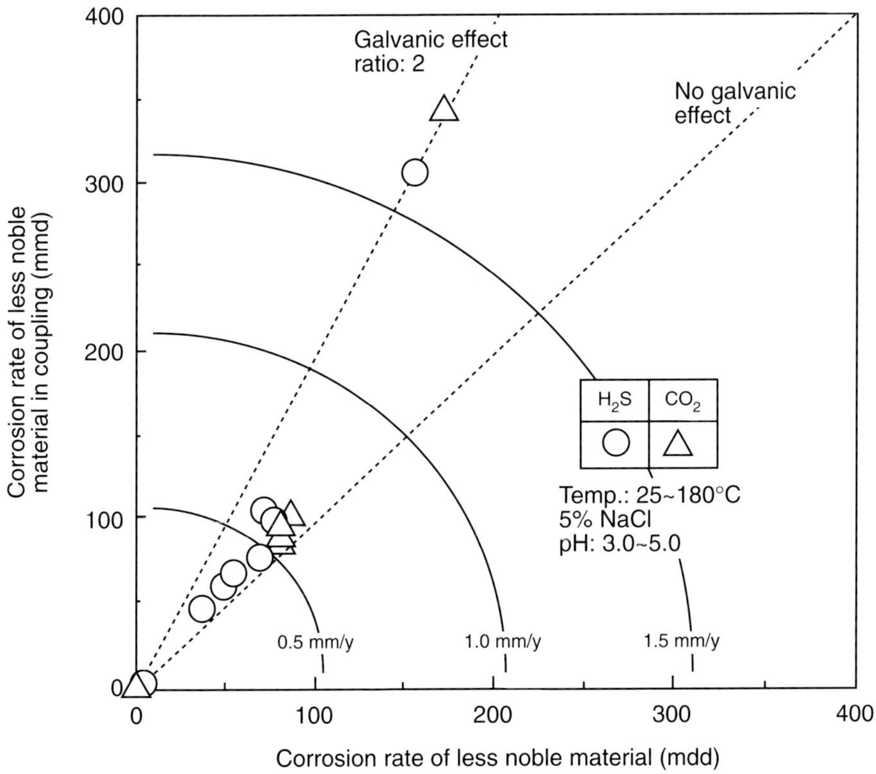

Fig. 6 *Relationship between corrosion and rate of less noble material and that of less noble material in coupling.*

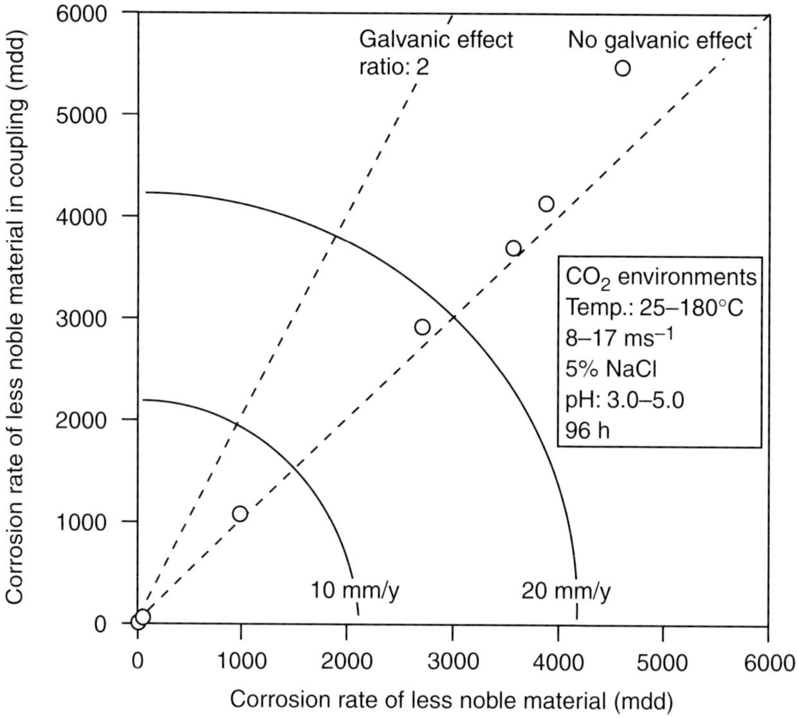

Fig. 7 *Relationship between corrosion rate of less noble material and that of less noble material in coupling.*

corrosion rate itself, the effect on GEI was very small. Galvanic test specimens corroded uniformly and did not selectively corrode near the contact site.

The GEI in the immersion test ranged from 1.0 to 2.0 in sweet and sour environments. On the other hand, the GEI in the loop test was in the range from 1.0 to 1.2 in the various environments. That is, the GEI in the loop test was smaller than that in the immersion test.

3.3. Influence of Surface Ratio

The corrosion rates of 'uncoupled specimens' and each sample of 'coupled specimens' with different surface ratios of 1:1, 1:2 and 1:10 were measured. For example, the corrosion rate of the low alloy steel (SP), which was the anode, increased with an increase in surface ratio when coupled with the 13Cr steel (T0) in a sour environment at 70°C. On the other hand, in the case of the modified 13Cr steel (T3) coupled with the duplex stainless steel (D) in a sour environment at 180°C, the GEI, irrespective of surface ratios, did not change.

Generally, the relationship between GEI (i) and surface ratio (R) is indicated as follows [7]:

$$\log i = B + \log R \quad (1)$$

where B is a constant.

The influence of surface ratio on GEI at 25 to 180°C is indicated in Fig. 8. GEI increases with increasing surface ratio and the logarithm of GEI is proportional to the logarithm of surface ratio. This slope is, however, not equal to 1 but is smaller than 1. In the sour environments at 25°C and 70°C, the slope was 0.61, which was the maximum value. In the sweet environments at 25°C and 70°C, the slope was 0.28. In Fig. 9, the influence of temperature on the slope is plotted as a function of temperature. This slope becomes smaller as temperature rises and no pronounced effect of surface ratio is observed at temperatures above 120°C. The reason is presumed to be as follows. Because an adherent corrosion film such as magnetite (Fe_3O_4) tends to form in high temperature environments, anodic dissolution is thought to be the rate-controlling process. Therefore, the corrosion rate of an anode site is probably not affected by the cathode surface area.

3.4. Effect of Distance

In the immersion and loop tests, the morphology of galvanic corrosion was uniform and the galvanic test specimens did not corrode selectively near the contact site. The reason for the uniform corrosion of galvanic test specimens in the laboratory test was further examined. Since actual tubing is known to be as long as 1000 m or longer, the effect of distance on galvanic corrosion was investigated using Masuko's equation [8] (see Appendix).

The case where the anode is passive. The modified 13Cr steel (T3) was the anode and the duplex stainless steel (D) was the cathode when these steels were coupled in 0.001 MPa

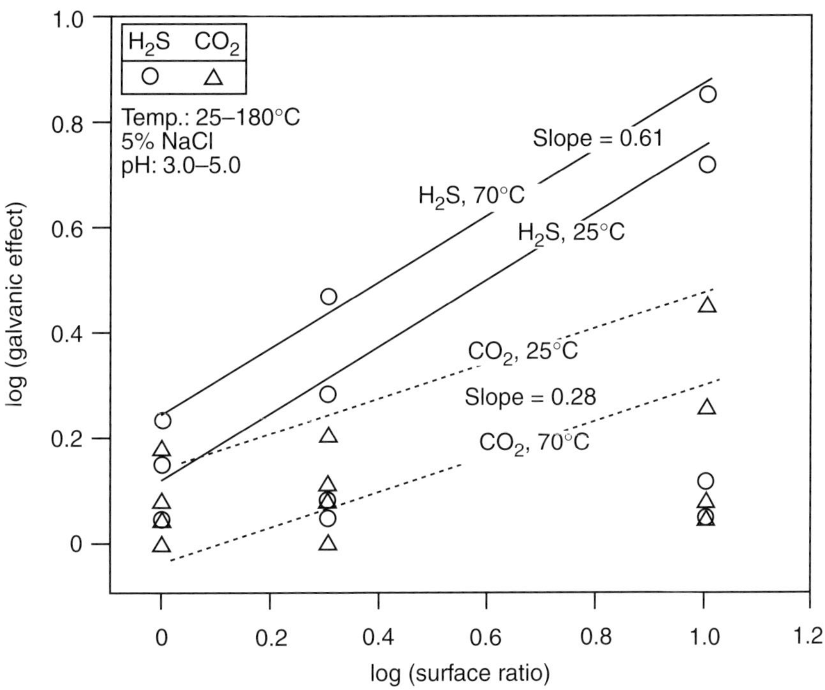

Fig. 8 Relationship between surface ratio and galvanic effect.

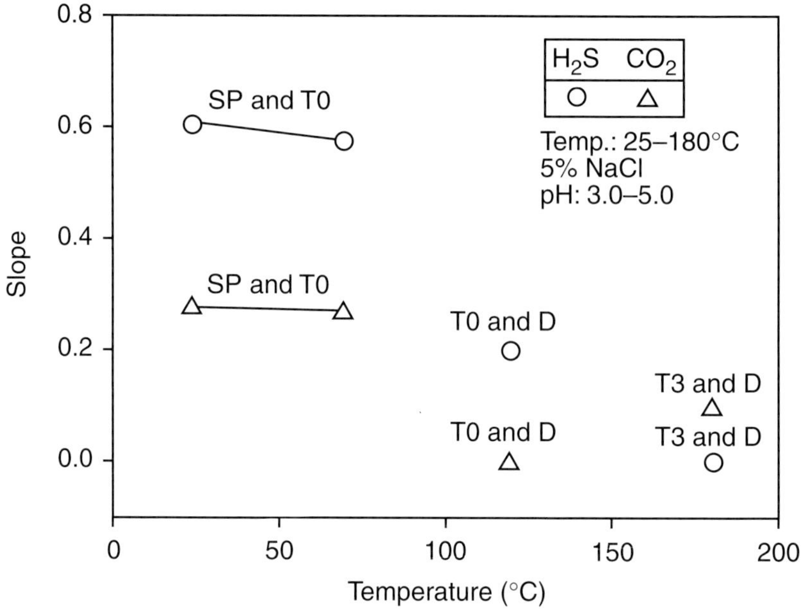

Fig. 9 Relationship between temperature and slope.

H_2S and 4 MPa CO_2 environment at 120°C. The modified 13Cr steel (T3) was passive in this environment. The corrosion potential for the modified 13Cr steel (T3) was –400 mV (Ag/AgCl (0.1M KCl)) and that for duplex stainless steel (D) was –380 mV before coupling. The current density immediately after these steels were coupled was 0.5 μA cm^{-2}. Figure 10 shows the results calculated using eqn (2) in the Appendix for the case where the difference of these corrosion potentials was 20 mV. This value of 0.5 μA cm^{-2} obtained from measurement of the coupled current density corresponded to the current density of 0.7 μA cm^{-2} at the contact site calculated using eqn (2). Galvanic coupled current density (*Ig*) was almost constant irrespective of distance (*L*).

The case where the anode is active. The low alloy steel (SP) was the anode and the 13Cr steel (T0) the cathode on coupling these steels in 4 MPa CO_2 environment at 120°C. The low alloy steel (SP) was active in this environment. The corrosion potential for the low alloy steel (SP) was –695 mV and that for the 13Cr steel (T0) –533 mV before coupling. The current density after coupling these steels was 733 μA cm^{-2}. Figure 11 shows the results calculated using eqn (2) when the difference of these corrosion potentials was 162 mV. The value obtained from the experiment almost coincided with the current density of 600 μA cm^{-2} at the contact site as calculated using eqn (2). The galvanic current *Ig* remained constant up to a distance of 1 m.

From the results of Figs 10 and 11, it appears that galvanic corrosion is uniform in oil and gas environments because of the high conductivity of the solution as indicated by the fact that galvanic corrosion was uniform in the laboratory test. Furthermore, the present study has shown that galvanic corrosion in oil and gas environments is not serious and low corrosion rates occur because galvanic corrosion is negligibly small.

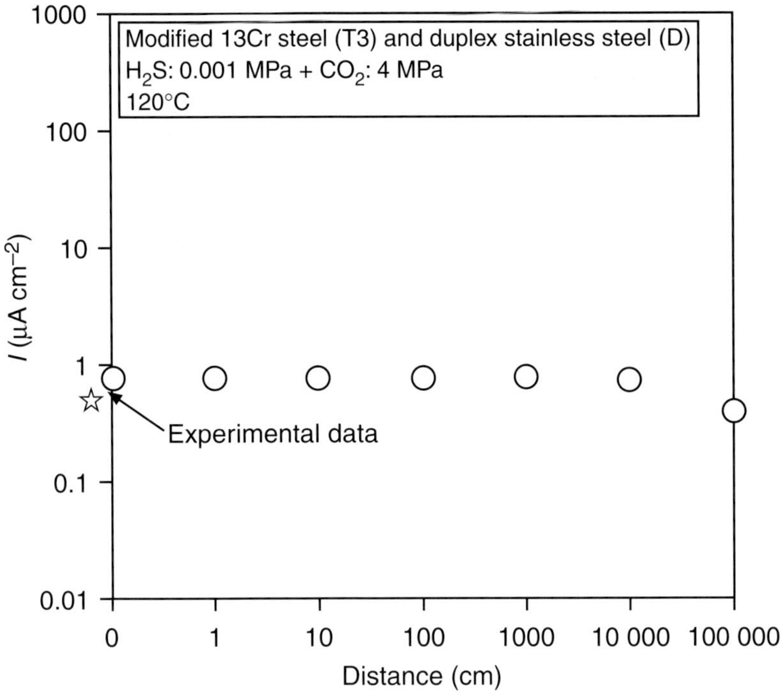

Fig. 10 *Effect of distance on galvanic coupled current density.*

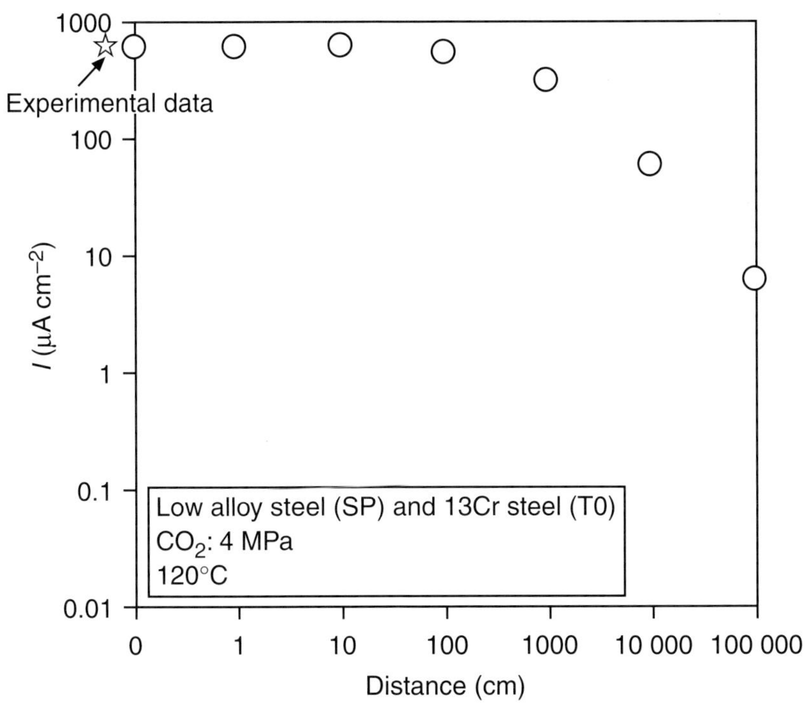

Fig. 11 *Effect of distance of galvanic coupled current density.*

4. Conclusions

The galvanic corrosion behaviour of tubing materials was investigated using immersion and loop tests and electrochemical measurements in typical oil and gas environments.

1. The galvanic effect index (GEI) increases with increase in the corrosion rate of 'uncoupled specimen', and is at most 2.0 when the surface ratio is 1:1 in the immersion test and loop test.

2. GEI increases with an increase in surface ratio, but the effect of surface ratio on GEI is negligibly small at temperatures higher than 120°C.

3. Galvanic current was almost constant up to a distance of 1 m because of the high conductivity of the solution. These results agree with experimental results from laboratory tests showing that galvanic corrosion was uniform.

4. The present study indicated that galvanic corrosion is not serious in oil and gas environments.

References

1. C. Su, *Corros. Prev. Control*, 1989, **36**, (1), 22.
2. C. Su, *Corros. Prev. Control*, 1989, **36**, (4),109.
3. S. M. Wilehelm, *Corrosion*, 1992, **48**, (8), 691.
4. M. A. S. Z. Farooqui and S. Holland, *SPE Production Engineering*, 1992, pp.149.
5. R. L. Martin, *Corrosion*, 1995, **51**, (6), 482.
6. J. Kolts and J. French Scott, *Corrosion '88*, Paper. No. 216. NACE International, Houston, Tx, 1988.
7. H. H. Uhlig, *Corros. Corros. Control*, 2nd edn, pp.100, 1974.
8. N. Masuko, Bousyoku-gizyutu, *Corros. Prev.*, 1978, **27**, 371.

APPENDIX

According to Masuko [8], galvanic coupled current density Ig is expressed by eqn (2)

$$Ig = (E_1 - E_2)/(h_1/A_1 + h_2/A_2 + L\rho/A_2) \quad (2)$$

where E_1 and E_2 are the corrosion potentials of the respective metals with surface areas A_1 and A_2 ($E_1 > E_2$), h_1 and h_2 are the polarisation resistances of the respective metals, L is the distance between the two metals; ρ is the resistance of the solution. Equation (2) is transformed into eqn (3):

$$Ig = (E_1 - E_2)/[(\rho/A_2) \cdot \{L + (h_1 \cdot A_2)/(A_1 \cdot p) + h_2/\rho\}] \quad (3)$$

Figure 1 shows the change of galvanic coupled current density *Ig* as a function of distance *L*. From this figure a maximum value of $E_1–E_2/(h_1/A_1 + h_2/A_2)$ appears at the contact site of the anode and cathode. Galvanic coupled current density *Ig* decreases with increasing distance *L*. The attenuation of the galvanic coupled current density *Ig* with distance *L* increases with increase in the resistivity of the solution ρ. The resistivity of the solution ρ is given by eqn (4)

$$\rho = 1/\kappa \cdot S \qquad (4)$$

where *S* is cross-sectional area, κ is the conductivity of the solution. In this case *S* is the cross-sectional area of the tubing. The conductivity of the solution in oil and gas environments is assumed to be that of ocean water ($\kappa = 0.05\ \Omega^{-1}\ \text{cm}^{-1}$). The polarisation resistance *h* is given by eqn (5)

$$h = K/i_o \qquad (5)$$

where *K* is a constant and i_o is the corrosion current density.

The constant *K* is given by

$$K = (b_a \times b_c)/\{2.303 \cdot (b_a + b_c)\} \qquad (6)$$

where b_a and b_c are the Tafel slopes for anode and cathode respectively.

The Tafel slope b_a for the anode is infinity for the case when the anode is passive, therefore constant *K* becomes $b_c/2.303$. Assuming that the Tafel slopes for the cathode are 0.1 V/decade, the constant *K* will be 0.04. On the other hand, if the Tafel slopes for anode b_a and cathode b_c are 0.1 V/decade respectively for the case where the anode is active, the constant *K* becomes 0.02. The values obtained from experimental data are substituted in eqns (2)–(6) and the effect of distance examined, assuming that the diameter of the pipe is 0.1 m (4 in.) and the total length of the pipe is 1000 m.

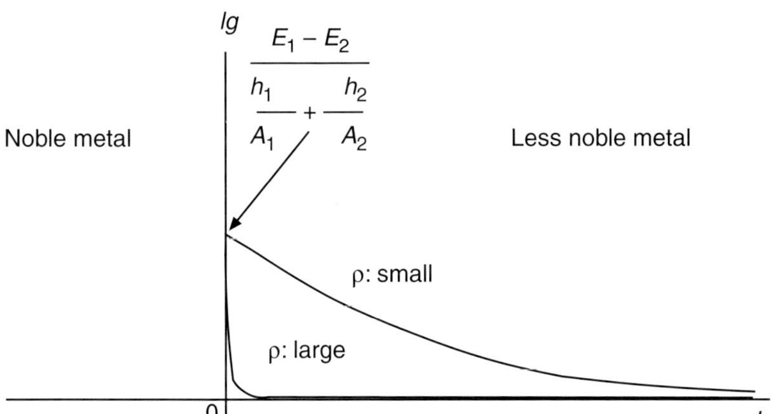

Appendix Fig. 1 *Effect of distance and resistivity of the solution on galvanic coupled current density.*

Part 6

Corrosion Inhibitors

41
Adsorption Isotherms for an Amine Based Fatty Acid Corrosion Inhibitor for Carbon Steel in CO_2-Saturated Solutions

J. BUCHWEISHAIJA and G. HAGEN
Department of Electrochemistry, Norwegian University of Science and Technology, NTNU, N-7034 Trondheim, Norway

ABSTRACT
The adsorption and corrosion protection properties of an amine based fatty acid corrosion inhibitor (KI384) have been investigated for non-corroded and 2 days pre-corroded carbon steel electrodes in CO_2-saturated formation water at 35°C. The inhibitor adsorption/desorption process can be characterised by a Langmuir isotherm by assuming that the inhibitor coverage can be calculated from corrosion currents at different inhibitor concentrations. The corrosion rate $i_{corr,i}$ in the presence of the KI384 inhibitor was then found to be related to the inhibitor concentration C by $i_{corr,i} = i_{corr,u}/(1 + KC)$, where K is the equilibrium constant for the adsorption/desorption process and $i_{corr,u}$ is the corrosion current without inhibitor. A corrosion protection of 99% was obtained with 10 ppm inhibitor concentration for electrodes without pre-corrosion, while more than 50 ppm was needed to obtain the same protection for 2 days pre-corroded specimens.

1. Introduction

Corrosion of carbon steel caused by the presence of dissolved carbon dioxide is a serious problem in the oil and gas production industry [1]. The corrosion problem can, however, to a certain extent be mitigated by the use of organic inhibitors [2]. In many corrosion systems, particularly in acidic solutions, inhibition by organic molecules is assumed to be due to the formation of a protective layer of adsorbed molecules at the metal surface [3]. The adsorption of an inhibitor *Inh* from the solution phase *aq* onto the metal surface *M* may be described by the following reaction

$$Inh(aq) + M \underset{k_d}{\overset{k_a}{\rightleftharpoons}} M\text{–}Inh(ads) \tag{1}$$

where k_a and k_d are the rate constants for the adsorption and desorption process, respectively.

The equilibrium situation for this process can be described by an adsorption isotherm such as

$$\theta/(1-\theta) = KC\exp(-g\theta) \qquad (2)$$

where $K = k_a/k_d$ is the equilibrium constant for reaction (1), C is the concentration of inhibitor in the aqueous phase, while θ is the fraction of the metal surface which is covered by adsorbed inhibitor molecules. The isotherm (2) is usually referred to as a Frumkin isotherm [4], where the g factor indicates lateral interactions between the adsorbed molecules. Thus $g > 0$ or $g < 0$ indicates repulsive or attractive interactions respectively. If $g = 0$ the isotherm is referred to as a Langmuir isotherm where there are no lateral interactions between the adsorbed molecules [4].

In this work we have investigated the influence of pre-corrosion of carbon steel electrodes on the performance of a commercial amine based fatty acid corrosion inhibitor (KI384) in CO_2-saturated aquous solutions.

2. Experimental

The experiments were carried out in an autoclave system with a rotating cylinder electrode assembly The electrodes were made of API 5L X60 carbon steel with the following composition in weight percentage: 0.07 C, 0.25 Si, 1.30 Mn, 0.02 Cr, 0.01 Ni, 0.018 Al, 0.01 Cu, 0.10 Mo, 0.039 V, bal. Fe.

The 1.0 cm long cylindrical specimens had a diameter of 2.5 cm, giving an exposed area of 7.85 cm^2. Four specimens were mounted on the rotating shaft and the segments were separated from each other by 1.0 cm long Teflon spacers. Two adjacent specimens were used as working electrodes, and the other two were used for weight loss measurements and surface analysis. Electrical connections to the working electrodes were made through the rotating shaft by mercury contacts.

The carbon steel electrodes were abraded with silicon carbide paper down to 1200 grit, ultrasonically cleaned with acetone, and finally rinsed with ethanol. The test solution was a simulated formation water with composition (in gL^{-1}): $NaHCO_3$ 0.72, CH_3COONa 1.3, $MgSO_4 \cdot 7H_2O$ 0.023, $CaCl_2 \cdot 2H_2O$ 0.46, $MgCl_2 \cdot 6H_2O$ 1.23, KCl 0.35 and NaCl 21.62. The solution was prepared from analytical grade reagents in deionised and distilled water. The inhibitor used was a commercially available water soluble amine based fatty acid compound (called KI384) produced by Dyno Oilfield Chemicals, Norway.

The test solution (4.5 L) was purged with CO_2 gas and the oxygen concentration in the solution was measured by an Orbisphere probe to be below 5 ppb. The solution pH was monitored by a pH sensor installed in the autoclave and observed to be 5.5 ± 0.2. Experiments were conducted at 35°C, and with inhibitor concentrations ranging between 0.1 and 100 ppm. The electrode rotation speed for all experiments was set at 1000 rpm. The electrodes were either pre-corroded for 2 days before inhibitor injection or non-corroded prior to injection of inhibitor (immediate injection). Cathodic and anodic polarisation curves were recorded on separate electrodes at a sweep rate of 1 mVs^{-1} from the corrosion potential, using a computer controlled AUTOLAB PGSTAT20 potentiostat from ECO CHEMIE, Netherlands. All polarisation curves were recorded at the end of the experiment, that is, 2 days after the inhibitor was injected into the solution. A 70 cm^2 plate of stainless steel was used as a counter electrode, and a saturated calomel electrode was used as reference. The reference electrode was placed outside the autoclave and connected to the test solution by a porous wooden plug. Electrochemical impedance measurements were performed at various time intervals using a computer assisted frequency response analyser (AUTOLAB/FRA) in the frequency range from 10 kHz to 10 mHz with a perturbation amplitude of 10 mV and 10 points per decade. The electrochemical impedance spectra were measured between two identical carbon steel electrodes mounted on the same rotating shaft.

3. Results and Discussion

The influence of pre-corrosion of carbon steel electrodes in CO_2 solutions containing various amounts (0 to 100 ppm) of the KI384 corrosion inhibitor was investigated by polarisation curves and electrochemical impedance spectroscopy (EIS). Figure 1 shows polarisation curves where the inhibitor has been injected into the test solution immediately after the immersion of the electrodes (no pre-corrosion), while polarisation curves for electrodes which had been pre-corroded for 2 days prior to

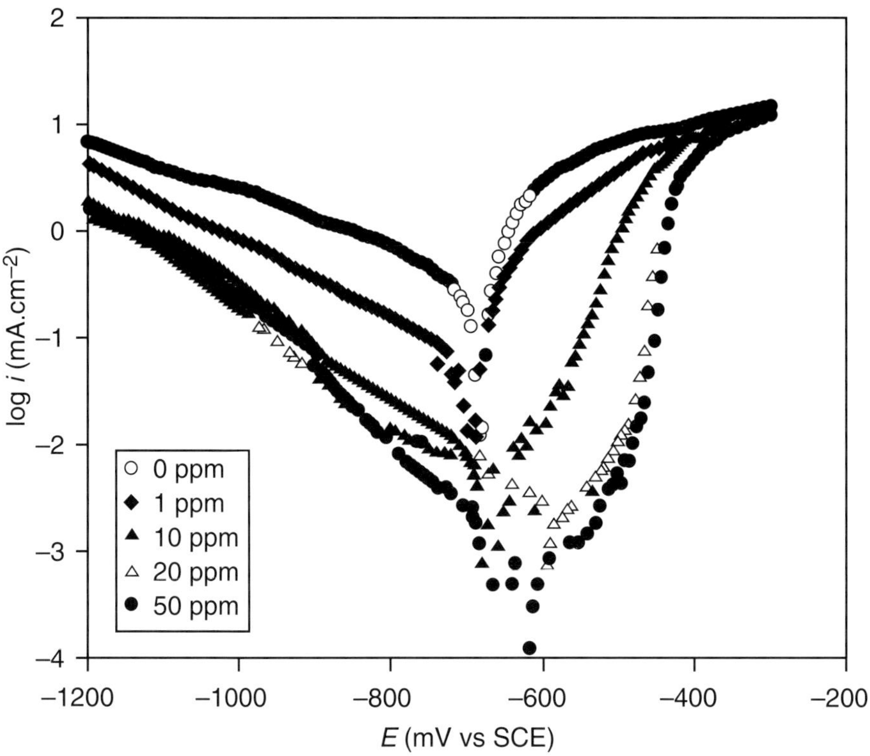

Fig. 1 Polarisation curves for non-corroded electrodes, recorded 48 h after inhibitor injection in the concentration range 0–50 ppm.

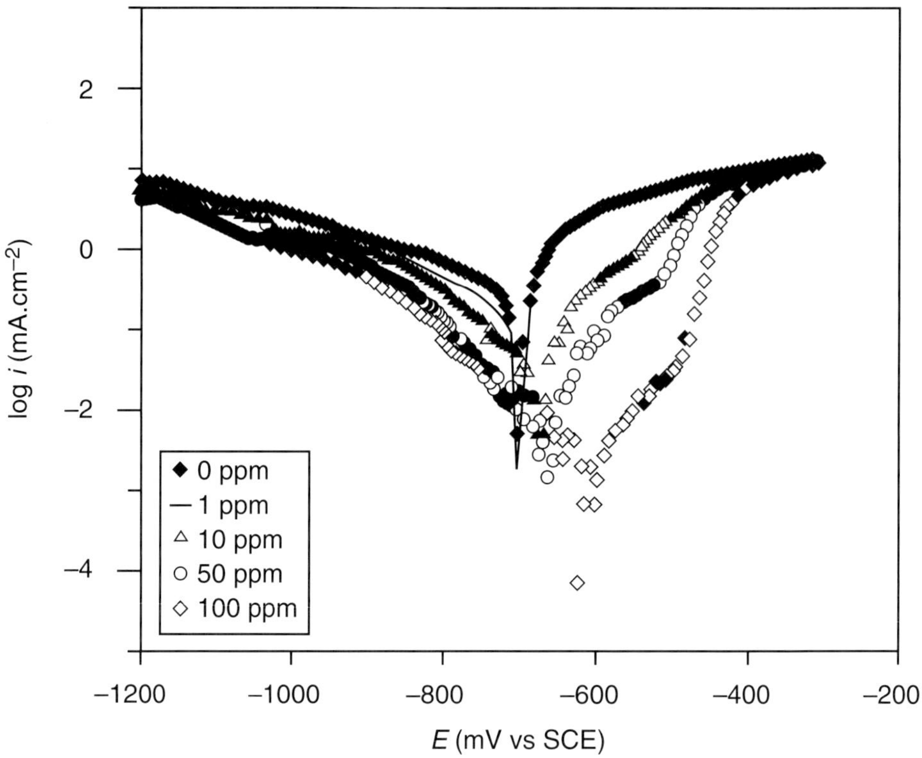

***Fig. 2** Polarisation curves for electrodes which have been pre-corroded for 2 days prior to inhibitor injection (0–100 ppm). The polarisation curves were recorded 48 h after inhibitor injection.*

inhibitor injection are shown in Fig. 2. In all cases the polarisation curves were recorded 48 h after the injection of inhibitor. Figures 1 and 2 show that the KI384 inhibitor has a significant influence on both cathodic and anodic polarisation curves and may be characterised as a mixed inhibitor. However, closer inspection of the curves shows that the corrosion potential is shifted towards anodic potentials, and the KI384 is therefore predominantly an anodic inhibitor. One can also see that the inhibitor has a significantly better protection performance on non-corroded electrodes (Fig. 1) compared to electrodes which have been pre-corroded (Fig. 2).

Tables 1 and 2 summarise electrochemical parameters, such as Tafel constants, corrosion potentials and corrosion currents densities, as well as calculated inhibition efficiencies obtained from the polarisation curves in Figs 1 and 2. The inhibition efficiency ϕ has been calculated from the corrosion current densities as follows:

$$\phi = 1 - i_{corr,i} / i_{corr,u} \tag{3}$$

where $i_{corr,i}$ and $i_{corr,u}$ are the corrosion current densities with and without inhibitor, respectively. The corrosion current density without inhibitor is referred to the system just before the injection of inhibitor. As can be seen from Tables 1 and 2 a corrosion protection of 99% is obtained by only 10 ppm inhibitor concentration for a non-corroded electrode, while nearly 100 ppm is needed to obtain the same protection

Table 1. *Tafel constants, corrosion current densities, corrosion potentials, and inhibitor efficiencies ϕ obtained from polarisation curves for rotating carbon steel electrodes (1000 rev/min) in CO_2-saturated formation water at 35°C with different inhibitor concentrations. KI384 inhibitor was* ***injected immediately (no pre-corrosion)*** *and polarisation curves were recorded 48 h after inhibitor injection*

$C_{inh.}$ (ppm)	b_a (mV/dec.)	b_c (mV/dec.)	i_{corr} (μA cm^{-2})	E_{corr} (mV vs SCE)	ϕ (%)
Blank/1 h	37	–120	280	–715	—
0.1	40	–120	170	–696	39.3
1	37	–218	54	–696	80.7
10	40	–194	2.84	–654	99.0
20	40	–177	2.23	–577	99.2
50	40	–141	0.55	–600	99.8
100	50	–117	0.42	–600	99.9

Table 2. *Tafel constants, corrosion current densities, corrosion potentials, and inhibitor efficiencies ϕ obtained from polarisation curves for rotating carbon steel electrodes (1000 rev/min) in CO_2 saturated formation water at 35°C with different inhibitor concentrations. The electrodes were* ***pre-corroded for two days*** *before inhibitor injection and polarisation curves were recorded 48 h after (KI384) inhibitor injection*

$C_{inh.}$ (ppm)	b_a (mV/dec.)	b_c (mV/dec.)	i_{corr} (μA cm^{-2})	E_{corr} (mV vs SCE)	ϕ (%)
Blank/48 h	40	–120	200	–700	—
1	39	–120	116	–700	42.0
4	54	–139	36.1	–688	82.0
10	54	–113	27.0	–673	86.5
20	41	–98	3.7	–661	98.2
50	40	–94	3.2	–654	98.4
100	50	–104	1.5	–600	99.3

for electrodes pre-corroded for 2 days. Similar results, were also calculated from EIS data (not shown here), and showed generally good agreement with results from polarisation curves.

The inhibitor protection efficiency, as defined by eqn (3), is here assumed to be equal to the degree of inhibitor coverage θ at the metal surface (θ = ϕ), and adsorption isotherms have been calculated for the KI384 inhibitor in the concentration range 0.1–100 ppm. The adsorption characteristics of the KI384 inhibitor were analysed according to the isotherm given in eqn (2). A Langmuir adsorption isotherm ($g = 0$) was found to give the best fit of the experimental data. Illustrations of the dependence of the log (θ/1–θ) vs log C for both non-corroded and 2 days pre-corroded rotating carbon steel electrodes are shown in Fig. 3. The plots for each system were found to yield a fairly straight line, with a slope close to 1, and the data were fitted to a straight line with a correlation coefficient of *ca.* 0.97 in all cases. From the results of the linear regressions, values of the slopes together with the calculated equilibrium constant K for the adsorption process are given in Table 3.

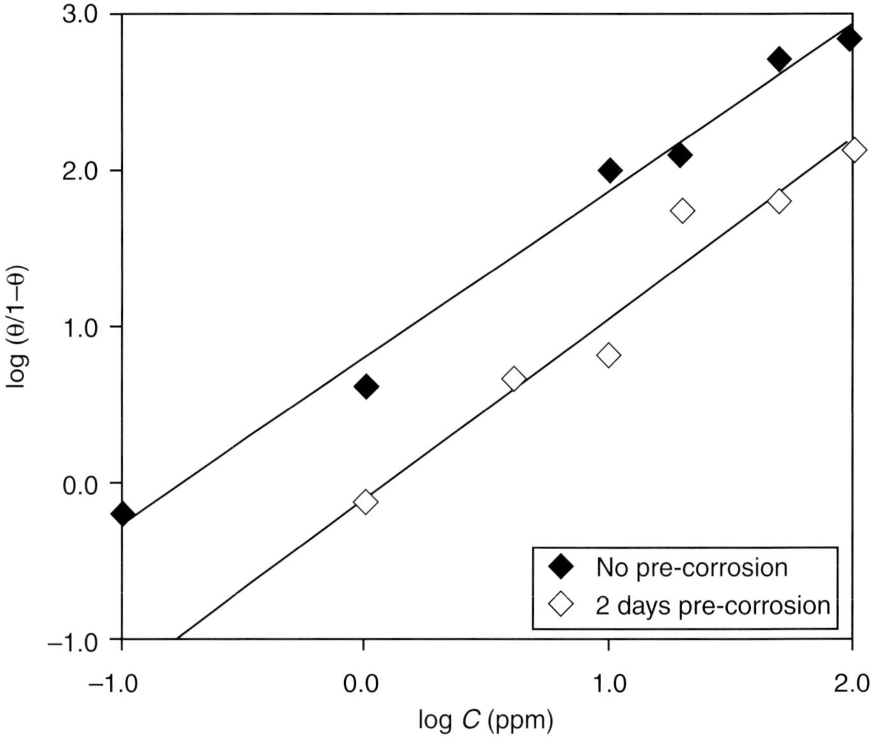

Fig. 3 Langmuir isotherms (from polarisation curves) for the KI384 inhibitor at 35°C with non-corroded electrodes and with electrodes which had been pre-corroded for 2 days before inhibitor injection.

The inhibitor coverage (= efficiency) as calculated from the polarisation curves using eqn (3) (see Tables 1 and 2) are plotted in Fig. 4 vs log *C* together with simulated data based on the Langmuir isotherm (Table 3). The simulated data points (solid line) in Fig. 4 show fairly good agreement with the experimental data points. The agreement is somewhat better for the non-corroded system compared to the pre-corroded system. Figure 4 shows that an inhibitor concentration of 1 ppm will give a

Table 3. *Equilibrium constant* K *calculated from eqn (2) assuming a Langmuir isotherm for the adsorption of the KI384 inhibitor on carbon steel in CO_2-saturated solutions. Data from polarisation curves and from EIS are fitted to the straight line, log (θ/1–θ) = log* K + *log* C

Pretreatment	**Polarisation results**		**EIS results**	
	Slope	***K* (ppm^{-1})**	**Slope**	***K* (ppm^{-1})**
No pre-corrosion	1.04	6.22	1.02	9.62
2 days pre-corrosion	1.14	0.79	1.23	0.81

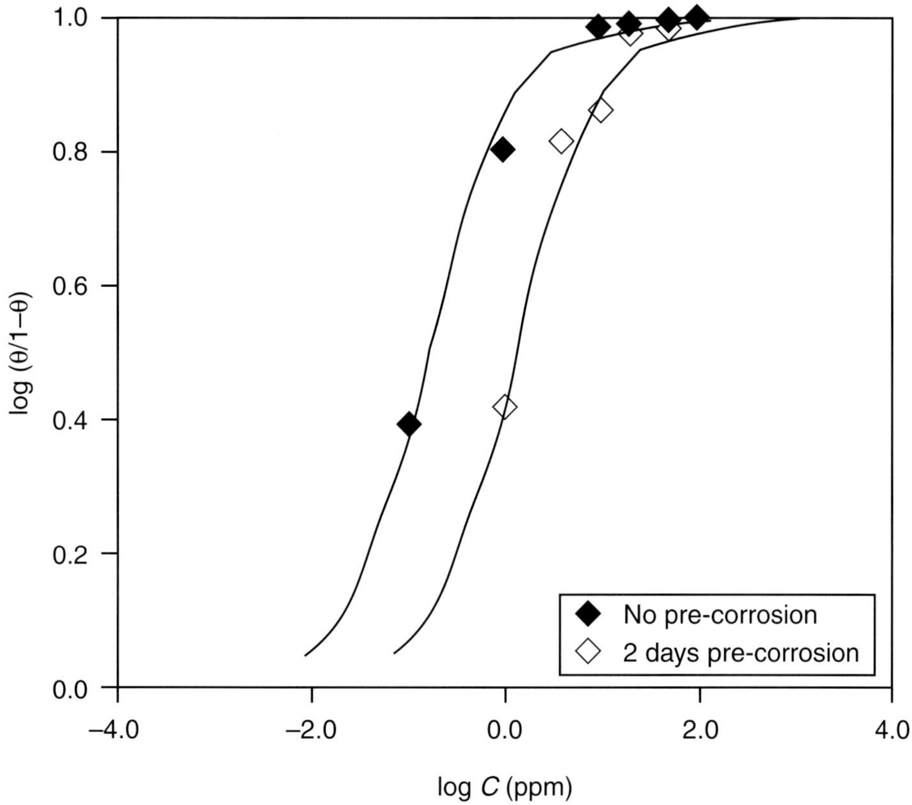

Fig. 4 *Simulated (solid lines) Langmuir isotherms (electrode coverage) and experimental electrode coverage calculated from polarisation curves for non-corroded and for 2 days pre-corroded electrodes.*

protection (coverage) of more than 80% for a non-corroded specimen but only 40% for a 2 days pre-corroded specimen.

As can be seen from Table 3, the *K* value for a non-corroded electrode is about 8–10 times higher than the *K*-value for an electrode which has been pre-corroded for 2 days prior to inhibitor injection, indicating a stronger inhibitor adsorption on the bare metal surface compared to the surface which is covered by corrosion products.

If the inhibitor is assumed to follow a Langmuir isotherm ($g = 0$) the corrosion current may be expressed as a simple function of the inhibitor concentration by combining eqns (2) and (3),

$$i_{corr,i} = i_{corr,u} / (1 + KC) \tag{4}$$

where the *K*-values are given in Table 3 and *C* is in ppm units. Typical values of corrosion rates before inhibitor $i_{corr,u}$ are 200 μAcm^{-2} for 2 days pre-corroded electrodes and 280 μAcm^{-2} or electrodes which have been exposed for 1 h in the solution.

The rate of the inhibitor adsorption process has been monitored by EIS, and Fig. 5 shows the time dependence of the inhibitor efficiency as calculated from the charge

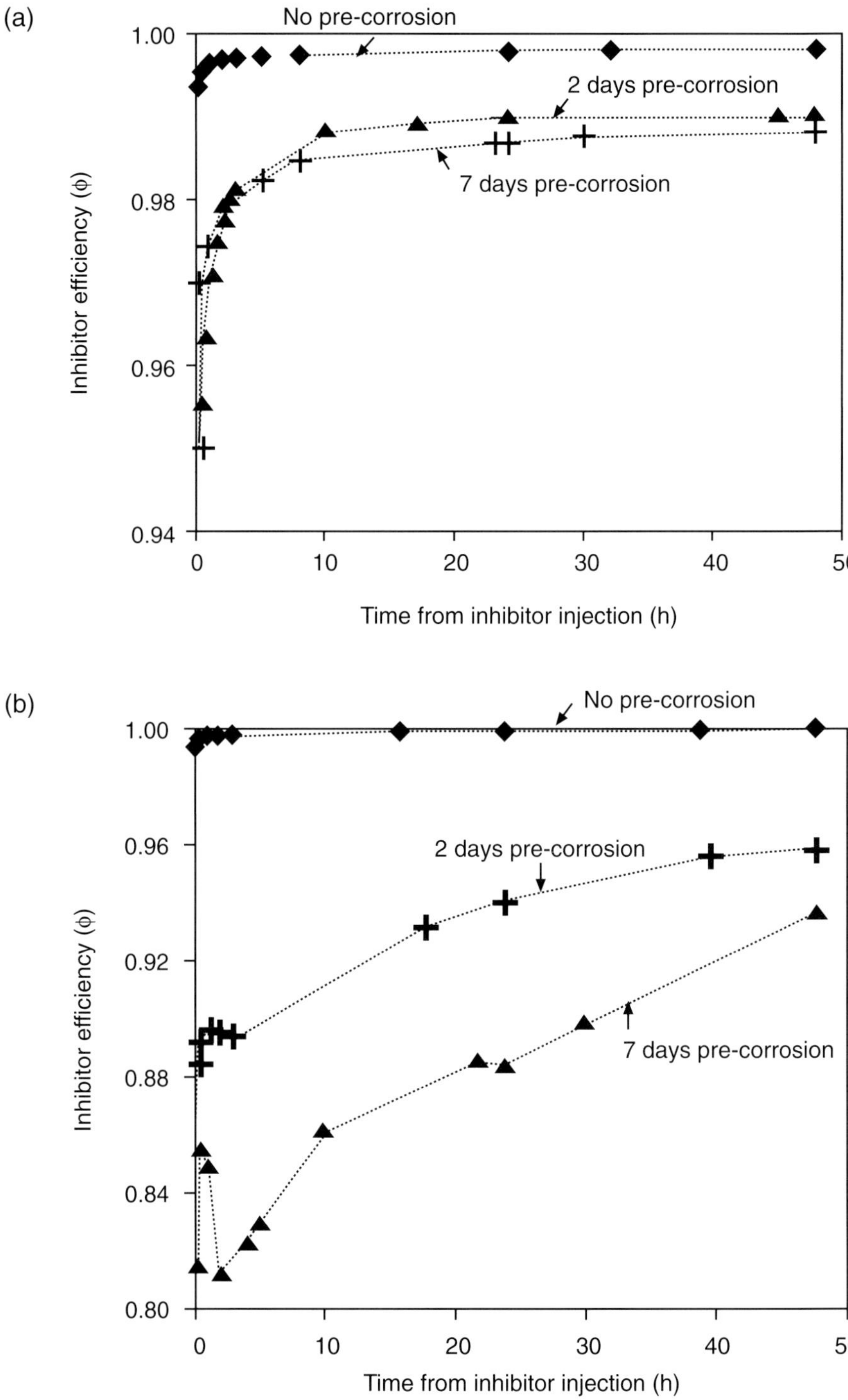

***Fig. 5** Inhibitor efficiency vs time of exposure for rotating carbon steel electrodes (1000 rev/min) in CO_2 saturated formation water at (a) 35°C and (b) 80°C with 50 ppm KI384 inhibitor.*

transfer resistance for a system with 50 ppm KI384. It can be seen from Fig. 5(a) that 99.8% efficiency has already been reached within the first hour of exposure for non-corroded electrodes with immediate inhibitor injection, while more than 10 h of exposure is needed in order to obtain 99% efficiency for electrodes which have been pre-corroded for 2 days at 35°C. The hindrance of the inhibitor adsorption on pre-corroded specimens is even more pronounced on electrodes which have been pre-corroded for 2 and 7 days in solutions at 80°C (Fig. 5b). This is most probably due to the formation of an iron carbonate corrosion product at the electrode surface which is more pronounced at temperatures above 60°C [5].

4. Conclusions

- A commercial amine based fatty acid inhibitor, KI384, has shown excellent corrosion protection properties for rotating carbon steel electrodes in a simulated CO_2 saturated formation water at 35°C. A protection efficiency of 99% has been obtained with only 10 ppm inhibitor concentration, while 100 ppm gave 99.9% protection for electrodes which had not been pre-corroded prior to inhibitor injection.

- The formation of corrosion products on the metal surface during pre-corrosion prior to inhibitor injection has a significant influence on the inhibitor performance, especially at temperatures above 60°C.

- The adsorption of the KI384 inhibitor has been found to follow a Langmuir isotherm both for non-corroded and pre-corroded electrodes, giving an equilibrium constant K for the adsorption/desorption process of 6.2 ppm^{-1} and 0.8 ppm^{-1} for non-corroded and 2 days pre-corroded electrodes respectively.

- By assuming a Langmuir isotherm for the adsorption/desorption process the concentration dependence of the corrosion rate, $i_{corr,i}$, in the presence of the KI384 inhibitor can be expressed by a simple formula, $i_{corr,i} = i_{corr,u} /(1 + KC)$, where $i_{corr,u}$ is the corrosion rate without inhibitor.

References

1. G. Schmitt, *Corrosion '83*, Paper No. 43. NACE International, Houston, Tx, USA, 1983.
2. R. H. Hausler, D.W. Stegmann and R. F. Stevens, *Corrosion*, 1989, **45**, (10), 857–870.
3. I. L. Rozenfeld, *Corrosion Inhibitors*, p. 130. McGraw-Hill, NY, 1981.
4. E. Gileadi, *Electrode Kinetics for Chemists, Chemical Engineers and Materials Scientists*. VCH, NY, 1993.
5. A. Dugstad, *Corrosion '92*, Paper No. 14. NACE International, Houston, Tx, USA, 1992.

42
The Effect of Diamines on CO_2 Corrosion of Low Carbon Steel

T. BURCHARDT, T. VÅLAND* and J. KVAREKVÅL†
Centre for Material Science, UiO, Gaustadallen 21, 0371 Oslo, Norway
*Agder College, Grooseveien 36, 4890 Grimstad, Norway
†Marintek, Norway

ABSTRACT

Although CO_2 corrosion inhibitors have been extensively studied the interpretation and understanding of inhibitor mechanisms is unclear, causing the need for more thorough investigations. Inhibitors of CO_2 corrosion are often organic molecules consisting of amines. A systematic study of diamines ranging from 1,2-ethyldiamine to 1,10-decyldiamine, has been performed using electrochemical techniques and molecular modelling. The electrochemical experiments were conducted in glass cells at 25°C and pH 3.9. A change in the adsorption mechanisms was observed with increasing carbon chain length. The large diamines reduced the corrosion rate the most effectively. A molecular modelling programme was used, minimising the quantum mechanical forces between a magnetite surface and the diamines. These calculations also showed a change in the adsorption mechanism when the carbon chain length between the diamines was varied.

1. Introduction

Corrosion inhibitors in oil and gas production are often molecules that contain amines, amides or imidazolines. The polar groups adsorb to the metal surface. In order to increase the persistency and the stability of the inhibitor films, long hydrocarbon chains are attached [1]. The understanding of inhibitor mechanisms is unclear, causing the need for more fundamental investigations [2]. Experiments have been conducted on a series of diamines, all having one amine group at each end of the molecule. The carbon chain length was altered in the range from 2–10 atoms. All the diamines were water soluble, giving a transparent solution. However, the solubility decreases with increasing carbon chain length. The diamines react with water according to the following acid-base reactions:

$$H_2N(CH_2)_n\ NH_2 + H_2O \Leftrightarrow H_2N(CH_2)_n\ NH_3^+ + OH^- \quad (1)$$

$$H_2N(CH_2)_n\ NH_3^+\ H_2O \Leftrightarrow H_3N(CH_2)_n\ NH_3^{2+} + OH^- \quad (2)$$

in which n indicates the number (2–10) of carbon atoms in the carbon chain. Solutions with pH values between 11 and 12 were obtained for the various amines. Since the pH of the electrolyte is adjusted to about 4, the ammonium ions dominate.

2. Experimental

The experiments were conducted at atmospheric pressure in a glass cell with a volume of 2 L. All the components of the cell were made from borosilicate glass. The electrolyte consisted of doubly distilled water containing 0.5M NaCl and CO_2 gas was continuously bubbled through the cell. During the experiments the temperature was kept constant at 25°C and pH at 3.9. The corrosion process was studied with electrochemical techniques using a computer controlled potentiostat. A three electrode set-up was used in all electrochemical experiments. The working electrode was of low carbon steel. An epoxy resin was used to seal the electrode into glass leaving an exposed area of approximately 2cm^2. A platinum wire was used as the counter electrode and was immersed in the middle of the cell. A saturated calomel electrode served as the reference electrode and was connected to the cell through a salt bridge filled with (0.5M NaCl) with a porous membrane. The diffusion of oxygen through the membrane was assumed to be insignificant. The pH of the solution was measured before and after the experiments. The values were compared with theoretical values under the given conditions. The temperature was monitored with an ETS-D2 probe, which also served as an input for the temperature control system consisting of a hot plate combined with a magnetic stirrer. The temperature could be kept stable at a given value with a precision of ±1 K. The amount corresponding to a concentration of 100 ppm of the required diamine was dissolved in distilled water. Before introducing the diamines into the cell, the pH was adjusted to that of the electrolyte using HCl. To ensure that no oxygen entered the cell, the diamine solution was injected one hour before the working electrode.

Prior to immersion, the low carbon steel surface was polished with 1000 mesh SiC paper, etched in concentrated HCl and washed in distilled water. For corrosion rate measurements the linear polarisation resistance (LPR) method was used with the sample being polarised ± 10 mV at a scan rate of 0.1 mVs^{-1}. At various exposure times linear polarisation scans were performed. Tafel scans were also conducted, scanning 100 mV in both directions from the corrosion potential, at a scan rate of 0.1 mVs^{-1}. Preliminary experiments with different scan rates were made to assure that the Tafel lines were unaffected by the scan rate. Cyclic polarisation scans were taken to ensure that the Tafel scans would not change the corrosion mechanisms.

3. Results and Discussion

In the experiments the presence of diamines caused a slight decrease in the corrosion rate. It should be noted that the diamines used are not complete inhibitors, but are components of many amine-based inhibitors and were used in an attempt to increase the understanding of adsorption mechanisms. For all samples, a decrease in the corrosion current was observed during the first hours, followed by a stabilisation. The diamines with short aliphatic chains produced a stable corrosion current of the same order of magnitude as the solution without diamine. The corrosion current with the longer diamines was significantly lower, the transfer region being around 1,7-heptyldiamine since experiments with this diamine gave corrosion currents of both magnitudes.

The correlation between the change in corrosion current and corrosion potential with time can be seen from a $E_{corr} - ln\ i_{corr}$ plot (Fig. 1). A gradual change in behaviour is observed with increasing carbon chain length of the diamines. For the short diamines represented (in Fig. 1 by 1,6 diamine) the decrease in the corrosion current is followed by an increase in the potential. For the long diamines the reduction in the corrosion is combined with a decrease in the potential. The reduction in the corrosion rate during the first few hours is probably due to the formation of a thin protective layer. This layer is only slightly affected by the short diamines as the corrosion rate is of the same order of magnitude as in the solution without diamine. When the larger diamines are present the behaviour of this layer is significantly altered as a pronounced reduction in the corrosion rate is observed.

Figures 2, 3 and 4 show Tafel scans taken after 24 h of corrosion. The LPR measurements are shown in the same figure. From these plots one can distinguish between the effect of the anodic and the cathodic reactions on the corrosion rate. In Fig. 2 the Tafel lines are extended so that the corrosion current and potential can be found from the intersection point. As can be seen from Fig. 2, the LPR measurements correlate well with the results found from the Tafel scan — remembering that the Tafel scan is taken immediately after the 24 h LPR experiment. Each LPR measurement represents the intersection between the anodic and the cathodic Tafel line. In that way, the LPR measurements indicate how the anodic and the cathodic reactions change with time.

For the solution with short diamines (1,2–1,6), an initial reduction in both the cathodic and the anodic reaction can be seen, since the corrosion rate decreases at a stable corrosion potential (Fig. 3). For the solution without diamine (Fig. 2), the cathodic reaction initially decreases faster than the anodic reaction. The difference between no diamine and the short diamines is small. After approximately 2–3 h, a change in the E_{corr}–$ln\ i_{corr}$ curve can be detected both for the samples with short

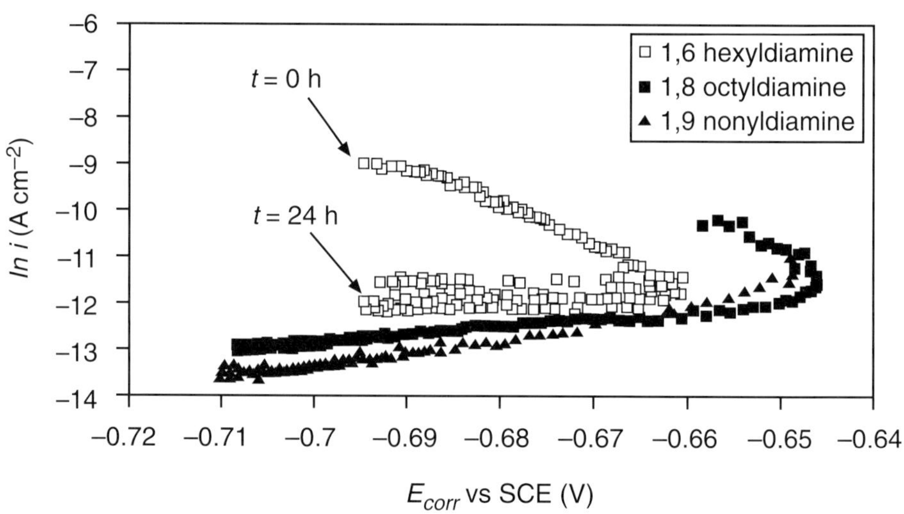

Fig. 1 *Corrosion potential vs ln corrosion current plot of low carbon steel at 25°C with pH 3.9 1 bar CO_2, 0.5M NaCl.*

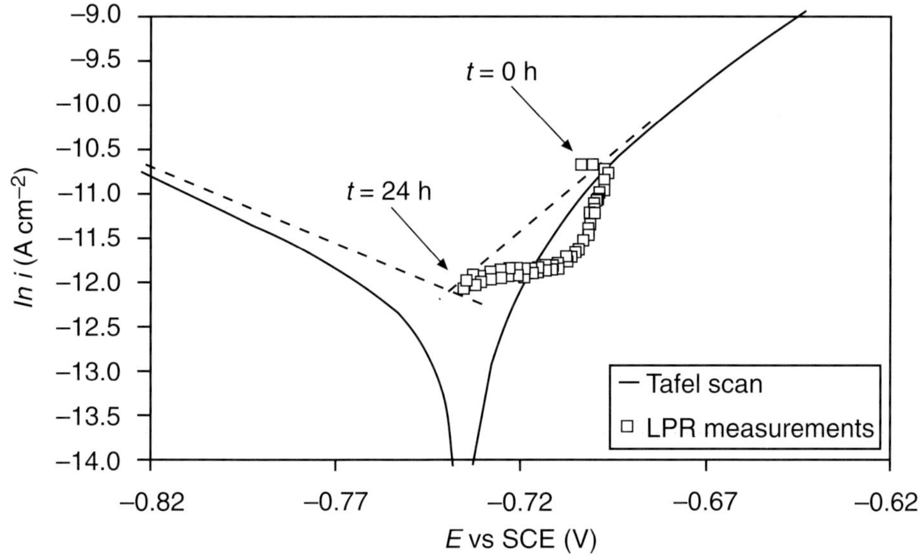

Fig. 2 Potential vs current density plot for sample without diamine; low carbon steel at 25°C and pH 3.9. 1 bar CO_2, 0.5M NaCl.

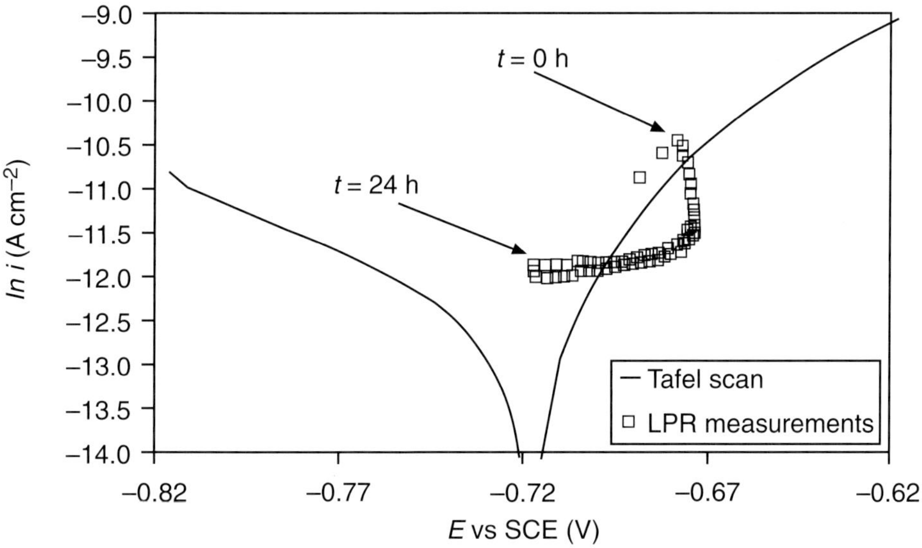

Fig. 3 Potential vs current density plot for sample with 1,6-hexyldiamine; low carbon steel at 25°C and pH 3.9. 1 bar CO_2, 0.5M NaCl.

diamines and the pure CO_2 sample. The corrosion rate becomes constant while, the corrosion potential continues to decrease. One explanation for the initial reduction in the anodic reaction can be the nucleation of a protective layer on the iron dissolution active sites. The initial reduction in the cathodic reaction can then either be due to the diffusion of reductive species through the produced layer, or to nucleation on the cathodic sites. For the short diamines reductions in both the anodic and the cathodic reactions are observed during the first 2–3 h, indicating the formation of a layer which decreases both reactions. After 2–3 h the changes in the corrosion current are small, although the potential drifts towards lower values. A reduction in the potential at a stable corrosion current indicates an increase in the anodic reaction, followed by a decrease in the cathodic reaction. One explanation for this can be that the inhibition of cathodic sites influences the inhibition of the anodic sites. If the inhibition of cathodic sites generates anodic sites it would cause the anodic reaction to increase proportionally to the decrease in the cathodic reaction.

For the large diamines the anodic reaction remains unchanged throughout the experiment, and the reduction in the corrosion rate is only due to a change in the cathodic reaction (Fig. 4). The reduction in the corrosion current is then probably caused by diffusion of oxidants through a rapidly formed protective layer. The properties of the protective layer formed on the low carbon steel surface are altered significantly when the large diamines are present. This is probably due to a difference in the adsorption mechanisms between the long and the short diamines.

4. Molecular Modelling

In the attempt to understand more about the adsorption processes of the diamines, a molecular modelling program was used. In addition to a visual picture of the metal surface and the diamines, this also provides the possibility of calculating the intermolecular interactive energies between them. A detailed explanation of how

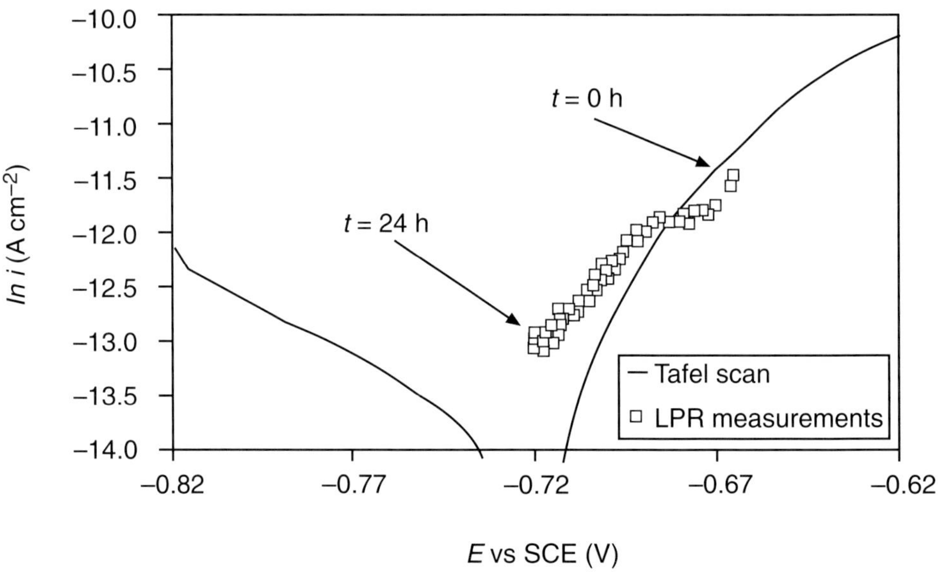

Fig. 4 *Potential vs current density plot for sample with 1,9-nonyldiamine. Low carbon steel at 25°C and pH 3.9. 1 bar CO_2, 0.5M NaCl.*

the total interaction energy is calculated can be found in the theory manual [3]. By minimising the interaction energies between a molecule and a crystal surface an idealised picture of how the diamines will adsorb onto a metal surface can be given. It must however be emphasised that this is an idealised picture. The simplifications made might not be justified. It can, however, be helpful in deducing some ideas about the adsorption processes.

From the electrochemical experiments a change in the adsorption mechanisms was observed with increasing aliphatic carbon chain length. Calculations were therefore made with different diamines. Using crystallographic data a magnetite (Fe_3O_4) crystal surface was constructed and used in the calculations. Magnetite is used since an oxidised surface is expected to be present. The interactions between a magnetite crystal surface and different diamines were calculated (Figs 5–7). All the calculations show that the two charged NH_3^+ groups present in the molecules try to come as close as possible to the opposite charged surface. At the same time, the aliphatic part of the surface attempts to get as far away from the surface as possible. The result produces the situations described in Figs 5–7. It can be seen, that for 1,3 diamine and 1,9 diamine both NH_3^+ groups are attached to the surface, for 1,6 diamines, on the other hand, only one group manages to be attached. From these calculations it should be expected that 1,9 diamine reduces the current to the largest extent as it covers a large surface area with both NH_3^+ groups attached to the surface. The 1,6 diamine also covers a large surface area but only one NH_3^+ group is attached, giving a weaker binding; 1,3 diamine has both groups attached but it does not cover such a large surface area. This correlates well with the electrochemical experiments. Although providing good correlation, the pictures are simplified compared to reality. Thus, no attention has been paid to the fact that the adsorption takes place in a liquid phase. Also the assumption of an ideal magnetite crystal surface is not probable. However, the use of molecular modelling visualises the adsorption process and therefore can increase the understanding.

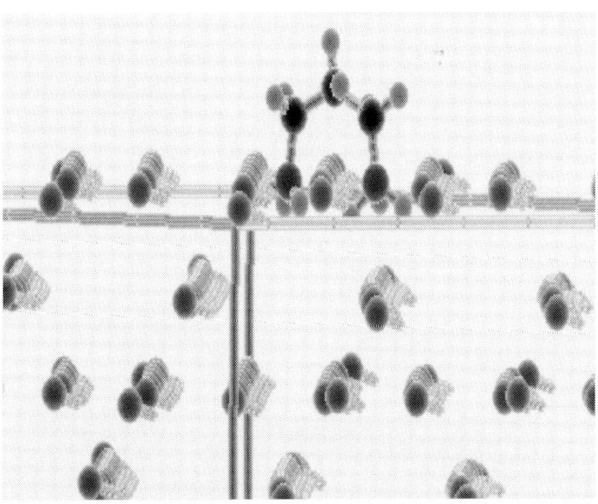

Fig. 5 *1,3 propyldiamine adsorbed at a magnetite surface.*

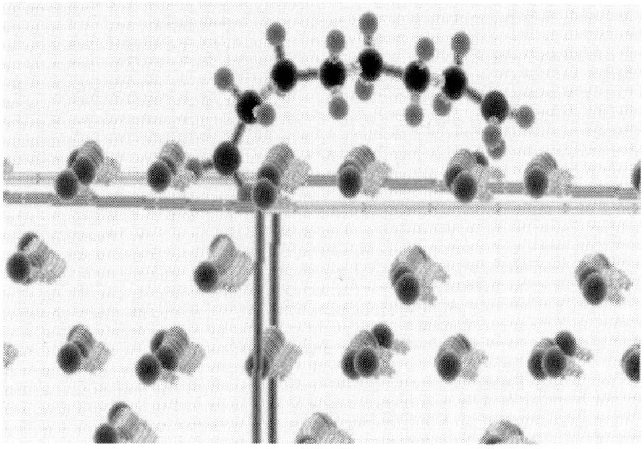

Fig. 6 *1,6 hexyldiamine adsorbed at magnetite surface.*

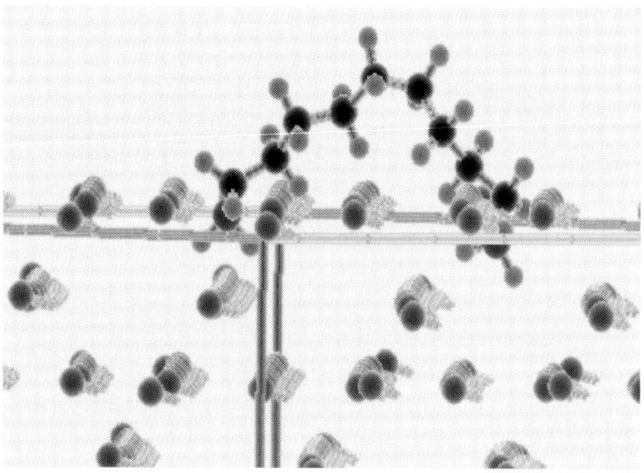

Fig. 7 *1,9 nonyldiamine adsorbed at a magnetite surface.*

5. Conclusions

The long diamines cause a relatively significant drop in the corrosion rate compared to a solution without diamine, whereas for the short diamines only a slight reduction in the corrosion rate could be observed. The experiments indicate that a difference in the adsorption mechanism occurs when the carbon chain length exceeds 7.

Molecular modelling simulations support this finding, as only one of the amine groups is adsorbed for the small diamines, whereas for the large diamines both groups adsorb. One exception is 1,3-propyldiamine where, according to the simulations, both groups adsorb.

References

1. A. K. Dunlop, *Corrosion Inhibitors*, C. C. Nathan, ed. NACE, Houston, Tx, USA, 1973, p.78–88.
2. G.Trabanelli, *Corrosion*, 1991, **47** (6), 410–419.
3. SYBYL. Theory manual, SYBYL Molecular Modelling Software, Version 6.1, Tripos Inc., 1994.

43

Effect of Pre-Corrosion on the Performance of Inhibitors for CO_2 Corrosion of Carbon Steel*

E. GULBRANDSEN, B. SUNDFÆR[†], S. M. HESJEVIK[†], S. SKJERVE[†], S. NESIC[‡] and T. BURCHARD[§]
Institute for Energy Technology (IFE), Norway
[†]Statoil Research Centre, Norway

ABSTRACT

Six commercially available, water soluble corrosion inhibitors for CO_2 corrosion have been tested in the laboratory on carbon steel specimens that were corroded for 3–18 days prior to inhibitor addition. The results show that the inhibitor performances were impaired, often dramatically, with increasing pre-corrosion time and increasing temperature. The corrosion was localised with deep pits. The detrimental effect of pre-corrosion is determined both by the nature of the steel and the inhibitor composition.

1. Introduction

Corrosion inhibitors are often tested on freshly ground specimens. However, in the field inhibitors meet steel surfaces that are covered with different kinds of corrosion products. These products may affect the performance of the inhibitor. Little work has been done on the effect of long time corrosion prior to inhibition (pre-corrosion). The problem of pre-corrosion is relevant to oil field applications of inhibitors, where a pipeline may have been in operation for several years before increasing water-cut necessitates inhibition. As part of a larger inhibitor testing programme carried out by Statoil and IFE, the effect of pre-corrosion has been studied experimentally.

2. Experimental

Six different commercially available inhibitors from three different suppliers were tested. The inhibitors are listed in Table 1. The pre-corrosion effect was examined on three different steels, here denoted as X65, 0.5%Cr and St52. The element analyses, microstructures and complete designations of the steels are given in Table 2. The steel St52 is a typical construction steel, while X65 and 0.5%Cr are used in pipelines.

---

*This paper was presented at NACE *Corrosion '98* conference: NACE International is the copyright holder.
[‡]Present address: University of Queensland, Department of Mechanical Engineering, Australia.
[§]Present address: University of Oslo, Department of Chemistry, Norway.

Table 1. *Inhibitors for CO_2 corrosion tested for pre-corrosion effect. All inhibitors were water soluble*

Inhibitor	Active compound	Inhibitor	Active compound
A	Imidazoline	D	?
B	Imidazoline	E	Amine
C	Amine	F	Amine

The inhibitor tests were conducted under natural convection conditions in glass cells and autoclaves, and under straight pipe flow conditions in a glass corrosion loop. The design of these facilities has been described in detail previously, (see e.g. [1]). Cylindrical specimens (3.1 cm^2) were used in the glass cells, and pipe wall flush mounted specimens (2.9 cm^2) in the loop. The specimens were pre-treated by grinding to 1000 grit with SiC paper, degreasing in acetone, and rinsing in ethanol before immersion. The test solutions were deoxygenated and CO_2 saturated by bubbling CO_2. The solution pH was adjusted with HCl or $NaHCO_3$. The concentration of dissolved O_2 was held below 10 ppb. The Fe^{2+} concentration was monitored by chemical analysis, and kept at the desired level by changing the electrolyte solution at regular intervals. The details of the experimental conditions are given in the Figure captions.

The corrosion rates were determined electrochemically [2] by linear polarisation resistance (LPR) measurements (± 5 mV vs E_{corr}, 0.1 mVs^{-1}, and in a few cases by electrochemical impedance (EIS) measurements. Mass loss measurements [3] were conducted to verify the LPR data.

3. Results and Discussion

The results are presented according to the type of corrosion layer that was formed during the pre-corrosion.

Table 2. *Carbon steels used in the inhibitor tests*

Steel	Designation	Microstructure	Shape	IFE ref. no.
X65	API 5L X65	Ferrite–pearlite	Plate t = 25 mm	56
0.5% Cr	API 5L X65	Ferrite–Wiedmanstätten	Pipe t = 19 mm	50
St 52	St. 52–3	Ferrite–pearlite	Rod dia. = 12 mm	53

t = thickness.

Steel	C	Si	Mn	S	P	Cr	Ni	V	Mo	Cu	Al	Sn	Nb
X65	0.064	0.26	1.55	0.001	0.012	0.05	0.04	0.035	0.01	0.04	0.041	0.002	0.041
0.5% Cr	0.072	0.17	0.89	0.002	0.014	0.6	0.02		0.01	0.01	0.038	0.001	
St 52	0.13	0.38	1.29	0.008	0.015	0.07	0.09	0.035	0.01	0.34	0.05	0.015	

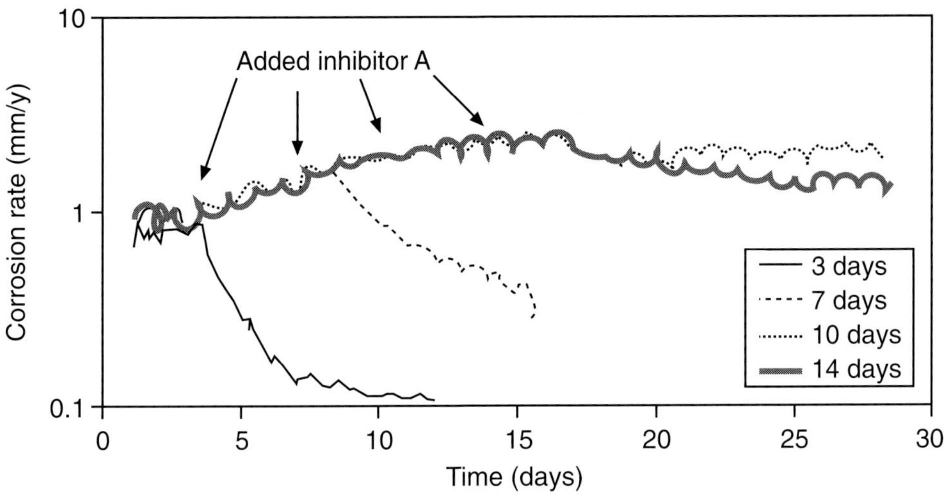

***Fig. 1** Corrosion rate of St52 specimens vs time in experiments with Inhibitor A. The pre-corrosion times are given in the legend, and the arrows show the points of time when inhibitor was added to the different glass cells. Experimental conditions: 20°C, pH 5, 1 bar CO_2, 1% NaCl, natural convection.*

3.1. Cementite Layers

The effect of increasing pre-corrosion time is well illustrated by the results from a test of Inhibitor A on St52 steel (Fig. 1). Two important trends are evident. Firstly, the corrosion rate increased during the pre-corrosion period, probably due to the formation of a porous cementite layer, which is believed to act as a cathode. Secondly, the inhibitor efficiency decreased strongly with increasing pre-corrosion time. The inhibitor (20 ppm) worked efficiently on the specimen that was pre-corroded for 3 days, giving a final corrosion rate of 0.1 mm/y. However, the inhibitor worked much less efficiently on the specimen with 7 days pre-corrosion, a corrosion rate of 0.4 mm/y being obtained 1 week after inhibitor addition. Specimens given 10 and 14 days pre-corrosion were practically not protected at all. In a similar set of experiments with X65 steel, it was found that Inhibitor A worked efficiently even after 14 days pre-corrosion. Obviously, the effect of pre-corrosion is influenced by the properties of the steel. This fact is further illustrated in Fig. 2, which shows the results from a test of Inhibitor B in the corrosion loop. The specimens were pre-corroded for 15 days (20°C, pH 5). Addition of 25 ppm inhibitor had little effect on the St52 specimen, whereas it provided better inhibition for the X65 (74% efficiency) and 0.5%Cr (94% efficiency) specimens. An X65 specimen with rust and mill scale at the surface prior to the pre-corrosion period was also tested in this experiment. This specimen obtained about the same degree of inhibition as the 0.5%Cr specimen. Improved inhibition was achieved for all specimens, except St52, by raising the inhibitor concentration to 55 ppm after 18 days. A similar experiment was performed with Inhibitor C. This inhibitor performed slightly better at 25 ppm, but the relative ranking between the different steels and surfaces with respect to corrosion rate and inhibitor efficiency was the same.

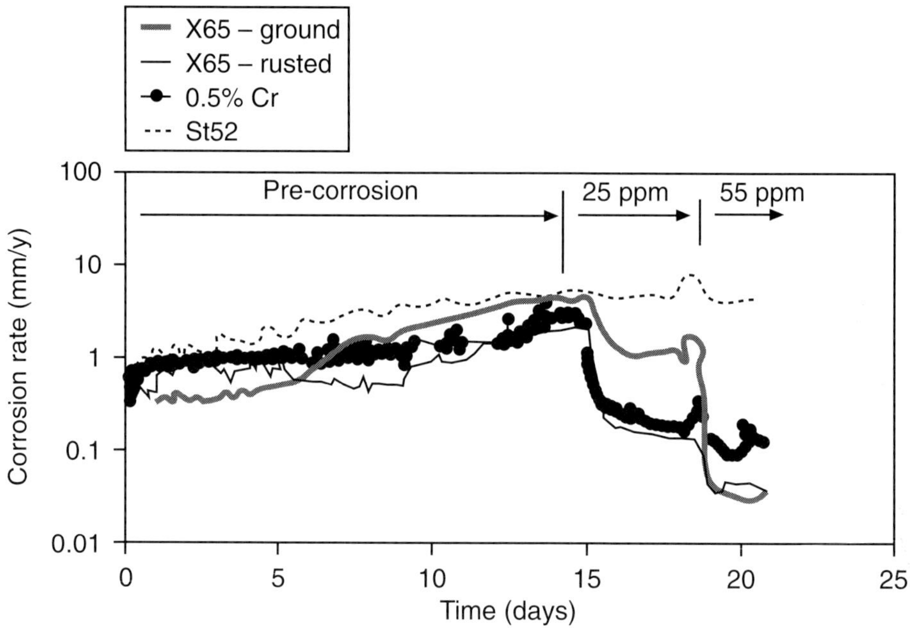

Fig. 2 Corrosion rate vs time in experiments with Inhibitor A. Experimental conditions : 20°C, pH 5, 1 bar CO_2, 1% NaCl, pipe flow 1 ms^{-1}.

The pre-corrosion effect became more severe when the temperature was increased. Inhibitor C and Inhibitor F were tested at 50°C, pH 5 in the loop. The specimens were pre-corroded for 18 days, and the experiment was continued 45–50 days after inhibitor addition. The results for Inhibitor F, which are typical, are shown in Fig. 3. None of the steels obtained stable corrosion rates below 1 mm/y. Freshly ground specimens were also inserted during this period. The inhibitor worked efficiently on these specimens, showing that thermal or chemical breakdown of the inhibitors could not explain the poor efficiency on the pre-corroded specimens. The X65 specimen was taken out of the solution after 60 days (see arrow in Fig. 3), and the corrosion products were removed by careful brushing before the specimen was re-inserted. The corrosion rate decreased by more than half a decade. Examination of the exposed specimens following the experiment with Inhibitor C revealed localised attack (Fig. 4). Parts of the specimen surface were protected, but at certain spots the corrosion continued at high rates, giving deep circular pits. More open pits were observed at 0.5%Cr, whereas in the case of St52 so small a fraction of the surface was protected (*cf*. Fig. 3) that the attack was nearly uniform.

3.2. Iron Carbonate Layers

Inhibitor A and Inhibitor F were tested on X65 specimens that had been pre-corroded for 14 days at 65°C, pH 5.5 and 2 bar CO_2. The specimens became covered with a 30–50 μm thick carbonate layer. The solution was replaced with a new solution at 50°C, pH 4 before inhibitor addition. A freshly ground specimen (Fig. 5) showed an initial corrosion rate of 7 mm/y, which decreased rapidly to 0.1 mm/y after addition

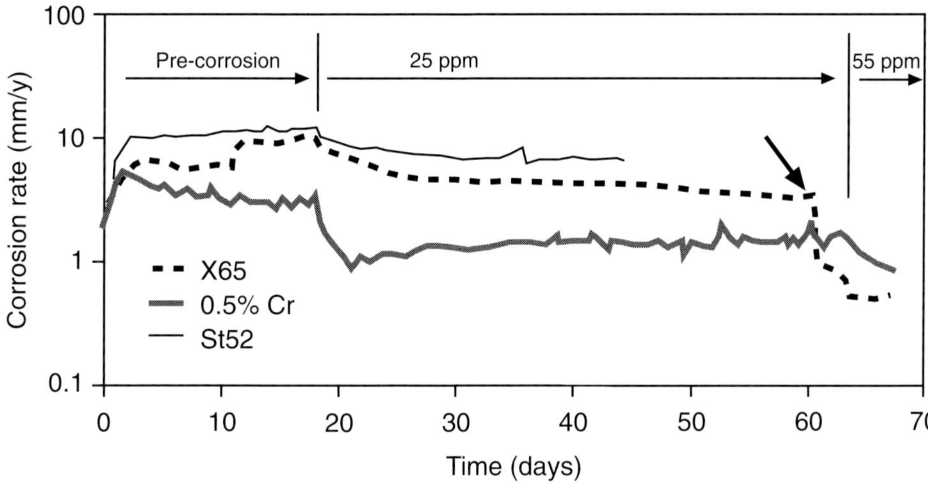

Fig. 3 *Corrosion rate vs time in experiments with Inhibitor F. Experimental conditions: 50°C, pH 5, 1 bar CO_2, 3% NaCl, pipe flow 1 ms^{-1}.*

of 30 ppm Inhibitor A. The pre-corroded specimens started at a corrosion rate of 0.1–0.2 mm/y. The rate increased rapidly towards 1–4 mm/y due to carbonate film dissolution. The inhibitors were clearly not able to protect the specimens. Inhibitor A and Inhibitor D were tested in a similar experiment (80°C, pH 5.5) after 1–4 days pre-corrosion at 60°C with very high Fe^{2+} concentration (steel wool was present in the pre-corrosion cell). The corrosion rate decreased to 0.1 mm/y when 20–30 ppm inhibitor was added (results not shown). This shows that the inhibitors performed better, when a shorter pre-corrosion time was applied, similar to the trend observed on cementite layers above.

3.3. Rust Layers

The results obtained for the specimen with mill scale and rust in Fig. 2. indicate that the rust layer did not adversely affect the inhibitor performance compared to that

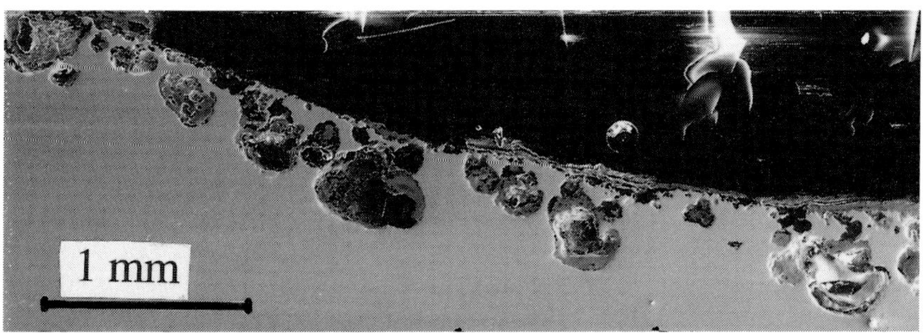

Fig. 4 *Scanning electron micrograph of X65 specimen from pre-corrosion experiments with Inhibitor C. Experimental conditions: see Fig. 3.*

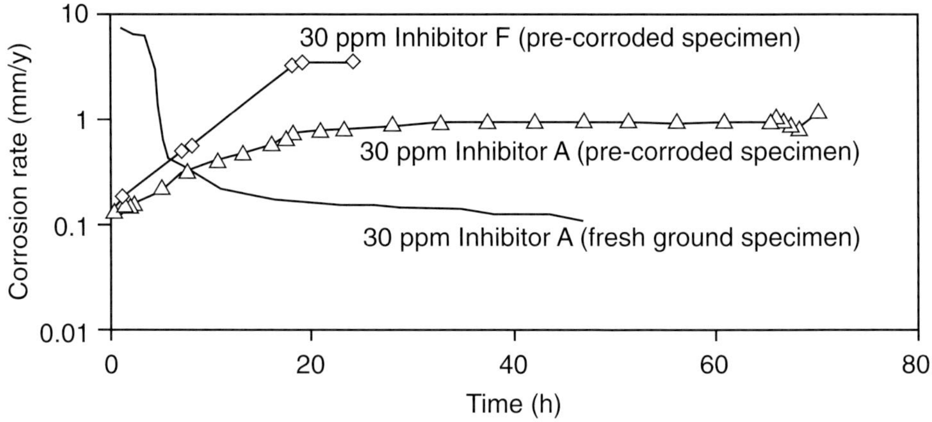

Fig. 5 *Corrosion rate of X65 specimens vs time after addition of Inhibitor A and F. Experimental conditions: 50°C, pH 4, 2 bar CO_2, 1% NaCl, natural convection. The specimens had been pre-corroded for 14 days at 65°C, pH 5.5, 2 bar CO_2, 1% NaCl, natural convection.*

with ground specimens. Inhibitor A was tested on a X65 specimen that had been pre-corroded in an air saturated solution. In this case the inhibitor worked efficiently, despite the rust layer that was formed.

The present results show that the performances of the tested inhibitors were impaired when the specimens had been pre-corroded for many days. The poor inhibitor efficiency is clearly connected to localised attack. One may suspect that this pre-corrosion effect is due to the accumulation of corrosion products at the surface (e.g. by way of 'steric' or chemical interactions with the inhibitor film). When the corrosion product layer was removed (Fig. 3), the corrosion rate decreased. The pre-corrosion effect occurs under conditions where mainly 'intrinsic' corrosion products (cementite, alloying elements) are present. Previously formed rust layers did not seem to contribute separately to the pre-corrosion effect. The same may be true for carbonate corrosion layers; the effect seen in Fig. 4 may well be due to the corrosion prior to the carbonate film formation. The cementite may accordingly have a detrimental effect on the inhibitor film. In addition, the large cathode surface provided by the cementite can contribute to maintaining a high corrosion rate in the unprotected parts (pits) of the steel surface. Both of these explanations are consistent with the observation of a decreased corrosion rate when the corrosion layer was removed. It may be noted that St52, which was the most severely affected steel amongst those tested, also had the highest content of carbon. However, other alloying elements may be co-determining factors for the reported pre-corrosion effect.

4. Conclusions

The experimental results show that inhibitor performances were impaired, often dramatically, with increasing pre-corrosion time and increasing temperature. The corrosion was localised with deep pits. The detrimental effect of pre-corrosion is determined both by the nature of the steel and the inhibitor composition. The pre-corrosion effect may be caused by the cementite layer or by alloying elements present in the steel. The effect of long term pre-corrosion should be given more attention when inhibitors are tested.

5. Acknowledgement

This work was financed by Statoil. The authors appreciate the permission granted by Statoil to publish these results.

References

1. S. Nesic, G. Solvi and J. Enerhaug, "Comparison of the rotating cylinder and pipe flow test for flow sensitive CO_2 corrosion", *Corrosion '95*, Paper No. 130. NACE International, Houston, Tx, USA, 1995.
2. ASTM Standard Practice G 102–89, "Standard Practice for Calculation of Corrosion Rates from Electrochemical Measurements", Annual Book of ASTM Standards Vol. 03.02., P. Fazio, Ed. American Society for Testing and Materials, West Conshohocken, Pa, pp. 406–411, 1992.
3. ASTM Standard Practice G 1–90, "Standard Practice for Preparing, Cleaning, and Evaluating Corrosion Test Specimens", Annual Book of ASTM Standards Vol. 03.02., P. Fazio, Ed. American Society for Testing and Materials, West Conshohocken, Pa, pp. 35–41, 1992.

44
Field Trials Using a New Generation of Electrical Resistance Probe for the Optimisation of Chemical Corrosion Inhibitors for Oilfield Applications

B. RIDD, R. JOHNSEN* and D. QUEEN†
CorrOcean Ltd, Norwich, UK
*CorrOcean a.s, Trondheim, Norway
†Shell UK Exploration & Production, Lowestoft, UK

ABSTRACT
In today's competitive world the use of carbon steels is still prevalent and the use of corrosion inhibitors to mitigate against internal pipework corrosion continues to offer an economic alternative to corrosion resistant alloys. Laboratory evaluation of corrosion inhibitor performance provides screening data; however, optimisation of the chemical can best be achieved through field trials which provide the ultimate test to evaluate the effectiveness of an inhibitor under actual operating conditions. CorrOcean have developed a new generation of electrical resistance probes to allow highly sensitive and immediate response to changes in corrosion rates on the internal environment of production pipework. The probe, because of its quick response to changes within the corrosive medium, provides a highly effective method of optimising the use of inhibitor chemicals resulting in confidence in corrosion control and minimising detrimental effects from the over-application of chemical products.

1. Introduction

In oil and gas production systems carbon steel has been widely employed as a construction material in many applications such as downhole tubulars, process facilities and transmission pipelines. When carbon dioxide is present in these systems it dissolves in the aqueous phase forming carbonic acid which subsequently produces an environment which is corrosive to carbon steel. To control the problems of CO_2 corrosion, corrosion resistant alloys (CRAs) can be used, however, the increased cost of these materials means that the use of carbon steels in conjunction with corrosion inhibitors is a practical and flexible alternative. Effective corrosion control by the use of inhibitors therefore provides an economic option for the design and operation of oil and gas fields.

The need to produce and transport oil and gas under increasingly more severe operating conditions is placing a major strain on the performance of conventional corrosion inhibitors used for this service. Furthermore, environmental legislation is pushing to reduce the damaging effects of conventional oil soluble inhibitors. Consequently, laboratory and field testing have become critical steps in the qualification, specification and optimisation of alternative inhibitors for oilfield applications.

Typically corrosion inhibitors are used to mitigate against the corrosion of production tubulars, flowlines, process pipework, interfield and export pipelines and each application entails different sets of operating conditions. The effect of these conditions must be considered carefully during the inhibitor selection process. The fluid characteristics are the primary consideration, i.e. temperature and pressure, fluid velocity, carbon dioxide and hydrogen sulfide content, oxygen concentration, water chemistry, condensate, multiphase or single phase flow must all be considered when selecting an inhibitor for a particular application. Other considerations include the compatibility with other chemicals which may be applied to the process fluid such as glycol, hydrate inhibitors, scale inhibitors, surfactants and oxygen scavengers which can produce processing problems through the formation of fluid interface emulsions. The method of application must also be considered, i.e. whether the chemical is to be applied as a bulk treatment or as a specific control measure. To ensure effective treatment, inhibitor application rates, concentrations and methods must also be considered.

Field trials provide the best means of proving the suitability of a particular chemical for a specific application and permit optimisation of the inhibition programme.

2. Field Corrosion Inhibitor Optimisation

The use of new 'environmentally friendly' inhibitors requires strict screening and performance testing if they are to provide the same level of performance traditionally achievable from oil-soluble inhibitors. Screening and performance testing must however reflect the intended operating conditions.

The primary selection and screening of candidate inhibitors is undertaken in the laboratory using various methods which provide a good indication of inhibitor performance over relatively short time scales. These tests should ensure that the operating process conditions are mimicked as closely as possible. Field trials, which monitor the ultimate performance and provide the means to tune the inhibition programme require equipment more suited to the offshore environment. A number of traditional methods to determine inhibitor performance and effectiveness are listed below:

- Weight loss coupons.
- Solution analysis to determine iron (Fe^{2+}) and inhibitor concentration.
- NDT methods to determine wall thickness loss.
- Traditional ER probe measurement.
- Various electrochemical measurements.
- Field Signature Method.

These methods are generally unsuitable for inhibitor field trials since they are either insufficiently responsive to changes in process conditions and hence inhibitor performance or prohibitively expensive for short term deployment.

Injection of inhibitor into the process stream also assumes that the inhibitor will form a continuous film at all locations downstream of the injection point. Typically a single injection rate is selected to suit all operating conditions and may result in under or over injection of chemicals since optimisation of chemical usage with traditional corrosion rate monitoring equipment has been extremely difficult to achieve. The production of a robust, highly sensitive and economical corrosion monitoring tool allowing the corrosivity of a process fluid to be monitored over short time intervals was therefore seen as a necessary development. Inhibitors identified from laboratory trials can now be tested under field conditions to determine the optimum injection level and identify any effect of local conditions.

3. Equipment Development

From the above, it can be seen that an effective field method of determining inhibitor efficiency provides a potent way of optimising inhibitor regimes. CorrOcean have designed such equipment based on technology developed during the design of its corrosion and erosion monitoring tools. The basis of this development is centred on thin element probe technology. The processes involved in the manufacture of probe elements has significantly improved in recent years and a new generation of probes with very thin measuring elements (less than 100 μm) coupled with the appropriate length to width ratio is now achievable. The sensitivity of such a probe has been measured at ±25 nm. In addition a new generation data logger unit, MultiLog™, has been developed to handle data from the thin element probe.

In selecting a probe for a particular monitoring application it is essential to consider the corrosivity of the process fluid and the quality of the data required. For long term monitoring of corrosive fluids it is prudent to select a probe with a relatively thick sensing element which will provide information over a long time period. The sensitivity of the probe is proportional to the probe thickness and length to width ratio. In situations where the corrosivity of the process fluid is expected to be relatively low, a thinner, longer element probe which provides more sensitive information is necessary.

Where the information required is to be gained over very short time scales, e.g. during an inhibitor optimisation programme where very sensitive changes in the operating environment are critical, it is essential to have a very sensitive and responsive means of determining changes in the corrosivity of the process fluid. This may be achieved by the introduction of the thin element probe which can provide quantitative readings almost instantaneously due to its sensitivity.

4. Advantages of Thin Element Probes

The main advantages of the sensitive thin element probe may be summarised as follows:

- Faster response achievable without reducing probe life.
- Probe lives may be increased without losing accuracy/response times.
- Active probe surface area may be increased thus providing a higher correlation between measured corrosion rates with actual corrosion rates.
- Enhanced sensitivity to process change and hence superior response times when compared with traditional ER probes.

Developments in datalogging capabilities have provided a useful tool to gather real time data which can identify accurately points in time when important events have occurred within the process fluid, e.g. increase in production rate, increase in water-cut or a change of inhibitor concentration. The logging time may be set as frequently as desired to provide the level of data required by the particular application.

CorrOcean also have the technology to put all monitoring probes on-line so that information may be viewed in the control room at all times. Alternatively, for installations that are not normally manned, it is possible to retrieve data remotely via the use of various data interfaces between the datalogger system and the control system network or dedicated modem links.

5. Summary of Inhibitor Trial

A new water soluble corrosion inhibitor identified from laboratory trials was introduced to the process stream on a Shell Expro gas field platform. A flush mounted thin element ER probe with a 100 μm sensing element was installed in a production flowline at a location just upstream of the common production header. The trial was performed to ascertain both the effectiveness of the new inhibitor to control CO_2 corrosion and to reduce the environmental impact compared with using the traditional oil-soluble product. The inhibitor trial was conducted over a period of 120 days at rates varying from 0 to 2 Lh^{-1}. Datalogging of the resultant corrosion rates was set for readings every six hours. Corrosion rate data was recorded for 3 periods during the trial, viz.:

- existing oil-soluble inhibitor (graphical data not shown)
- no inhibitor (freely corroding); and
- new water soluble inhibitor.

The various monitoring stages of the trial and associated results which were recorded are illustrated in Figs 1 and 2.

Corrosion rates during application of the existing chemical were recorded at a value of 0.01 mm/y. During the period of no inhibitor injection, corrosion rates increased to 0.8 mm/y with depletion of the existing inhibitor film. Recommencing inhibition with the new chemical reduced the corrosion rates down to an acceptable rate of <0.05 mm/y.

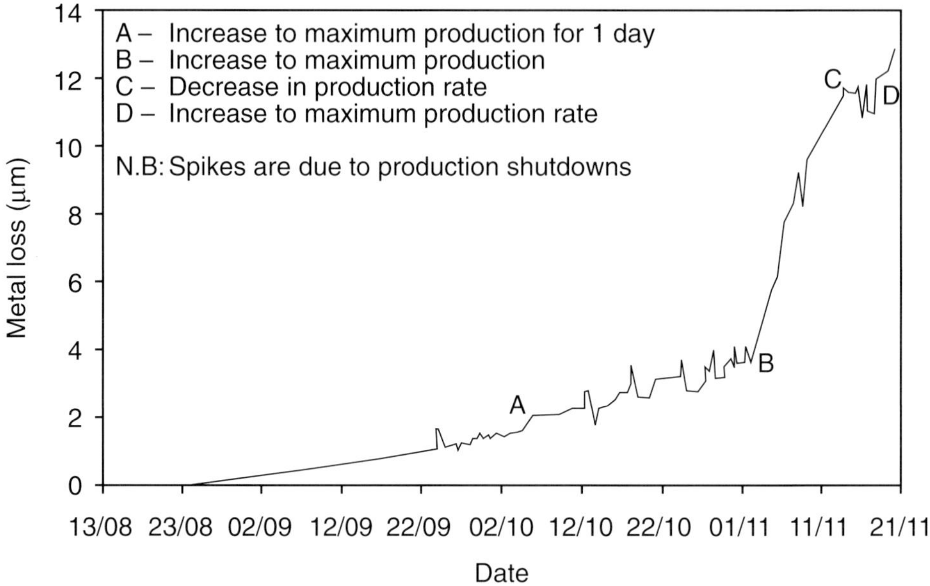

Fig. 1 ER probe metal loss results with no corrosion inhibitor injection at low and high production rates (time scale continued on Fig. 2).

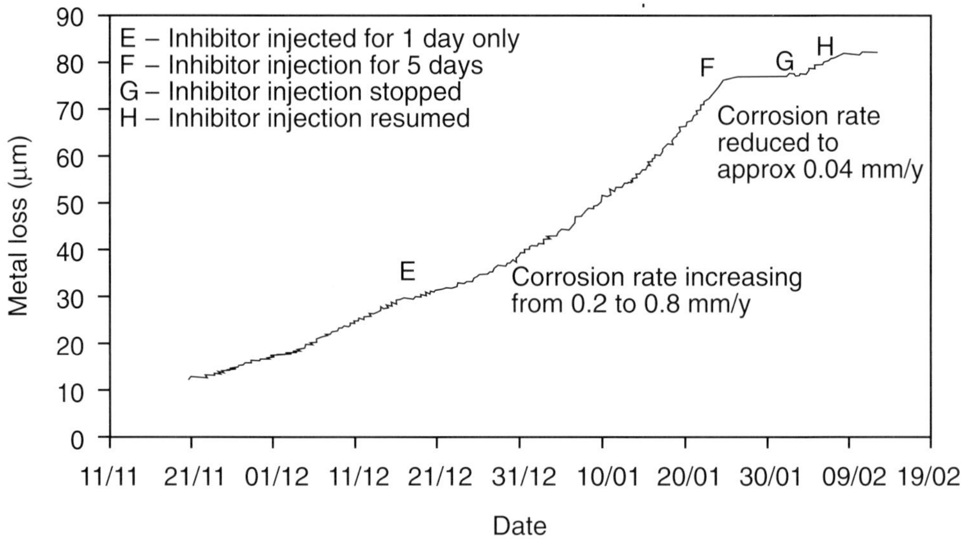

Fig. 2 ER probe metal loss results with the introduction of new water soluble corrosion inhibitor at maximum production (following Fig. 1).

Optimisation of the chemical injection system was achieved at a rate of 0.8 Lh^{-1}, which was comparable to the injection rate specified for the original chemical although it was believed that over injection of the existing chemical had occurred in the past.

Water quality improved visually and process problems associated with the formation of emulsions also reduced significantly.

6. Conclusions

It is clearly demonstrated from the results of the inhibitor replacement trial conducted under this project that the thin element ER probe can clearly and rapidly identify changes in corrosion rates due to changes in the production process. This information can then be readily used to optimise the use of corrosion inhibitor chemicals thus providing confidence in both control of the corrosion process and control over the volumes of chemical being used.

The thin element probe may also be suitable in a number of other field applications as illustrated above for the optimisation of inhibitor usage and in other applications such as the monitoring of subsea pipelines or flowlines to provide high quality corrosion data in the initial days of operation or for example for the detection of water in export gas. It is clear that CorrOcean's advancement in ER probe element design for this traditional monitoring technique will provide a useful tool for the monitoring of a wide range of applications in future.

45
Inhibition of Vapour Phase Corrosion in Gas Pipelines*

R. L. MARTIN
Petrolite Corporation, St. Louis, MO 63119, USA

ABSTRACT
Corrosion in gas pipelines takes many forms. One of these is corrosion in vapour space areas. This paper describes results of tests designed to understand better the process in sour gas lines and how to inhibit this corrosion. Corrosion in the vapour space above sour fluids was found to proceed in an unsteady manner. The presence of liquid hydrocarbon lowered rates, whereas oxygen contamination increased rates. Methanol did not increase uninhibited rates but did diminish inhibitor effectiveness. The most effective inhibitors were found to be blends of imidazolines with light amines at particular ratios. Field confirmation of inhibition was demonstrated.

1. Introduction

The subject of vapour space or 'top of the line' corrosion in gas pipelines has been a concern for a number of years [1–6]. The concern is both in sour (H_2S-containing) systems [1,2,6] and in sweet (all CO_2) systems [3–5]. Other corrosion phenomena can be interrelated and must be considered; examples are hydrogen related problems [6–9], corrosive water accumulation at sags in low velocity lines [10–12], and bacteria [13]. Incidentally, hydrogen entry into steels can be used as a corrosion monitoring technique in many of these lines [8,9,14,15]. Logistic conditions such as cold climates [7,10] offshore location [16], and special production conditions such as the presence of elemental sulfur [17,18] must also be considered. The importance of flow regime [2,10,19] as well as hydrate control additives such as methanol [2,5,7] and glycol [4,20] are now recognised. A factor not treated in literature to the author's knowledge is the corrosive consequences of oxygen dissolved in methanol [21], although once inside the system the consequences of oxygen are well known [22,23].

A review of this literature suggests that proper initial design would give a flow regime which would minimise corrosion problems [10–12]. Proper selection of the pipeline steel can also be important in sour lines [6–9]. However, in many situations, pipelines are in use which were originally designed for conditions different from present use. Sometimes production rates have declined, sometimes the nature of transported fluids has changed. We must therefore address these conditions; in many cases corrosion inhibitors are the economical answer.

Even though the primary aim of this work is to understand better corrosion and inhibition in the vapour space of sour gas pipelines, the above factors cannot be ignored. Therefore, a second portion of this study considered inhibition in the aqueous phase and hydrogen entry properties; a portion of the hydrogen entry effort has been reported earlier [6].

2. Experimental

Laboratory experiments were conducted in magnetically stirred glass resin kettles of 2 L and 4 L capacity. Weight loss, linear polarisation and electrochemical hydrogen detection methods were used for monitoring. All coupons except for hydrogen detection were round AISI 1020 (UNS G 10200), 4.4 cm long and 0.6 cm dia.

For most of the vapour space experiments, 2 L resin kettles were half filled with deaerated 0.5% NaCl and sparged with H_2S. Coupons were vertically mounted on Teflon-coated rods so that the coupon end was 3 cm above the liquid level. Test periods were 24 h at laboratory temperature; in these tests corrosion rates were determined by weight loss. Inhibitor concentrations were usually 2000 ppm. A few vapour space experiments were conducted in 4 L kettles (same diameter, twice the height) one-quarter filled with deaerated, H_2S sparged 0.5% NaCl. Procedures were identical as for 2 L kettle experiments except one set of coupons was 5 cm above the liquid level and one set was 10 cm above the liquid level liquid.

The liquid phase tests were conducted in 2 L kettles totally filled with deaerated, H_2S sparged 0.5% NaCl. Monitoring was by weight loss, linear polarisation, and back-side hydrogen measurements on totally exposed coupons. Details of hydrogen detection measurements are given in a report of earlier work [6]. Inhibition concentration in these tests was usually 250 ppm.

In field tests, weight loss coupons, ultrasonic inspection, and electrochemical hydrogen probes were used for monitoring. The characteristics of the systems were examined and any failure samples were examined. The final result was a corrosion history of the pipelines before treatment with designed inhibitors compared with the period after treatment.

3. Discussion of Results

The vapour space experiments performed in this study were conducted with no external condensation. Results would likely be different with a condensation step. For example, our experiments with CO_2 rather than H_2S (shown in Table 1) showed no corrosion whereas other investigations have developed corrosive conditions above CO_2 saturated solutions [3–5]. Under our test conditions corrosion was appreciable above all sour solutions. Rates above 0.5% NaCl were somewhat greater than above 7% NaCl. This observation must be related to evaporation since the environment in condensing droplets must have been similar in the two cases.

Attempts were made to collect condensing water for analysis but were unsuccessful. Information on pH and inhibitor concentration would have been informative. Evidently only very small quantities of water were associated with the

Table 1. *Unihibited corrosion above (vapor space) and immersed in 0.5% NaCl (except as noted) in various conditions*

	Corrosion rate (mpy)	
Condition	Vapour space at 3 cm	Immersion
H_2S saturated	86	50
H_2S sat. 7% NaCl	43	45
H_2S/O_2 saturated	160	350
CO_2 saturated	0	40
H_2S, 20% analytical grade MeOH	88	40
H_2S, 20% industrial grade MeOH	88	42
H_2S, 5% kerosene	4	50
H_2S, kerosene, MeOH	28	45

metal surface. Additional conclusions, however, can be drawn from the results in Table 1. First, the addition of deaerated methanol (reagent or plant grade) makes no difference to uninhibited H_2S corrosion. Methanol affects inhibition efficiency and oxygen loading and will be discussed later. Second, a layer of kerosene placed on the stirred brine greatly reduces vapour space corrosion. This observation is qualitatively confirmed by experience. Lines with considerable quantities of layered hydrocarbon condensate experience little or no vapour space damage, presumably because the condensible vapour is then hydrocarbon, not water [21]. However, our laboratory results indicated that the hydrocarbon benefit was reduced when methanol was added to the system. Finally, oxygen added to the system significantly accelerated vapour space corrosion. This also agrees with field experience and is not unexpected in view of the corrosive behaviour in liquids [22,23].

Another observation is that vapour space corrosion is unsteady; data in Table 2 illustrate this point. Unsteadiness has also been observed in vapour space corrosion in sweet systems [3]. At some times vapour space steel corrosion progressed at a faster rate than if the coupon had been immersed in a corresponding liquid. During

Table 2. *Unhibited vapour space corrosion*

	Corrosion rate (mpy)				
Height over liquid	Days 1	2	3	4	6
(Immersion)	45	40	40	39	35
3 cm	80	40	40	50	60
3 cm, d*	80	0	40	80	80
5 cm	30	80		20	
5 cm, d*	30	130		0	
10 cm	20	10		10	
10 cm, d*	20	0		10	

*d = implied rate since previous period.

other periods, the vapour space rate was almost zero. Since changes in corrosivity of condensing droplets seems unlikely, the answer must lie in corrosion product type or adhesion. Evidently the corrosion product offered corrosion protection but periodically spalled to expose fresh metal.

The magnitude of corrosion was influenced by height over liquid in our tests. This may simply be related to the quantity of water which condensed on the metal surface. The periodicity of rise and falls in corrosion rate also appeared to be influenced by height above liquid. These two observations may be interrelated if a critical thickness of corrosion product layer must be achieved before spalling begins.

The corrosion rates of completely immersed coupons in the same media suggest no such periodicity (see also Table 2). The rates started slightly higher but soon reached a nearly steady rate. This steady state rate was of the same order of magnitude as the average rate for the lower mounted vapour space coupons. These observations could be explained by the formation of the same type of surface film in both cases but with the spalling process hastened by the immersion condition.

Inhibition of corrosion in the vapour space was addressed by the next series of experiments. Tables 3–5 and Fig. 1 show these results. A series of light amines, imidazoline salts, and proprietary inhibitors show a variety of responses. No obvious correlation appears between vapour space inhibition and boiling point, molecular structure, or immersion inhibition (Table 3). More discussion of the inhibitor blends used in the field will follow.

As shown by results in Table 4 and Fig. 1, it was possible to blend components and achieve far better vapour space inhibition than with each component alone. Not only the components, but the blending ratio was found to be important (Fig. 1). These results are consistent with the formation of azeotropes of the blends to develop an inhibiting concentration in the vapour. The strongest synergism is between certain imidazolines (average molecular weights of about 500) and low boiling amines. For proprietary reasons, the components will not be further identified. Figure 1 data also suggested that the optimum ratio for vapour space effectiveness varies depending

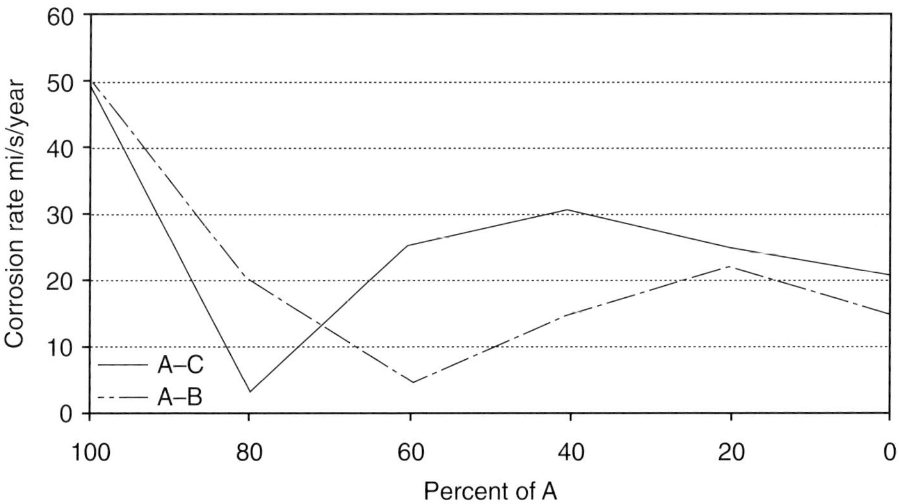

Fig. 1 *Inhibition by blends of inhibitors.*

Table 3. *Inhibited corrosion*

	Corrosion rate (mpy)	
Inhibitor	**Vapour space at 3 cm**	**Immersion**
None	86	50
Diethylamine	58	4.8
Methoxypropylamine	27	7.9
Ethylenediamine	25	16
Triethylenetetramine	16	10
Crude morpholine	14	4.2
Dimer salt of imidazoline	17	2.5
Phosphate ester salt of imidazoline	51	3.1
Pipeline inhibitor A	2	1.6
Pipeline inhibitor B	10	3.2
Pipeline inhibitor C	78	5.4
Pipeline inhibitor D	16	1.6

on the starting components. Commercial inhibitors A and D were developed using these tests.

For completeness, other corrosion related properties of potential pipeline inhibitors were examined. These results are shown in Table 5. Inhibitors A and D are both oil soluble and both specially blended to combat vapour phase corrosion. Inhibitor B is water soluble; inhibitor C is oil soluble. Inhibitor D was also designed to minimise hydrogen entry [6].

As shown by results in Table 5, inhibition in the vapour space was always degraded by methanol but to varying degrees. The weight loss was usually about doubled. Inhibitors which gave good results without methanol (for example, Inhibitor A) gave poorer but still acceptable results with methanol. Inhibitors which gave poor results in the vapour space without methanol (for example, Inhibitor C) gave unacceptable results in the presence of methanol. In fact, vapour space corrosion seemed to be accelerated for the case of Inhibitor C in the presence of methanol. For this reason Inhibitor C would never be applied in conditions where vapour space corrosion is suspected. Hydrogen entry into steel in the vapour space of sour pipelines can be significant [6]. Therefore, the hydrogen entry properties of these inhibitors were examined as shown in Table 5. Since Inhibitor D was designed to minimise vapour space hydrogen entry, it is not surprising that it was the most effective in these tests.

Table 4. *Vapour space (3 cm height) inhibition by mixtures*

Inhibitor	Corrosion rate (mpy)
Imidazoline acetate	61
Quaternary ammonium	27
Mix of above two	16
Pyridine salt	10
Quatenary ammonium	27
Mix of above two	8

Table 5. *Inhibition by pipeline inhibitors*

Inhibitor	**Corrosion rate**	
	Vapour space at 3 cm without MeOH (mpy)	**Vapour space hydrogen entry (μAcm^{-2})**
None	88/86	3.7
Dimer salt of imidazoline	40/17	0.64
Pipeline inhibitor A	5.6/2	0.34
Pipeline inhibitor B	30/10	1.3
Pipeline inhibitor C	129/78	2.8
Pipeline inhibitor D	29/16	0.27

Field observations were made in five fields which had proven by failure examination to be experiencing vapour space corrosion. Two of these were examined with electrochemical hydrogen probes. Compared with the previous inhibitor, hydrogen opposite the vapour space was lowered by more than a factor of two with Inhibitor D. In each of these cases, Inhibitor D had been in the system from 2 to 4 years with no leaks experienced due to internal corrosion. One particularly troublesome line had leaked four times in the past eight years, two leaks coming in the last two years. Failure samples from the last two both showed significant vapour space damage as well as some stepwise cracking. In the two years since the change to Inhibitor D, no internal leaks have occurred.

In a few of these cases, methanol was being injected in the winter season. Hydrogen probe readings indicated higher readings during these periods from steel opposite both liquid and vapour spaces. A special oxygen inhibitor [23] dissolved in the methanol resulted in lower hydrogen readings opposite both liquid and vapour phases. These inhibitors (special oxygen inhibitor and Inhibitor D) are used together in some of the five test systems and many other field systems.

4. Discussion and Conclusions

Corrosion in the vapour space liquids can be significant in gas pipelines. Our tests indicated the following trends for uninhibited corrosion:

1. Worse in sour than in sweet conditions.
2. Worse if oxygen is present.
3. Better if liquid hydrocarbon is present.
4. No change in the presence of anaerobic methanol.
5. Methanol weakens the protective influence of liquid hydrocarbon.
6. Water phase composition may have an effect, higher brines may give lower rates.

7. Steel further from the water phase experiences lower rates.

8. Penetration is discontinuous but averages about the same as for immersion in the corresponding liquid.

Corrosion in the vapour space can be inhibited and our tests indicated the following:

1. Some potentially inhibitive compounds provide inhibition of vapour space corrosion.

2. Inhibition is not directly related to boiling point of the inhibitor.

3. Blends of heavy inhibitive molecules with lower boiling point nitrogen compounds give best inhibition.

4. The activity of blends is ratio specific and could be explained by azeotrope formation.

5. New pipeline inhibitors have been identified by this testing.

6. Methanol diminished inhibitor effectiveness; final rates were about a factor of two greater with methanol.

7. Hydrogen entry can also be inhibited.

8. In laminar flow systems, the proportioning behaviour of inhibitor can influence corrosion inhibition in the liquid phase.

Many of these conclusions have been borne out in at least a qualitative way in the field:

1. Some of the worst vapour phase corrosion occurs in sour lines which contain some oxygen.

2. Methanol seems to contribute to increased vapour phase corrosion; this may be due to an inhibition influence or to oxygen contamination.

3. Inhibitors are effective at controlling weight loss corrosion in vapour spaces.

4. Inhibitors are effective at controlling hydrogen problems — delamination and stepwise cracking — in vapour spaces.

References

1. L. A. Jean, *H_2S Corrosion in Oil and Gas Production.* NACE International, Houston, Tx, USA, 1981, p.826.

2. D. F. Ho-Chung-Qui and A. I. Williamson, "Corrosion experiences and inhibition practices in wet sour gas gathering systems", *Corrosion '87*, Paper No. 46. NACE International, Houston, Tx, USA, 1987.
3. S. Olsen and A. Dugstad, "Corrosion under dewing conditions", *Corrosion '91*, Paper No. 472. NACE International, Houston, Tx, USA, 1991.
4. R. Nyborg, A. Dugstad and L. Lunde, "Top-of-the-line corrosion and distribution of glycol in a large wet gas pipeline", *Corrosion '93*, Paper No. 77. NACE International, Houston, Tx, USA, 1993.
5. T. R. Anderson and A. Valle, "Prediction of top-of-the-line corrosion in pipelines carrying wet CO_2 containing gas", *Corrosion '93*, Paper No. 73. NACE International, Houston, Tx, USA, 1993.
6. R. L. Martin, *Corrosion*, 1993, **49** (8), 694.
7. V. B. Larson, C. Duncan and R. S. Treseder, *Mater. Perform.*, 1988, **27** (4), 10.
8. M. G. Hay, "Correlation of laboratory hydrogen induced cracking test environments with field sour gas environments using hydrogen permeation measurements", *Corrosion '91*, Paper No. 13. NACE International, Houston, Tx, USA, 1991.
9. C. Christensen, H. Arup and P. B. Mortensen, "Hydrogen permeation/concentration effects in pipe line and pressure vessel steels as a result of wet sour corrosion", *Corrosion '92*, Paper No. 40. NACE International, Houston, Tx, USA, 1992.
10. B. R. D. Gerus and J. N. Cassin, *H_2S Corrosion in Oil and Gas*. NACE International, Houston, Tx, USA, p. 809.
11. K. van Gelder, L. van Bodegom and A. Visser, *Mater. Perform.*, 1988, **27** (4), 17.
12. N. J. Grahmann and J. L. Pikas, "Corrosion inhibition of a west gas pipeline – A case history", *Corrosion '93*, Paper No. 101. NACE International, Houston, Tx, USA, 1993.
13. O. C. Dias and M. C. Bromel, *Mater. Perform.*, 1990, **29** (5), 53.
14. R. G. Asperger and P. G. Hewitt, *Mater. Perform.*, 1986, **25** (8), 47.
15. R. L. Martin and E. C. French, *J. Petrol. Technol.*, 1978, **20** (11), 1566.
16. M. W. Joosten, E. Buck, J. Kolts, D. Erickson and M. Mai, "Monitoring and control of corrosion in offshore, wet gas-condensate pipelines", *Corrosion '92*, Paper No. 9. NACE International, Houston, Tx, USA, 1992.
17. A. Dougherty, "Corrosion inhibition of wet, sour gas lines carrying elemental sulfur", *Corrosion '92*, Paper No. 2. NACE International, Houston, Tx, USA, 1992.
18. J. A. Dougherty, B. A. Alink, D. F. Ho-Chung-Qui and K. van Gelder, *Mater. Perform.*, 1995, **34** (1), 41.
19. M. Bopal, A. Kaul and W. P. Jepson, "Mechanisms contributing to enhanced corrosion in three phase slug flow in horizontal pipelines", *Corrosion '95*, Paper No. 105. NACE International, Houston, Tx, USA, 1995.
20. K. van Gelder, M. J. J. Simon Thomas and C. J. Krose, *Corrosion*, 1986, **42** (1), 36.
21. J. A. Svetgoff, Private communication.
22. Y. Wu, K. E. McSperitt and G. D. Harris, *Corrosion '88*, Paper No. 365. NACE International, Houston, Tx, USA, 1988.
23. R. L. Martin and R. R. Annand, *Corrosion*, 1981, **37** (5), 295.

Part 7

Non-Metallics

46
Service Experience with Glass Reinforced Epoxy Pipelines and the Way Forward

S. R. FROST, M. R. KLEIN*, S. J. PATERSON† and G. E. SCHOOLENBERG§

Shell International E & P, Volmerlaan 8, P.O. Box 60, 2280 AB Rijswijk, The Netherlands
*Nederlandse Aardolie Maatschappij (NAM), De Vosholen 66, 9611 TD Sappemeer, The Netherlands
†Nederlandse Aardolie Maatschappij, Beekweg 33, 7761 PK Schoonebeek, The Netherlands
§Shell Research & Technology Centre Amsterdam, Badhuisweg 3, 1031 CM Amsterdam, The Netherlands

ABSTRACT

Glass reinforced plastic epoxy (GRP/GRE) pipes have been used successfully in a wide range of applications from offshore piping systems to onshore process plant piping and buried pipelines. Typically service conditions are pressures up to 50 bar, temperatures up to 50°C with fluids transported, being either water, oil, or water/oil mixtures.

The cost benefit of using GRP is that life-cycle costs are lower than other competing pipeline material options. The excellent corrosion resistance and easy installation imply that although the capital expenditure (Capex) of GRP is greater than that of carbon steel, the operating expenditure (Opex) costs are significantly lower. The combined cost over a 20 year design life makes GRP an extremely cost effective material selection.

The economic attractiveness of GRP implies that its use is constantly being widened, and extending the application envelope in terms of either pressure, temperature or service condition is a natural progression.

This paper addresses the issue of extending the application of GRP both to water/condensate and to gas service. It includes a review of two specific projects and summarises the required technical support. Furthermore, new design guidelines and qualification procedures are introduced. Finally, conclusions are drawn with recommendations made for the way forward relevant to both suppliers and other potential users.

1. Introduction

Glass reinforced plastic (GRP) pipes and piping systems have been used successfully in a wide range of applications from offshore piping systems to onshore process plant piping and buried pipelines. Typically service conditions are pressures up to 50 bar, temperatures up to 60°C with transported fluids being either water, oil or water/oil mixtures.

The cost benefit of using GRP (or GRE, Glass Reinforced Epoxy) is that life-cycle costs (LCC) are lower than other competing pipeline material options. The excellent corrosion resistance and easy installation imply that although the capital expenditure (Capex) of GRP is greater than that of carbon steel, the operating expenditure (Opex) costs are significantly lower. The combined cost over, for example, a 20 year design life makes GRP a very cost effective material selection.

The economic attractiveness of GRP means that its use is constantly being widened, extending the application envelope in terms of either pressure, temperature or service condition is a natural progression.

This paper addresses the issue of extending the application of GRP both to WACO (Water Gas Condensate) and to gas service. It includes a review of two specific projects and summarises the required technical support. Furthermore, new design guidelines and qualification procedures are introduced. Finally, conclusions are drawn with recommendations made for the way forward relevant to both suppliers and other potential users.

2. Qualification Procedures and Design Standards

The design and qualification of GRP pipes is achieved through performance based procedures. This effectively means that the emphasis placed on the suppliers of GRP is to demonstrate, through testing, that their products will satisfy the design specification. No prescriptive measures defining fibre or matrix type or construction geometry are required. The benefit of such an approach is that the suppliers of GRP products are free to optimise their material selection and construction procedures. The drawback is that a substantial database of test data, for both pipes and fittings, is required before a product family can be considered qualified.

To promote the safe yet efficient use of GRP, Shell developed a Design and Engineering Practice (DEP) document for the use of GRP pipes and piping systems. The policy within Shell concerning design standards is to write amendments to existing accepted standards and guidelines. The latest version of the DEP is based on the UKOOA (United Kingdom Offshore Operators Association) guidelines for the offshore application of GRP [1] with reference also made to API 15 HR [2] and the NORSOK (Norsk Sokkels Konkuranseposisjon) standard on GRP pipes [3]. There are 5 chapters or sections within the DEP, namely:

- Philosophy and scope
- Qualification (components and manufacture)
- System design
- Fabrication/installation
- Operation.

The important aspects of the DEP are summarised as follows:

- The scope of the document is broader than any other standard and encompasses onshore pipelines and piping systems, and offshore piping systems. The range of service conditions is also broad covering water, hydrocarbon and wet/dry gas transport.

- Qualification procedures have been simplified and manufacturers are now encouraged to perform 1000 hour tests on all components to verify (qualify) long term regression data (ASTM D 2992) for their products. The large amount of testing required to qualify a product family has been reduced but still remains significant. This issue still requires further discussion and development before a satisfactory outcome is reached. The nominal pressure rating (NPR) is defined as the lower confidence limit (LCL) of the regression curve at the required design life multiplied by a safety factor, f_2. Currently, $f_2 = 0.67$.

- Previous international standards have linked the NPR to system design loads. In this DEP the NPR is solely defined in terms of internal pressure capacity. This implies that the maximum allowable operating pressure for a specific design could be lower than the NPR. The system design procedure is based on measured long term failure envelopes. If the sum of all the specific system stresses lies within the failure envelope, then the design is safe. If the total stresses lie outside then a thicker walled (or higher NPR) pipe must be chosen and the design process repeated.

- Fabrication, installation and operational issues are broadened to cover buried pipelines and piping systems (both onshore and offshore). Maintenance of GRP pipes is considered but the emphasis of inspection is placed on quality assurance during manufacture and installation, i.e. getting it right before and during installation to minimise operational problems.

For critical applications (where critical is defined as system integrity not strength), extra requirements have been set. Primarily critical applications involve wet or dry gas transport. These extra criteria include:

- Air testing (at low pressure), and

- Microstructural sampling and characterisation.

At the qualification stage of the project (i.e. before pipe manufacture), air tests are specified for the pipe plus joint. Also a microstructural investigation of the pipe wall in two locations is required. The locations are in the pipe wall body and at the 'connection' end of the pipe. The microstructural investigation involves sectioning the pipe, removing samples, polishing, and preparing micrographs from a high-resolution optical microscope. Specific interpretation and analysis details are contained in Section 5. These micrographs form the base case for the quality assurance process with comparison to produced pipe to ensure consistency throughout the manufacturing process.

During manufacture, air tests are performed and micrographs are taken at specified intervals and compared with the base case micrographs. By adopting this rigorous approach it is intended that the consistency in the quality of manufacture is such that GRP pipes can be safely used with confidence in gas and water/condensate service applications. Sections 3 and 4 describe specific applications within the NAM, The Netherlands, where these procedures were applied.

3. Experience with the Installation of an 8 in. GRE Water Injection Pipeline

For the transport of water with traces of gas condensate, NAM Business Unit Groningen has replaced an existing 8 in. dia., 8 km carbon steel pipeline between the water/condensate separation unit at Delfzijl and the water injection well at location Borgsweer with an 8 in. dia. GRE pipeline.

The carbon steel pipeline had suffered from extensive pitting corrosion due to the presence of oxygen in the water. Glass reinforced epoxy was selected because of its excellent corrosion resistant properties and previous good experience with WACO in NAM. The design pressure of this buried GRE pipeline is 50 bar and the minimum/maximum design temperatures are 10 and 50°C, respectively. Hydrostatic testing of the GRE pipes at the manufacturing plant was performed at 75 bar. The GRE pipes (and fittings) with integral thread connections had to comply with the requirements specified in the previous DEP (based on API 15HR).

After delivery of the containers severe damage was observed on the GRE pipes and fittings. Transport damage occurred because insufficient precautionary measures had been taken to support the GRE pipes during transport at sea. Visual inspection of all GRE pipes and fittings was performed to determine whether or not damaged GRE components could be properly repaired. All GRE components, which could not be repaired were rejected (approximately 5%).

At the start of pipeline construction it was found that the gripping tools supplied by the GRE manufacturer for assembly of the GRE pipes caused unacceptable damage to the outer surface of the pipe and the development of new tools was required. Furthermore, installation procedures provided by the GRE manufacturer were not sufficiently clear for the pipe-laying contractor. Additional assembly instructions were issued to solve this problem. Although representatives from the GRE manufacturer carried out supervision during construction, assembly of the threaded pipes was troublesome and resulted in a low production rate with a relatively high number of leaking connections. After special supporting equipment had been developed and fabricated to facilitate proper alignment, assembly of the GRE pipes improved and the installation rate reached an acceptable level.

During field pressure testing with air at 7 bar after each day's production, several connections showed leakage through the GRE wall at the flat female end of the threaded connection. Microstructural investigations were performed on the female ends of the leaking GRE pipes to find the cause of leakage. The investigation showed that cracks (typically 1–2 mm length in the root of the threads) and poorly wetted areas were present at the inner surface of the pipe wall. The cracks did not extend beyond the first layer of reinforcement, and on investigation were found to have been caused during pressure test at the manufacturer's factory. Furthermore, a relatively high number of 'standard' voids (clusters with a void 'diameter' of 2 to 5 times the fibre diameter, e.g. Fig. 1) and 'non-standard' large elongated voids (void 'diameter' of 15 to 60 times the fibre diameter, e.g. Fig. 2) were present at the inner surface of the pipe wall at the female thread end.

Although a leak path could not be physically identified, it was postulated that the leak path starts at either the poorly wetted areas or at the cracks in the root within the female end connection, then flows through the laminate by means of the clusters

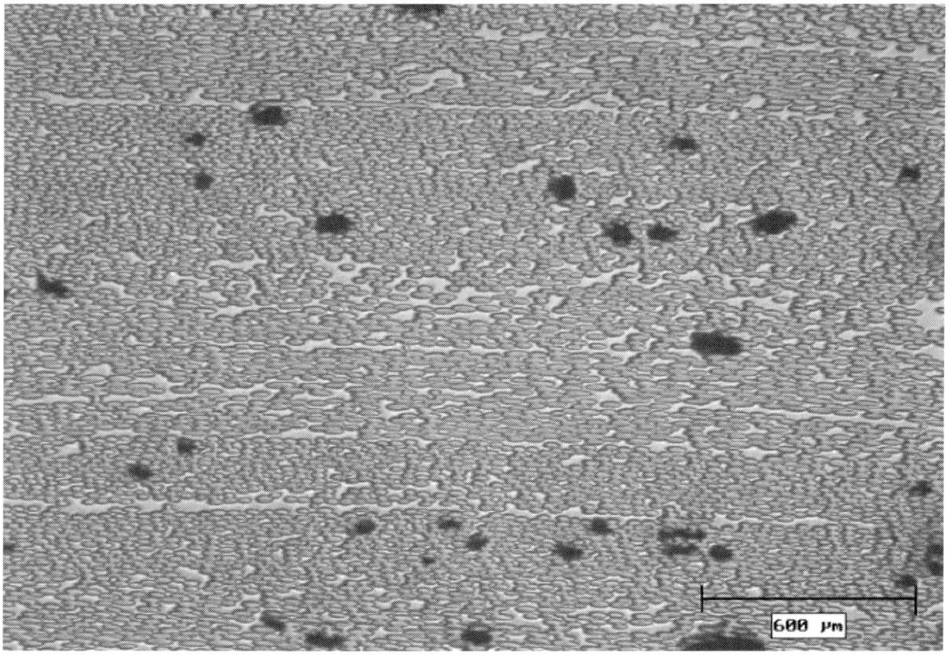

Fig. 1 *Microstructural photograph of small spherical voids.*

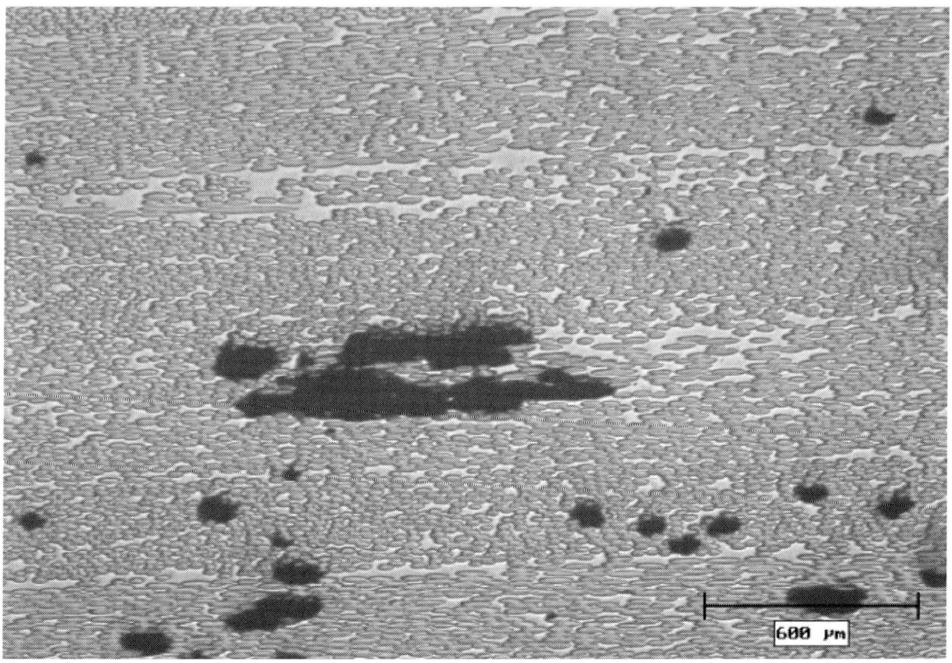

Fig. 2 *Microstructural photograph of large elongated voids.*

of 'standard' voids and large voids. Based on these results it was concluded that the quality of the delivered GRE pipe material was not satisfactory and the technical integrity of the GRE pipes could not be guaranteed during the service lifetime of the pipeline. Therefore all the GRE pipes were rejected and new GRE pipes were ordered with additional requirements concerning void size and void content. Because agreement with the manufacturer regarding these additional requirements could not be reached, additional air testing at the manufacturing plant with a pressure of 7 bar was specified in order to reduce the risk of leakage through the GRE wall at the female end during installation.

Installation of the 8 in. dia. GRE injection water pipeline was completed in July 1997 and the pipeline has been in operation since then without problems.

The experience with this GRE pipeline demonstrated that the manufacture of relatively large diameter GRE pipes for high pressure applications requires improved fabrication procedures, additional test requirements to verify the material quality, and adequate procedures for transport, handling and assembly.

4. Experience with the Installation of a 6 in. GRE Gas Pipeline

The first GRE pipeline for gas transport in NAM was installed in Drenthe in 1997. The 6 in. dia., 5.5 km GRE pipeline with integral thread connections has a dual purpose. It will be used initially for the transport of sour gas with a design pressure of 40 bar for a period of 3–4 years, and subsequently for injection water transport with a design pressure of 80 bar and temperature of 60°C. Normally carbon steel with corrosion inhibition would have been used for transporting the wet sour gas, but this is not suitable for sour injection water because problems with oxygen control can result in rapid and deep pitting corrosion. Glass reinforced epoxy was an obvious choice for the injection water but it had never been used before in NAM for gas transport.

To prove the viability of GRE for this application a detailed risk assessment was conducted in accordance with the requirements of the Netherlands standard NEN-3650 [4]. Based on this risk assessment, and also recognising that once installed the greatest chance of damage to the pipeline is from third party intervention, the following measures were employed to reduce the risk to a comparable level of a steel pipeline:

- Large margin between design and operating pressure (20 bar max.).
- Greater depth of soil cover: 1.8 m instead of the usual 1.5 m.
- Use of mantle pipes at road and waterway crossings.
- Markers at ditches, road and waterway crossings to indicate differential soil subsidence.
- Use of concrete slabs at vulnerable points such as ditches.
- Regular pipeline surveys.

- Use of a dedicated installation contractor.

- Training of installation crews and supervision by the GRE pipe supplier.

As a result of the experience with the GRE pipeline in Groningen the following additional measures were specified for the supply of the GRE pipe which this time came from a different manufacturer:

- More extensive material qualification requirements including additional gas tightness testing for straight pipe, elbows and metal/GRE connections and microstructural examination of the 1000 h test pipe.

- Strict quality assurance and control during manufacture and installation. In addition to the normal testing requirements for GRE pipe, air-testing at 7 bar was conducted on 10% of the order during manufacture.

- Microstructural examination of samples from the first ten pipes from the production line and retention of all pipe end cut-off rings during manufacture for subsequent examination if necessary.

- Careful packing and handling of the GRE pipe.

Microscopic examination of the samples from the production pipe revealed a similar microstructure to the sample from the 1000 h qualification test. The voids were regular and evenly distributed with a size of 3–5 times the fibre diameter. The overall impression was of a well manufactured GRE pipe.

Despite the emphasis on proper packing and handling the delivery inspection revealed that damage had occurred to the outside surface of the female ends of many pipes due to movement during shipping. Two damaged pipes selected from the consignment successfully passed an air test at 7 bar for 30 min. One of the pipes sectioned for microscopic examination showed that small microcracks had extended approximately 5 mm from the damaged areas. Subsequent tests in the laboratory showed that this damage could be removed by light grinding and that the absence of cracks could be confirmed by dye penetrant inspection. In order to confirm these observations another pipe with an impact on the external corner of the female end which had a white 'frost-like' appearance was examined. This showed that a crack had extended by 17 mm at approximately 45° from the damaged area. The rest of the consignment was inspected for this type of damage and where repair was not possible the pipes were rejected.

During the examination of this last pipe cracks were also observed in the threads in the female end of the pipe. These cracks were located in the base of the 5 to 6 outer threads of the female end of the pipe and extended in the axial direction towards the open end. The cracks were clearly visible by a much lighter appearance of the thread.

The supplier was able to demonstrate that the thread cracking had occurred during hydrotesting in their manufacturing plant and was generally limited to the first few threads which were used as 'leaders' for the sealing area of the thread. The cracks occurred during pressure testing because the steel end plugs had not been sufficiently made up, causing distortion and cracking in the threads. The extent of cracking was reduced by adapting the make-up procedure. It was agreed that provided no cracks were present within the sealing area the pipes were fit for service. All the pipes were re-inspected for thread damage and were rejected if any damage was present beyond the first two 'leader' threads.

Overall approximately 20% of the pipes had to be replaced because of damage that had occurred either during hydrotesting in the manufacturing plant or during shipment. To reduce the risk of this occurring with future orders more specific requirements for inspection, packing and handling of GRE pipe have been incorporated into the latest version of the DEP.

During installation of the pipeline the gas tightness of the threaded joints was ensured by using a proprietary thread sealant compound and was checked by air-testing at 2.5 bar and visual inspection at the end of each day. At the end of each week the whole length was hydrotested at 120 bar for 1 h. After completion of the installation a hydrotest was carried out at 120 bar for 48 h, after which the pipeline was de-watered and filled with nitrogen ready for service. The GRE pipeline has been in service since November 1997 and is currently operating at a pressure of 15 bar.

5. Microstructural Characterisation

Based on recent experiences in NAM likely causes of failures or leaks in threaded GRE pipes are:

- Damage during manufacture or transport, and
- A deficient microstructural quality of the GRE pipe.

With performance based testing poor microstructural quality is not normally an issue unless it results in direct leakage during the test. Field experience has shown that possible leak paths are not always detected during quality assurance procedures in the factory, even though gas tightness tests and hydrotests are performed. Apparently, some leaks only occur if the pressure and/or duration of the test are sufficient to highlight these defects. In some cases even when the field air test is successful, leaks still occur during commissioning. This raises the question whether there are suitable procedures for assessing the microstructural quality achieved during manufacture.

5.1. Possible Leak Paths

The microstructural investigations indicated that possible causes of leak paths are:

1. Poor impregnation around deficiencies in the filaments, e.g. knots.
2. Poor temperature control during resin curing.

3. Poor wetting of the fibres, leading to a leak path from the pipe interior to the flat female end of the pipe.

Deficiencies due to poor impregnation are considered to be 'one off' production faults which should be detected during visual inspection in the manufacturing plant.

Resin hardening is an exothermic reaction which may become unstable resulting in overheating. This can introduce voids into the laminate. Accurate temperature control is required to prevent this process occurring.

The poor wetting of fibres is of a more systematic nature and could be monitored by taking samples at regular intervals to identify dips in production quality below a certain acceptance criterion.

5.2. Need for Fibre Wetting

The need for fibre wetting is evident when levels of wetting are so poor that a leak path is present directly after manufacture or after installation of the pipe. Even if this is not the case, poor wetting has negative effects on the long term performance of the pipes in either fatigue and static service. The long term failure mode of GRE pipes is weepage, where a pattern of cracks develops within the pipe wall to create a leak path from the bore to the external surface. These cracks generally develop parallel to the fibres and propagate along the fibre matrix interface. When fibre wetting is poor, the rate of formation, i.e. initiation and propagation of such cracks will increase significantly. Apart from early leak prevention, knowing the level of wetting is relevant for assessing the long term performance.

5.3 Spherical vs Elongated Voids; Difficulty of Defining Acceptance Criteria

Poor wetting of the fibre results in voids in the material. Methods to measure void content e.g. through density are readily available. However, all GRE materials contain voids and it was demonstrated that it is not the void content itself that should be used as a measure for risk of leakage. It is the shape and distribution of the voids that is important. Void content provides an average measure of voids and should only be used as a quality assurance check. Small spherical voids (Fig. 1), e.g. due to matrix shrinkage, are not harmful as long as their quantity is not excessive or their distribution is random. It is the large elongated voids that are indicative of the possibility of a leak path (Fig. 2).

Pipes were cross-sectioned in different areas and an acceptance criterion through visual observation was attempted. During this exercise it became clear that, within a population of pipes where several leaks were found, some of the microstructural cross-sections looked acceptable. Those microstructural cross-sections that looked particularly poor were usually taken from leaking pipes. Also there was not a gradual variation in the production quality. Subsequent pipes from the same production batch could have greatly varying degrees of quality. However, comparing microstructural cross-sections from one population of pipes which did not leak with another population which did leak, showed that there was a distinct difference in quality, (Fig. 3).

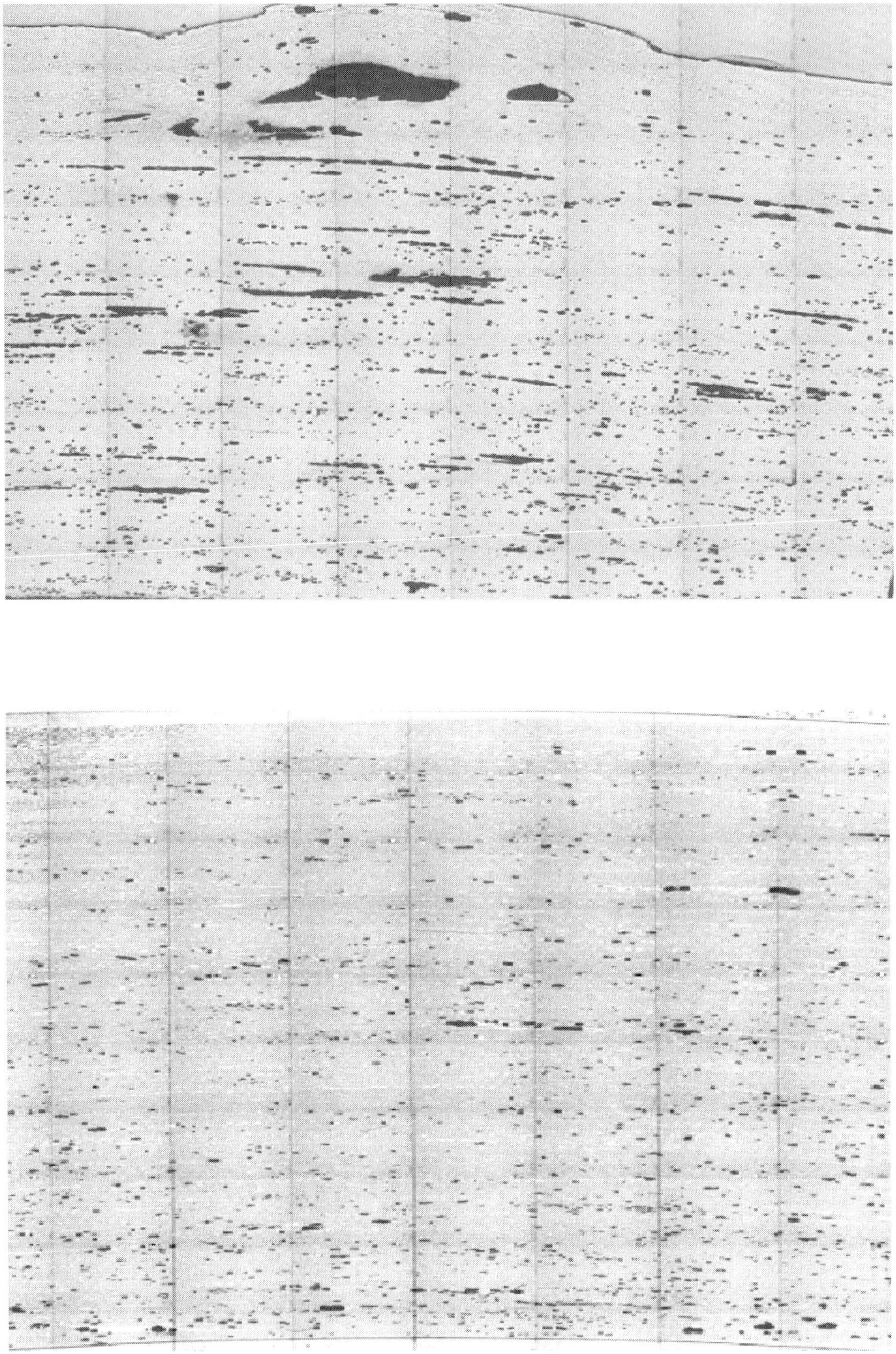

***Fig. 3** Photographs comparing microstructures: leaking pipe* (top) *and non-leaking pipe* (bottom).

To summarise, if the quality of production is borderline, microstructural investigations cannot be used as a suitable tool to define an acceptance criterion. However, as a rough tool to discriminate between 'good' and 'poor' quality, microstructural analysis is a suitable method.

5.4. End Porosity Test

The end porosity test provides a measure of the rate of flow of gas through a small sample or section of a GRE pipe. However, this rate of flow is dependent on many variables and can vary from supposedly equivalent samples by significant margins. Therefore, as it currently stands this test cannot be used as an acceptance/rejection criterion for the quality of pipe wall microstructure during production. Defining a limit to what is an acceptable level of flow is not possible. However, if this test could be developed to minimise the scatter in results and if an investigation into defining the minimum acceptable flow rate through the sample as a function of pipe wall microstructure were carried out then end porosity testing could form a basis for acceptance/rejection criteria.

6. Conclusions and Recommendations

1. A new Shell standard (DEP) has been produced based on the UKOOA guidelines for GRP pipes and fittings.

2. Glass reinforced epoxy pipes have been successfully applied in both water/condensate and gas service applications within the NAM.

3, Project experience has shown that improved inspection, packing and handling procedures are required for GRE pipes during transport to prevent damaged pipes arriving at site.

4. For critical GRE pipe applications air testing and microstructural analysis are specified. Microstructural analysis cannot be used as an acceptance/rejection criterion. However, it can be used as a check on the quality of produced pipes, when referenced to the microstructure of the qualification pipe tests. End porosity testing could form the basis for an acceptance/rejection criteria but the test needs significant development.

5. Generally performance testing and strict quality assurance procedures will form the basis for qualification of GRE pipes. Inspection during operation will be limited until tools are developed for GRE pipes and fittings.

6. Suppliers will be continually encouraged to broaden their test database. Particular emphasis will be placed on 1000 h test results for pipes, joints and fittings. Suppliers will also be actively encouraged to become involved in the future of standards development.

7. Acknowledgement

The authors wish to thank H. van Zummeren for the microstructural work used in this paper.

References

1. Specification and recommended practice for the use of GRP piping offshore UKOOA, First Edition, March 1994.
2. Specification for High Pressure Fiberglass Line Pipe, API 15 HR, 1996.
3. Guideline for NDT of GRP piping systems and tanks, NORSOK M-CR-622.
4. NEN-3650: Requirements for steel pipeline transportation systems.

47
New Ceramic–Metallic Materials for Choke Valves in Oil Production

C. H. AHLEN, E. BARDAL*†, L. MARKEN and T. SOLEM*

Statoil R&D Centre, Trondheim, Norway
*SINTEF Corrosion and Surface Technology, Trondheim, Norway
†Norwegian University of Science and Technology, Trondheim

ABSTRACT

To increase the lifetime for choke valves in oil and gas production a number of materials for choke trim components, such as plug, cage and seat, have been investigated using various erosion test methods. The study also includes corrosion testing and the synergistic effect of erosion and corrosion. Hardness testing, fracture toughness testing and microscopic examination complete the investigation. The life time of choke valves can be increased several times by making a better selection from the special cemented tungsten carbide grades existing on the market. Further improvements of the erosion and the erosion–corrosion resistance can probably be made by optimising the microstructure and the chemical composition of the binder phase.

1. Introduction

Traditionally, choke valves in oil production systems have been used only on offshore production platforms or onshore. Because of high erosion rates, usually combined with corrosion, the life time of the valves have often been very short. This problem has been tolerated since replacement of worn parts under these conditions is relatively simple and the maintenance cost moderate.

The development of subsea production has changed the situation. Replacing subsea valves is very expensive, and there is a high economic potential in improving valve materials for this service.

This potential is particularly valid for many of the new field installations where the choke valves are designed to considerably reduce the pressure in order to obtain a lower design pressure for pipelines or improved flow conditions.

Two joint projects between Statoil, Norsk Hydro and SINTEF [1–4] have been carried out to establish a better basis for materials design and selection and to qualify materials with better erosion and erosion–corrosion resistance. During the last two years Saga Petroleum has also taken part in this co-operation. For sand-producing wells with a high pressure drop in the valve, these projects have indicated that a cost reduction of 240 000 GBP – 400 000 GBP (£ sterling) per year may be obtained for each well.

The present paper gives a short review of the research work and the results used as a basis for the indicated cost reduction. Only tungsten carbide grades are dealt with in the paper since earlier erosion testing has shown that this type of material has a suitable combination of fracture toughness and erosion resistance. Pure ceramic materials are still considered too brittle and all metallic materials are too soft for this service.

2. Experimental

The following methods have been used in the tests reported in the present work.

(a) Sand erosion and corrosion of cylindrical specimens fixed to a rotating disk. The method and the equipment have been described elsewhere [1,4–7]. By means of weight loss and electrochemical measurements both total material loss, pure corrosion, pure erosion and synergistic material loss could be measured. Peripheral velocities from 18 to 30 ms^{-1}, sand concentrations 0.25 and 2.5 wt% and a temperature of 54°C were used.

(b) Air–sand and water–sand jet impingement at various high velocities and different impact angles at 20°C. The results in the present publication were obtained with jet velocities 190 ms^{-1} (air–sand) and 90 ms^{-1} (water–sand) and with impact angles of 30° and 90° to the test specimen surface. Specimen weight loss was measured and converted to volume reduction, and related to the amount of sand used during the representative test period [2,3].

The testing with method (a) was carried out at SINTEF Corrosion and Surface Technology and tests (b) were run at Statoil R & D Centre. The water used was simulated well formation water saturated with CO_2 in the SINTEF tests. Ordinary tap water was used in the tests at Statoil. The sand used was Skanør silica sand with average grain size 200/230 μm in tests (a) and Baskarp (more angular) silica sand with average grain diameter 150 μm in tests (b).

To obtain a better understanding of how structural properties of the materials affect erosion and erosion–corrosion resistance, the materials have been studied by transmission electron microscopy (TEM), and scanning electron microscopy (SEM). The composition of the binder phase has also been investigated.

The applicability of the materials depends also on fracture properties, and a simplified method has been used to determine the fracture toughness, K_{IC}. This was determined by means of Vickers pyramid indentations on a flat specimen, which generated cracks from the corners of the indentations. The value of K_{IC} is calculated from the average crack length, the apparent hardness, the modulus of elasticity and the applied force [8].

3. Results and Discussion

Results from the erosion–corrosion experiments at SINTEF are shown in Figs 1 and 2 at two different levels of erosion.

With relatively low levels of erosion (Fig. 1) there is a large synergistic effect of erosion and corrosion for most of the materials. This is typical for materials with a metallic phase of pure Co but occurs also for some of the materials with Cr and Ni in the metallic phase. A large synergistic effect also leads to high rates of total material loss. The two best materials in this test (M13 and M19) are characterised by a low synergy effect. It is interesting to notice that these two materials have respectively the largest and the smallest content of metallic phase of all the materials tested. Material M13 (with 9.5% metal phase) has only a small content of Co (according to

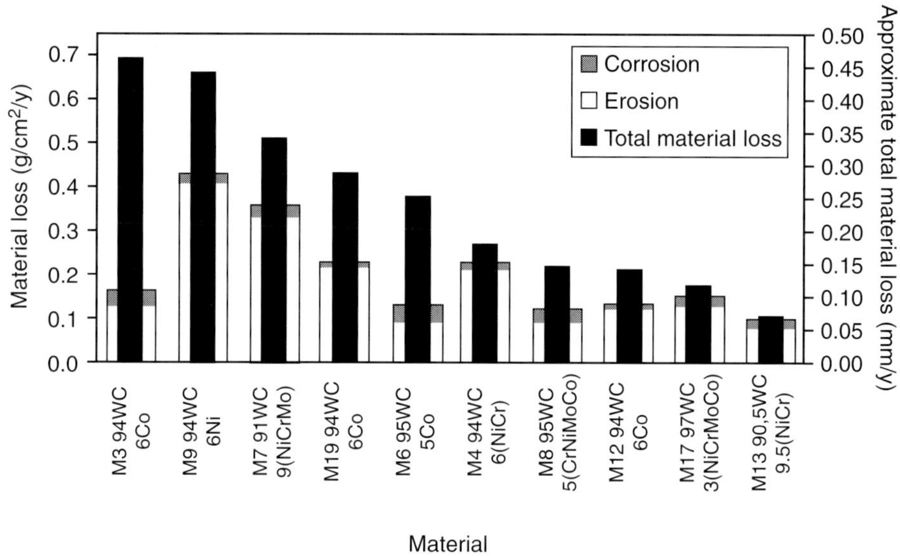

Fig. 1 *Rates of corrosion, erosion and total material loss at peripheral velocity 18.3 ms*$^{-1}$ *and sand concentration 0.25 wt%.*

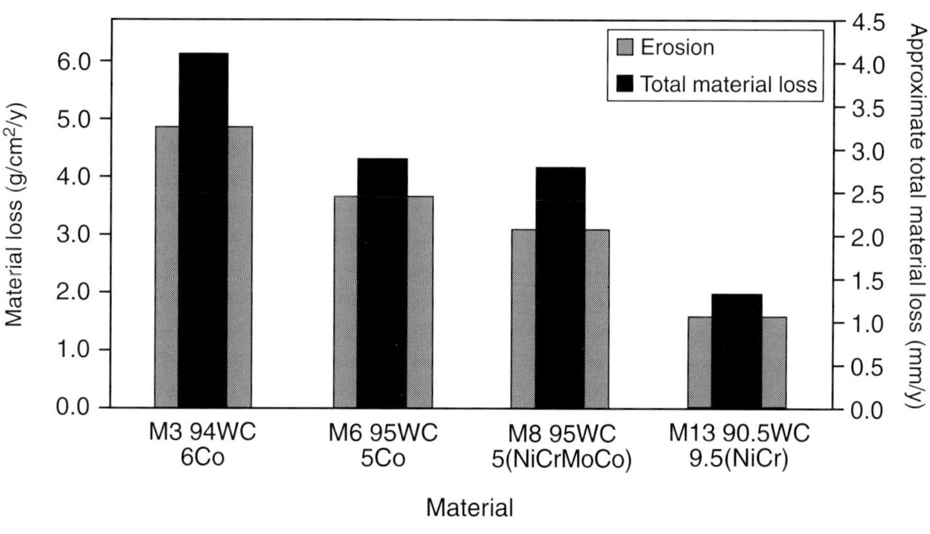

Fig. 2 *As Fig. 1 but with velocity 29.3 ms*$^{-1}$ *and sand concentration 2.5 wt%.*

microanalysis) and this may contribute to the high corrosion resistance and to the good total result for this material. The tests have also shown that there are large differences in material loss between materials with similar compositions, e.g. between M12, M6, M19 and M3.

These results show that the material loss is not a simple function of metal binder content. It is obvious that both the amount and the composition of the metal phase are of importance but also the microstructure has to be considered. The best materials are characterised by small WC grains (typically 0.5–0.8 μm). The very best material in these tests (M13) is special: it contains a considerable fraction of very fine grains, i.e. with grain size as low 0.1–0.2 μm [4]. In addition, the metallic binder phase in this material seems to have a high corrosion resistance under erosive conditions.

A relatively good material with a Co matrix (M12) has a metal phase mainly concentrated in the space between groups of WC grains. On the other hand, a WC–Co material with inferior erosion properties, M19, has a fraction of very coarse WC grains in addition to the normal small grains [4].

Results from tests with much higher erosion are shown for four of the materials in Fig. 2. As would be expected, pure erosion makes up a larger part of the total material loss in this case. The synergy effect is much smaller and the differences between the total material loss of the various materials are smaller. However, the ranking order of the four materials is the same as at the lower erosion level represented in Fig. 1.

Selected results from the jet impingement with water–sand at Statoil are given in Fig. 3. The four most erosion resistant materials in these tests correspond to the four best in the SINTEF tests represented in Fig. 1. The ranking order of these materials is also the same with the exception of the two best materials. In the Statoil test, the material M17 with only 3% metallic phase has shown the best results. This is also the case in the air–sand jet impingement tests, as shown in Fig. 4. The ranking order in Fig. 4 is somewhat different from those in Fig. 1 and Fig. 3. The air–sand erosion seems to be much more dependent on the amount of metallic phase than the other types of erosion investigated. Another difference between air–sand and water–sand

Table 1. *Material data*

Specimen	Composition*	Grain size WC*	HV30* (kg mm^{-2})	Densities (g cm^{-3})	Matrix (%)	ρ_{matrix} (gcm^{-3})
M3	94WC 6Co	Fine (0.8 μm)	1780	14.95	6	8.81
M4	94WC 6(NiCr)	Fine (2.0 μm)	1800	14.95	6	7.87
M6	95WC 5Co	Fine	1750	15.00	5	8.81
M7	91WC 9(NiCrMo)	Fine	1375	14.60	9	8.77
M8/M16	94.9WC 5.1 (NiCrMoCo)	Extra fine	1800	15.00	5.1	8.80
M9	94WC 6Ni	Fine (2.0 μm)		14.93	6	8.90
M12	94WC 6Co			14.80	6	8.8
M13	91WC 9(NiCr)			14.50	9.5	8.4
M17	97WC 3(NiCrMoCo)			15.10	3	8.9
M19	94WC 6Co			15.00	6	8.8

*Given by materials manufacturer.

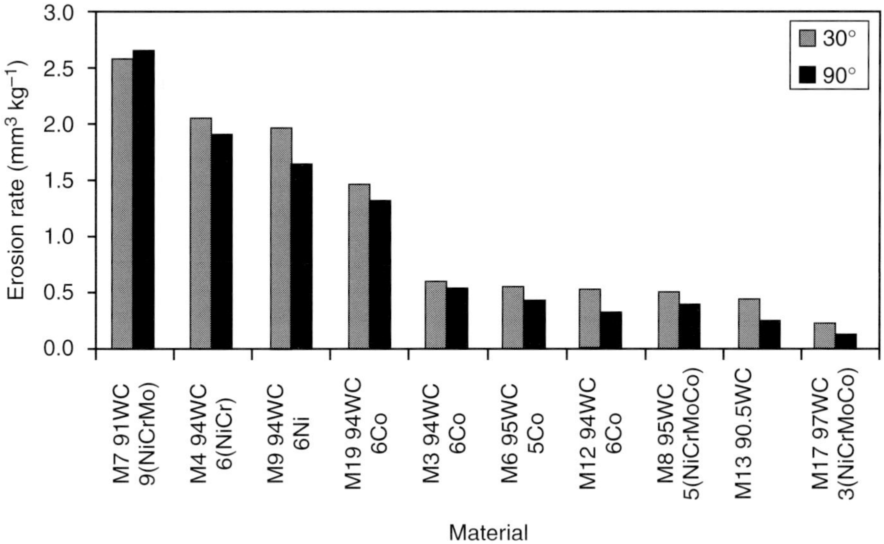

Fig. 3 Erosion rates in water–sand jet impingement tests with jet velocity 90 ms^{-1} and impact angles of 30° and 90°.

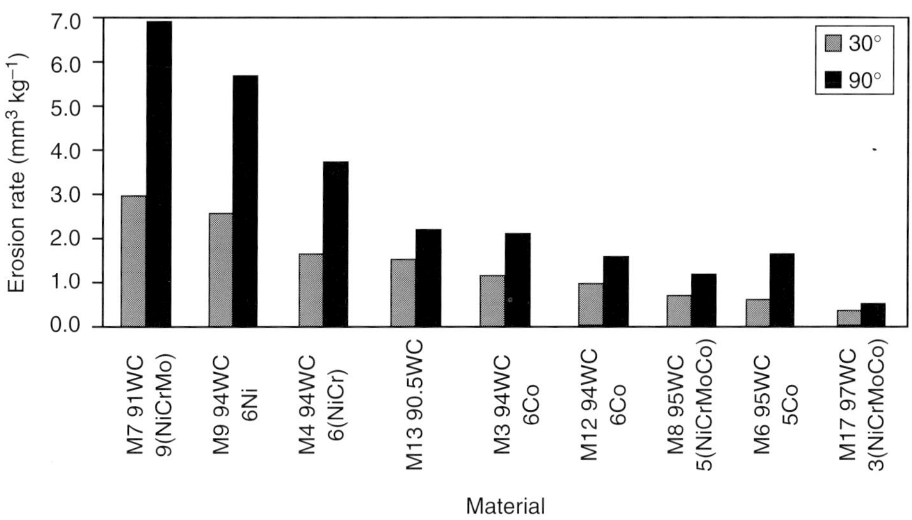

Fig. 4 Erosion rates in air–sand jet impingements tests with velocity 190 ms^{-1} and angles of 30° and 90.

is the effect of impact angle (compare Fig. 3 and Fig. 4.). An explanation of this difference is that there is some deflection of the sand particles in the water jet before they hit the specimen surface, resulting in reduced real impact angles, particularly for a nominal angle of 90°.

A typical material loss in service may often be in the order of 1mm/year or less. The erosion levels of the tests represented in Fig. 1 is therefore probably the most realistic. In extreme cases material losses similar to those in Fig. 2 may possibly occur. The jet impingement tests are considerably accelerated with respect to time but the velocities of the particles and the erosive wear per amount of sand are considered to be realistic.

From the above considerations it seems that the four best materials in Fig. 1 and Fig. 3 are the best suited of the materials tested with respect to application in choke valves in service. For a safe material choice the fracture properties should also be taken into consideration. Table 2 shows the decrease of fracture toughness resulting from a decreasing amount of metallic binder phase.

Materials like M3 and M19 have frequently been used in choke valves. Figures 1–3 indicate that the degradation rates of the materials can be reduced several times by selecting one of the best materials. The life time may be increased by factors of, for example, 3–6. Further improvement may be obtained by systematic research on composition, structure and production method.

4. Conclusions

- The life time of choke valves in oil and gas production can be increased several times by the use of the best materials investigated compared to the materials which have been used frequently until now. Systematic R & D is expected to give further improvements.

- The erosion–corrosion resistance of the cemented tungsten carbides is related to the amount and composition of metallic binder phase and the grain structure of the carbides. A very fine WC grain structure and high corrosion resistance of the binder phase under erosive conditions are particularly important factors. Reduced content of metal binder phase improves the erosion properties, but this possibility may be limited due to decreasing fracture toughness.

Table 2. *Hardness and fracture toughness for some of the materials tested**

Material	Average Vickers hardness HV_{60} (kg mm^{-2})	Vickers hardness (GPa)	Fracture toughness K_{IC} (MPa $m^{1/2}$)
M12 94WC 6Co	1979	19.4 –0.8 +0.9	11.4 ± 0.1
M13 90.5WC 9.5NiCrCo	1769	17.4 –0.2 +0.4	13.0 –0.4 +0.5
M16 94.9WC 5.1NiCrCo	2139	21.0 –0.8 +0.5	10.1 –0.5 +1.1
M17 97WC 3NiCrMoCo	2242	22.0 –0.9 +1.0	8.5 ± 0.3

*Data determined by hardness indentations leading to cracks.

5. Acknowledgements

The research work has been financed by the Norwegian Research Council, Norsk Hydro, S and Saga Petroleum as part of the projects, 'Materials selection and lifetime of components exposed to multiphase flow'. The determination of fracture toughness has been carried out by Hans Iver Lange, and TEM and microanalysis by Asgeir Bardal, both at SINTEF Materials Technology. We wish to thank the sponsors as well as personnel at SINTEFF, Statoil and Norsk Hydro who have contributed in the project work.

References

1. E. Bardal and T. Solem, "Materialvalg og levetid for komponenter utsatt for flerfasestrøm," (New ceramic-metallic materials for choke valves in oil production) SINTEF report STF24 A95254, February, 1995.
2. C. H. Ahlén and E. Willmann, "Materialvalg og levetid for komponenter utsatt for flerfasestrøm. Resultater fra erosjonstesting med luft-sand og vann-sand" (Material selection and lifetime for components exposed to multiphase flow. Results from erosion testing with airborne sand particles and water mixed with sand) Statoil 95.03.03.
3. C. H. Ahlén and L. Marken, "Erosion testing of tungsten carbide grades for use in choke valves", Statoil 97.01.20.
4. E. Bardal, T. Solem and A. Bardal, "Materialvalg og levetid for komponenter utsatt for flerfasestrøm II", (Material selection and lifetime for components exposed to multiphase flow. Phase II) SINTEF report STF24 F97216, Jan. 1997.
5. E. Bardal, T. G. Eggen and Aa. Stølan Langseth, "Combined erosion and corrosion of steels and hard metal coatings in slurries of water and silica sand", *12th Scand. Corros. Congr. EUROCORR '92*, Espoo, Finland, 1992.
6. E. Bardal, M. Bjordal, T. G. Eggen, T. Rogne and T. Solem, "Prevention of erosion and corrosion in slurries using various inorganic coatings". *Progress in the Understanding and Prevention of Corrosion*, J. M. Costa and A. D. Mercer, eds. The Institute of Materials, London, 1993, pp. 158–164.
7. E. Bardal, C. H. Ahlén, L. Nøkleberg and T. Solem, "Erosion and Corrosion of Ceramic-Metal Materials", *13th Int. Corros. Congr.*, Melbourne, Australia, Nov. 1996.
8. G. R. Anstis, P. Chanticul, B. R. Lawn and D. B. Marshall, *J. Amer. Ceramic Soc.*, 1981, **64**, (a) 533–538.

List of Abbreviations

The following abbreviations occur in the text and in the Index of contents.

APB	Anaerobic Acid Producing Bacteria
API	American Petroleum Institute
BDWT	Battelle Drop Weight Test
BHP	Bottom Hole Pressure
BHT	Bottom Hole Temperature
BOPD	Barrels of Oil Per Day
BPP	Bubble Point Pressure
CA	Corrosion Allowance
CAMP	Current Advances in Materials and Processes
Capex	Capital Expenditure
CCI	CO_2 Corrosion Index
CDTM	Controlled Depth Tow Method
CER	Contact Electric Resistance
CEV	Vallourec Research Centre
CL	Constant Load
CLR	Crack Length Ratio
CORPOS	Positional Corrosion Model
CR	Controlled Rolling also Corrosion Rate
CSR	Crack Sensitivity Ratio
CT	Compact Tension
CTOD	Crack Tip Opening Displacement
CTR	Crack Thickness Ratio
DO	Dissolved Oxygen
E_c	Corrosion potential
E_{oc}	Open circuit potential
E_p	Pitting potential
E_{pp}	Passivity potential
E_t	Transpassivity potential
ECSC	European Coal and Steel Community
EEW	Erndtebrücke Eisenwerk
EIS	Electrochemical Impedance Spectroscopy
EN	Electrochemical Noise
EP	Europipe GmbH Gerrnany
EPMA	Electron Probe Microanalysis
ETPM	ETPM International S.A. France
FAC	Free Available Chlorine
FEM	Finite Element Method
FILC	Flow Induced Localised Corrosion
FTE	Flow Through Electrode
GEI	Galvanic Effect Index
GLT	Groningen Long Term (project)
GMA	Gas Metal Arc
GOR	Gas Oil Ratio
GRE	Glass Reinforced Epoxy
GRP	Glass Reinforced Plastic
GTA	Gas Tungsten Arc
HAZ	Heat Affected Zone
HFI	High Frequency Induction
HIC	Hydrogen Induced Cracking
HSLA	High Strength Low Alloy Steels
i_{cp}	critical current density
i_p	passive current density
i.d.	internal diameter
IE	Impingement Electrode
IFE	Institute for Energy Technology
IIW	International Institute of Welding
IRC	Instrumented Restraint Cracking

JIP Joint Industry Profect

K_{ISSC} Critical stress intensity for SSC

LCC Life Cycle Costs
LCL Lower Confidence Limit
LPR Linear Polarisation Resistance
LW Laser Welding

MAPS Multiphase Autoclave Pipeline Simulator
MAR Maximum Acceptable Risk
m.p.y. mils (thousanths of an inch) per year
MRI Mannesmann Research Institute
MMSCF Millions of Standard Cubic Feet
MUSD Million US Dollars

NACE National Association of Corrosion Engineers
NAM Nederlandsee Aardolie Maatschappij
NKK Nippon Ko Kan
NORSOK Norsk Sokkels Konkuranseposisjon
NPR Nominal Pressure Rating
NPV Net Present Value

OCTG Oil Country Tubular Goods
o.d. outside diameter
OLGA A computer based information program
OMAE Offshore Mechanics and Arctic Engineering
Opex Operating Expenditure

psia pounds per square inch atmosphere
PAG Prior Austenite Grain
POD Probability of Detection
PREN Pitting Resistance Equivalent Number
PTFE Polytetrafluoroethylene
PVDF Polyvinylidene difluoride
PWHT Post Weld Heat Treatment

QT Quenched-and-tempered

R_{ct} Charge Transfer Resistance
RHE Reversible Hydrogen Electrode
RN Noise Resistance

SAC Surface active compounds
SAW Submerged Arc Welding
SCC Stress Corrosion Cracking
SEM Scanning Electron Microscope
SD Superduplex
SHPE Soil Hydrogen Permeation Electrode
SIMS Secondary Ion Mass Spectroscopy
SM Supermartensitic
SMAW Shielded Metal Arc Welding
SMYS Specified Minimum Yield Strength
SOHIC Stress Oriented Hydrogen Induced Cracking
SORM Second Order Reliability, Method
SPE Society of Petroleum Engineers
SRB Sulfate Reducing Bacteria
SS Stainless steel
SSC Sulfide Stress Cracking
SSRT Slow Strain Rate Test
STTP Surface Tension Transfer Process
SZE Static Zone Electrode

TEM Transmission Electron Microscopy
TFL Towed Flow Line
TTF Time to Failure
TMCP Thermomechanically Controlled Processing

UKOOA United Kingdom Offshore Operators Association

VLCC Very Large Crude oil Carriers

WACO Water Condensate
WT Wall Thickness

XRD X-ray Diffraction

Index